Pathophysiology and Rational Pharmacotherapy of Myocardial Ischemia

Gerd Heusch (Ed.)

Pathophysiology and Rational Pharmacotherapy of Myocardial Ischemia

With contributions by

P. B. Corr · J. Dirschinger · K. Egstrup · K. P. Gallagher
B. D. Guth · J. Herlitz · B. Heusch · G. Heusch · A. Hjalmarson
Th. Hohlfeld · J. C. Kaski · M. Kobayashi · F. Kraus · A. Malliani
A. Maseri · H. A. McCann · L. H. Opie · S. M. Pogwizd
W. Rafflenbeul · G. Reiniger · J. Ross, Jr. · W. Rudolph · J. Schaper
W. Schaper · R. J. Schott · K. Schrör · M. Schwaiger · F. Waagstein
H. G. Wolpers

Springer-Verlag Berlin Heidelberg GmbH

The Editor:
Prof. Dr. med. Gerd Heusch
Abteilung für Pathophysiologie
Zentrum für Innere Medizin
Universitätsklinikum Essen
Hufelandstraße 55
D-4300 Essen, FRG

CIP-Titelaufnahme der Deutschen Bibliothek

Pathophysiology and rational pharmacotherapy of myocardial ischemia / Gerd Heusch (ed.). With contributions by P. B. Corr ... — Darmstadt : Steinkopff ; New York : Springer, 1990

   ISBN 978-3-642-54135-3   ISBN 978-3-642-54133-9 (eBook)
   DOI 10.1007/978-3-642-54133-9

NE: Heusch, Gerd [Hrsg.]; Corr, Peter B. [Mitverf.]

Copyright © 1990 by Springer-Verlag Berlin Heidelberg
Originally published by Dr. Dietrich Steinkopff Verlag GmbH & Co. KG, Darmstadt in 1990.
Softcover reprint of the hardcover 1st edition 1990

Medical Editorial: Sabine Müller — English Editor: James C. Willis — Production: Heinz J. Schäfer

Foreword

Ischemic heart disease is still the most frequent cause of death in the western world. There have been significant achievements in diagnostic procedures as well as in the medical, invasive, and surgical treatment of ischemic heart disease in recent years. A variety of drugs are available for the pharmacotherapy of ischemic heart disease, particularly nitrates, β-blockers, and calcium-antagonists which are used as monotherapy or in various combinations. However, the selection of patients for a certain treatment, as well as the optimization of an individual treatment are still largely empirical. On the other hand, the recent advances in experimental cardiology emphasize the extremely complex and dynamic scenario of ischemic heart disease, involving endothelial damage, coagulation processes, metabolic and morphologic derangements, coronary constrictor mechanisms, blood flow redistribution, arrhythmogenesis, contractile dysfunction during ischemia and reperfusion, and finally lack or presence of pain perception. Therefore, it appears desirable to close the gap between experimental and clinical cardiology and, thus, to provide a pathophysiological basis for rational clinical decisions with respect to diagnostic and therapeutic procedures.

The idea for this book arose during the preparation of a seminar series on experimental cardiology, when I found it difficult to collect the pertinent information from textbooks of cardiology, physiology, pathology, and pharmacology, as well as from numerous review and original articles on specific topics. I am now very grateful that expert clinical and experimental colleagues from around the world have joined me in the effort to provide a comprehensive textbook on the pathophysiology of myocardial ischemia and its rational pharmacotherapy.

Finally, I would like to thank the publisher, in particular Ms. Sabine Müller for the pleasant cooperation and constructive support in editing this book.

Essen, spring 1990   Gerd Heusch

Contents

A Brief History of Angina Pectoris:
Change of Concepts and Ideas

Beate Heusch, Brian D. Guth, and Gerd Heusch

Abt. für Pathophysiologie, Universitätsklinikum Essen and Dept. of Medicine, University of California, San Diego, USA

The history of angina pectoris may be considered in three developmental periods. The first extended from antiquity into the late middle ages during which numerous descriptions of clinical symptoms appeared that, in retrospect, were likely angina pectoris but at the time were not associated with a disease of the heart. The period from the late middle ages to the early 20th century brought about the correlation between angina pectoris and a pathological state of the coronary vessels. Finally, since the 1930s there has been the development of pathophysiological concepts of angina pectoris which now focus on an impaired relation between the work performed by the heart and the oxygen supply to the heart through the coronary circulation.

The earliest description of a disease which may be characterized as angina pectoris is found in the Egyptian Papyrus Ebers (Fig. 1) which originates from 1500 B.C. The content of the Papyrus Ebers dates back to an even older dynasty (2500–2000 B.C.) [8]: "If you find a man whose chest hurts and who suffers pain from his shoulder and his stomach, you shall say that death is approaching". Here already is the association of angina-like chest pain radiating into the shoulder with a fatal result. These symptoms were, however, related to a disease of the stomach, consequently herbs to elicit vomiting and diarrhea were the recommended treatment.

The term "angina" is first found in writings ascribed to the Greek physician Hippocrates (460–370 B.C.) [10, 14]. Chest pain radiating into the shoulder and arm, also in connection with dyspnea, is called κυνάγχη. This term is composed of κυών (the dog) and ἄγχω (to narrow, to compress). Apparently Hippocrates regarded angina pectoris and angina tonsillaris – diseases of the throat associated with a typical "barking" voice – as the same disease. A well known letter from Seneca to Lucilius (originating 60–65 A.D.) describes pectanginal symptoms in himself, symptoms which were called "meditatio mortis" (preparation of death) by the physicians. In this context Tacitus's description of Seneca's suicide is interesting. He could only bleed to death after he was carried into a warm bath, because his extremities were hardened – an indication of generalized atherosclerosis. Galen called angina pectoris "Kardialgia" [25]. The designation of both the upper portion of the stomach and the heart by the term "Kardia" is probably not accidental. Galen knew that the vagal nerves innervate both the stomach and the heart and attributed heart burn and tachycardia to a "sympathetic" radiation from the stomach to the heart. Apart from the fact that Galen anticipated modern ideas about autonomous reflexes, even today a clinical differential diagnosis between an acute, painful disease of the stomach and an anginal attack may be difficult. This ambivalent nosological relation between heart and stomach persists also into today's nomenclature, in that ventricle means "little stomach".

1

Fig. 1. Papyrus Ebers

Until the late middle ages medicine was committed to Galen's ideas, including those about angina pectoris. However, graphics by Leonardo da Vinci (1452–1519) (Fig. 2) and Andreas Vesalius (1514–1564) prove that the anatomy of the heart and the coronary vessels were known in great detail. According to the prevailing idea, however, the coronary veins functioned to bring nutritive blood from the liver to the heart [22].

This concept persisted until 1628 when W. Harvey published his "Exercitatio Anatomica de Motu Cordis et Sanguinis" (Fig. 3), in which he not only described the circulation of blood through the lungs and the body, but also described the supply of blood to the heart by the coronary arteries. Thus, the initial period in the understanding of angina pectoris ended with a knowledge of its symptoms but ideas about the genesis of this disease and its relation to the heart remained unclear.

In 1761, Morgagni characterized the pathology of typical coronary atherosclerosis at autopsy. Progress in the understanding of angina pectoris exceeding the morphology of coronary atherosclerosis was provided by British clinicians in the late 18th century. The surgeon John Hunter suffered from angina pectoris and studied the symptoms carefully in himself. He noticed the provocation of anginal attacks by physical exercise. In 1793 he died during an anginal attack after he became very excited in a discussion during rounds in St. George's Hospital. John Hunter is, thus, a classic example for the acute precipitation of angina pectoris by physical and mental stress. At his autopsy a marked ossification of the coronary arteries was revealed [1]. In 1768 W. Heberden gave his classic lecture on angina pectoris to the Royal College of Physicians which was published 4 years later. On the basis of numerous clinical observations, he described the typical pain sensation, the provocation of angina pectoris by physical exercise and opulent meals, and the higher incidence in men, particularly in those older than 50 years. Heberden, whose name is used today to designate the poststenotic myocardial ischemia on the basis of coronary atherosclerosis,

2

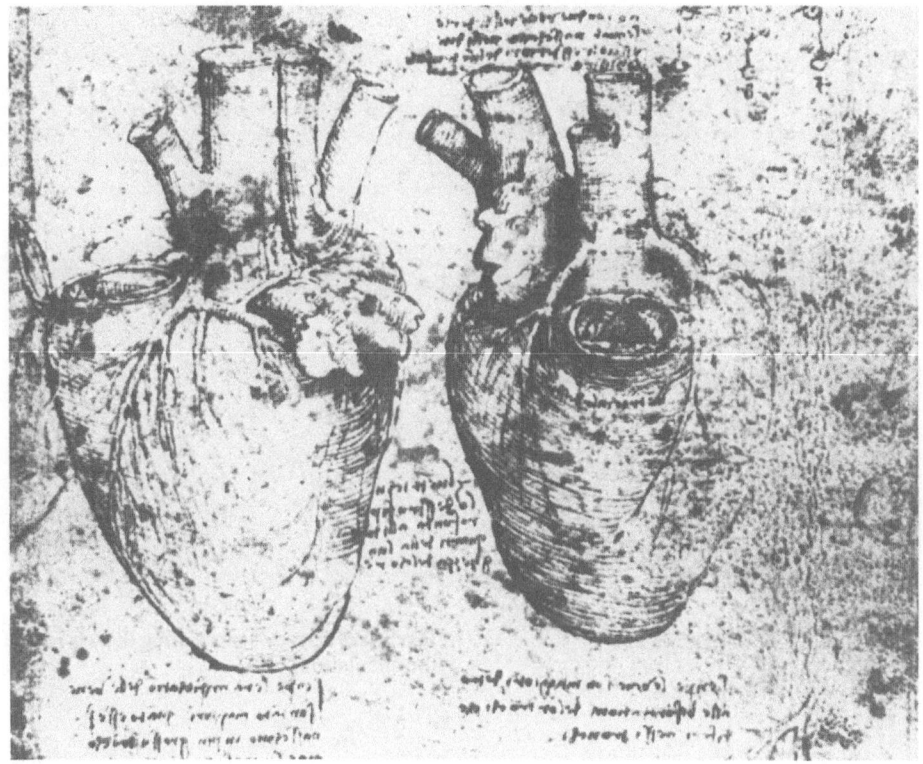

Fig. 2. Illustration of the heart by Leonardo da Vinci

considered coronary spasm as the cause of angina pectoris. Consequently, he recommended opium and alcohol for therapy.

In 1809 A. Burns was the first to develop a pathophysiological concept for the relation between the clinical anginal symptoms and the coronary atherosclerosis at autopsy. He argued that the diseased heart, like an extremity with a tourniquet, could not increase its function without an adequate increase in blood flow, and that cardiac failure would result from the excessive demand. This concept remained hypothetical until it was later confirmed by experimental investigations in the course of the nineteenth century. In 1881 the German pathologists Cohnheim and v. Schultheiss-Rechberg, using animal experiments, characterized the coronary arteries as functional end-arteries, such that their occlusion leads to irreversible loss of the perfused myocardial area. They attributed the loss of cardiac function and the initiation of ventricular fibrillation after coronary ligation, not to the lack of oxygen, but to the accumulation of metabolic byproducts due to the lack of washout.

Crucial for the development of more precise ideas about myocardial ischemia was the development of appropriate experimental models. In 1912 Morawitz and Zahn [16] developed a coronary sinus cannula which first permitted the measurement of coronary blood flow in situ. Progress in the diagnosis and therapy of the clinical symptoms occured in 1867 as L. Brunton first described in "Lancet" the therapeutic

3

Fig. 3. W. Harvey's pioneering description of the circulation

effect of amylnitrite in angina pectoris [4]. Brunton attributed the effect of amylnitrite to a reduction in arterial tension in analogy to hemorrhagia, a still valid idea about the antianginal effect of nitrates by a reduction in pre- and afterload. In 1912 J. B. Herrick provided the classic clinical and morphological demonstration of myocardial infarction as the result of a thrombotic coronary occlusion [12]. In 1918 G. Bousfield first described characteristic ECG-changes during angina pectoris, in particular disturbances of repolarization − the discordant T − and their normalization by nitrates [2].

The modern period in the understanding of angina pectoris in which more precise pathophysiological concepts about the genesis of myocardial ischemia were developed was initiated by H. Rein, a physiologist in Freiburg and later in Göttingen. Rein developed a thermoflowprobe which first permitted the simultaneous measurement of cardiac output and coronary blood flow in the dog in situ. In 1931 he demonstrated in such experiments that coronary blood flow is not matched to arterial blood pressure, but to cardiac output (Fig. 4) [19, 20]. Based on his physiological experiments, Rein postulated that also under pathological conditions, the relation between coronary blood flow and myocardial performance should be the only criterion for sufficiency or insufficiency of coronary blood flow. According to this idea coronary insufficiency is a coronary blood flow which is inadequate in relation to the instanteous myocardial performance. The physiological ideas of Rein were immediately taken up by the pathologist F. Büchner who transferred them to pathological conditions and confirmed them morphologically [5, 6]. During autopsy

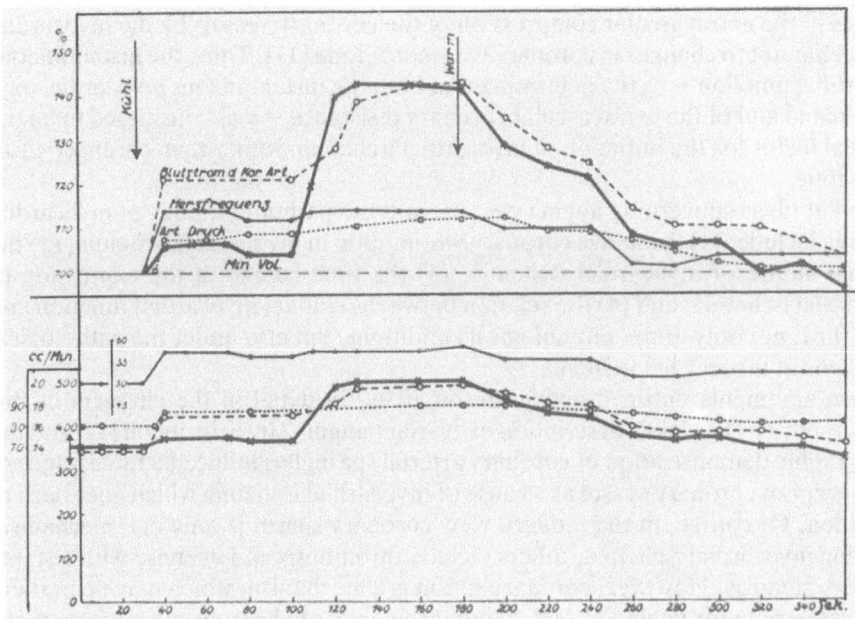

Fig. 4. Representative recording from H. Rein's experiments. Cardiac output (solid line) and coronary blood flow (dashed line) are increased in parallel by exposure to cold (first arrow) and decreased by exposure to warm air (second arrow).

of patients who had suffered from angina pectoris, Büchner found atherosclerotic coronary stenoses and disseminated necroses and fibroses in the myocardium which could be experimentally reproduced by a combination of a reduction in coronary supply (anemia, hypoxia, CO-poisoning) with physical exercise. In Büchner's view these experiments proved that these myocardial lesions were caused by a discrepancy between coronary oxygen supply and myocardial oxygen demand.

An imbalance between the coronary supply and the myocardial demand for blood or oxygen is found in modern textbooks to be the essential pathophysiological explanation of myocardial ischemia [3, 11, 21]. The term "myocardial ischemia" is regularly used synonymous with the term "coronary insufficiency".

The experimental observations of Rein, however, were confined to physiological conditions and those of Büchner to the morphological result of myocardial ischemia. They both viewed coronary insufficiency as a homogeneous event affecting the whole heart, i.e., they related total myocardial performance to total coronary blood flow. The regional nature of myocardial ischemia was not taken into consideration, although Büchner did explicitly describe the inhomogeneous distribution of myocardial necroses preferentially in the subendocardium of the left ventricle. Also, the time-course for the ultimate expression of myocardial ischemia as tissue necrosis was not taken into consideration. Finally, as a consequence of Rein's and Büchner's ideas the terminal coronary vascular bed was assumed to be maximally dilated during myocardial ischemia. Changes in coronary blood flow were only attributed to

5

changes in the extravascular compression of the coronary vessels by the myocardial muscle, but not to changes in coronary vasomotor tone [11]. Thus, the instantaneous myocardial function – as the determinant of both the instantaneous myocardial oxygen demand and of the extravascular coronary resistance – was considered to be the essential factor for the initiation of myocardial ischemia, rather than coronary vasomotor tone.

Current ideas concerning angina pectoris and the pathophysiology of myocardial ischemia include: (1) the active coronary vasomotion in myocardial ischemia, (2) the regional nature of myocardial ischemia, (3) the time course of the expression of myocardial ischemia, and (4) the relation between regional myocardial function and blood flow, not only under physiological conditions, but also under the pathological conditions of myocardial ischemia.

These arguments outlined above are discussed in detail in the chapters of this monograph: 1) The classic description of "variant angina" by Prinzmetal [17] and the angiographic demonstration of coronary arterial spasm has induced a renaissance of the concept of coronary spasm as a cause of myocardial ischemia which goes back to Heberden. Of course, in the modern view coronary spasm is only one mechanism initiating myocardial ischemia, others include thrombosis and exercise with a stable coronary stenosis. However, coronary spasm is a mechanism which may be particularly responsible for angina at rest. Neither the epicardial coronary arteries nor the resistive vessels are maximally dilated during myocardial ischemia. The complex active coronary vasomotion is discussed in the chapters 9 by Heusch and Guth and 10 by Kaski and Maseri.

2) The regional nature of myocardial ischemia as an essential pathophysiological feature has been respected only after the development of adequate techniques for the measurement of regional myocardial function (sonomicrometry) and blood flow (tracer microspheres). The interaction between ischemic and nonischemic myocardial areas with respect both to regional function and regional blood flow and its collateral and transmural distribution is covered in the chapters 6 by Gallagher, 9 by Heusch and Guth, 12 by Guth and Heusch, and 16 by McCann and Ross.

3) The time course of the expression of myocardial ischemia is difficult to understand from the point of view of an energetic imbalance. In the phase immediately following the onset of ischemia, contractile function rapidly decreases but without any evidence of a reduction in myocardial ATP, a phenomenon referred to as "early hypoxic failure" [15]. Also, despite a contractile dysfunction at rest there remains an inotropic reserve during early ischemia which can be recruited by postextrasystolic potentiation [7] or catecholamine stimulation [24]. Contractile dysfunction may serve as a protective downregulation of myocardial work and oxygen consumption during the early phase of reversible ischemia [13, 23]; potentially this downregulation is unmasked in the persistent dysfunction of reperfused myocardium (see chapter 8 by W. Schaper et al.). The energetic conditions involved in the transition from reversible to irreversible ischemic damage are also as yet unclear. Potentially ischemic myocardium can "hibernate" at reduced function for prolonged periods of time before it undergoes irreversible damage [18].

4) The analysis of regional function (demand) and blood flow (supply) in ischemic myocardium during a hemodynamic steady state reveals no discrepancy between flow and function since the demand (as indicated by contractile function) does not

6

exceed the actual supply (as indicated by myocardial blood flow) as would be indicative of an energetic imbalance. Instead, regional myocardial function is reduced in proportion to the available oxygen supply (blood flow). Thus, in the steady-state there is no "relative" ischemia with a characteristic supply-demand imbalance, but rather an "absolute" ischemia characterized by a new balance between contractile function and blood flow at a reduced level [9].

It must be emphasized, however, that the concept of a discrepancy between myocardial demand and coronary supply as the underlying cause of myocardial ischemia, which was proposed by H. Rein and F. Büchner, is not simply refuted by the lack of evidence for a discrepancy between ischemic regional function (demand) and ischemic regional blood flow (supply) during a hemodynamic steady state. The strength of this imbalance theory is certainly the ability to separate ischemia from autoregulatory changes of blood flow and function where metabolically mediated blood flow adjustments follow changes in myocardial function and oxygen consumption. Such separation, however, is also performed by the definition of myocardial ischemia as a reduction in absolute blood flow per heart beat (see chapters 6 by Gallagher and 16 by McCann and Ross). The normalization of blood flow on a per-beat basis is supposed to take into account that measures of regional contractile function (systolic segment shortening or wall thickening) describe a single cardiac cycle and do not indicate the number of contractions per time interval, a major determinant of oxygen consumption. On the other hand, it may be argued in defense of the imbalance theory that blood flow per heart beat reflects already a ratio of supply (blood flow) and demand (number of beats) and that contractile function adjusts to this balance (see chapter 6 by Gallagher). Nevertheless, at any given heart rate regional myocardial blood flow and function are proportionally reduced during myocardial ischemia.

In conclusion, it appears necessary to distinguish the demand for the integrity and viability of myocardial cells from the demand for contractile function. Thus, there may then be discrepancy between demand and supply on the cellular level, but not on the hemodynamic level. The disseminated myocardial necroses in patients with angina pectoris which are regularly found at autopsy support this hypothesis. It is completely unclear when and why the experimentally confirmed balance (steady state) between function (demand) and blood flow (supply) turns into an imbalance threatening the cellular integrity and viability. The investigation of these complex interactions is currently limited by a lack of adequate methods. In principle, the simultaneous analysis of function, blood flow, metabolism and ultrastructure of ischemic myocardium on a regional level would be necessary. Currently, however, only regional contractile function can be measured continuously. Even in animal experiments, regional blood flow can only be measured during a limited number of steady state conditions. Nondestructive continuous methods for the analysis of regional myocardial metabolism are just now being developed (NMR, PET). A nondestructive, continuous analysis of the morphology of in situ ischemic myocardium appears impossible at this moment.

Apart from these theoretical considerations on the pathophysiology of myocardial ischemia, the characterization of myocardial ischemia just as a hypoperfusion of the myocardium appears reasonable from a pragmatic point of view. All therapeutic interventions from surgical revascularization to thrombolysis and to pharmacotherapy can then be summarized independently from myocardial demand in that they all increase blood flow to the ischemic myocardium and thereby permit an increased function.

The occurence of angina pectoris in man then appears to predate even the earliest written records. Despite this long association and an ever increasing research effort to understand the pathophysiology involved, we still do not have an adequate theoretical concept of the disease. We can hope that the advent of improved technology will provide better insight into these problems and form a background on which we may base an improved therapeutic approach.

Summary: The history of angina pectoris may be subdivided into three periods. From antiquity into the late middle ages there are numerous descriptions of clinical symptoms which may be characterized as angina pectoris in retrospect, but were not associated with a disease of the heart then. In modern times angina pectoris is recognized as a heart disease; initial connections between pathological alterations of coronary vessels and clinical symptoms are drawn. It is only since the thirties of this century that in essence pathophysiological concepts of angina pectoris are developed which focus on an impaired relation between the work performed by the heart and the supply of energy to the heart by the coronary vessels. In recent years the view of myocardial ischemia as an energetic imbalance of myocardial oxygen demand and supply has been questioned. In the hemodynamic steady state of ischemia regional myocardial function (demand) and regional blood flow (supply) are rather adequately reduced. It is still unclear – particularly because of methodological reasons – when and why the steady state of reversible myocardial ischemia is turned into an imbalance which threatens myocardial integrity and viability.

References

1. Bing R (1964) Circulation of the Blood. Men and ideas. Coronary circulation and cardiac metabolism. Oxford Univ Press: Fishman A P, Richards D W, pp 199–263
2. Bousfield G (1918) Angina pectoris: changes in electrocardiogram during paroxysm. Lancet I: 457–458
3. Braunwald E, Sobel BE (1988) Coronary blood flow and myocardial ischemia. In: Braunwald E (ed) Heart disease. Saunders Comp., Philadelphia London Toronto, pp 1191–1221
4. Brunton TL (1867) Use of nitrite of amyl in angina pectoris. Lancet:97–98
5. Büchner F (1932) Über Angina pectoris. Klin Wochenschr 42: 1737–1739
6. Büchner F (1939) Die Koronarinsuffizienz. Steinkopff, Dresden Leipzig, pp 1–92
7. Crozatier B, Franklin D, Theroux P, Tomoike H, Sasayama S, Ross Jr (1977) Loss of regional ventricular postextrasystolic potentiation after coronary occlusion in dogs. Am J Physiol 233:H392–H398
8. Deines H, Grapow H, Westendorf W (1958) IV. D. Magen-Herz. Berlin: Grundriß der Medizin der alten Ägypter, Akademie Verlag, Berlin, pp 88–100
9. Gallagher KP, Matsuzaki M, Osakada G, Klemper WS, Ross J Jr (1983) Effect of exercise on the relationship between myocardial blood flow and systolic wall thickening in dogs with acute coronary stenosis. Circ Res 52:716–729
10. Harris CRS (1973) The heart and the vascular system in ancient greek medicine. Oxford Univ. Press, pp 432–455
11. Hellige G (1981) Koronardurchblutung. In: Krayenbühl HP, Kübler W (Hrsg) Kardiologie in Klinik und Praxis. Thieme, Stuttgart New York, 8.1–8.12
12. Herrick JB (1912) Clinical features of sudden obstruction of the coronary arteries. J Am Med Assoc 59:2015–2020

13. Jacobus WE, Pores IH, Lucas SK, Kallmann CH, Weisfeldt ML, Flaherty JT (1982) The role of intracellular pH in the control of normal and ischemic myocardial contractility: a 31P nuclear magnetic resonance and mass spectrometry study. In: Intracellular pH: its measurement, regulation and utilization in cellular functions. Liss, New York, pp 537−565
14. Katz AM, Katz PB (1962) Diseases of the heart in the works of Hippocrates. Br Heart J 24:257−264
15. Kübler W, Katz AM (1977) Mechanism of early "pump" failure of the ischemic heart: possible role of adenosine triphosphate depletion and inorganic phosphate accumulation. Am J Cardiol 40:467−471
16. Morawitz P, Zahn A (1912) Über den Koronarkreislauf am Herzen in situ. Zbl Physiol 26:465−470
17. Prinzmetal M, Kennamer R, Merliss R, Wada T, Bor N (1959) Angina pectoris (I) A variant form of angina pectoris. Am J Med 1:375−388
18. Rahimtoola SH (1989) The hibernating myocardium. Am Heart J 117:211−221
19. Rein H (1931) Die Physiologie der Herz-Kranz-Gefäße. I. Mitteilung. Z Biol 92:101−114
20. Rein H (1931) Die Physiologie der Koronardurchblutung. Untersuchungen des Koronarkreislaufes am intakten Organismus. Verh Dtsch Ges Innere Med 43:247−262
21. Roskamm H (1982) Koronarerkrankungen. In: Roskamm H, Reindell H (Hrsg) Herzkrankheiten. Springer, Berlin Heidelberg New York, pp 529−548
22. Schadewaldt H (1980) Herz und Herzkranzgefäße. Jahrbuch der Universität Düsseldorf 3:192
23. Schipke JD, Burkhoff D, Schäfer J (1986) Änderung des ventrikulären mechanischen Wirkungsgrades bei reduzierter Koronardurchblutung. Z Kardiol 75 [Suppl 4]:43
24. Schulz R, Miyazaki S, Miller M, Thaulow E, Heusch G, Ross Jr J, Guth BD (1989) Consequences of regional inotropic stimulation of ischemic myocardium on regional myocardial blood flow and function in anesthetized swine. Circ Res 64:1116−1120
25. Siegel RE (1968) Galen's system of physiology and medicine. Heart Disease 3. Angina pectoris. Karger Basel New York, pp 344−365

Authors' address:
Prof. Dr. Gerd Heusch
Abt. für Pathophysiologie
Zentrum Innere Medizin
Universitätsklinikum Essen
Hufelandstr. 55
4300 Essen
F.R.G.

9

Myocardial Ultrastructure in Ischemia

Jutta Schaper

Max-Planck-Institute, Department of Experimental Cardiology, Bad Nauheim, FRG

Introduction

Concern about the increasing number of patients afflicted by ischemic heart disease has focussed research on the pathophysiology of ischemia in an attempt to elucidate the various mechanisms behind ischemic injury. Electron microscopy is a useful tool for this type of investigation because it not only identifies irreversible injury, but it also reveals different stages of reversible injury due to ischemia. Reversible injury, however, is the most interesting phase of ischemia because it allows for therapeutic measures such as delay of occurrence of myocardial necrosis and for reperfusion by thrombolysis or angioplasty to induce recovery of the tissue.

In earlier studies [10] we determined four different stages of ischemic injury by comparing the ultrastructural appearance of the myocardium with metabolic changes, as well as with the degree of functional recovery after induction of reperfusion. The standardization of ischemic injury thus obtained has proved useful in various investigations of the effects of either global or regional ischemia and has been adopted by other groups as well.

This "calibration" of ultrastructural alterations by electron microscopic evaluation can be used for the determination of the degree of injury under experimental conditions and gives detailed information about the actual injury at any given time of biopsy removal. It is furthermore hoped that the information from ultrastructural studies of ischemic tissue may add to our knowledge on pathogenetic mechanisms leading to cell death.

Myocardial Subcellular Organelles in Ischemia

In the course of ischemia all subcellular components of the myocardial cell undergo changes, but there is a distinct difference in the susceptibility of the various organelles and, therefore, the time sequence of alterations varies from mitochondria to nuclei to contractile material.

The mitochondria are those subcellular organelles that are the most sensitive to ischemia. The first ultrastructural sign is the disappearance of the normal matrix granules which occurs already at 5–10 min after onset of ischemia. At later stages, the matrix becomes more and more electron-lucent and the cristae are fragmented. The outer mitochondrial membrane remains structurally intact during reversible injury and only appears to be interrupted at late stages of irreversible injury (Fig. 1). These two principal phases of ischemic damage, the reversible vs the irreversible

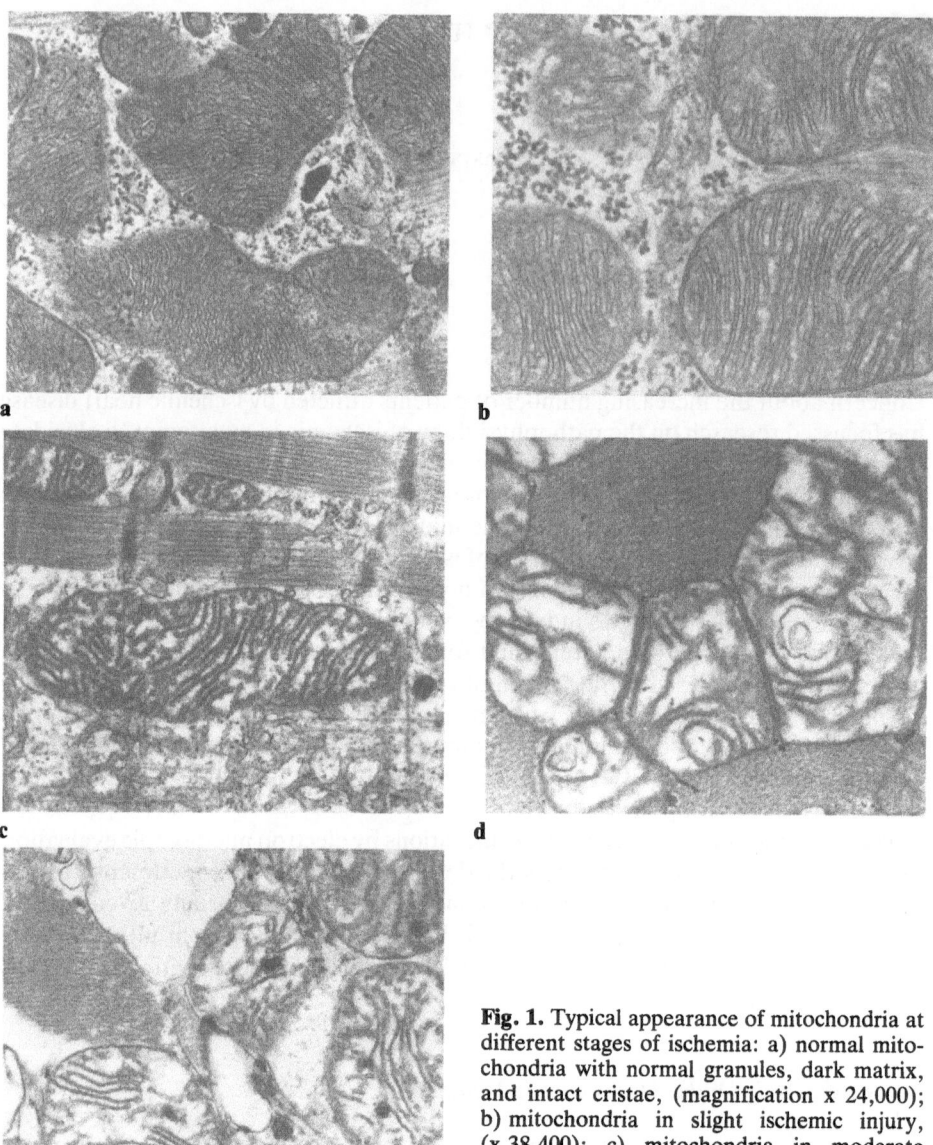

Fig. 1. Typical appearance of mitochondria at different stages of ischemia: a) normal mitochondria with normal granules, dark matrix, and intact cristae, (magnification x 24,000); b) mitochondria in slight ischemic injury, (x 38,400); c) mitochondria in moderate ischemic injury, (x 16,100); d) mitochondria in severe ischemic injury, (x 27,200); e) mitochondria in irreversible injury (x 19,200).

states, can be well differentiated by the observation of amorphous or flocculent densities within the mitochondrial matrix. These depositions most probably consist of an insoluble accumulation of proteins, lipids, and Ca^{++}.

The nuclei are only altered by ischemia when the mitochondria already exhibit slight injury. The nucleoplasm is increasingly electron-lucent and the chromatin is

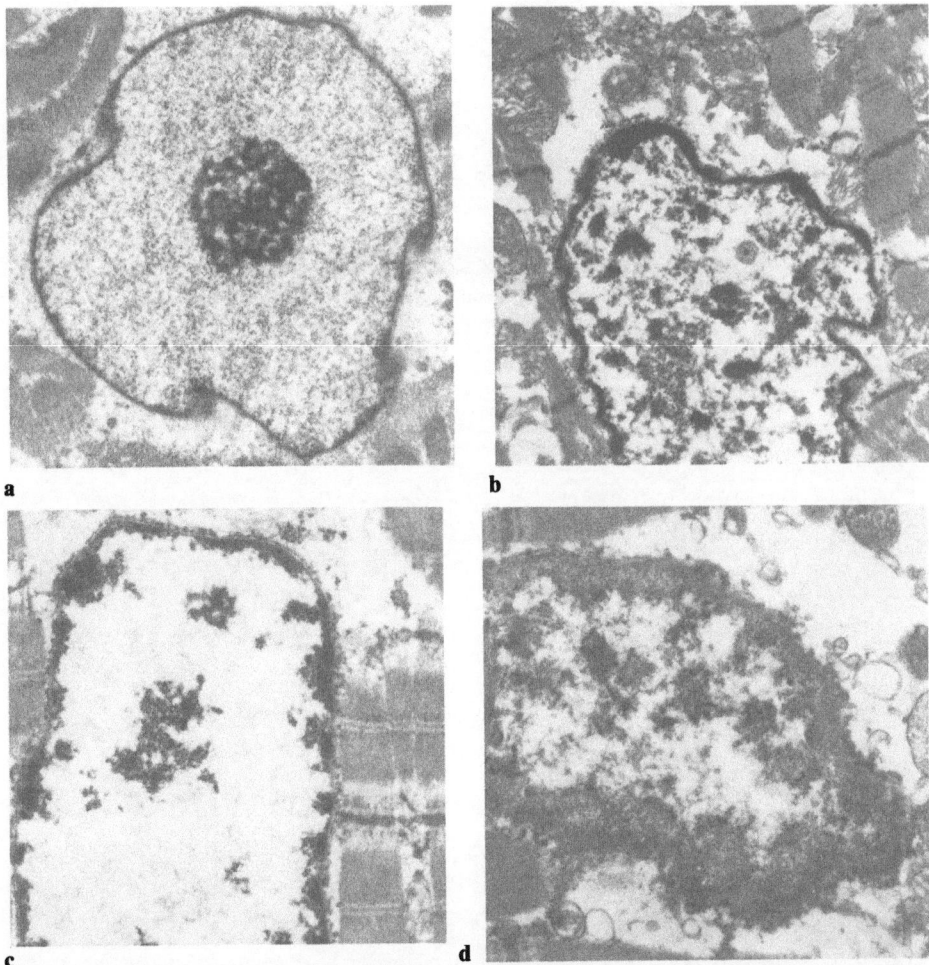

Fig. 2. Typical appearance of nuclei in different stages of ischemia: a) normal nucleus, (magnification x 12,100); b) clumping of chromatin and clearing of nucleoplasm, (x 8,800); c) very light nucleoplasm and margination of chromatin, (x 16,700); d) pyknosis and beginning of disappearance of nuclear membrane (x 21,500).

more clumped with the persistence of ischemia. The nuclei show pyknosis, and at late irreversible stages the outer nuclear membrane disappears and finally the nucleus dissolves (Fig. 2).

The sarcomeres appear ultrastructurally unchanged during the development of ischemic injury except that relaxation with prominent I-bands prevails at later stages. With the progression of irreversible injury a N-band appears at the level of the I-band, the M-line becomes very prominent and the Z-bands gradually disappear (Fig. 3). Contracture bands may be present at all stages of ischemia.

The sarcolemma appears to be intact in reversible injury and only in irreversible injury it is disrupted, showing defects while the basement membrane seems to be

13

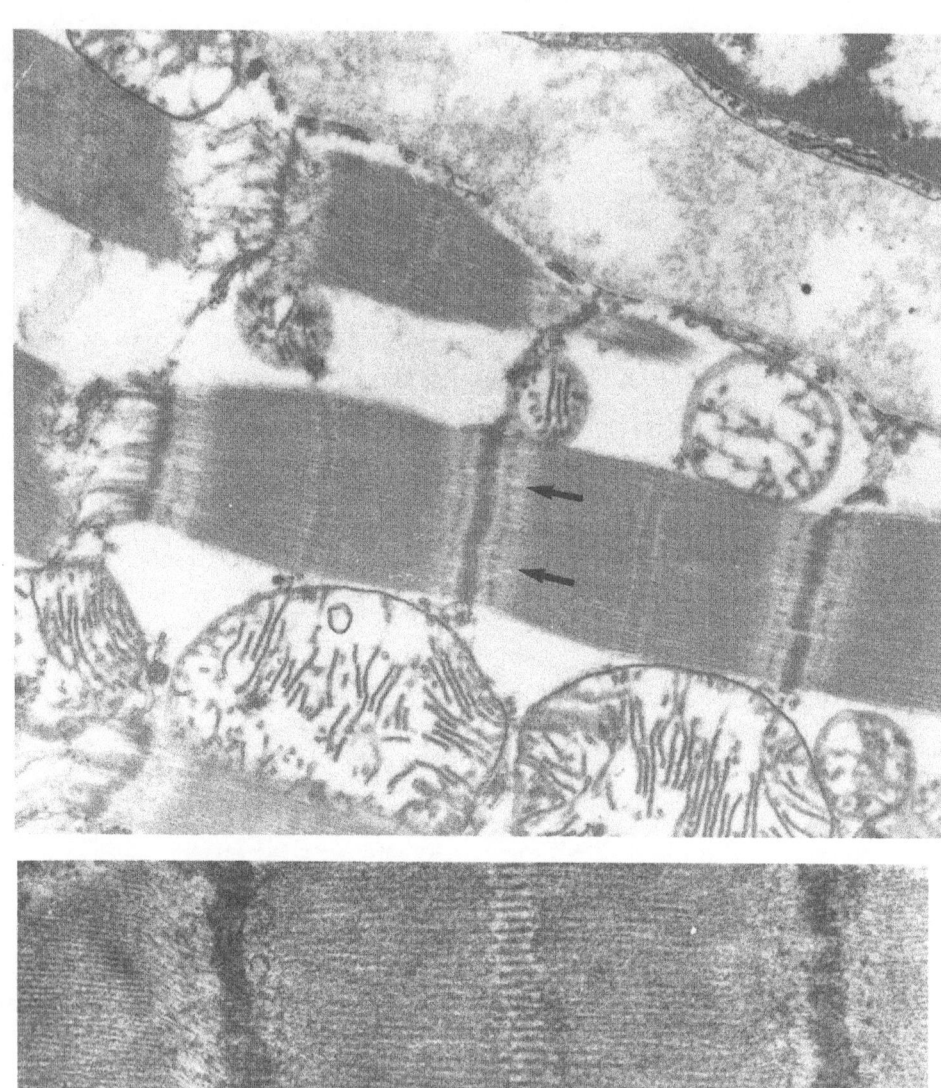

Fig. 3. Alterations in the sarcomeres: a) appearance of an N-band (arrow) within the I-band in sarcomeres in relaxation, (magnification x 29,400); b) very prominent M-band in sarcomere that also exhibits a distorted I-band (x 54,300).

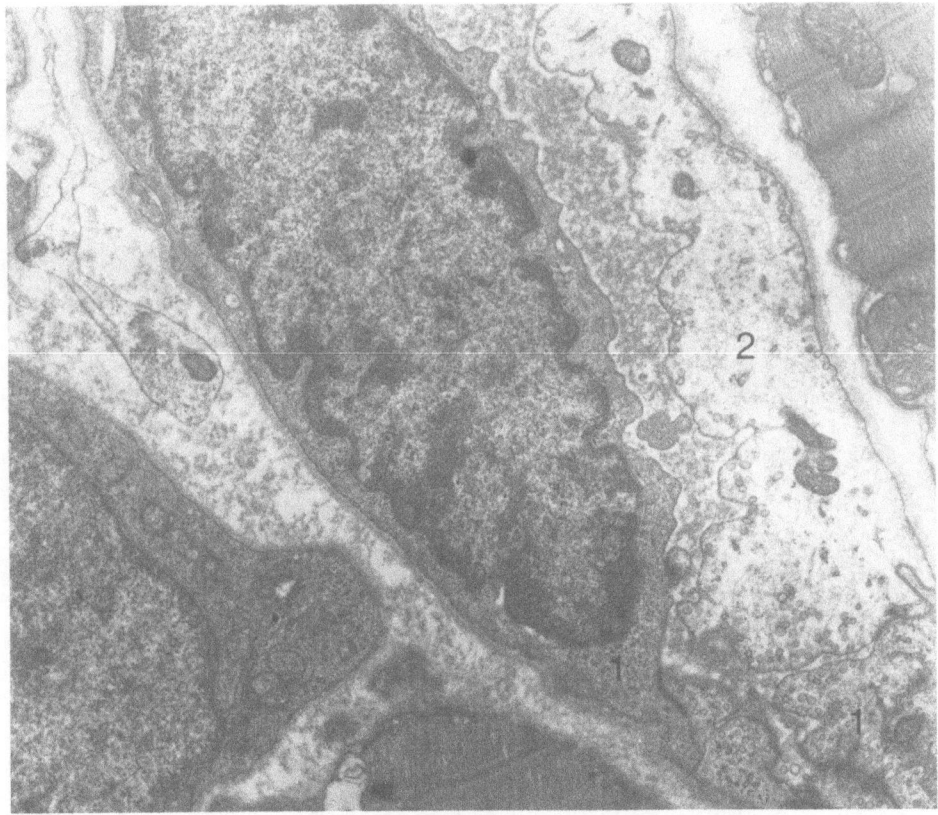

Fig. 4. A capillary shows endothelial cells with different reactions to ischemia. One cell appears to be normal (1), the other is significantly swollen (2) (magnification x 11,300).

unchanged. The sarcoplasmic reticulum and the T-tubular system may show dilation as is also the case for the rough endoplasmic reticulum and the Golgi apparatus. In irreversible injury these subcellular organelles gradually disappear as is also the case for lysosomes. Phospholipid accumulation and myelin figures persist during all stages of ischemia. Glycogen is reduced in amount during ischemia and the cells are usually completely depleted of this finely granular material in the more advanced destruction of the tissue.

The extracellular space often exhibits widening and evidence of edema. The endothelial cells of capillaries, venules and arterioles are either swollen or very thin structures but many remain of almost normal appearance. It is interesting that several endothelial cells from one particular blood vessel show different alterations due to ischemia as if each cell had an individual tolerance to ischemia (Fig. 4). Generally, blood vessels seem to be more resistant to ischemia than myocardial cells, i.e. they lag behind with regard to the time of occurrence of the point-of-no return when cell death occurs.

15

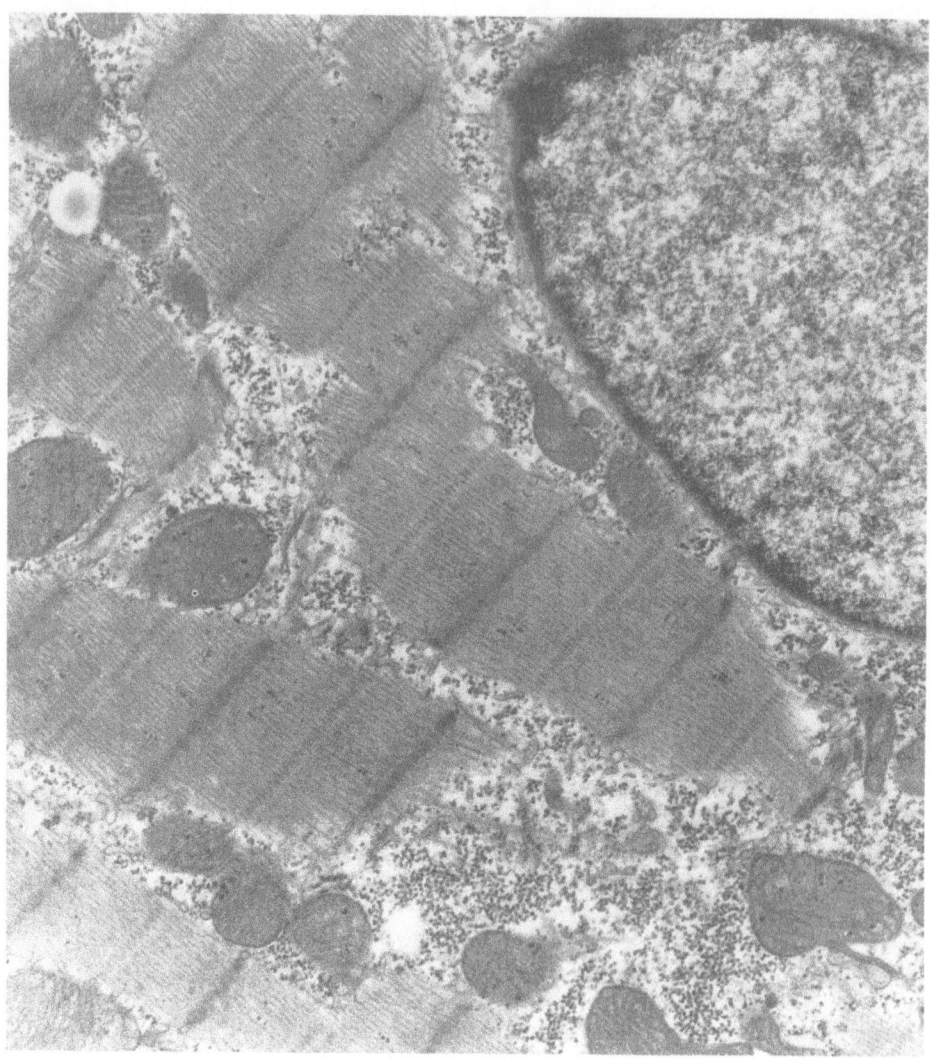

Fig. 5. Normal myocardium (magnification x 20,600).

On the basis of a careful evaluation of all ultrastructural changes in ischemic myocardium and by comparing these morphological findings to metabolic and functional data from the same hearts it was possible to standardize the different stages of ischemic injury as mentioned in the Introduction. Compared to the state of normal myocardium, reversible injury was differentiated (subdivided into slight, moderate, and severe injury) and additionally irreversible injury was characterized. The different phases of ischemic injury with their typical ultrastructural appearance are illustrated in Figs. 5−9 and in Table 1 all alterations typical of each degree of injury are schematically presented.

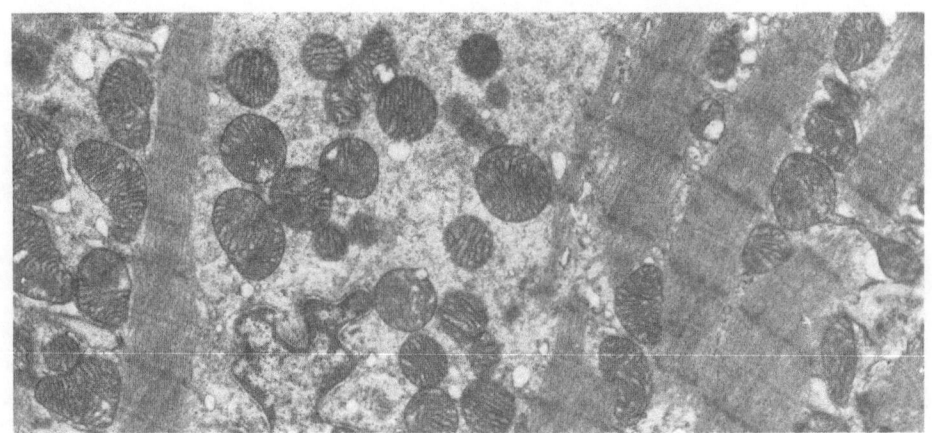

Fig. 6. Slight ischemic injury (magnification x 10,500).

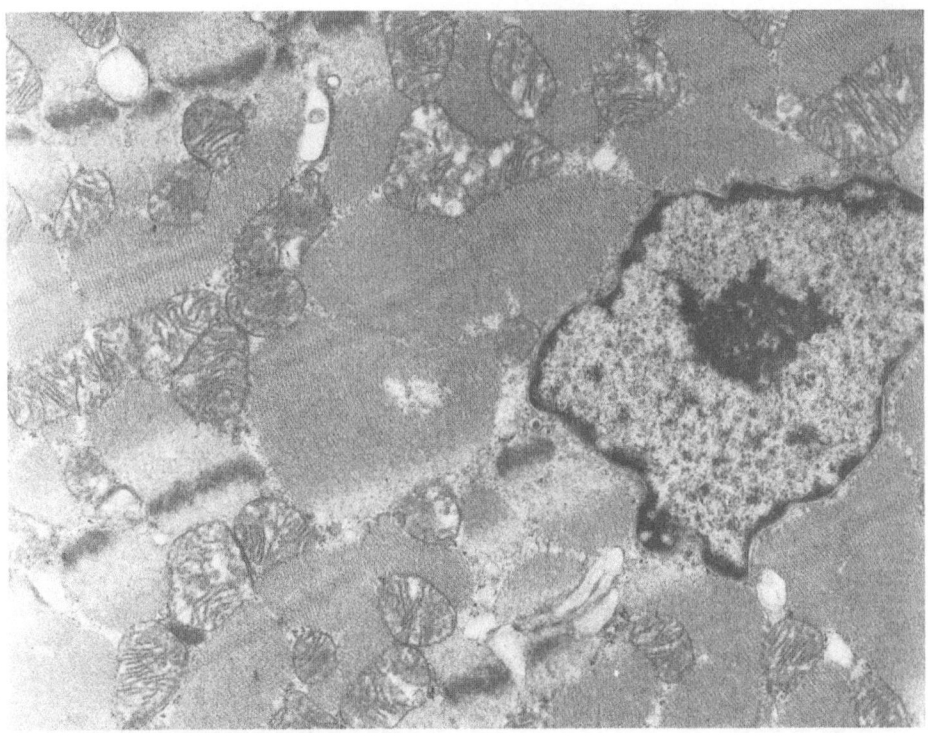

Fig. 7. Moderate ischemic injury (magnification x 15,700).

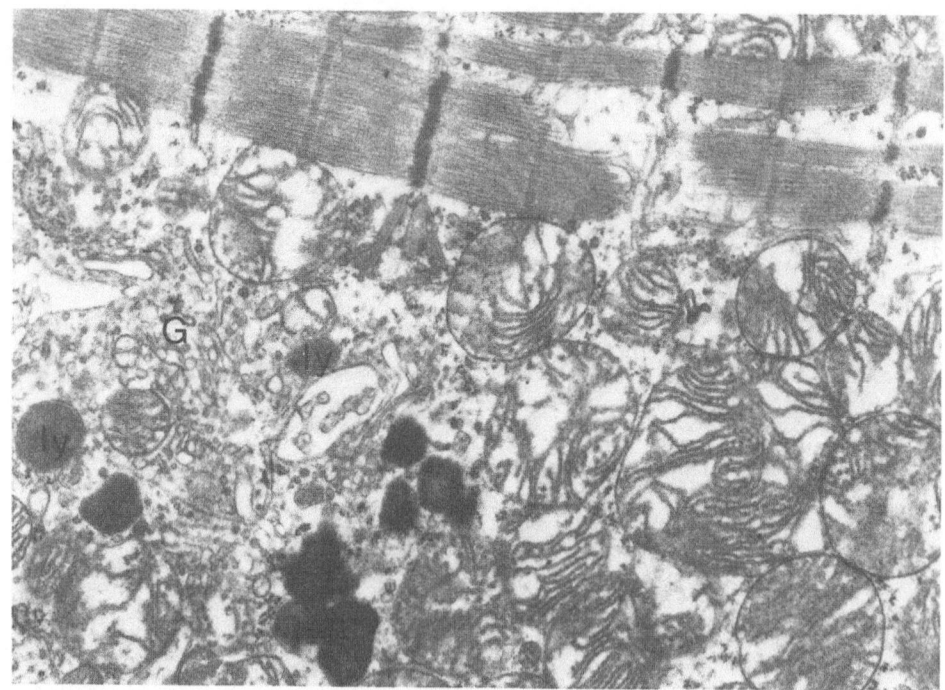

Fig. 8. Severe ischemic injury. Persistence of lysosomes (ly) and Golgi-apparatus (G) (magnification x 16,300).

Fig. 9. Irreversible injury (magnification x 9,700).

18

Table 1. Calibration of ischemic injury

Myocardium		Degree of injury				
		none = normal	slight	mod.	sev.	irre.
Myocytes						
Sarcolemma	Normal	+	+	+	+/−	+/−
Oedema	Present	−/+	−/+	+	+	−/+
+ or −						
Nucleus						
+ or −	Normal	+	+	−	−	−
+, ++, +++	Clearing[a]	−	−/+	+/++	++	+++
+, ++, +++	Clumping[a]	−	−	+/++	++	+++
+ or −	Margination	−	−	+	+	+
Mitochondria						
+ or −	Normal granules	+	−	−	−	−
	Flocculent densities	−	−	−	−	+
	Broken cristae[a]	−	−/+	+/++	++/+++	+++
	Clearing[a]	−	+	+/++	++/+++	+++
Myofilaments						
	Contraction	+	+	+	+	+
	Relaxation	−	−	+/−	+	+
+ or −	Contracture	(+)	+	+	+	+
	Destruction	−	−	+/−	+	+
Cellular organelles and content						
	SR and T systems	+	+/−	+/−	−	−
	Golgi, ER, ribosomes	+	+	+	−	−
	Lysosomes	+	+	−/+	−/+	−
	Phospholipid	+	+	+	+	+
+ or −	Myelin figures	−	−/+	−/+	+	+
	Glycogen	+	−/+	+	−	−
	Lipid	+/−	+/−	+/−	+	+
	Protein	−	−	−/+	−/+	−/+
Microvessels						
Endothelium						
	Normal	+	−/+	−/+	−/+	−/+
	Swollen	−	−/+	−/+	−/+	−/+
+ or −	Thinned	−	−	−/+	−/+	−/+
	Protrusions	−	−/+	−/+	−/+	−/+
Endothelial nuclei						
	Normal	+	−/+	−/+	−/+	−/+
	Clearing	−	−/+	+	−/+	−/+
+ or −	Clumping	−	−/+	−/+	−/+	−/+
	Margination	−	−/+	−/+	−/+	−/+
Extracellular space						
	edema	−/+	−/+	−/+	−/+	−/+

+ present, − absent; [a] grading (+, ++, +++) for nuclei and mitochondria. Each column indicates ultrastructural symptoms typical of one particular degree of ischemic injury. The occurrence of flocculent densities in the mitochondrial matrix marks irreversible injury.

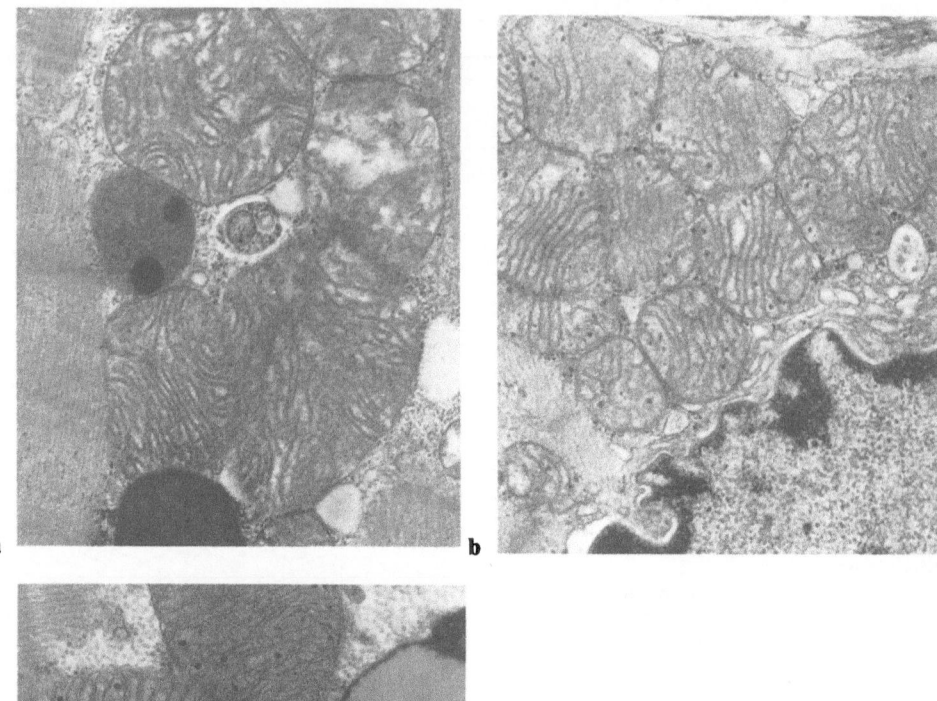

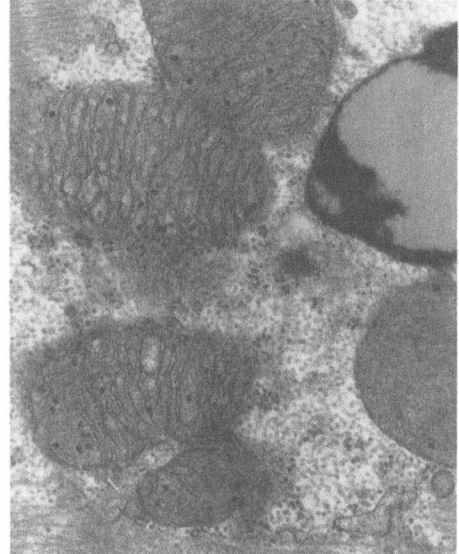

Fig. 10. Recovery of structural integrity during reperfusion of reversibly injured tissue. a) moderately injured mitochondria at the end of ischemia (magnification x 18,500); b) reappearance of normal granules and matrix density at early reperfusion (magnification x 22,500); c) complete reconstitution of the mitochondrial structure at late reperfusion (magnification x 16,500).

Ultrastructural Changes in Reperfused Myocardium

Upon restoration of blood flow reversibly injured myocardial cells begin to recover, structurally as well as functionally. A certain completion of structural restoration seems to be the prerequisite for functional recovery, i.e., the latter lags behind the occurrence of structural integrity. The mitochondria, which are those cellular organelles that are most susceptible to ischemic injury show the first signs of structural recovery by reappearance of the dark matrical granules. The matrix itself

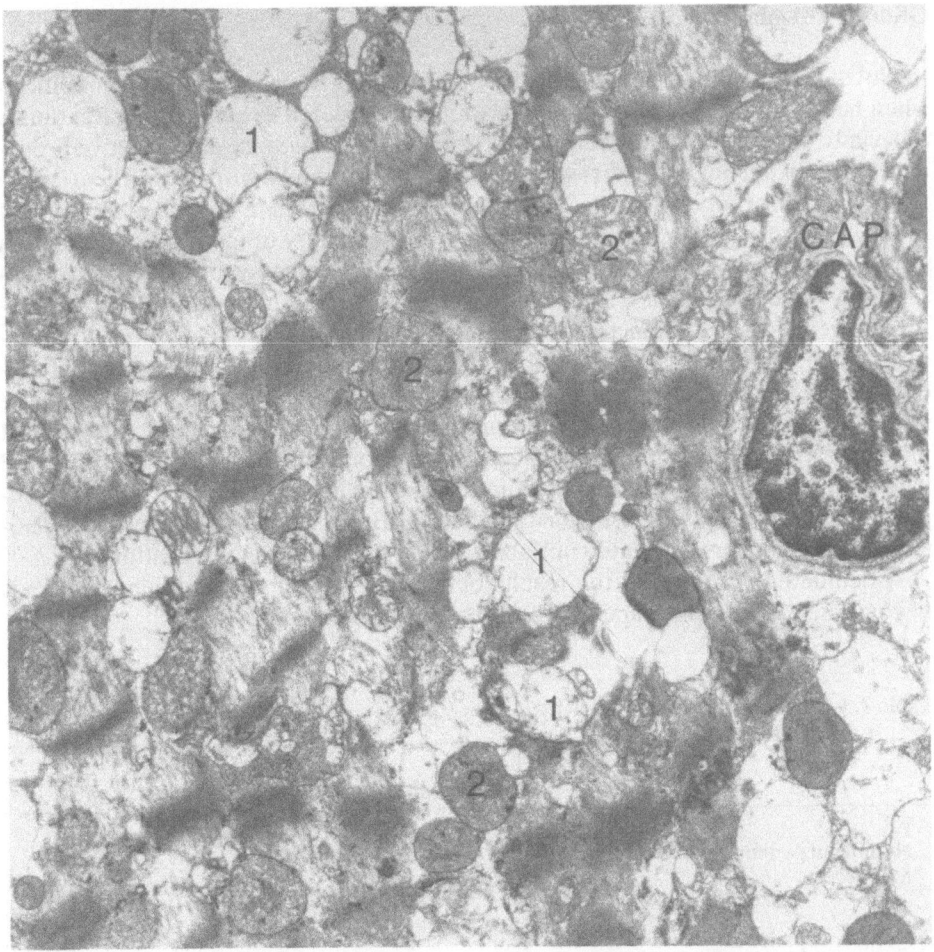

Fig. 11. Reperfusion of irreversibly injured tissue leads either to more severe mitochondrial damage (1) or to a temporary increase in matrix density with persisting amorphous densities (2). Contracture bands are evident, the sarcolemma is destroyed while the basement membrane persists. Note the intact capillary (cap) next to the irreversibly injured cell (magnification x 8,000).

becomes increasingly electron-dense and the cristae more numerous. The nucleus exhibits an increased electron-density and the chromatin is again homogeneously distributed with a slight margination at the nuclear membrane (Fig. 10).

Reperfusion of irreversibly injured myocardium leads to further structural deterioration, i.e., the nuclei show pyknosis, the mitochondria exhibit an increased electron-density while the amorphous densities persist, and the most obvious feature are numerous contracture bands (Fig. 11).

Reperfusion of irreversibly injured tissue very often results, especially at late stages, in hemorrhagic necrosis after ischemia of several hours duration. Functional recovery, of course, is absent in hearts that are reperfused after irreversible injury due to ischemia.

21

Global Ischemia

In global ischemia the whole heart is ischemic. Apart from experimental conditions when hearts in the Langendorff preparation are rendered ischemic, this situation is especially important in daily routine in cardiac surgery because all corrective procedures are carried out in the arrested, i.e., non-beating and non-perfused heart.

The classification of ischemic injury in different degrees as described above is of practical importance for the determination of the quality of myocardial protection during induced cardiac arrest in cardiac surgery. By evaluating intraoperative biopsies removed during and after ischemia it is possible to clearly identify the various factors influencing the tolerance to global ischemia such as: time, myocardial temperature, mode of inducing arrest and the quality and composition of the different cardioplegic solutions used during maintenance of arrest. It is now evident [16] that the ischemic tolerance of patients with coronary heart disease is superior to that of patients with severely hypertrophied hearts. Two hours of ischemia are tolerated well when local as well as total body hypothermia are applied, and three different cardioplegic solutions provide satisfactory protection (Bretschneider, Hamburg, St. Thomas). These data are presented in Table 2 taken from a recent publication [16].

A detailed description of the results obtained by our group and many others is beyond the scope of this paper (for further information see [2, 9]).

Table 2. Ischemic injury with different cardioplegic solutions or intermittent fibrillation.

	Kirsch	Bret- schneider	Hamburg	St. Thomas' Hospital	Intermittent fibrillation	Total
AVR						
Number of patients	26	25	12	17		80
Mean score	2.8	2.0	2.0	1.8		
CABG						
Number of patients	10	123	24	56	17	230
Mean score	2.7	1.5[a]	1.6[a]	1.6	2.6	

AVR-aortic valve replacement; CABG-coronary artery bypass grafting. Score: 0 − normal, 1 − slight, 2 − moderate, 3 − severe ischemic injury; mean scores were calculated from individual scores for each patient. The Bretschneider, Hamburg, or St. Thomas cardioplegic solutions provide satisfactory myocardial protection. Patients with coronary heart disease possess a better tolerance to ischemia. [a] $p < 0.05$

Regional Ischemia

Following occlusion of a coronary artery, regional ischemia develops; it differs from global ischemia because the ischemic zone is surrounded by well perfused, contracting myocardium and this difference is reflected in the ultrastructural appearance of the tissue.

In principle, the three different stages of reversible ischemia may again be separated from that of irreversible injury, i.e., the evaluation system described in the beginning of this text is applicable in regional ischemia as well. The changes occurring in mitochondria as well as in nuclei are identical with those in global ischemia. Alterations of the sarcomeres are somewhat different. In global ischemia, the sarcomeres exhibit more pronounced relaxation with increasing severity of the ischemic insult. In regional ischemia, however, the situation is more complex because of the mechanical stretch of the contracting myocardium acting on ischemic tissue and because the infarcted area is usually only partially, but not totally ischemic due to collateral blood flow. Therefore, the existence of relaxed myofibrils next to contracted myofibrils in the same cell and the concurrent incidence of contracture bands are not surprising.

It is interesting to observe in irreversible ischemic injury persisting for a longer period of time that the dissolution of the sarcomeres starts at the I-band, i.e., the actin filaments dissolve first, and at later stages the Z-line disappears. Myosin, i.e., the A-band persists even after the disappearance of nuclei, mitochondria, as well as the sarcolemma (Fig. 12).

The basement membrane persists even in far advanced irreversible injury so that the myocytes are still recognizable as a cellular unit.

Table 3 shows how irreversible injury proceeds and what the typical ultrastructural signs are of this deterioration process. The classification of different stages of irreversible injury has been useful for the determination of the time of occurrence of necrosis in experiments with a rather long duration of ischemia, such as 24 h. It also is useful in studies concerned with the effects of induced reperfusion by thrombolysis in human myocardium.

Loss of glycogen and accumulation of lipid droplets are ultrastructural symptoms indicative of severe disturbances in lipid and carbohydrate metabolism of ischemic myocardial cells (Fig. 13). Fine granular material of medium electron-density is frequently observed in many regions of ischemic myocytes. This amorphous material is believed to originate from the degradation of all kinds of cellular organelles, especially from mitochondria and myofilaments, as well as from the denaturation of cytoplasmic proteins by ischemia. The amount of amorphous proteins increases significantly with increasing severity of ischemia and is particularly prominent in late stages of irreversible injury.

Many blood vessels are damaged and show swelling or shrinkage of the endothelial cells. As in global ischemia, it is interesting to observe an individual tolerance to ischemia in individual endothelial cells. In the canine heart capillaries can survive tissue necrosis and still carry blood cells reflecting the remaining collateral flow. In these capillaries neutrophil granulocytes are observed to stick to the endothelium and later to penetrate into the extravascular space where they closely adhere to and engulf irreversibly injured cells (Fig. 14). In the porcine heart where blood flow during ischemia is minimal because of the absence of collaterals, most capillaries are damaged and later destroyed by ischemia. Occasionally, however, capillaries are seen to maintain their integrity, indicating the increased tolerance of endothelial cells to ischemia as compared to the tolerance of myocytes (Fig. 14).

The extracellular space occasionally is edematous but this is not a very regular phenomenon. Extravasation of blood cells is common as is deposition of fibrin and proteinous material. Cellular debris is present and necrosis and concomitant proliferation of interstitial cells – indicating the beginning of repair processes – are obvious.

In the stage of ischemia when irreversible injury begins to develop, inhomogeneity of the degree of injury of neighboring cells is a typical feature (Fig. 15). This fact is most probably due to the collateral flow distributed heterogeneously which permits some cells to survive longer than others. In contrast, in global ischemia all cells are uniformly damaged. Another ultrastructural sign typical of regional myocardial ischemia is the occurrence of numerous lipid droplets in the myocardial cells. This may be indicative of the inability of the ischemic myocardial cell to metabolize triglycerides or free fatty acids that are carried to the ischemic area by collateral flow in the dog heart. In the pig heart, these lipid droplets are absent from ischemic myocardium which is easily explained by the absence of collateral flow. This lipid accumulation is also absent in global ischemia because there is no resting flow present.

The ultrastructural characteristics of regional ischemia in the myocardium have been discussed in detail in several recent papers from our group [13, 14, 17]. We believe that ischemia is a multifactorial event involving all subcellular components of the various cell types present in the myocardium. We further believe that the cause of cell death should not be sought in alterations of one particular cellular organelle alone, such as, for example, the mitochondrion or the sarcolemma, but that changes in all these different structures contribute to ischemic injury and, finally, to cell death.

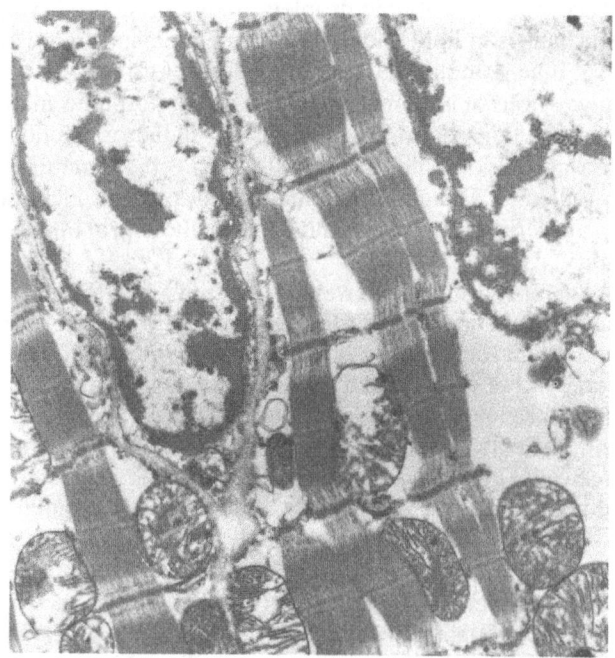

a

Fig. 12. Progression of irreversible injury. For further information see Table 3. a) early irreversible injury (magnification x 18,000); b) stage 3 in irreversible injury: beginning of destruction of the Z-line and the I-band (x 13,300); c) stage 4 in irreversible injury: only the A-band (myosin filaments) are still present (x 12,600).

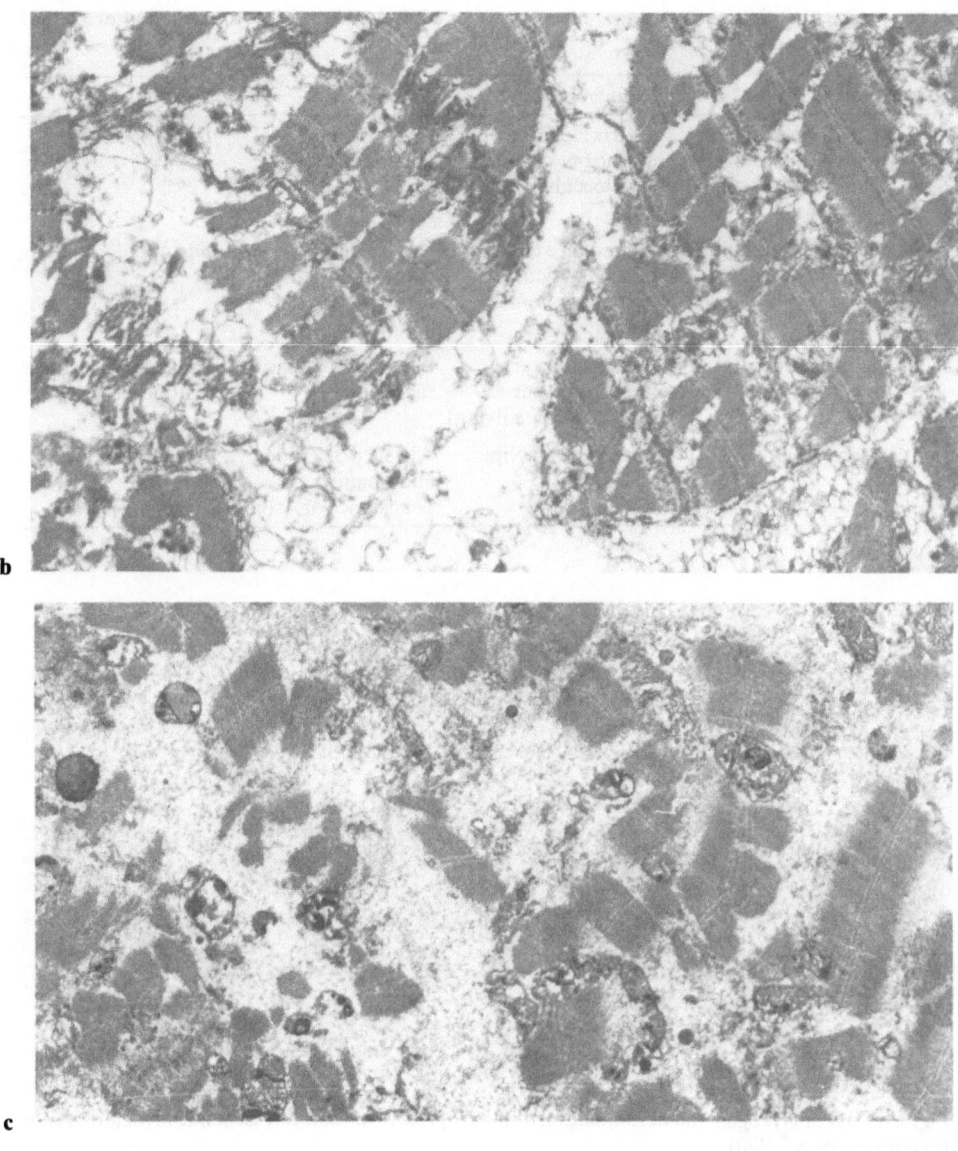

b

c

As has been shown earlier, the occurrence of irreversible injury is not only dependent on the duration of ischemia, but it also depends on the amount of collateral blood flow and on the level of oxygen consumption at the time of occlusion [19]. Species differences are another factor influencing the speed of development of myocardial ischemic injury and infarction [18]. This difference, however, may be reduced to differences in oxygen consumption as well as in collateral flow. Rats possessing a high heart rate and no collaterals develop an infarct (90% of area at risk at 45 min), i.e.,

Table 3. Classification of four stages of progression of irreversible injury observed after 24 h of regional ischemia

Stage	Nuclei	Mitochondria	Myofilaments	Sarcolemma
1.	Clearing of nucleo-plasm; clumping of chromatin; swelling; pyknosis	Matrix cleared; cristae broken; few flocculent densities	Contraction; contracture bands;	Intact or disrupted
2.	Clearing; clumping; swelling; pyknosis	Abundance of flocculent densities; outer membrane intact	Contraction; contracture bands; disappearance of Z-lines	Disrupted
3.	Loss of structural integrity	Loss of structural integrity (outer membrane dissolving)	Absence of Z-lines; relaxation	Progressive disruption
4.	Absent	Lysis; phospholipid inclusions	Absence of Z-lines and I-bands; persistence of A-bands	Absent

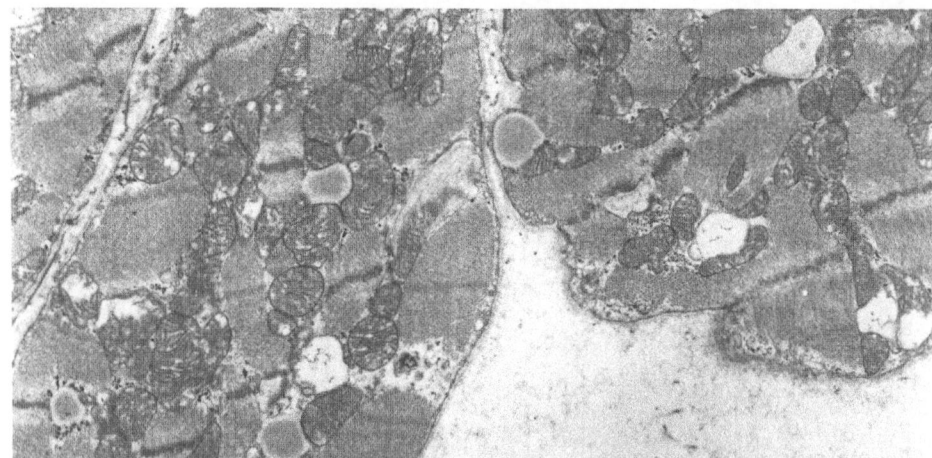

Fig. 13. Loss of glycogen and accumulation of fat droplets in moderately ischemic myocardial cells (magnification x 8,900).

irreversible injury, within 1 h of occlusion, whereas cats show a much slower time course of myocardial necrosis (12% of the area at risk at 90 min, 70% at 6 h). Dogs, and most probably humans, fall between these two extremes of the curve [17, 18].

Reperfusion of regionally ischemic myocardium leads to recovery of the structural integrity of the tissue as described above, as long as reversible damage exists. This knowledge is implemented in many cardiology centers where reperfusion following acute coronary artery occlusion is induced in patients by thrombolysis or by angio-

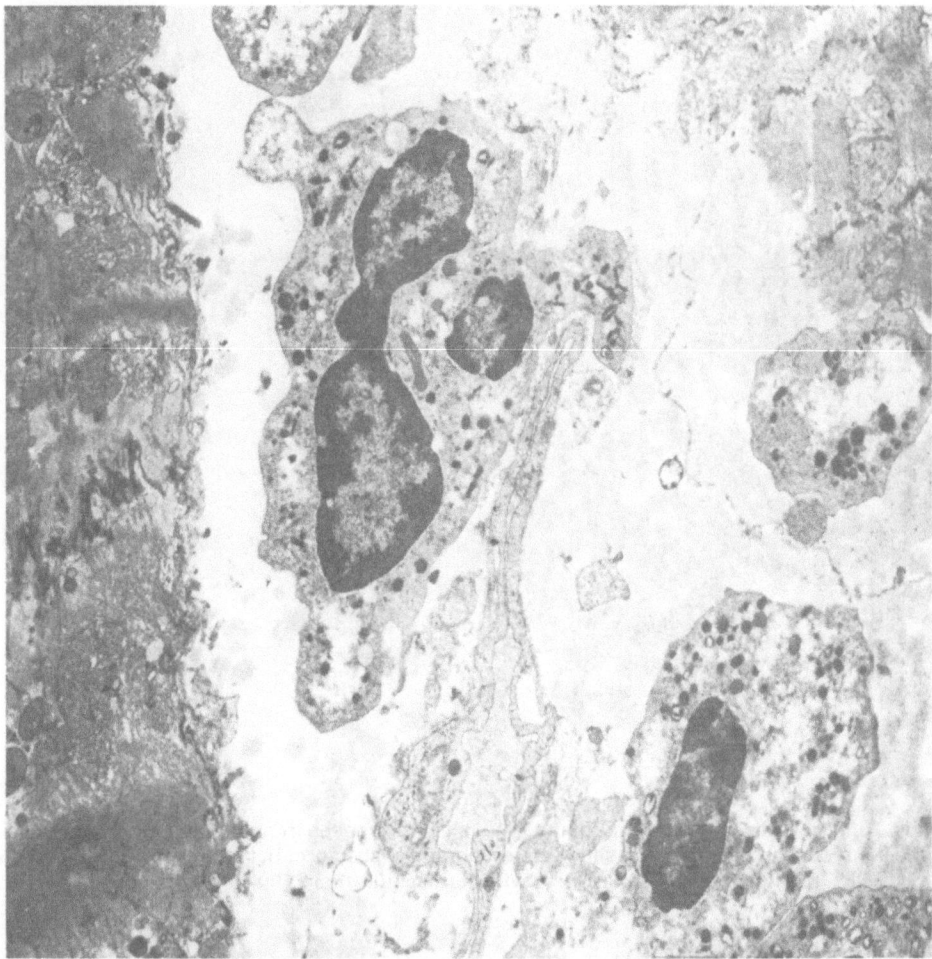

Fig. 14. Extravasation of neutrophil granulocytes into the extravascular space, most probably from the microvessel in the center of the picture. Two irreversibly injured myocardial cells are present. At the right side of the picture a neutrophil has penetrated the basement membrane and adheres to the destroyed cell (magnification x 11,000).

plasty. Time plays a critical role in this procedure and presently a 4 h time lag is accepted as the upper limit for successful reperfusion after the onset of symptoms [1].

Effects of Multiple Ischemic Events on the Myocardial Ultrastructure

Patients undergoing coronary bypass surgery because of stenosis or occlusion of one or several coronary arteries have had multiple episodes of angina pectoris over their preoperative period of perhaps many years. Even in patients lacking any evidence of infarction the left ventricular wall motility is regionally reduced and that may be

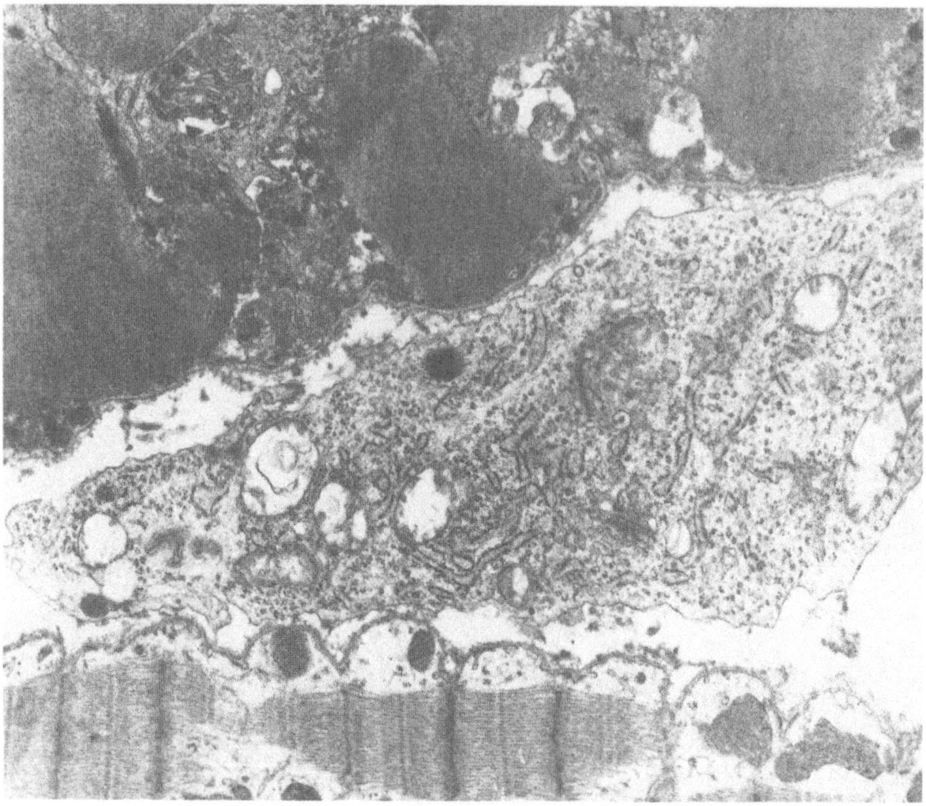

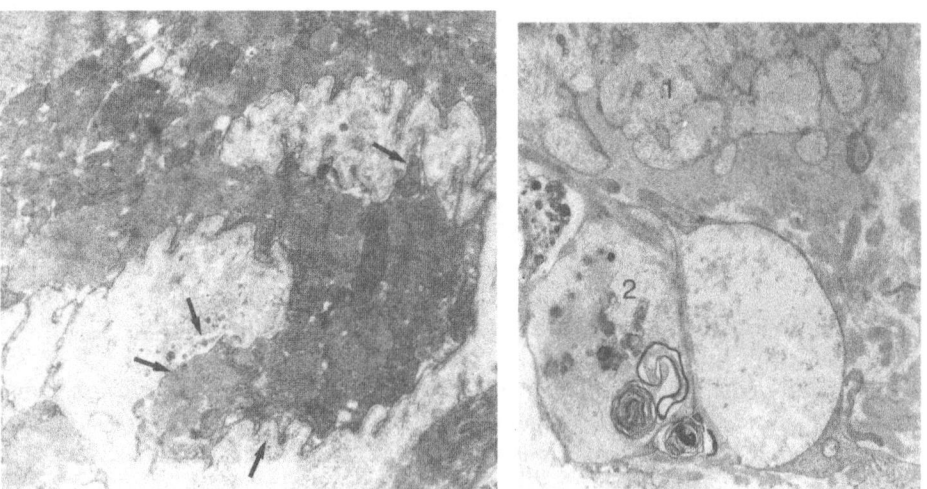

Fig. 15. Heterogeneity of ischemic injury. The cell in the upper part of the picture shows irreversible injury, that in the lower part shows almost normal appearance. In between these two myocardial cells a proliferating interstitial cell is evident (magnification x 10,800).

a

b

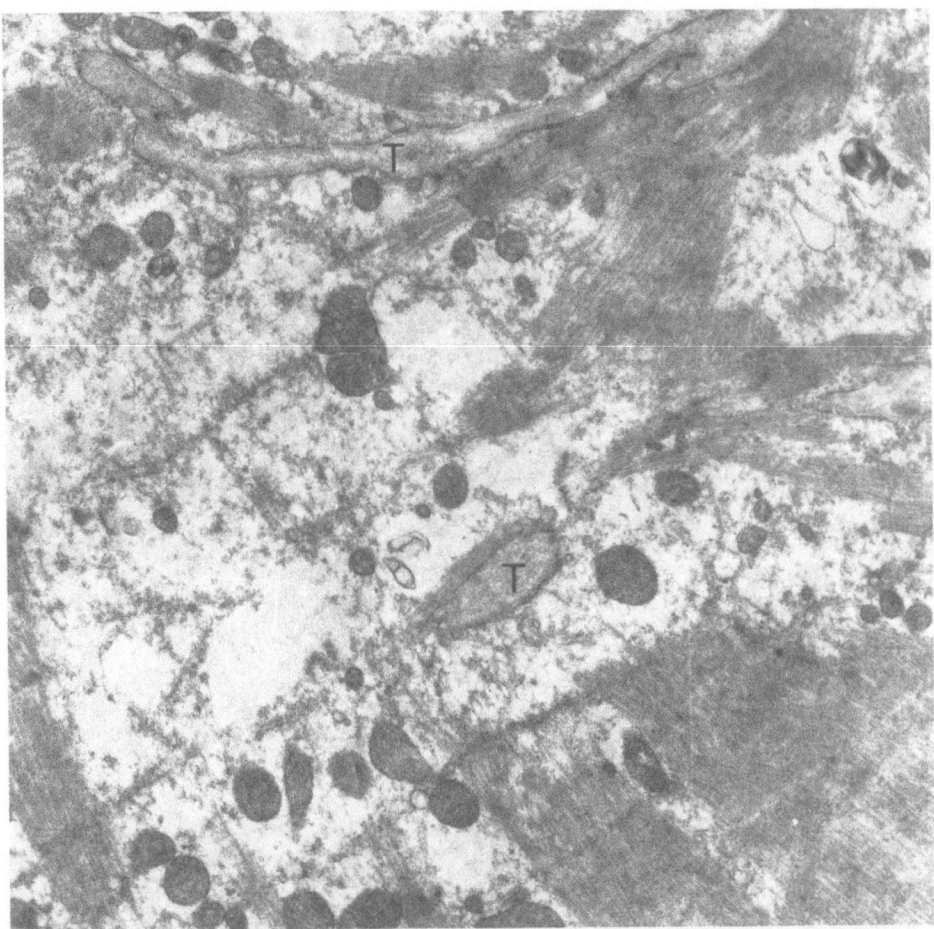

Fig. 17. Human myocardium from a patient with coronary heart disease. Area of myofibrillar lysis in a myocyte of normal size. Note also the dilated and irregular T-tubular system (T) (magnification x 10,800).

defined as hypokinesia or akinesia. In an attempt to study the morphological correlate of reduced regional myocardial function intraoperative needle biopsies were obtained from the diseased areas during surgery and then studied by electron microscopy. All biopsies were removed while the hearts were still beating, before the induction of cardiac arrest.

Numerous morphological changes were observed by light and electron microscopy which will be described in a detailed manner in the following text.

Fig. 16. Human myocardium from patients with coronary heart disease. a) Myocardial cells of very irregular shape that show sequestration (arrows) into the extracellular space. Contractile material is still present in this atrophic cell; b) absence of contractile material in a cell showing sequestration (1). Remnant of myocardial cell containing myelin figures (2) (magnification x 5,900).

29

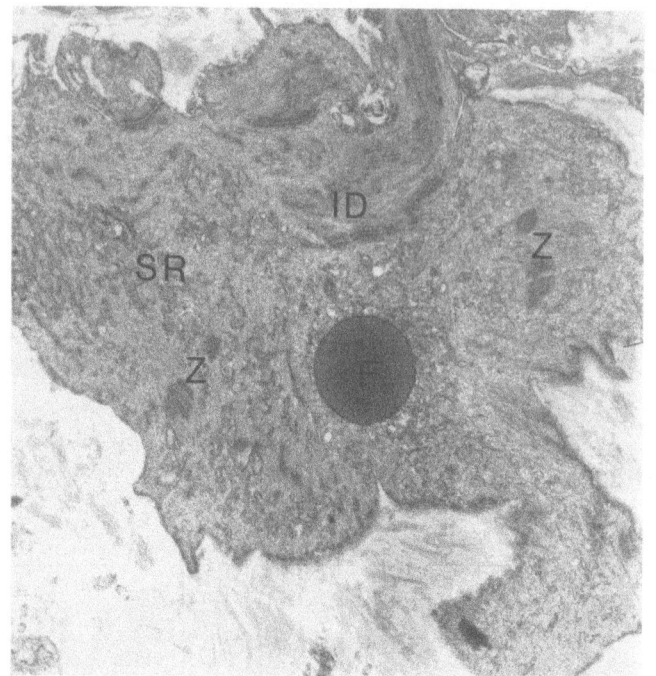

a

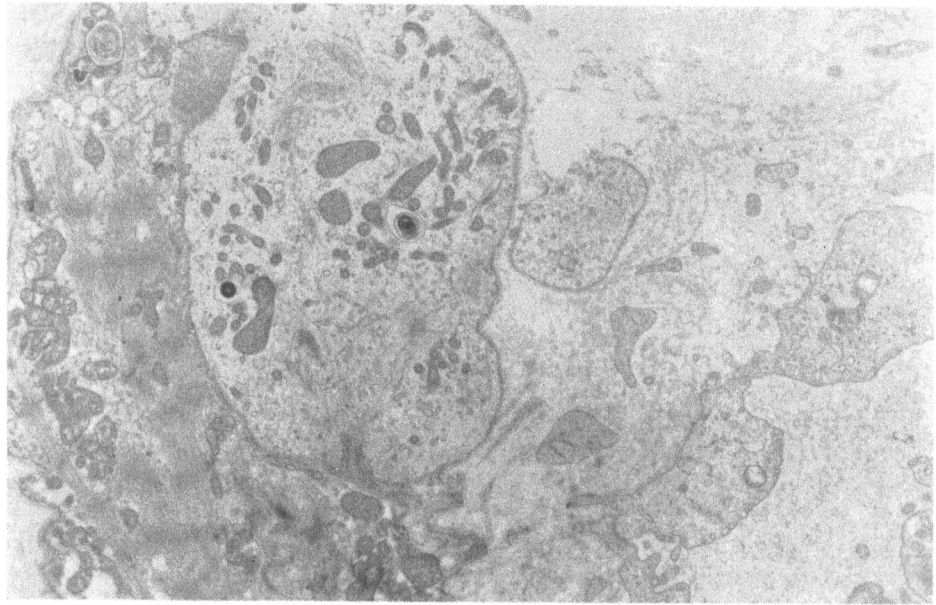

b

Fig. 18. Human myocardium from patients with coronary heart disease. Loss of contractile elements in atrophic cells: a) absence of regular sarcomeres in the presence of disorganized myofilaments showing also Z-line changes (Z). Sarcoplasmic reticulum is abundant (SR), the intercalated disc is degenerated (ID). F = fat droplet (magnification x 6,500); b) almost complete absence of myofilaments in an atrophic cell also showing sequestration. The neighboring cell contains several sarcomeres but is also smaller than normal (magnification x 7,800).

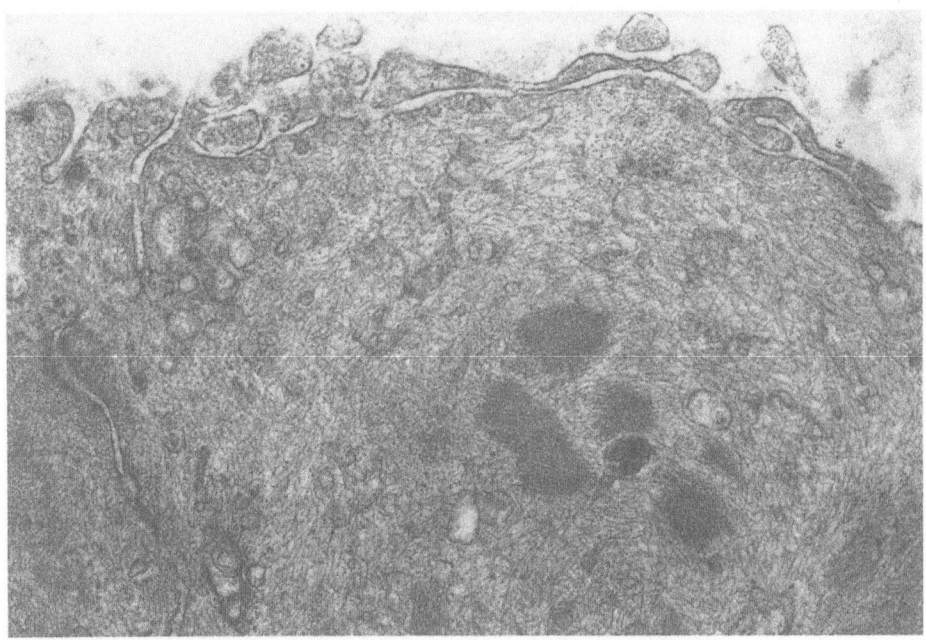

Fig. 19. Human myocardium from a patient with coronary heart disease. Accumulation of Z-line material (magnification x 16,200).

By light microscopy, an increased amount of connective tissue was present and this amount was different in different biopsies ranging from 12% to 37% with a mean value of 19% to 25% as measured by morphometry. These values are in agreement with those reported by Thiedemann [21] and Flameng [3]. Normal values for the content of interstitial tissue is 11% in human hearts [11, 12] which is in accordance with the values reported by Knieriem [5] and Schoenmackers [20] but is higher than those reported by Krayenbühl [6, 7] and Unverferth [22]. Many myocardial cells were of abnormal shape and size showing either a reduction in dimension resulting finally in atrophy or an increase in size, i.e., hypertrophy. The irregular shape was confirmed by electron microscopy which revealed highly irregular cell borders that seemed to sequester parts of the cell (Fig. 16). In contrast to normal cells that contain about 60% of contractile material [12] a significant loss of myofilaments was found in diseased human myocardium. Values ranged from 52.0 ± 8.0% in the subepicardium to 48.0 ± 8.0% in the subendocardium in 40 patients investigated by our group [12]. Kunkel [8], and Frenzel [4] reported similar reductions of contractile material in patients with dilated cardiomyopathy, and our own study in patients with longstanding hypertrophy due to aortic valve disease showed a similar significant reduction [11] which raises the question as to the specificity of the ultrastructural changes observed for any particular cardiac disease.

This reduction of myofilaments in myocardium from patients with CHD could be observed in all types of cells independent of their actual size. In Fig. 17 an area of myofibrillar lysis in an otherwise normal cell is shown and Fig. 18 exhibits extreme

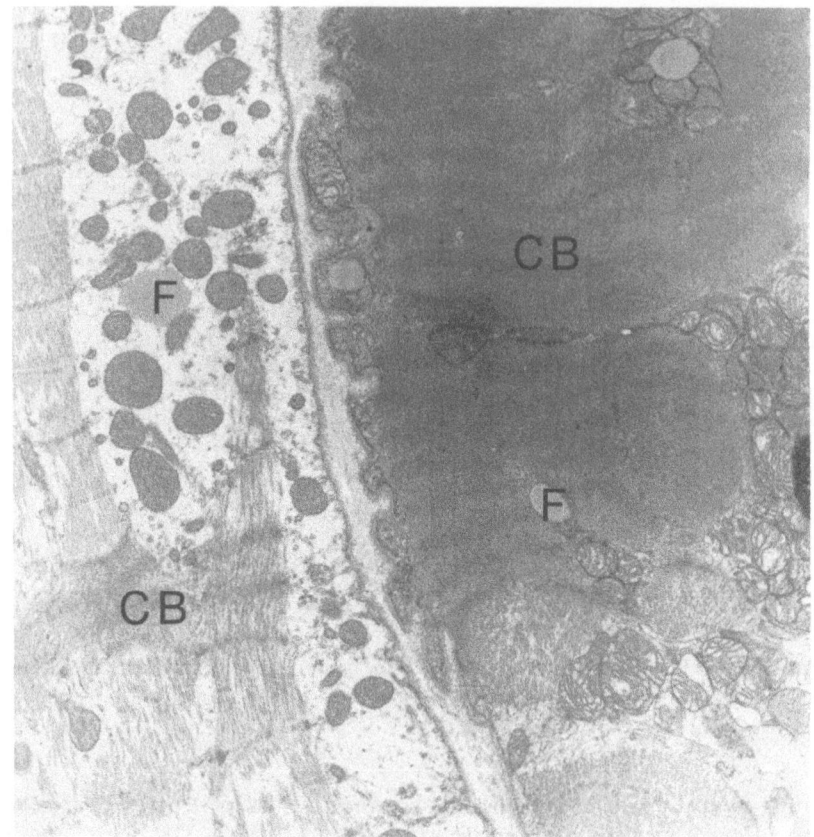

Fig. 20. Human myocardium from a patient with coronary heart disease. Contracture band (CB) in one cell, absence of sarcomeres in the other, with only a small contracture band visible. Note the fat droplets (F) (magnification x 7,400).

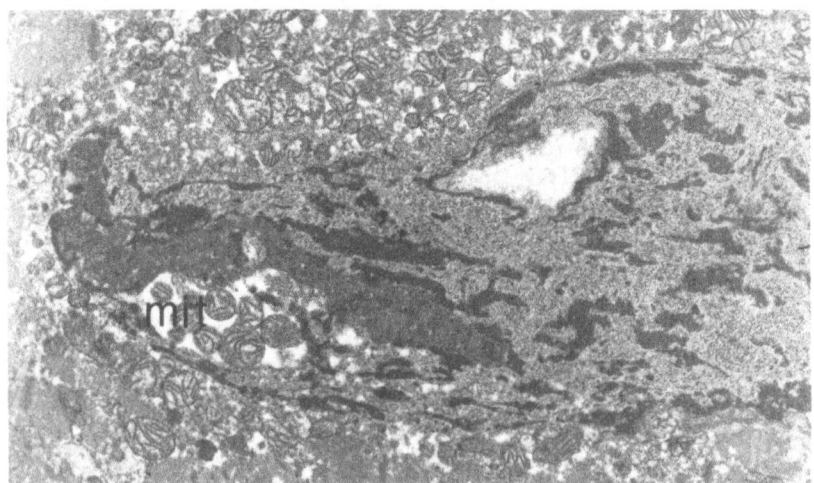

Fig. 21. Human myocardium from a patient with coronay heart disease. Nucleus of abnormal size and shape showing inclusion of mitochondria (mit) (magnification x 6,500).

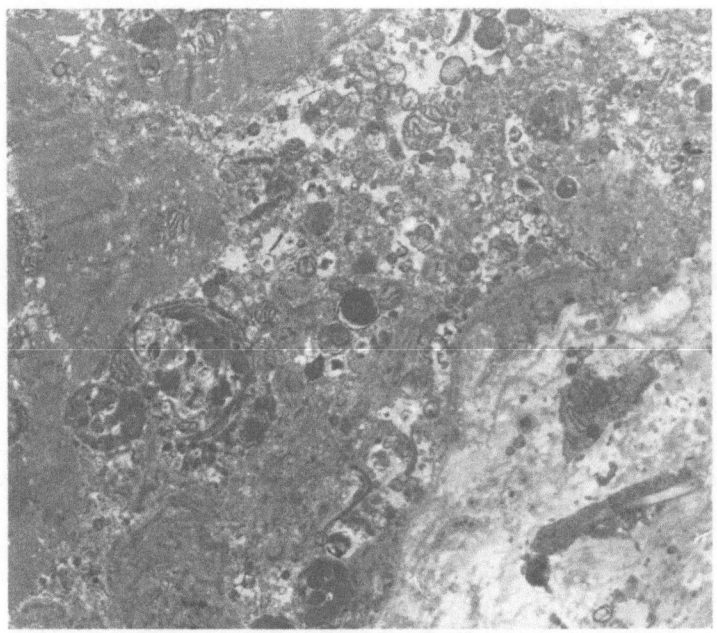

Fig. 22. Human myocardium from a patient with coronary heart disease. Occurrence of cellular breakdown products in a myocardial cell (magnification x 6,500).

loss of contractile material in an atrophic cell. Z-line changes were often observed and are indicative of disarrangement of sarcomeres (Fig. 19).

Contracture bands were often seen (Fig. 20); they are typical of cardiac tissue fixed by immersion in glutaraldehyde. Their occurrence in a frequency greater than in normal tissue may be an indication of an increased vulnerability of the tissue due to pathogenetic processes.

Mitochondria were of varying size and shape and also exhibited degenerative signs such as a reduction of the number of cristae (Figs. 17, 18b, 21).

The nuclei were frequently enlarged and exhibited cytoplasmic inclusions, a very irregular shape, and sometimes a prominent nucleolus (Fig. 21).

Other intracellular alterations included the occurrence of an increased amount of lipofuscin vacuoles, vacuoles filled with debris, myelin bodies, and lipid droplets (Figs. 16b, 18a, 22). Glycogen often was abundantly present (Fig. 23) and the T-tubular system showed dilated and proliferated elements (Fig. 17).

The interstitial space was widened in many instances and contained cellular debris, and macrophages in addition to the normally occurring fibroblasts, fibrocytes, and collagen fibrils (Fig. 24). This appears to be a reaction of the connective tissue elements to the degeneration of myocardial cells, i.e., to their partial sequestration producing cellular debris which is phagocytosed by macrophages as well as by fibroblasts.

Microvessels exhibited swollen endothelial cells in 30% of all biopsies and their basement membrane was frequently thickened.

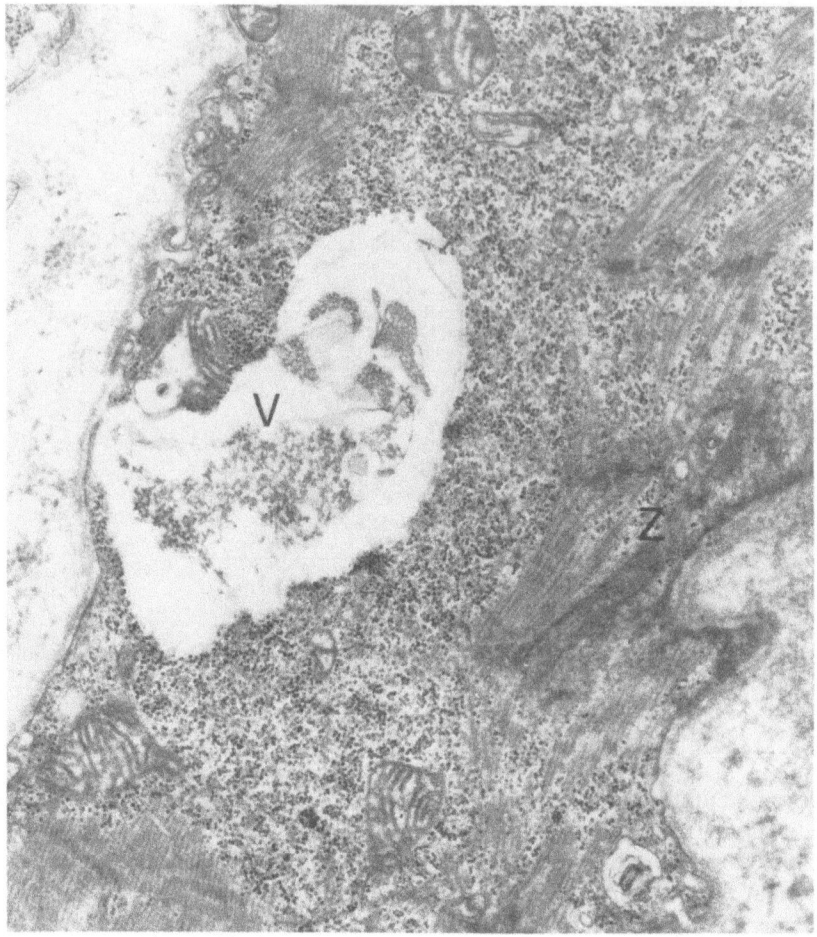

Fig. 23. Human myocardium from a patient with coronary heart disease. Abundance of glycogen in an atrophic cell. Note also the large vacuole (v) and the accumulation of Z-line material (Z) (magnification x 21,000).

In the same group of 40 patients in which the morphometric data cited above were obtained all the different qualitative changes were systematically evaluated as is shown in Table 4.

From this study in human myocardium it may be concluded that multiple ischemic events producing multiple episodes of reversible acute ischemic injury finally lead to slow cellular death by degeneration and a concomitant increase in the amount of fibrosis. These cellular changes are considered to be the morphological correlate to reduced left ventricular wall motility, i.e., hypokinesia in myocardium from patients with coronary heart disease.

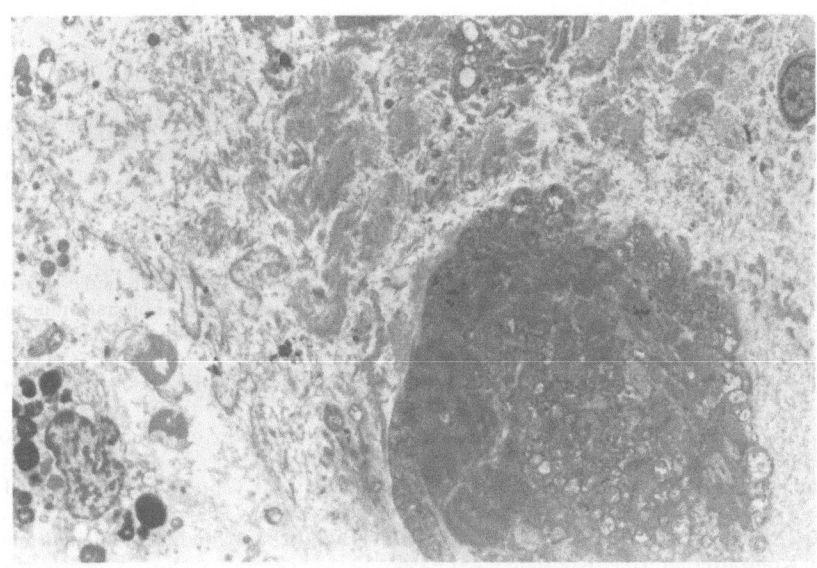

Fig. 24. Human myocardium from a patient with coronary heart disease. The interstitial space is widened and contains cellular debris, macrophages, and abundant collagen fibrils (magnification x 2,000).

Table 4. Qualitative ultrastructural alterations in patients with coronary heart disease

Sarcomeres	
contracture bands	77.5%
lack of contractile material	57.5%
Mitochondria	
irregular shape, very small	37.5%
Nuclei	
enlarged	45.0%
clumping of chromatin	35.0%
irregular shape	32.5%
cytoplasmic inclusions	12.5%
Other intracelluar alterations	
lipofuscin	80.0%
vacuoles	45.0%
lipid	35.0%
vacuoles with debris	32.5%
dilated T-system	32.5%
glycogen present	32.5%
myelin bodies	20.0%
atrophic degenerated cells	15.0%
Interstitial space	
proliferating fibroblasts	57.5%
cellular debris	35.0%
macrophages	10.0%
mast cells	7.5%
Microvessels	
swollen endothelial cells	30.0%
thick basement membrane	22.5%

References

1. Effert S (ed) (1986) Facts and hopes in thrombolysis in acute myocardial infarction. Stein-kopff, Darmstadt
2. Engelman RM, Levitsky S (eds) (1982) A textbook of clinical cardioplegia. Futura Publishing Company, Mount Kisco, New York
3. Flameng W, Suy R, Schwarz F, Borgers M, Piessens J, Thone F, van Ermen H, DeGeest H (1981) Ultrastructural correlates of left ventricular contraction abnormalities in patients with chronic ischemic heart disease. Determinants of reversible segmental asynergy post revascularization surgery. Am Heart J 102:846–857
4. Frenzel H, Kasper M, Kuhn H, Lösse B, Reifschneider G, Hort W (1985) Licht- und elektronenmikroskopische Befunde in Früh- und Spätstadien der Herzinsuffizienz. Untersuchungen an Endomyokardbiopsien von Patienten mit latenter und dilativer Kardiomyopathie. Z Kardiol 74:135–143
5. Knieriem HR (1964) Über den Bindegewebsgehalt des Herzmuskels des Menschen. Arch Kreislaufforsch 44:231–258
6. Krayenbühl HP, Hess OM, Schneider J, Turina M (1983) Physiologic or pathologic hypertrophy. Eur Heart J [Suppl A] 4:29–34
7. Krayenbühl HP, Hess OM, Schneider J, Turina M (1984) Left ventricular function and myocardial structure in aortic valve disease before and after surgery. Herz 5:270–278
8. Kunkel B, Schneider M, Kober G, Kaltenbach M (1984) Bioptische Befunde bei Patienten mit normalen Kranzarterien. Herz/Kreislauf 9:451–456
9. Roberts AJ (ed) (1987) Myocardial protection in cardiac surgery. Dekker Inc., New York Basel
10. Schaper J, Mulch J, Winkler B, Schaper W (1979) Ultrastructural, functional, and biochemical criteria for estimation of reversibility of ischemic injury: A study on the effects of global ischemia on the isolated dog heart. J Mol Cell Cardiol 11:521–541
11. Schaper J, Schwarz F, Hehrlein F (1981) Ultrastrukturelle Veränderungen im menschlichen Myokard bei Hypertrophie durch Aortenklappenfehler und deren Beziehung zur linksventrikulären Masse und Auswurffraktion. Herz 6:217–225
12. Schaper J, Meis M, Froede R (1985) Morphologische Untersuchungen am Myokard von Patienten mit koronarer Herzkrankheit. In: Mall G, Otto HF (eds) Herzhypertrophie. Springer, Berlin Heidelberg New York Tokyo, pp 89–111
13. Schaper J, Alpers P, Gottwik M, Schaper W (1985) Ultrastructural characteristics of regional ischaemia and infarction in the canine heart. Eur Heart J [Suppl E] 6:21–31
14. Schaper J (1986) Ultrastructural changes of the myocardium in regional ischaemia and infarction. Eur Heart J 7:3–9
15. Schaper J, Schaper W (1986) Morphological changes in myocardium from patients with coronary heart disease and cardiac hypertrophy. Adv Cardiol 34:16–24
16. Schaper J, Scheld HH, Schmidt U, Hehrlein F (1986) Ultrastructural study comparing the efficacy of five different methods of intraoperative myocardial protection in the human heart. J Thorac Cardiovasc Surg 92:47–55
17. Schaper J, Schaper W (1988) Time course of myocardial necrosis. Cardiovasc Drugs Ther 2:17–25
18. Schaper W (1984) Experimental infarcts and the microcirculation. In: Hearse DJ, Yellon DM (eds) Therapeutic approaches to myocardial infarct size limitation. Raven Press, New York, pp 79–90
19. Schaper W, Binz K, Sass S, Winkler B (1987) Influence of collateral blood flow and of variations in MVO2 on tissue-ATP content in ischemic and infarcted myocardium. J Mol Cell Cardiol 19:19–37
20. Schoenmackers J (1966) Über den Bindegewebsgehalt des Myokards der linken Herzkammer bei elastischer und unelastischer Koronarsklerose. Arch Kreislaufforsch 50:208–230
21. Thiedemann KU (1979) Ultrastructure in chronic ischemia-studies in human hearts. In: Schaper W (ed) The pathophysiology of myocardial perfusion. Elsevier, North-Holland Biomedical Press, Amsterdam New York Oxford, pp 675–716
22. Unverferth DV, Fetters JK, Unverferth BJ, Leier CV, Magorien RD, Arn AR, Baker PB (1983) Human myocardial histologic characteristics in congestive heart failure. Circulation 68:1194–1200

Authors' address: Prof. Dr. Jutta Schaper, Max-Planck-Institute, Department of Experimental Cardiology, Benekestraße 2, D-6350 Bad Nauheim, FRG

Myocardial Metabolism in Ischemia

Lionel H. Opie

Heart Research Unit, University of Cape Town
Medical School, South Africa

Introduction

Derangements of myocardial metabolism in ischemia are widespread, affecting especially carbohydrates and lipids. Depending on the duration of ischemia such changes may be reversible or irreversible; in the latter case they are deemed to constitute myocardial infarction. The emphasis of this chapter will be on the early myocardial metabolic changes following the onset of coronary occlusion; in addition, the question of reperfusion damage will be briefly covered.

Definition of Ischemia

Myocardial ischemia is a reversible oxygen imbalance of the myocardium, caused by an inadequate coronary blood flow. Coronary artery disease, generally obstructive in nature, is the usual cause, yet additional or alternate causes of myocardial ischemia include: 1) coronary artery spasm or functional obstruction added to organic obstruction, and 2) an increased myocardial oxygen demand resulting from myocardial hypertrophy.

The word "ischemia" is derived from two Greek words, *isch* meaning "too little" plus *aemia* meaning "blood".

Central to all definitions of ischemia is a *lack of an adequate oxygen supply to the mitochondria* with a consequent fall in the energy available to the cytoplasm. The rundown of high-energy phosphate compounds accelerates glycolysis and glycogenolysis, so that glycolytic flux is stimulated to a greater extent than its end-products, pyruvate and $NADH_2$, can enter the mitochondria for oxidation. The end result is production of lactate, the first recognized metabolic "gold standard" of ischemia. Attention has now shifted to the direct monitoring of enhanced glycolysis by the use of labelled glucose (F18-deoxyglucose) which can be visualized non-invasively with positron emission tomography.

Currently, continuous electrocardiographic monitoring of ischemia is increasingly frequently undertaken. ST-segment shifts result from transsarcolemmal loss of potassium ions and localized depolarization; potassium loss is very closely linked to early myocardial energy deterioration and egress of lactate and phosphate. Such episodes of ST-segment change can occur in the absence of any subjective pain to constitute *silent ischemia* which, however, may be accompanied by other features of "true ischemia" such as regional wall abnormalities, or volume changes.

Anoxia and Glycolysis

To understand the complex metabolic patterns found in ischemia, an analysis of the simplified situation found in anoxia is required. Yet there are important metabolic differences between anoxia and ischemia. The basic defect in anoxia is inadequacy of oxygen supply. In ischemia, the almost invariable existence of at least some collateral flow means that there is a state of not total anoxia but severe hypoxia. This basic defect is combined with inadequate washout of metabolites. For simplicity's sake, the metabolic pathways in two models will be considered, first anoxia without ischemia and secondly ischemia with residual oxidative metabolism.

Anoxia without Ischemia

When the isolated heart is perfused with an anoxic medium, then the fundamental pathophysiology is stimulation of glycolysis (the Pasteur effect) with washout of metabolites by the continued coronary flow. Pasteur [51] described that a reduced oxygen supply to micro-organisms (yeasts) increased the process of "fermentation"; thus anaerobiosis accelerated those metabolic processes which kept the organisms alive. In 1933, Evans et al. [16] showed the existence of the Pasteur effect in the dog heart-lung preparation when nitrogen rather than oxygen was used for respiration; the normal metabolic pattern was changed as glucose uptake increased and there was lactate output instead of uptake. The factors accelerating glucose uptake and glycolysis (from glucose-6-phosphate to pyruvate) during anoxia have been elucidated by the work of Morgan et al. [37], Newsholme and Randle [40] and Williamson et al. [71].

Phosphofructokinase is the key enzyme and its low activity normally inhibits the glycolytic rate. Phosphofructokinase activity is accelerated during anoxia as ATP (normally inhibitory) breaks down and as creatine phosphate falls and forms inorganic phosphate (the latter de-inhibiting phosphofructokinase). To keep the flow of six carbon units through glycolysis adequate, a greater input is achieved by: 1) increased glucose uptake, through a poorly understood mechanism, possibly related to a fall of high-energy phosphates, and 2) increased glycogen breakdown. The mechanism whereby glycogen is broken down depends on dual stimulation of the enzyme phosphorylase, firstly by products of ATP breakdown such as inorganic phosphate and AMP, and secondly formation of the active form of phosphorylase (phosphorylase A) in response to adrenergic stimulation and an increase of tissue cyclic AMP levels. Because glucose uptake and glycogen breakdown are stimulated more than the flow through phosphofructokinase, the intermediate compounds glucose-6-phosphate and fructose-6-phosphate accumulate. Probably also the newly described activator of phosphofructokinase, fructose 1,2-bisphosphate, increases [31].

Pyruvate, formed at the "exit" end of glycolysis, cannot enter the citrate cycle because the enzyme pyruvate dehydrogenase is inhibited by increasing levels of $NADH_2$ which is no longer oxidized, so that pyruvate is converted to lactate under the influence of activity of the enzyme lactate dehydrogenase. Alanine formation is a much more minor fate of pyruvate during anaerobic metabolism, occuring by trans-

Table 1. Anaerobic glycolysis, $NADH_2$ (NADH + H$^+$) and protons

---

A. *Changes in NAD/NADH$_2$*
    Glucose + 2 NAD$^+$ → 2 pyruvates + 2 NADH + 2 H$^+$
    2 pyruvates + 2 NADH + 2 H$^+$ → 2 lactates + 2 NAD$^+$

    Net: glucose → 2 lactates

B. *Allowing for charges and ATP breakdown*
    Glucose + 2 Mg ADP^{1-} + 2 Pi^{2-} → 2 lactate^{1-} + 2 Mg ATP^{2-}
    2 Mg ATP^{2-} → 2 Mg ADP^{1-} + 2 Pi^{2-} + 2 H$^+$

    Net: glucose → 2 lactate^{1-} + 2 H$^+$

Table 2. Basic equations for triglyceride-free fatty acid (FFA) cycle

---

A. *Formation of triglyceride*

    glucose + 2 Mg ATP^{2-} + 2 NADH → 2 glycerol 3-phosphate^{2-} + 2 Mg ADP^{1-} + 2 NAD^{1+}

    3 FFA^{1-} + 3 Mg ATP^{2-} + 3 CoA^{4-} → 3 fatty acyl CoA^{4-} + 3 AMP^{2-} + 6 Pi^{2-} + 3 H^{1+} + 3 Mg^{2+}

    glycerol 3-phosphate^{2-} + 3 fatty acyl CoA^{4-} → 1 triglyceride + 3 CoA^{4-} + Pi^{2-}

B. *Net equation for triglyceride synthesis*

    3 FFA^{1-} + 4 Mg ATP^{2-} + NADH + 1/2 glucose → 1 triglyceride + Mg ADP^{1-} + 3 AMP^{2-} + 7 Pi^{2-} + 3 H^{1+}
    $\qquad\qquad\qquad\qquad\qquad\qquad\qquad\qquad\qquad\qquad$ + 3 Mg^{2+} + NAD^{1+}

C. *Triglyceride hydrolysis*

    1 triglyceride + 3 H$_2$O → 3 FFA^{1-} + 3 H^{1+} + glycerol

D. *Net equation for triglyceride-FFA cycle*

    1/2 glucose + 4 Mg ATP^{2-} + NADH + 3 H$_2$O → glycerol + 6 H^{1+} + Mg ADP^{1-} + 3 AMP^{2-} + 7 Pi^{2-}
    $\qquad\qquad\qquad\qquad\qquad\qquad\qquad\qquad\qquad\qquad$ + 3 Mg^{2+} + NAD^{1+}

For basic equations, see [18]

amination. $NADH_2$ formation in anoxia is not as simple as might be supposed. The process of converting pyruvate to lactate also reconverts each mole of $NADH_2$ produced by the triose phosphate dehydrogenase step of glycolysis to NAD (Table 1). *Thus the end-product of anaerobic glycolysis is lactate, not $NADH_2$,* thereby keeping the cytoplasmic $NADH_2$/NAD ratio normal. However, it is known that the ratio $NADH_2$/NAD rises in both the cytosol and mitochondria, which must reflect sources of protons other than glycolysis.

Proton formation is multifactorial in origin. The major source of protons is probably continued breakdown of ATP formed by anaerobic glycolysis and to a lesser extent from ATP-turnover cycles such as the triglyceride-free fatty acid (FFA) cycle (Table 2). Other sources are 1) inhibition of $NADH_2$ oxidation in the mitochondria, and 2) reactions indirectly producing protons such as continued production and accumulation of CO_2 (Table 3). Although ATP-wastage cycles may theoretically

Table 3. Sources of production of protons in anoxia-ischemia

Process	Mechanism of generation	Comment
Inhibition of mitochondrial oxidation of $NADH_2$	Inhibition of mitochondrial metabolism	$NADH_2$ formed by anaerobic glycolysis is regenerated to NAD by conversion of pyruvate to lactate; other processes must be responsible for increased cytosolic $NADH_2$/NAD ratio in ischemia.
Anaerobic glycolysis	ATP breakdown	Anaerobic glycolysis results in no proton production; protons form during breakdown of ATP.
Increased tissue CO_2	1) Continued residual respiration 2) Poor washout	Only in ischemia.
Triglyceride-FFA cycle	Continued breakdown and resynthesis of TG; ATP lost with proton production	3 ATP used per cycle, 6–7 protons produced per cycle.
Glycogen turnover	Excess recycling uses ATP and produces protons	1 ATP, 1 UTP, and 1 proton per cycle.
Mitochondrial uptake of calcium	1) Countertransport of protons with calcium 2) ATP breakdown	Uptake of calcium by mitochondria utilizes ATP and therefore produces protons.

contribute to energy loss during ischemia [18], the measured rate of the triglyceride-FFA cycle is in fact low (only about 2.5% of the total energy production in one model [62]).

In well perfused but anoxic preparations, tissue lactate, protons and inorganic phosphate all leave the anoxic cells. Thus there is only a mild accumulation of lactate to about 2 μmol/g fresh weight, reflecting the washout of lactate at nearly the same rate as its formation. The intracellular pH does not fall markedly with anoxia because the intracellular buffers are capable of coping with the mild accumulation of protons.

Anaerobic Energy Production in Anoxia

Because in anoxia mitochondrial energy production has virtually ceased, the output of lactate must reflect the rate of anaerobic ATP production (there is relatively little accumulation of lactate in the tissue). That anaerobic rate can meet the energy needs of the anoxic potassium-arrested heart, but becomes inadequate when the heart starts to contract even though it is not working (Langendorff perfusion), and is grossly inadequate when the energy requirements of the external pumping heart are considered [42]. Maximum rates of anaerobic glycolysis are achieved in the presence of insulin.

Lipid Metabolism in Anoxia

Normally the major fate of FFA taken up lies in oxidative metabolism. During anoxia, the end-products of the oxidative metabolism of FFA such as acetyl CoA and citrate

40

Table 4. Some tissue metabolic differences between anoxia and ischemia

	Anoxia	Mild ischemia	Severe ischemia
Glucose uptake	++	+	−
Tissue acidosis	+/−	+	++
Tissue lactate	+/−	+	++
Tissue Pi	+/−	+	++
Residual oxidative metabolism	0	+/−	0

Abbreviations: Pi = inorganic phosphate; + = increased compared with control; − = decreased compared with control; ++ = considerable increase compared with control; 0 = inhibition compared with control.

must decrease [9]. As the rate of oxidative metabolism of FFA falls, so does the uptake, which is increasingly converted to non-oxidative fates such as tissue FFA, phospholipids, and triglycerides. According to data obtained with the use of gas-liquid chromatography, the tissue concentrations of FFA are increased in ischemia, especially the arachidonic and linoleic fractions, suggesting enhanced breakdown of endogenous phospholipids [64]. It is proposed that in anoxia there is a "triglyceride-FFA cycle" (Table 2) whereby as increased tissue FFA are formed they are converted to triglyceride which in turn, under the stimulus of anoxia, is increasingly broken down to FFA [43]. This energy-wasting cycle uses up ATP and produces six protons and seven inorganic phosphates per molecule of glycerol broken down (Table 2).

Ischemia without Total Anoxia

There are important metabolic differences between ischemia and anoxia, and between mild and severe ischemia (Table 4). The major metabolic changes found in regional ischemia (with some collateral flow) but not in anoxic perfused hearts are as follows: 1) reduced provision of oxygen without total anoxia, 2) continued residual respiratory production of CO_2, and 3) decreased removal of metabolic end-products such as lactate, protons, and CO_2. According to the severity of the ischemia, the glucose uptake might either increase or decrease. During mild ischemia the Pasteur effect dominates with increased glycolysis (Fig. 1), whereas in severe ischemia glycolysis is inhibited as the end-products thereof, such as lactate, protons and $NADH_2$, inhibit crucial enzymes of glycolysis (phosphofructokinase and glyceraldehyde 3-phosphate dehydrogenase).

Chance's Model for Mitochondrial Activity in Ischemia

Can mitochondria function at "half-speed"? Chance [11] has presented powerful data suggesting that they cannot. According to him the Km of the mitochondria for O_2 is so low (10^{-10}M) that even minute amounts of oxygen would allow full-functioning of oxidative phosphorylation. Two points arise: ischemia is highly focal, and even hypoxia does not result in uniform oxygen lack throughout the heart. Rather, micro-foci of anoxia result. Thus the cellular pattern of a lack of oxygen in ischemia is likely

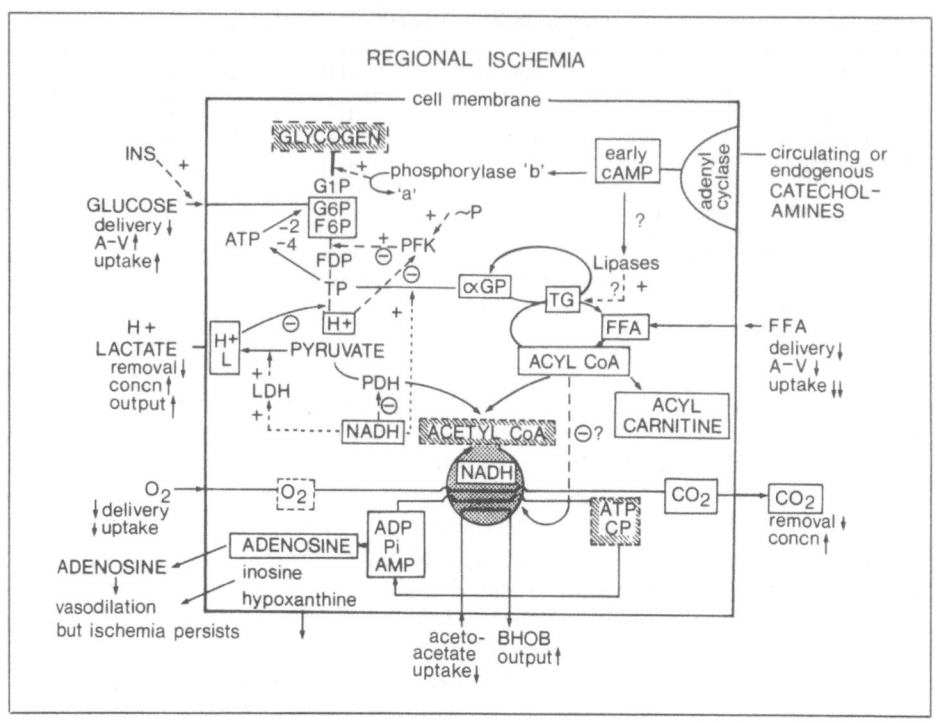

Fig. 1. Regional ischemia (mild) without total anoxia, i.e., developing myocardial infarction with collateral flow. The major differences from anoxia without ischemia are: a residual oxidative metabolism resulting from a decreased but not absent tissue O_2, increased tissue CO_2, definite intracellular acidosis, a much increased tissue lactate, and a definite increase in acyl CoA. Decreased blood flow results in decreased delivery of oxygen, glucose, and FFA; at circulating FFA levels of about $500-700$ μEq/liter, the glucose arteriovenous (AV) difference is increased relative to that of FFA. PFK activity is relatively inhibited (compare anoxia) and tissue contents of hexose monophosphate (G6P and F6P) rise rather than fall as in anoxia. Glycolytic flux is not stimulated to the same extent as in anoxia. Glycolytic H^+ is produced via ATP breakdown (bottom right corner of figure) and intracellular acidosis may both contribute to and be caused by the increased tissue pCO_2. The intracellular (IC) acidosis inhibits PFK ($-$) which receives conflicting stimuli from changes in high-energy phosphate ($\sim P$) and from H^+ accumulation. αGP and CoA increase; the latter inhibits the adenine translocase transferring ATP across the mitochondrial membrane. Values of acetyl CoA are not known in regional ischemia but tend to fall in uniform ischemia. Values to TG are not reported, and the roles of cAMP and catecholamine activity need clarification. In global ischemia, catecholamine release may stimulate lipase activity. There is evidence for an increased turnover of "triglyceride-FFA cycle" which would "waste" ATP and produce H^+.

Abbreviations and symbols are as follows: $\square$ = increased tissue content of metabolite; $\boxtimes$ = decreased tissue content of metabolite; x = absence of metabolite; $\longrightarrow$ = pathways of metabolism; $\longrightarrow$ = accelerated pathways; $------\rightarrow$ = factors altering rates of enzyme activity; $\otimes$ = citrate cycle activities; INS = insulin activity; PFK = phosphofructokinase; TP = triose phosphates, including dihydroxyacetone phosphate; αGP = α-glycerophosphate; L = lactate; H^+ = protons; LDH = lactate dehydrogenase; PDH = pyruvate dehydrogenase; cAMP = cyclic AMP; TG = triglyceride; BHOB = β-hydroxybutyrate; G1P = glucose 1-phosphate; G6P = glucose 6-phosphate; F6P = fructose 6-phosphate; FDP = fructose 1, 6-diphosphate; FFA = free fatty acids.

Modified from Opie [44] with permission of Circulation Research and the American Heart Association.

to be even more focal. In relation to the size of the myocyte, mitochondria are small. Perhaps within one cell some mitochondria are not working at all and others are working excessively. The net effect would still be a hypoxic but not an anoxic cell, and on an average, the mitochondria might indeed appear to be functioning at "half-speed" [44].

Collateral Flow in Ischemia

Clearly the persistence of collateral flow would make a crucial difference to the patterns of metabolic pathways occuring during regional ischemia. In ischemic tissue without any collateral flow at all, residual oxidative metabolism is going to be very low. In a situation closer to human coronary artery disease, at least some collateral flow is likely. The result is at least some residual oxidative metabolism so that there is competition between glucose and FFA for this residual oxygen uptake. Citrate cycle activity decreases, approximately in proportion to the decreased blood flow [44]. The rise of pCO_2 in tissue and in coronary venous effluent may reflect either retention of respiratory CO_2 and/or evolution of CO_2 from bicarbonate as the result of proton accumulation. The arteriovenous differences of glucose are increased relative to those of FFA so that glucose accounts for an increased percentage of the residual oxygen uptake. The basis for this observation is that in mild to moderate ischemia, glycolysis is accelerated, whereas FFA metabolism is relatively inhibited [44], as these two metabolic pathways compete for residual mitochondrial metabolism.

Lipid Metabolism in Regional Ischemia

During ischemia, the tissue content of acyl CoA increases, as a result of ischemic inhibition of oxidative phosphorylation, decreased citrate cycle activity, and lessened β-oxidation (Fig. 1). The end result is that extramitochondrial acyl carnitine as well as acyl CoA also accumulate (both these long-chain lipids have detergent properties and contribute to membrane breakdown in ischemia).

Lysophosphoglycerides are membrane-derived fatty acids released from the phospholipids of the cell membranes (including the sarcolemma) during ischemia. The enzyme phospholipase is activated by ischemia (probably as a result of increased cytosolic calcium levels), to split-off fatty acids and hence to produce lysophosphatidyl cholines, which are normally resynthesized into phosphatidyl choline. In ischemia, this process is inhibited by the acyl carnitine which accumulates during ischemia. Lysophosphatidyl cholines are also converted to glycerophosphoryl choline, a process which is inhibited by the acidosis associated with ischemia. The accumulated lysophosphatidyl cholines (LPC) are thought to have arrhythmogenic properties [14], probably acting by their detergent effect on the membranes.

Triglycerides are broken down in ischemia as in anoxia, and the triglyceride-FFA cycle also operates, albeit at a low rate [62]. Increased uptake of glucose *relative* to that of FFA provides the glycerol-3-phosphate required for re-esterification of the tissue FFA formed from 1) triglyceride hydrolysis, and 2) residual uptake of FFA

43

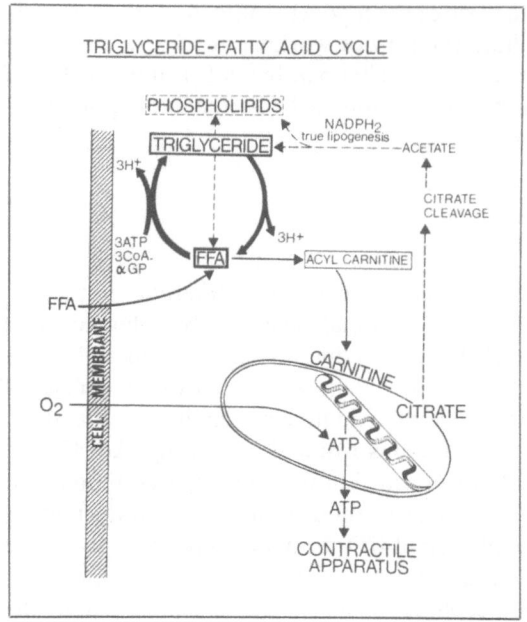

Fig. 2. Triglyceride-fatty acid cycle. Intracellular free fatty acids (FFA) when converted to triglyceride produce three protons (H$^+$) with a further three protons produced when triglyceride is broken down. Activation of this cycle occurs in ischemia (see text). (Copyright by L. Opie).

(Fig. 2). The hydrolysis of triglyceride appears, unexpectedly, to take place (at least in part) in the *lysosomes* which can take up triglyceride and release FFA, a process enhanced by decreasing tissue pH [57].

High-Energy Phosphate Compounds in Ischemia

In regional ischemia, breakdown of ATP and CP produces increased tissue levels of ADP, AMP, and inorganic phosphate. Ultimately, adenosine, a coronary vasodilator, forms and there are increased coronary venous values of inosine and hypoxanthine, further end-products of adenosine breakdown. Although there is an early and progressive fall of the tissue levels of ATP, it is really the decrease of creatine phosphate which is much more striking. The energy transfer between creatine phosphate and ATP occurs under the influence of the enzyme creatine kinase which catalyzes the following reaction:

$$CP \longrightarrow creatine + P_i + free\ energy$$

The fall of ATP and the rise of ADP at the onset of anoxia shows that the rate of ATP hydrolysis by ischemia exceeds the rate at which ATP is formed from creatine phosphate. The reason why creatine phosphate falls so much more rapidly than ATP is that the specific process transferring ATP in and out of the mitochondria (adenine nucleotide transferase) is inhibited by an accumulation of acyl CoA (see earlier) during ischemia. Therefore, as the waning contractile activity continues to use up ATP, CP must be used to replenish the ATP stores in the cytoplasm, whereas mitochondrial ATP continues to be stored.

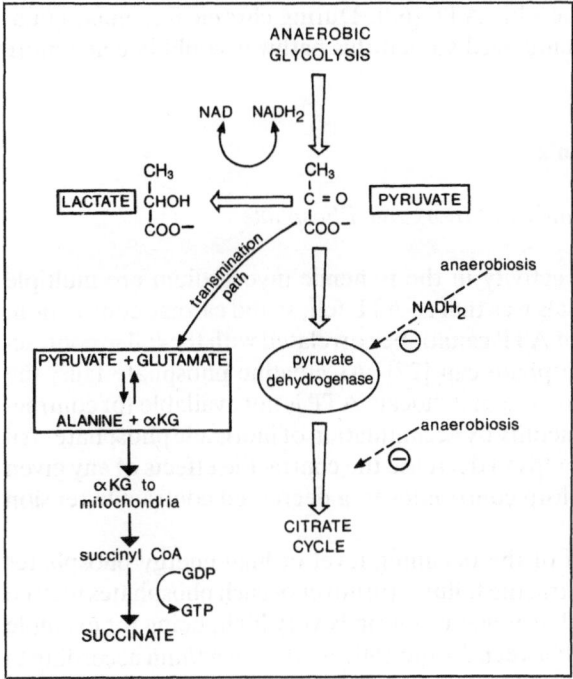

Fig. 3. Possible fates of pyruvate. During ischemia, conversion to acetyl CoA is inhibited and formation of lactate becomes the major fate. However, a variable percentage undergoes transamination to alanine and α-ketoglutarate (2-oxoglutarate) which can enter the mitochondria by countertransport with malate, and intramitochondrial conversion to succinyl CoA. The latter is converted to succinate with a substrate level phosphorylation of GDP to GTP; the latter in turn can be converted to ATP.

ATP "Overspending"

Schaper et al. [55] introduced this dramatic term to describe the drop of ATP in ischemic tissue soon after the onset of ischemia. Their reasoning is that as ischemia causes contractile activity to diminish, oxygen demand falls and although the turnover of ATP must fall, its actual level should stay constant. Hence the fall of tissue levels of ATP reflects mismatching of ATP utilization and synthesis. Possible explanations are: 1) the initial trigger to impaired contractility is different from the fall in high-energy phosphates, 2) contractility is stimulated more than it should be, for example, by release of catecholamines in anoxia, and 3) there are ATP-wasting cycles, such as the triglyceride-FFA cycle and others (Table 2).

Sources of Energy during Ischemia

During ischemia, energy may be provided by continued residual oxidative metabolism. In addition, there are three non-oxidative sources: 1) breakdown of CP and ATP, 2) anaerobic glycolytic ATP production, and 3) substrate level phosphorylation, by conversion of succinyl CoA to succinate. It is generally assumed that anaerobic glycolysis is the major anaerobic energy source and that succinate formation can be discounted. However, in severely hypoxic papillary muscles succinate formation by this pathway (Fig. 3) can account for 16% of anaerobic ATP [59] and after

45

coronary ligation for 18% of anaerobic ATP [68]. During chronic ischemia, glutamate uptake may be adaptively enhanced so that this pathway could become more
active.

Contractile Impairment in Ischemia

High-Energy Phosphate Compounds and Inorganic Phosphate

The causes of fall of contractile activity in the ischemic myocardium are multiple
(Table 5). One of the first proposals was that as ATP fell, so did cardiac contraction.
Although the fall in tissue levels of ATP cannot be correlated with the fall in contractile activity, that of creatine phosphate can [27]. As creatine phosphate falls, the
creatine phosphate shuttle decreases so that "local" ATP is not available for contraction. CP fall can also inhibit contractility by accumulation of inorganic phosphate. An
increase of inorganic phosphate helps to decrease the contractile effects of any given
cytosolic calcium level and therefore contributes to a decreased contractile tension
[29].
Another proposal is that instead of the declining level of high-energy phosphates
being important in relation to contractile failure, turnover of such phosphates may be
more important. The rate of ATP turnover is normally very high, being for example
in the isolated rat heart anything between 20 and 150 μmol/g/wet wt/min according to
the heart rate (Tables 4, 5 in [45]). Compared with an average tissue level of say
5 μmols ATP/g, the turnover of each molecule is something like 4- to 30-times per
min. Hence the turnover of ATP could decrease considerably before the total level
falls. Such a decrease of ATP turnover undoubtedly occurs [7], but the problem is
that, in general, it is the *tissue levels* of metabolites that act as metabolic signals rather
than *turnover* rates of compounds.
   A third proposal is that the free energy of hydrolysis of ATP is decreased, at least
in hypoxia [26]. The basis of this proposal is that the phosphorylation potential of the
myocardial cell determines the free energy of hydrolysis, where:

$$\text{phosphorylation potential} = (ATP)/(ADP)(P_i).$$

   Thus any increase in ADP and especially P_i will decrease free energy of hydrolysis,
as will a decrease in ATP. Thus the marked increase of tissue inorganic phosphate

Table 5. Possible causes of impaired contractility in ischemia

Proposed mechanism	Reference
1) Intracellular acidosis with displacement of Ca^{2+} from intracellular binding sites	[28], [1]
2) Accumulation of inorganic phosphate	[29]
3) Decreased turnover of ATP	[7]
4) Decreased ATP in a "contractile" compartment	[41], [20]
5) Decreased free energy change of ATP hydrolysis	[26]
6) Reversed "garden hose" or "erectile" effect	[35], [65]
7) Accumulation of lactate	Table 6

concentration during severe "supply" ischemia may contribute to loss of contractile function (see section "Supply vs Demand Ischemia").

Retention of CO_2, Protons and Contractile Activity

Accumulation of protons and CO_2 may contribute to impaired contractile activity soon after coronary artery ligation, partially as a result of continued residual respiration, partially as a result of poor washout, and also partially by liberation of CO_2 from bicarbonate by protons. Thus, respiratory acidosis (pH 6.6) with a high pCO_2 produced a rapid fall of left ventricular pressure in the isolated working rat heart; the depressive effect could be counteracted by an increase of calcium in the perfusion fluid [58, 71]. Protons could be exerting an effect by competing with calcium for sites on the contractile proteins [28] or at more superficial sites [71].

These earlier speculations have now been proved by direct measurement. As protons are added to an isolated preparation the contractile activity decreases, even though the cytosolic calcium increases [1].

Lactate Accumulation

It is not generally known that an accumulation of lactate (Table 6), which is a well-established abnormality in ischemia, has certain effects which are detrimental to the heart. Yet this proposal goes back about 45 years to Tennant [60] who found that an infusion of neutral lactate decreased contractile activity in the ischemic zone. In addition, recent data show that lactate can promote mitochondrial damage [6], decrease the action potential duration [46, 72], and inhibit glyceraldehyde-3-phosphate-dehyd-

Table 6. Reported effects of buffered lactate relevant to the development of ischemic injury

Reference	Lactate concentration	Preparation	Effect
[60]	About 40 mM	Dog with coronary artery ligation; lactate infused via peripheral branch of coronary artery.	Lactate alone could inhibit contractions.
[72]	5,9–19,0 mM	Sheep Purkinje fiber	Shortened action potential duration; enhanced phase 4 depolarization.
[6]	33 mM	Dog heart slices	Increased mitochondrial damage.
[36]	10 mM	Isolated perfused rat heart	L-lactate but not D-lactate inhibits glycolytic flux.
[10]	5 mM	Isolated perfused rat heart with coronary ligation	Doubling of enzyme release when compared with glucose; slight fall in cardiac output and power production.
[46]	2.5–10.0 mM	Guinea-pig papillary muscle	Shortened action potential duration, relieved by addition of glucose.

rogenase [54]. It may be supposed that with the addition of large amounts of lactate small amounts of lactic acid could be formed. Yet the dissociation constant of lactate is such that even millimolar concentrations in the tissue would be associated only with micromolar concentrations of lactic acid. The mechanism of these lactate effects is not clarified and may speculatively include extracellular calcium-binding, inhibited glycolytic flux, and in the case of glyceraldehyde-3-phosphate dehydrogenase, a stereospecific interaction with the enzyme [36]. High lactate concentrations should alter the cytoplasmic ratio $NADH_2/NAD$ in favor of $NADH_2$; such cytosolic changes can in turn be communicated to the mitochondrial $NADH_2/NAD$ system by the malate-aspartate shuttle, and the intramitochondrial increase of $NADH_2$ (really $NADH + H^+$) can increase the rate of calcium uptake into mitochondria by the H^+/Na^+ exchange followed by Na^+/Ca^{2+} exchange.

Bearing in mind that in developing pig heart infarcts tissue lactate levels can rise to $30-50$ µM/g/fresh weight [38], these effects of high lactate levels may well be relevant to the overall mechanism of ischemic damage. However, it must be stressed that all the experimental data have been derived with the addition of external lactate which needs a transport system for its uptake, so that the results of such extracellular addition are not necessarily the same as an increase of internal lactate.

Early Ionic and Electrophysiological Changes

Potassium

Potassium loss from the ischemic myocardium was apparently first described by Harris et al. [19] who linked the egress of potassium to early ventricular arrhythmias. As potassium is lost, the ionic difference across the myocardial sarcolemma is diminished so that there is depolarization, which in turn helps to diminish the activity of the sodium channel, so that there is a decreased upstroke in addition to depolarization. The action potential duration is shortened through mechanisms not completely understood but which may include lack of cytoplasmic ATP. All these changes explain many of the characteristic electrophysiological changes that develop in the action potential pattern in myocardial hypoxia or ischemia.

The mechanism of the potassium loss is not well understood (Fig. 4). The three major theories are: 1) inhibition of the sodium-potassium pump, 2) co-ionic loss as the potassium cation is thought to leave the myocardial cell in company with negatively charged lactate and phosphate ions, and 3) opening of the ATP-inhibited potassium channel as a result of cytosolic ATP deficiency. The latter theory is currently receiving considerable attention (for review, see [49]).

Calcium

Abnormalities of calcium regulation in ischemia have long been suspected and an increased tissue level has been found in reperfused tissue. A role for early cytosolic calcium accumulation in ventricular arrhythmias has been proposed. There is multiple evidence for this proposal. First, calcium channel antagonists such as diltiazem

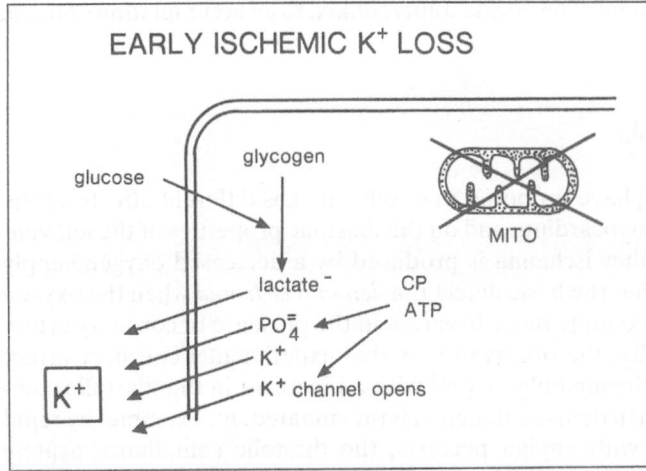

Fig. 4. Schematic diagram of early ischemic postassium loss, depending on cotransport of potassium with negatively charged ions such as lactate and phosphate. (Copyright, LH Opie.)

[12] or verapamil isomers [61] have delayed or inhibited the onset of ventricular fibrillation following the onset of myocardial ischemia. Recently, the elegant measurements of Clusin's group [32] and of Allen [1] have confirmed the suspected early rise of cytosolic calcium in ischemia. Such a rise occuring soon after the onset of ischemia could in turn activate calcium-dependent enzymes such as the phospholipases (see earlier). Harmful electrophysiological consequences are possible, including a contribution to ischemic depolarization, especially during tachycardia [12]. In addition there may be a calcium-activated inward current (I_{ti}) which could help provoke early ischemic arrhythmias, provided that there is still sufficient ATP left to allow rhythmical uptake and discharge of calcium from the sarcoplasmic reticulum [48].

Adrenergic Activation and Cyclic AMP

Ischemia is associated with adrenergic activation, as shown by the rise of tissue cyclic AMP in the ischemic zone. Initially there is a very early rise in the cyclic AMP level within minutes, detected by Wollenberger and his group [30]. A later rise, approximately 10–20 min after coronary occlusion, was first shown by Podzuweit et al. [52] in the baboon and by Muller and her associates in the pig [38]. Further evidence of adrenergic activation is the release of catecholamines from ischemic tissue as early as 10 min after the onset of total global ischemia [70]. After 20–30 min of ischemia, extracellular noradrenaline concentrations are estimated to be about 10^{-6}M [56].

Closely linked to tissue accumulation of cyclic AMP is the development of ventricular fibrillation [33, 47]; the mechanisms are multiple and may include: increased automaticity in Purkinje fibers, activation of phospholipases, an increased intracellular cytosolic calcium level, and the induction of slow responses in depolarized tissue.

The metabolic consequences of adrenergic activation include: 1) enhancement of glycogen breakdown, 2) stimulation of glycolysis, 3) stimulation of the sodium pump

[70], and 4) an increased membrane permeability, linked to an accumulation of tissue cyclic AMP [22, 47].

Supply vs Demand Ischemia

Apstein and Grossman [4] have proposed that ischemia has different effects on the contractile activity of the myocardium and on the diastolic properties of the left ventricle, depending on whether ischemia is produced by a decreased oxygen supply *(supply ischemia)* or whether the basic defect is a *demand ischemia* when the oxygen demand is increased, for example by tachycardia in the presence of coronary artery disease (Table 7). Basically, the observation is that experimental coronary artery ligation results in local ischemic bulging [69] with an increase in the diastolic compliance [4]. In contrast, when demand ischemia is precipitated, for example, by rapid atrial pacing in patients with angina pectoris, the diastolic compliance acutely decreases so that there is stiffening and a failure of the ischemic tissue to relax. The corresponding animal model is as follows. When a globally underperfused isolated rabbit heart is paced, then there is a rapid fall in left ventricular chamber compliance with an increase in left ventricular diastolic pressure [23].

The basic proposal of Apstein and Grossman [4] is that a complete coronary occlusion is basic to supply ischemia. The result is a marked tissue acidosis and accumulation of inorganic phosphate − and both factors are thought to decrease the availability of intracellular cytosolic calcium with decreased contractile activity (Table 5; [2]). In contrast, during demand ischemia intracellular calcium is thought to increase without a significant acidosis and therefore an ischemia-induced loss of compliance takes place. The basic cause of the rise of intracellular cytosolic calcium in demand ischemia is not clear. Of interest is the proposal that continued glycolysis has a role in the maintenance of intracellular calcium homeostasis so that as glycolysis is inhibited by ischemia there is an increase in diastolic tension [50]. In isolated rat ventricu-

Table 7. Some proposed differences between "supply" and "demand" ischemia [4]

	Supply ischemia	Demand ischemia
Animal model	Coronary artery ligation	Low-flow global ischemia with added pacing
Human condition	Coronary occlusion; acute myocardial infarction	Angina of effort or of pacing
Diastolic compliance	Increases	Falls
Cytosolic calcium	Increases	Increases
Cause of increased cytosolic calcium	Tissue acidosis	? inhibited glycolysis
Factors opposing effects of rise of cytosolic calcium	Tissue acidosis, increased tissue inorganic phosphate	None
Coronary arteries in ischemic zone	Collapsed; reversed "garden hose" or erectile effects	Arteries normally distended

lar cells, metabolic blockade causes the development of contracture which is dependent on an increase of cytosolic calcium [15].

The anatomical state of the coronary arteries may also be of importance. During supply ischemia, the coronary arteries are occluded and collapsed with a reversed "erectile" effect (Table 7). In contrast, in demand ischemia the arteries stay filled with blood which contributes to the tissue "stiffening".

Chronic Reversible Ischemia: Hibernating Myocardium

Ischemia is frequently thought of as an acute event which either resolves on its own or progresses to an irreversible state of myocardial necrosis. Recently there have been fragments of information suggesting that "chronic" ischemia may occur. One important metabolic hallmark of mild ischemia in isolated hearts is an increased glucose uptake – yet this characteristic can persist between attacks in patients with unstable angina pectoris [5]. Metabolically, it is not impossible to conceive of a situation in which the various factors regulating demand vs supply ischemia actually balance each other so that the tissue could be "preserved" in an ischemic state for a chronic period. Moderate ischemia with a flow reduction to 20–70% of control values could decrease and "freeze" ATP levels for ½ to 5 h [39]. The degree of tissue acidosis and increase of inorganic phosphate may be just enough to inhibit contractile activity in the ischemic zone, whereas the degree of coronary flow might be just enough to maintain some oxidative metabolism. There could thus be a balance between these factors. This rather new concept of chronic ischemia corresponds to Rahimtoola's [53] "hibernating myocardium" where there is chronic myocardial ischemia in which the heart responds to a reduced myocardial blood flow through shutting off its contractile activity. Why this "smart heart" outwits its fellows who undergo ischemic necrosis in somewhat similar circumstances is not clear. Presumably "smart hearts" have only minor degrees of myocardial ischemia with which to cope.

Point of No Return

Sustained ischemia (total no-flow) causes irreversible damage rather rapidly, within 20–40 min, unless specific steps are taken to reduce the myocardial energy demand, such as hypothermia and rapid potassium arrest. However, in regional ischemia produced by coronary artery occlusion, survival can be much longer, depending on the extent of collateral flow and the oxygen demand placed on the ischemic tissue. Thus irreversibility may not occur until at least 6 h after the onset of ischemia [55].

Five chief theories for the development of irreversible ischemic damage are: 1) loss of a critical amount of ATP, 2) mechanical effects resulting from intracellular edema and sodium retention including cell swelling and membrane rupture, 3) metabolically induced membrane damage, 4) formation of free radicals, and 5) calcium overload.

A *critical level of ATP* can be excluded as the sole cause of irreversibility because of the very different values for the "critical" level given by different authors and because of the established concept of ATP compartmentation (mitochondrial vs

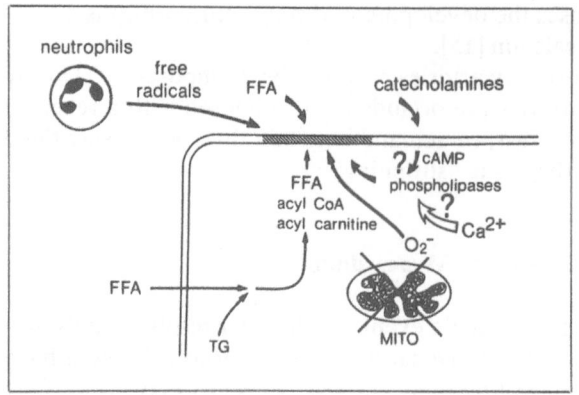

Fig. 5. Schematic diagram of role of neutrophils and fatty acid metabolites in causing membrane damage. For proposed role of neutrophils, see [34].
(Copyright, LH Opie.)

cytoplasmic), so that it is impossible to say which compartment is depleted by measurements of overall ATP. However, the ATP theory is being revived and extended with the recent evidence that ATP produced by glycolysis may have a specific role in the prevention of membrane-related ischemic events such as calcium influx [50] and potassium loss [67].

Mechanical consequences of ischemia may also promote cellular irreversibility. Jennings et al. [24] have suggested that ischemia causes an abnormality in cell volume regulation with swelling of the cell and breaks in the sarcolemmal membrane which could account for the greatly increased enzyme release as a feature of irreversibility.

Metabolically induced membrane damage is multifactorial in origin (Fig. 5), including accumulation of FFA inside and outside the ischemic cells, the increased amounts of acyl CoA and acyl carnitine, and the self-perpetuating circle whereby part of membrane damage results from the action of phospholipases which break down membrane lipids, with formation of lysophosphoglycerides which, in turn, further promote membrane damage.

Oxygen free radicals, derived in part from neutrophils and formed particularly during the reperfusion phase of ischemic damage, may contribute to membrane damage by lipid peroxide formation. However, rates of free radical formation are much lower in ischemia than they are during early reperfusion.

The *calcium overload* concept of irreversibility has received special prominence in relation to conditions of massive calcium overload such as catecholamine stimulation or severe reperfusion damage or the very unusual experimental conditions of the calcium paradox when extracellular calcium is totally removed and then, when re-introduced, causes massive cellular damage. The basic concept of such severe degrees of calcium overload is that the mitochondria initially act as a "buffer" to take up calcium from the cytosol, thereby spending considerable energy. As a consequence, generalized cellular energy depletion is enhanced, the energy required for maintenance of ionic gradients becomes inadequate, and membrane integrity is lost. A modified form of this hypothesis has been introduced by Opie and Coetzee [48]. In ischemia, cytosolic calcium levels rise, possibly as a result of intracellular redistribution of calcium. Such cytosolic calcium increases can still activate phospholipases, increase resting tension, and provoke fatal arrhythmias.

52

Probably irreversibility depends on no single metabolic event but, like the death of a person, it is a complex phenomenon resulting from the simultaneous operation of many diverse mechanisms.

Reperfusion Damage

An important unresolved problem is whether the act of reperfusion induces specific damage that would not otherwise have occurred. Definitive evidence favoring this view is provided by the development of early reperfusion arrhythmias. The origin of such arrhythmias is still controversial, with the two main hypotheses being as follows. First, the effects of cytosolic calcium overload during the ischemic period are followed at the onset of reperfusion by excess recycling of calcium [13, 48]. Secondly, formation of free radicals, known to occur in a burst at the start of reperfusion, causes membrane damage which hypothetically predisposes to arrhythmias [21], although data are controversial [48]. Formation of free radicals has come to the fore as a possible factor in reperfusion damage and particularly in reperfusion "stunning", when the contraction of the myocardium is impaired due to unknown reasons. There is now no doubt that during reperfusion there is an increased formation of oxygen free radicals. At least some of the free radicals are derived from neutrophils re-entering the ischemic zone, which may explain the harmful consequences of neutrophil infiltration during developing infarction or reperfusion. In addition, it is known that a certain amount of free radical generation can actually occur during the ischemic period [66]. Therefore, free radical scavengers may have the potential of decreasing ischemic as well as reperfusion damage, although it seems clear that the greater accumulation of free radicals lies in the reperfusion period [3].

Whether, in addition to provoking arrhythmias, reperfusion can cause other types of damage remains controversial. For example, reperfusion "stunning" may be decreased by free radical scavengers [8], yet there are several contradictory studies. The proposal that free radical scavengers can decrease ultimate infarct size when regional ischemia is followed by reperfusion [25] has since been challenged by other negative studies [17, 63]. There are few studies in which the free radical scavengers were given at the time of reperfusion; mostly they had already been introduced during the period of ischemia when they could have acted to reduce the ischemic injury and, hence, indirectly improved reperfusion injury.

So the question of free radical mediated reperfusion damage must still remain an open question. Just as the mechanism of cellular ischemic damage is multiple, so probably is that of reperfusion damage. Increased cytosolic calcium, formation of free radicals, and other as yet unclarified factors may all play a role.

Metabolism of Ischemia – Summary

Two central aspects of the metabolic changes in ischemia are 1) a decreased mitochondrial energy production with a consequent rundown in the cytoplasmic ATP and CP stores, with acceleration of glycolysis and lactate formation; increased glycolysis has been shown in the human heart by positron emission tomography; and 2) accumulation of potentially harmful metabolites such as intracellular lactate, $NADH_2$ and protons, as well as unoxidized lipid long-chain intermediates. Decreased contractile function in early ischemia may be either systolic failure or an increased diastolic tension, reflecting, respectively, supply and demand ischemia. In supply ischemia, there is total coronary occlusion which decreases contractility as a result of an increase of inorganic phosphate and protons which interfere with the action of calcium. In demand ischemia, there is still some residual coronary flow, as when patients with coronary artery disease are paced. It is thought that decreased availability of ATP, and specifically that made by glycolysis, may impair intracellular calcium regulation with decreased tissue compliance and decreased myocardial compliance and an increased diastolic tension.

Many of the electrophysiological changes of acute ischemia can be explained in terms of ionic alterations including the following: first, localized potassium loss with hyperkalemia causes depolarization and changes in the pattern of the cardiac action potential thereby setting the stage for re-entry arrhythmias; furthermore, by creating a potential difference between the ischemic and non-ischemic zones, localized hyperkalemia is able to promote the formation of a potentially arrhythmogenic current between these two zones. Secondly, an additional arrhythmogenic change is the early increase in intracellular cytosolic calcium concentrations. The increase of cellular cyclic AMP levels is a reflection of enhanced adrenergic activity, which is susceptible to therapy by β-adrenergic blockade. Whether a specific reperfusion injury occurs is still controversial. While the case is proven for reperfusion arrhythmias, other aspects of reperfusion injury, such as "stunning" and possible reperfusion-induced cell necrosis, remain ill-understood. The chief mechanisms proposed for reperfusion injury are calcium overload and the formation of free radicals.

The "stunned myocardium", on its way to recovery but not yet there, is to be distinguished from the "hibernating heart" which suffers from chronic ischemia sufficiently mild so as not to lead to infarction, but sufficiently severe to lead to inhibition of contractile function.

The difference between reversible ischemic injury and irreversibility probably lies in multiple metabolic aberations, including depletion of ATP, intracellular edema and sarcolemmal rupture, metabolically induced membrane damage, formation of free radicals, and particularly, calcium overload.

References

1. Allen DG (1988) Does calcium play a role in early ischemic injury? J Mol Cell Cardiol 20 (Suppl V):S1
2. Allen DG, Lee JA, Smith GL (1988) The effects of simulated ischemia on intracellular calcium and tension in isolated ferret ventricular muscle. J Physiol 440:91 P
3. Ambrosio G, Weisfeldt ML, Jacobus WE, Flaherty JT (1987) Evidence for a reversible oxygen radical-mediated component of reperfusion injury: reduction by recombinant human superoxide dismutase administered at the time of reflow. Circulation 75:282–291
4. Apstein CS, Grossman W (1987) Opposite initial effects of supply and demand ischemia on left ventricular diastolic compliance: the ischemia-diastolic paradox. J Mol Cell Cardiol 19:119–128
5. Araujo LI, Camici P, Spinks TJ, Jones T, Maseri A (1988) Abnormalities in myocardial

metabolism in patients with unstable angina as assessed by positron emission tomography. Cardiovasc Drugs Ther 2:41–46

6. Armiger LC, Gavin JB, Herdson PB (1974) Mitochondrial changes in dog myocardium induced by neutral lactate in vivo. Lab Invest 31:29–33
7. Bittl JA, Balschi JA, Ingwall JS (1987) Contractile failure and high-energy phosphate turnover during hypoxia: ^{31}P-NMR surface coil studies in living rat. Circ Res 60:871–878
8. Bolli R, Patel BS, Jeroudi MO, Lai EK, McCay PB (1988) Demonstration of free radical generation in "stunned" myocardium of intact dogs with the use of the spin trap α-phenyl N-tert-butyl nitrone. J Clin Invest 82:476–485
9. Bowman RH (1966) Effects of diabetes, fatty acids and ketone bodies on tricarboxylic acid cycle metabolism in the perfused rat heart. J Biol Chem 241:3041–3048
10. Bricknell OL, Opie LH (1978) Effects of substrates on tissue metabolic changes in the isolated rat heart during underperfusion and on release of lactate dehydrogenase and arrhythmias during reperfusion. Circ Res 43:102–115
11. Chance B (1976) Pyridine nucleotide as an indicator of the oxygen requirements for energy-linked functions of mitochondria. Circ Res (Suppl I):31–38
12. Clusin WT (1987) What is the solution to sudden cardiac death: calcium modulation or arrhythmia clinics? Cardiovasc Drugs Ther 1:335–342
13. Coetzee WA, Dennis SC, Opie LH, Muller CA (1987) Calcium channel blockers and early ischemic ventricular arrhythmias: electrophysiological versus anti-ischemic effects. J Mol Cell Cardiol 19 (Suppl II):77–97
14. Corr PB, Gross RW, Sobel BE (1982) Arrhythmogenic amphiphilic lipids and the myocardial cell membrane. Editorial. J Mol Cell Cardiol 14:619–626
15. Eisner DA, Nichols CG, O'Neill SC, Smith GL, Valdeolmillos M (1989) The effects of metabolic inhibition on intracellular calcium and pH in isolated rat ventricular cells. J Physiol 411:393–418
16. Evans CL, de Graff AC, Kosaka T, Mackenzie K, Murphy GE, Vacek T, Williams DH, Young FG (1933) The utilization of blood sugar and lactate by the heart-lung preparation. J Physiol (Lond) 80:21–40
17. Gallagher KP, Buda AJ, Pace D, Gerren RA, Shlafer M (1986) Failure of superoxide dismutase and catalase to alter size of infarction in conscious dogs after 3 hours of occlusion followed by reperfusion. Circulation 73:1065–1076
18. Gevers W (1977) Generation of protons by metabolic processes in heart cells. J Mol Cell Cardiol 9:867–874
19. Harris AS, Bisteni A, Russell RA, Brigham JC, Firestone JE (1954) Excitatory factors in ventricular tachycardia resulting from myocardial ischemia. Science 119:200–203
20. Hearse DJ (1979) Oxygen deprivation and early myocardial contractile failure: a reassessment of the possible role of adenosine triphosphate. Am J Cardiol 44:1115–1121
21. Hearse DJ, Tosaki A (1987) Free radicals and reperfusion-induced arrhythmias: protection by spin-trap agent PBN in the rat heart. Circ Res 60:375–383
22. Horak AR, Opie LH (1983) Energy metabolism of the heart in catecholamine-induced myocardial injury. Concentration-dependent effect of epinephrine on enzyme release, mechanical function, and "oxygen-wastage". In: Chazov E, Saks V, Rona G (eds) Advances in Myocardiology, Vol 4. Plenum Publishing Corporation
23. Isoyama S, Apstein CS, Wexler LF, Grice WN, Lorell BH (1987) Acute decrease in left ventricular diastolic chamber distensibility during simulated angina in isolated hearts. Circ Res 61:925–933
24. Jennings RB, Reimer KA, Steenbergen C (1986) Myocardial ischemia revisited. The osmolar load, membrane damage, and reperfusion. J Mol Cell Cardiol 18:769–780
25. Jolly SR, Kane WJ, Bailie MB, Abrams GD, Lucchesi BR (1984) Canine myocardial reperfusion injury. Its reduction by the combined administration of superoxide dismutase and catalase. Circ Res 54:277–285
26. Kammermeier H, Schmidt P, Jüngling E (1982) Free energy change of ATP-hydrolysis: a causal factor of early hypoxic failure of the myocardium. J Mol Cell Cardiol 14:267–277
27. Kapelko VI, Kupriyanov VV, Novikova NA, Lakomkin VL, Steinschneider AYa, Severina MYu, Veksler VI, Saks VA (1988) The cardiac contractile failure induced by chronic creatine and phosphocreatine deficiency. J Mol Cell Cardiol 20:465–479

28. Katz AM, Hecht HH (1969) The early "pump" failure of the ischemic heart. Am J Med 47:497–502
29. Kentish JC (1986) The effects of inorganic phosphate and creatine phosphate on force production in skinned muscles from rat ventricle. J Physiol 370:585–604
30. Krause EG, Wollenberger A (1980) Cyclic nucleotides in heart in acute myocardial ischemia and hypoxia. Adv Cycl Nucl Res 12:49–61
31. Lawson JWR, Uyeda K (1987) Effects of insulin and work on fructose 2,6-bisphosphate content and phosphofructokinase activity in perfused rat hearts. J Biol Chem 262:3165–3173
32. Lee H-C, Smith N, Mohabir R, Clusin WT (1987) Cytosolic calcium transients from the beating mammalian heart. Proc Natl Acad Sci USA 84:7793–7797
33. Lubbe WF, Podzuweit T, Daries P, Opie LH (1978) The role of cyclic adenosine monophosphate in adrenergic effects on vulnerability to fibrillation in the isolated perfused rat heart. J Clin Invest 61:1260–1269
34. Lucchesi BR, Romson JL, Jolly SR (1984) Do leukocytes influence infarct size? In: Hearse DJ, Yellon DM (eds) Therapeutic Approaches to Myocardial Infarct Size Limitation. Raven Press, New York, pp 219–248
35. Marban E, Kitakaze M, Chacko VP, Pike MM (1988) Ca^{2+} transients in perfused hearts revealed by gated ^{19}F NMR spectroscopy. Circ Res 63:673–678
36. Mochizuki S, Kobayashi K, Neely JR (1978) Effects of L-lactate on glyceraldehyde 3-P dehydrogenase in heart muscle. In: Kobayashi T, Ito Y, Rona G (eds) Recent Advances in Cardiac Structure and Metabolism, Vol 12. Cardiac Adaptation. University Park Press, Baltimore, pp 175–182
37. Morgan HE, Henderson MJ, Regen DM, Park CR (1961) Regulation of glucose uptake in muscle. I. The effects of insulin and anoxia on glucose transport and phosphorylation in the isolated, perfused heart of normal rats. J Biol Chem 236:253–261
38. Muller CA, Opie LH, Hamm CW, Peisach M, Gihwala D, Steyn JM, Basset HM (1986) Prevention of ventricular fibrillation by metoprolol in a pig model of acute myocardial ischemia: absence of a major arrhythmogenic role for cyclic AMP. J Mol Cell Cardiol 18:375–387
39. Neill WA, Ingwall JS (1986) Stabilization of a derangement in adenosine triphosphate metabolism during sustained, partial ischemia in the dog heart. J Am Coll Cardiol 8:894–900
40. Newsholme EA, Randle PJ (1961) Regulation of glucose uptake by muscle. 5. Effects of anoxia, insulin, adrenaline and prolonged starving on concentrations of hexose phosphates in isolated rat diaphragm and perfused isolated heart. Biochem J 80:655–662
41. Opie LH (1969) Metabolism of the heart in health and disease. Part II. Am Heart J 77:100–122
42. Opie LH (1971/72) Substrate utilization and glycolysis in the heart. Cardiology 56:2–21
43. Opie LH (1975) Metabolism of free fatty acids, glucose and catecholamines in acute myocardial infarction. Am J Cardiol 36:938–953
44. Opie LH (1976) Effects of regional ischemia on metabolism of glucose and fatty acids. Relative rates of aerobic and anaerobic energy production during myocardial infarction and comparison with effects of anoxia. Circ Res 38:52–74
45. Opie LH (1984) The heart: physiology, metabolism, pharmacology and therapy. Grune and Stratton, London, p 354
46. Opie LH, Muller CA, Lubbe WF (1978) Cyclic AMP and arrhythmias revisited. Lancet II:921–923
47. Opie LH, Nathan D, Lubbe WF (1979) Biochemical aspects of arrhythmogenesis and ventricular fibrillation. Am J Cardiol 43:131–148
48. Opie LH, Coetzee WA (1988) Role of calcium ions in reperfusion arrhythmias: relevance to pharmacologic intervention. Cardiovasc Drugs Ther 2:623–636
49. Opie LH, Clusin WT (1990) Cellular mechanism for ischemic ventricular arrhythmias. Ann Rev Med (in press)
50. Owen, P, Dennis S, Opie LH (1990) Glucose flux rate regulates onset of ischemic contracture in globally underperfused rat hearts. Circ Res 66:344–354
51. Pasteur L (1876) Ganthier-Villars, Paris
52. Podzuweit T, Dalby AJ, Cherry GW, Opie LH (1978) Cyclic AMP levels in ischaemic and non-ischaemic myocardium following coronary artery ligation: relation to ventricular fibrillation. J Mol Cell Cardiol 10:81–94

53. Rahimtoola SH (1989) The hibernating myocardium. Am Heart J 117:211−221
54. Rovetto MJ, Lamberton WF, Neely JR (1975) Mechanism of glycolytic inhibition in ischemic rat hearts. Circ Res 37:742−751
55. Schaper W, Binz K, Sass S, Winkler B (1987) Influence of collateral blood flow and of variations in MVO_2 on tissue ATP content in ischemic and infarcted myocardium. J Mol Cell Cardiol 19:19−37
56. Schömig A, Dart AM, Dietz R, Mayer E, Kübler W (1984) Release of endogenous catecholamines in the ischemic myocardium of the rat. Part A. Locally mediated release. Circ Res 55:689−701
57. Schoonderwoerd K, Broekhoven-Schokker S, Hulsmann WC, Stam H (1990) Involvement of lysosome-like particles in the metabolism of endogenous triglycerides in the normoxic and ischemic rat heart. Uptake and degradation of triglycerides by lysosomes isolated from rat heart. Basic Res Cardiol (in press)
58. Serur JR, Urschel CW, Sonnenblick EH, LaRaia PJ (1976) Experimental myocardial ischemia: III. Protective effect of glucose on myocardial function. J Mol Cell Cardiol 8:521−531
59. Taegtmeyer H (1978) Metabolic responses to cardiac hypoxia. Increased production of succinate by rabbit papillary muscles. Circ Res 43:808−815
60. Tennant R (1935) Factors concerned in the arrest of contraction in an ischemic myocardial area. Am J Physiol 113:677−682
61. Thandroyen FT, Higginson LM, Opie LH, Yon E (1986) The influence of verapamil and its isomers on vulnerability to ventricular fibrillation during acute myocardial ischemia and adrenergic stimulation in isolated rat heart. J Mol Cell Cardiol 18:645−649
62. Trach V, Buschmans-Denkel E, Schaper W (1986) Relation between lipolysis and glycolysis during ischemia in the isolated rat heart. Basic Res Cardiol 81:454−464
63. Uraizee A, Reimer KA, Murry CE, Jennings RB (1987) Failure of superoxide dismutase to limit size of myocardial infarction after 40 minutes of ischemia and 4 days of reperfusion in dogs. Circulation 75:1237−1248
64. Van der Vusse GJ, Roemen ThHM, Prinzen FW, Coumans WA, Reneman RS (1982) Uptake and tissue content of fatty acids in dog myocardium under normoxic and ischemic conditions. Circ Res 50:538−546
65. Vogel WM, Apstein CS, Briggs LL, Gaasch WH, Ahn J (1982) Acute alterations in left ventricular chamber stiffness. Role of the "erectile" effect of coronary arterial pressure and flow in normal and damaged hearts. Circ Res 51:465−478
66. Watanabe H, Nagao B, Nishiyama T, Kamikawa T, Kobayashi A, Yamazaki N (1986) The changes of free radicals in myocardial mitochondria during ischemia and reperfusion. J Mol Cell Cardiol 18 (Suppl 1):114 (abstract)
67. Weiss J, Hiltbrand B (1985) Functional compartmentation of glycolytic versus oxidative metabolism in isolated rabbit heart. J Clin Invest 75:436−447
68. Wiesner RJ, Deussen A, Borst M, Schrader J, Grieshaber MK (1989) Glutamate degradation in the ischemic dog heart: contribution to anaerobic energy production. J Mol Cell Cardiol 21:49−59
69. Wiggers CJ (1934) Physiology in health and disease. Lea and Febiger, Philadelphia, p 568
70. Wilde AAM, Peters RJG, Janse MJ (1988) Catecholamine release and potassium accumulation in the isolated globally ischemic rabbit heart. J Mol Cell Cardiol 20:887−896
71. Williamson JR, Safer B, Rich T, Schaffer S, Kobayashi K (1976) Effects of acidosis on myocardial contractility and metabolism. Acta Med Scand 587 (Suppl):95−112
72. Wissner SB (1974) The effect of excess lactate upon the excitability of the sheep Purkinje fiber. J Electrocardiol 7:17−26

Authors' address:
Professor L. H. Opie, M.D., Ph.D.
Heart Research Unit
University of Cape Town
Medical School
Observatory 7925
South Africa

Metabolic Imaging of Ischemic Heart Disease by Positron Emission Tomography

Hans Georg Wolpers and Markus Schwaiger

University of Michigan, Medical Center, Department of Internal Medicine, Division of Nuclear Medicine, Ann Arbor, Michigan, USA

Introduction

Over the last decade many new invasive and non-invasive technologies have been developed to characterize ischemic heart disease. Coronary artery disease can be detected with high sensitivity and specificity by using non-invasive imaging approaches. At the same time, diagnostic coronary angiography has become a safe and widely accepted invasive procedure for characterizing the extent and severity of coronary artery disease. Regional and global assessment of function, perfusion, and coronary anatomy have been shown to provide important diagnostic and prognostic information in acute and chronic disease. However, diagnostic problems persist in efforts to define tissue viability in patients with advanced coronary artery disease. Since many large clinical studies have shown that patients with impaired left ventricular function benefit most from revascularization, the diagnostic differentiation between infarcted and ischemically compromised, but viable, myocardium has become of major clinical importance. Functional evaluation by radionuclide ventriculography or echocardiography does not allow for separation of irreversible and reversible tissue injury, especially early after myocardial infarction. Delayed functional recovery (stunned myocardium) [8] and chronically depressed function of hypoperfused myocardium (hibernating myocardium) [38] have been recently recognized as clinical entities. Thallium-201 stress imaging has been widely used to separate ischemic viable myocardium from scar tissue. However, recent reports indicate that this technique overestimates the extent of scar formation and is, therefore, of only limited value in selecting patients for revascularization [9]. Since tissue viability relies on residual metabolic activity for energy production, the use of radiopharmaceuticals that trace cardiac metabolism has been proposed for the definition of viable tissue. Such an approach has been shown to provide specific detection of viable myocardium when combined with positron emission tomography (PET) [42].

This review will address the role of metabolic imaging using PET to characterize the extent and severity of ischemic heart disease. Following a brief description of the technique, the principles and experimental validation of metabolic imaging, as well as its clinical application in acute and chronic stages of coronary artery disease will be discussed.

This work was done during the tenure of an established investigatorship from the American Heart Association (M. Schwaiger), and supported in part by the National Institute of Health (RO1HL41047-01) and the American Heart Association of Michigan (#88-0699-J1). Dr. H. G. Wolpers is recipient of a research fellowship by the Deutsche Forschungsgemeinschaft.

PET Imaging Principles

The imaging of myocardial tracer retention by PET is accomplished by the unique properties of positron decay. Positron-emitting radionuclides are characterized by a nuclear imbalance with an excess of protons. In order to restore stability to the nuclear structure, a proton is converted to a neutron and a positron is emitted from the nucleus. This energetic positron traverses a few millimeters through tissue until it combines with a free electron to form a two-particle atom-like entity called a positronium. The positronium quickly decays by annihilation (complete conversion of the positron and electron mass into energy), thereby generating a pair of photons. Conservation of energy and momentum of the positronium prior to annihilation requires that the photons travel in nearly opposite directions (180° apart and energy of 511 KeV each). It is this unique characteristic of positron annihilation that is exploited for image formation [21]. The photons from positron annihilation can be detected through the use of colinearly aligned detectors (Fig. 1). Spatial definition of the radioactive event is achieved by multiple detector pairs surrounding the body. This angular information is used to reconstruct tomographic images of body organs, such as the heart. The major objective of cardiac PET is to label physiologically active substrates with positron-emitting radionuclides and to acquire transverse sectional images of the radionuclide distribution in the heart following the injection of the tracer. Most state-of-the-art positron emission tomographs consist of multiple detector rings that make the simultaneous definition of tracer distribution over the entire heart possible. This multiplane data acquisition allows three-dimensional display of tracer distribution, that is especially useful in the definition of the extent of tracer

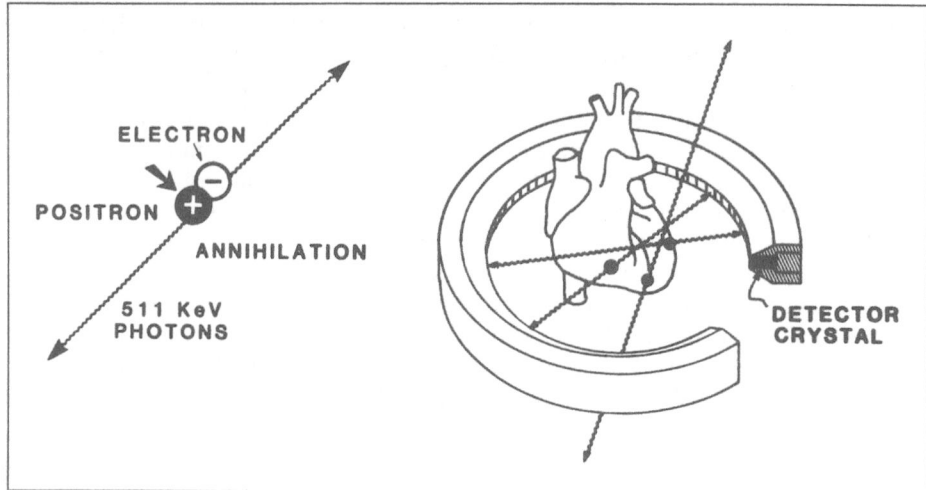

Fig. 1. Uptake of positron-emitting radiotracers by the heart results in pairs of photons being emitted at an angle of approximately 180° from the tissue. The twin 511 KeV photons arise when an emitted positron interacts with an electron (annihilation radiation). The photons are detected externally by multiple detector pairs positioned around the subject. Coincident circuits provide angular information of the radioactive events which is used for reconstruction of a quantitative image of the radionuclide distribution within the heart.

uptake abnormalities. The spatial resolution of current PET instrumentation ranges from 5 to 10 mm and is primarily determined by the size of the detectors in the tomograph as well as the distance the positron travels in tissue before it interacts with an electron. Due to these physical limitations, PET cannot provide the same spatial resolution as echocardiography or magnetic resonance imaging. Since the left ventricular wall thickness averages around 1 to 1.5 cm, currently available image resolution does not allow for separation of endo- and epicardial regions. However, PET excels by its extreme sensitivity. Small tissue concentrations of radiopharmaceuticals (picomolar) can be detected, which minimizes the patient's exposure to radiation and avoid the pharmacologic effects of most radiopharmaceuticals used for imaging (tracer technique). In addition, dynamic studies with high temporal resolution provide the delineation of regional activity changes that define tissue tracer kinetics. The use of such dynamic data acquisition in combination with tracer kinetic modeling allows the quantitative assessment of physiologic processes, such as myocardial blood flow or substrate utilization [4, 22, 45].

Evaluation of Myocardial Blood Flow

Myocardial blood flow is closely related to myocardial oxygen consumption [26]. Therefore, evaluation and quantification of regional myocardial perfusion provide indirect estimates of myocardial oxygen consumption. Since both the delivery and utilization of substrates define the metabolic state, the combined characterization of regional myocardial perfusion and metabolic activity allows the specific definition of pathophysiologic alterations occuring during myocardial ischemia. In animal experiments a number of techniques are available to measure blood flow (i.e., microspheres, flow probe). Although enabling sophisticated measurements, these techniques are invasive and cannot be transferred easily to the clinical situation. In comparison, non-invasive approaches of studying coronary flow in humans have been crude, providing primarily qualitative evaluation. However, the principles of PET allow precise non-invasive measurements of regional myocardial perfusion as shown in recent investigations [6, 29]. Most alternative clinical blood flow techniques, such as thermodilution and Doppler methods, assess flow velocity in coronary arteries, while the determination of regional tracer tissue concentration by PET permits the quantification of tissue perfusion per gram of myocardium.

Aggregated albumin microspheres labeled with Ga-68 and C-11 can be used to assess blood flow accurately in the human heart, but the injection site (left atrium, left ventricle) limits the clinical application. There are two types of blood flow tracers that can be administered intravenously: 1) indicators that are trapped in myocardium, at least transiently, in proportion to blood flow, and 2) freely diffusible tracers that accumulate and clear from myocardium as a function of flow, such as O-15 water and C-11 butanol (Table 1). The most commonly used tracers of the first type are rubidium-82 and N-13 ammonia, which have a high first-pass extraction fraction by the myocardium (about 80% for N-13 ammonia and 60% for rubidium). Rubidium-82 is a potassium analogue, sharing, at least partially, the sodium-potassium transport mechanism, while N-13 ammonia is converted to N-13 glutamine by the glutamine synthesis reaction. The development of tracer kinetic models has been shown to

Table 1.

Radionuclide	Half-life [min]	Compound	Myocardial blood flow	Myocardial metabolism
Cyclotron produced				
O-15	2.1	water	+++	−
N-13	10.1	ammonia	++	+
C-11	20.4	palmitate	+	++
		acetate	+	++
F-18	110	deoxyglucose	−	+++
Generator produced				
Rb-82	1.3	Rb-chloride	++	+
Cu-62	10	PTSM[a]	+++	−

[a] copper − (II) bi's (N-4-methylthiosemicarbazone)

allow quantification of regional myocardial blood flow using N-13 ammonia [23, 29]. New radiopharmaceuticals, especially lipophilic compounds, which have high extraction and retention by the myocardium, may improve the accuracy of blood flow measurements in the future. Copper-62 PTSM is currently being tested in animal studies and provides flow measurements over a wide flow range [56]; further validation, however, is required before this compound can be clinically applied.

O-15 water is the most widely used PET flow marker of the freely diffusible compounds. This tracer is metabolically inert and avidly extracted by myocardial tissue (first-pass extraction fraction close to 100%). Most studies have shown that this high extraction fraction is maintained over a wide flow range. This finding indicates that the myocardial distribution of O-15 water is limited by flow rather than by diffusion [6]. In addition to these favorable physiologic properties of O-15 water, the short physical half-life (120 s) also allows repetitive flow measurements within short intervals. However, the high concentration of O-15 water in the vascular space (cardiac chambers) limits the accurate definition of myocardial tissue concentration using static imaging approaches. Dynamic PET scanning allows the differentiation of tracer activities in blood and in tissue. Accurate quantification of regional blood flow using tracer kinetic approaches has been demonstrated in animal and clinical studies [5, 6].

Application of PET technology for the evaluation of blood flow in patients with coronary artery disease has shown that qualitative image evaluation yields high sensitivity and specificity for the detection of coronary artery disease. Quantitative measurements of coronary reserve following pharmacological coronary vasodilatation have become possible and provide a functional, non-invasive parameter (coronary reserve) for the assessment of the severity of regional coronary artery disease [6, 46, 52].

Evaluation of Cardiac Metabolism

Regional differences in myocardial blood flow at rest or during stress may not necessarily reflect myocardial ischemia. In clinical situations such as reperfusion, there may exist a transient dissociation between regional flow and metabolic requirements.

Since ischemia represents pathophysiology on a cellular level, assessment of cellular function by metabolic parameters may define myocardial ischemia more specifically than flow parameters.

Invasive characterization of cardiac metabolism by arterio-venous sampling is limited to global measurements of substrate extraction. However, coronary artery disease affects regional vascular territories; consequently, regional definition of tissue function is required for the assessment of severity and extent of disease. The three-dimensional data acquisition provided by PET permits the evaluation of regional changes of myocardial substrate utilization and, thus, overcomes the previously existing limitations of the in-vivo evaluation of cardiac metabolism.

Myocardial high energy production relies on an intact myocardial substrate metabolism. The normal heart utilizes various substrates, such as free fatty acids, glucose, lactate, pyruvate, acetate, keton bodies, and amino acids. Under fasting conditions, the heart derives most of its energy (approximately 70%) from oxidation of long-chain fatty acids, while carbohydrates account for about 30% of cardiac energy production [31]. Myocardial substrate selection is influenced by the concentrations of alternative fuels in arterial blood, by hormonal variations, cardiac workload, and the availability of oxygen. This metabolic versatility may complicate the interpretation of metabolic measurements, but may result in unique metabolic patterns associated with pathophysiologic conditions, which can be identified by PET in combination with metabolic tracers.

Definition of Metabolic Tracers

Two general strategies have been pursued for tomographic assessment of metabolism (Table 1, Fig. 2). The first is by incorporation of positron-emitting radionuclides into physiologic substrates, such as C-11 palmitate, C-11 glucose, and C-11 acetate. Since these radiopharmaceuticals behave biochemically as counterparts of the traced substrates, no specific tracer-related assumptions about transport or metabolic pathways are required. However, contamination by labeled metabolites may impede the quantitative interpretation of externally recorded tracer kinetics. The second approach is characterized by the use of radiolabeled substrate analogues. These analogues are selected to be refractory to enzymes acting on their physiological counterparts, such as F-18-2-deoxyglucose or 1-C-11-beta-methylheptadecanoid acid. This selection may facilitate assessment of specific aspects of glucose or fatty acid metabolism, but it entails the definition of correction factors for discrepancies between the behavior of the analogue and the natural substrate [33].

Clinical investigations with PET thus far have employed only a limited number of tracers. The most frequently used are C-11 palmitate, F-18-2-deoxyglucose, and C-11 acetate as tracers of fatty acid metabolism, exogenous glucose utilization, and overall oxidative metabolism of the heart, respectively (Fig. 2).

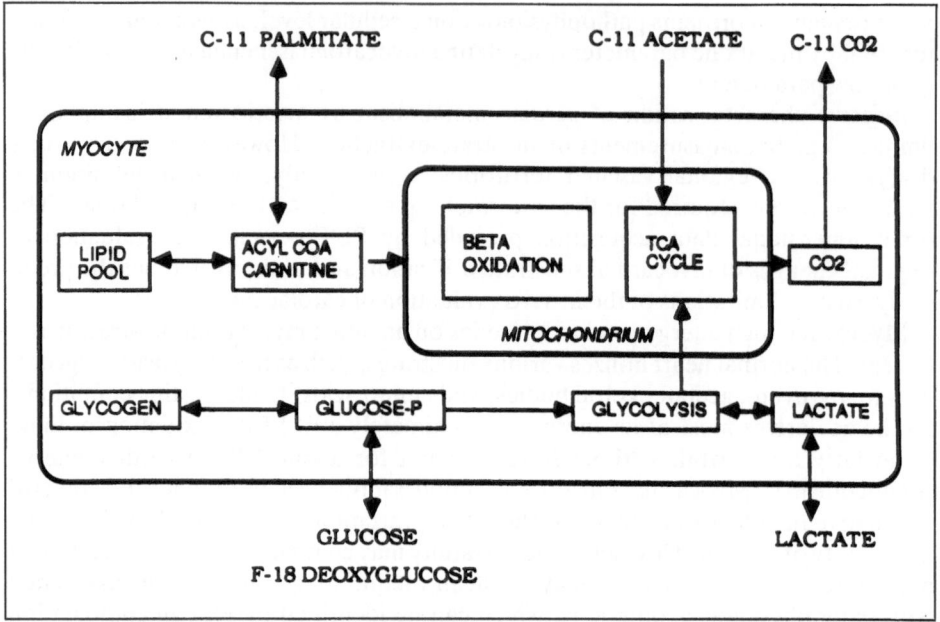

Fig. 2. Metabolic fate of various radiolabeled substances in the myocyte. After uptake into the myocyte C-11 palmitic acid either enters mitochondria via the "carnitine-shuttle" for immediate oxidation or is incorporated into triglyceride or phospholipid pools. After beta-oxidation the C-11 label is released as C-11 dioxide. C-11 acetate enters directly the TCA cycle and again the label clears from the myocardium as a function of C-11 dioxide production. FDG traces glucose uptake and phosphorylation by hexokinase, but does not enter glycogen-synthesis or glycolysis. FDG uptake reflects exogenous glucose utilization.

C-11 Palmitate

C-11 palmitate has been widely employed in experimental and clinical studies for the non-invasive characterization of cardiac fatty acid metabolism [18, 27, 43, 47, 54, 57, 58]. Under physiologic conditions, palmitate comprises 25–30% of circulating fatty acids and its oxidation accounts for about 50% of the myocardial energy production attributable to fatty acid oxidation [4, 40]. Cellular uptake is influenced by the concentration of free fatty acids in blood, the albumine binding ratio, hormonal influences, and the myocardial affinity for fatty acids with a specific chain length and saturation. The favored uptake mechanism includes passive diffusion and transfer from extravascular albumin to intracellular carrier proteins [60].

Because the extraction fraction of C-11 palmitate is high (>50%), initial uptake of the tracer primarily reflects blood flow. Clearance of C-11 from the myocardium is biexponential, reflecting the metabolic fate of the tracer. The early rapid clearance phase corresponds to the release of C-11 dioxide as the end-product of fatty acid oxidation. The size and rate of this clearance phase has been shown to reflect sensitively changes in cardiac work load and substrate utilization [19, 43, 44, 45]. In contrast, the late slow phase most likely relates to the deposition of tracer in endogeneous lipid pools, such as triglycerides and phospholipids [4, 47, 55].

The complexities of fatty acid metabolism, however, limit the quantitative evaluation of fatty acid oxidation based on PET-derived C-11 palmitate kinetics. This limitation is due to the contamination of the early tissue kinetics by back-diffusion of unmetabolized tracer into the vascular space [22, 47, 55]. In addition, the relative amount of tracer entering the endogenous lipid pool varies, depending on the intracellular availability of glycerophosphate and the activity of carnitine-palmitoyl transferase (CPT I) [65]. Nevertheless, the evaluation of C-11 palmitate kinetics has rendered valuable qualitative information on fatty acid metabolism in the normal and ischemic heart.

F-18-2-Deoxyglucose

F-18-2-deoxyglucose (FDG) is widely used for the non-invasive assessment of exogenous glucose utilization. This technique, initially developed by Sokoloff et al. for autographic imaging of brain metabolism, was subsequently adapted for measuring glucose utilization in the heart and other organs [37]. FDG traces the initial metabolic steps of glucose uptake and phosphorylation by hexokinase. The phosphorylated product is not further metabolized via glycolysis, glycogen synthesis, or the pentose phosphate shunt. Because dephosphorylation of FDG-6-phosphate is slow in the heart and the sarcolemma is largely impermeable for charged species, FDG-6-phosphate remains essentially trapped in the cytosol and accumulates proportionally to exogenous glucose utilization. In contrast to the brain, where glycogen synthesis is negligible, FDG uptake in the heart reflects the sum of both glycolysis and glycogen synthesis. Yet, under steady-state conditions, where the rates of glycogen formation and breakdown are in equilibrium, the net accumulation of myocardial FDG-6-PO_4 over a specific time period (e.g., 40 min) reflects the steady-state extraction and phosphorylation rate of exogenous glucose [28]. In combination with a three-compartment tracer kinetic model, FDG can be used for the quantification of exogenous glucose utilization rate [28, 37]. Using dynamic PET imaging, this approach has been successfully applied to the human heart [17, 39]. Since myocardial glucose utilization is low in the fasting state and increases in the postprandial state, quantitatively meaningful and reproducible measurements rely on standardized metabolic conditions, which are often difficult to control in the clinical environment.

C-11 Acetate

C-11 acetate has been recently introduced as a tracer for the assessment of oxidative metabolism. C-11 acetate is avidly taken up by the myocardial cell and converted to acetyl-CoA, which enters the tricarboxic acid (TCA) cycle. C-11 activity clears in the form of C-11 dioxide from the myocyte as a function of TCA cycle flux [12]. The use of this tracer offers several advantages over C-11 palmitate. C-11 acetate undergoes immediate metabolism, and only minimal activity is retained in the myocardium as TCA-cycle intermediates. Thus, the tracer clearance kinetics directly reflect myocardial oxidative metabolism. Several animal and clinical studies have demonstrated a close relationship of C-11 clearance halftimes and myocardial oxygen consumption

[3, 11, 20]. In contrast to C-11 palmitate and FDG, uptake and metabolism of C-11 acetate are little affected by substrate interaction. Therefore, TCA-cycle flux, which represents the common metabolic pathway of myocardial energy production, can be assessed by C-11 acetate independently of the overall metabolic state.

Current research focuses on the development of a tracer kinetic model for C-11 acetate for the accurate quantification of myocardial oxidative metabolism. Such an approach may be useful in assessing the effect of various therapeutic interventions on global and regional oxygen consumption. Furthermore, in combination with the measurement of cardiac work, the relationship of metabolic turnover and cardiac performance may be used to define the efficiency of external cardiac work.

Metabolic Changes in Ischemic Myocardium

Myocardial energy stores last for approximately 10 beats following experimental coronary artery occlusion, and myocardial contractile function ceases rapidly. The concomitant metabolic changes in ischemic myocardium include impaired oxidation of fatty acids, accelerated glycogen breakdown, and increased glycolytic flux. The mechanisms involved are described in detail in the preceding chapter by Opie. Cyclic adenosine monophosphate accumulates early in ischemic tissue, and the lack of oxygen impairs or inhibits oxidative fatty acid metabolism. This inhibition subsequently leads both to activation of the glycogen breakdown and to increased glucose uptake and glycolytic flux. The activity of phosphofructokinase is specifically accelerated by changes in the energy status of the cells, as reflected by the cytosolic concentrations of ATP, creatine phosphate and their breakdown products. Because of the inhibitory action of protons, lactate, and $NADH_2$ produced during non-oxidative glucose utilization, the glycolytic rate depends critically on the washout of detrimental metabolites in ischemic tissue. Accumulation of lactate leads to cessation of glycolysis and, subsequently, to cell death.

Metabolic Imaging in Animal Models of Coronary Artery Occlusion

The relationship between residual blood flow, anaerobic and aerobic metabolism, and cell viability has been studied with various metabolic tracers in animal experiments. Studies employing FDG have shown a striking dependence of FDG accumulation on residual blood flow in the center of ischemic tissue and in the transition zone between normal and ischemic tissue during acute occlusion of the left anterior descending artery (LAD) in the dog model [24, 41]. In contrast to the close correlation of glucose utilization and blood flow encountered in normal myocardium, FDG uptake was disproportionally increased relative to flow in ischemic zones with residual blood flows between 80% and 20% of baseline, thus suggesting accelerated glycolysis in mild to moderate ischemia (Fig. 3). However, FDG uptake was greatly diminished in regions with severe ischemia, i.e., residual blood flow below 20% of control. These data indicate the existence of a threshold value of blood flow for the preservation of glycolytic flux. The sensitive coupling between ischemic flow and preserved glucose utilization may reflect the washout of lactate discussed above or of

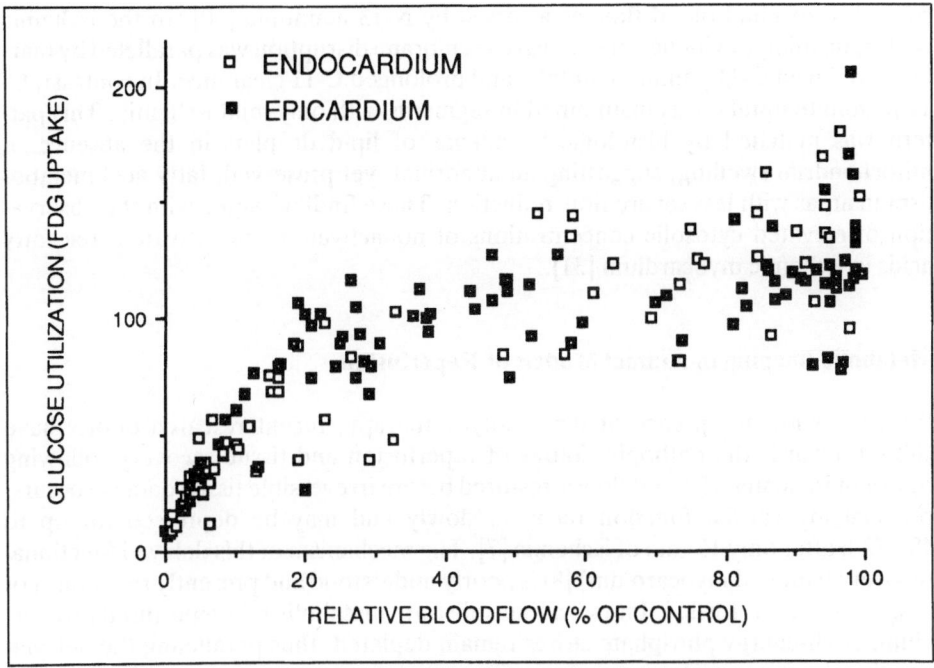

Fig. 3. Relationship between myocardial blood flow (microspheres) and FDG uptake (0.5 mCi i.v.) during acute occlusion of the LAD in six dogs. Data are derived from microsphere and F-18 activities (well counting) in tissue slices excised from the left ventricle and divided into 15 to 18 sequential cross-sectional endo- and epicardial strips. Both FDG uptake and myocardial blood flow are expressed as percentages of normal values in remote myocardium. Myocardial FDH uptake is maintained fairly constant, despite reduced residual blood flow, until flow falls below 20% of control. Thus, glucose utilization by the myocardium is increased relative to blood flow in regions of moderate flow reduction (between 80–20% of control).

other detrimental metabolites, thus preventing inhibition of glycolysis. On the other hand, the anaerobic, high-energy phosphate production in the presence of decreased exogenous glucose delivery may be too low to cover the energy requirements necessary for cell survival. The finding that FDG uptake and, hence, residual metabolic activity fall rapidly beyond a critical bllod flow level agrees with observations that small fluctuations in blood flow may be of utmost importance for maintaining tissue viability. In an animal model of myocardial stunning, the magnitude of residual collateral flow in the ischemic territory was shown to account for most of the variation of functional recovery during reperfusion [7].

The tracer kinetics and metabolic fate of C-11 labeled palmitate in ischemic myocardium have been characterized in several studies [47, 55]. C-11 activity clears more slowly from ischemic myocardium, indicating decreased C-11 dioxide production (Fig. 4B). Increased back diffusion of non-metabolized tracer into the vascular space occurs during ischemia and limits the specificity of the tracer clearance for fatty acid oxidation. However, the increased late tissue retention of C-11 activity in ischemic myocardium is consistent with increased deposition of fatty acids in endogenous lipid pools. Using a dog model with LAD occlusion, C-11 palmitate kinetics in the center and on the periphery of the ischemic territory have been evaluated by PET and related to residual blood flow as assessed by N-13 ammonia [50]. In the ischemic center, histologic evidence of extensive membrane disruption was paralleled by markedly reduced C-11 palmitate uptake and prolonged C-11 clearance. In contrast, C-11 palmitate uptake was maintained in segments with only mild ischemia. This pattern was matched by histological evidence of lipid droplets in the absence of mitochondrial swelling, suggesting an abnormal, yet preserved, fatty acid metabolism in areas with less severe flow reduction. These findings agree with the observation of elevated cytosolic concentrations of nonactivated and activated free fatty acids in ischemic myocardium [31].

Metabolic Imaging in Animal Models of Reperfusion

With the wide acceptance of thrombolytic therapy, recent research efforts have shifted towards the pathophysiology of reperfusion and tissue recovery following transient ischemia. If blood flow is restored before irreversible tissue damage occurs, regional myocardial function recovers slowly and may be depressed for up to 48–72 h after only 15 min of ischemia [7]. The mechanism of this delayed functional recovery (stunned myocardium) [8] is poorly understood and presently the subject of intense research. Little is known about substrate metabolism in reperfused myocardium. High-energy phosphate stores remain depleted, thus paralleling the delayed functional recovery. The decreased ATP content, however, does not exclude normal high energy phosphate turnover. The in-vitro assessment of mitochondrial function following short time periods of ischemia failed to demonstrate sustained abnormalities of oxidative metabolism in reversibly injured myocardium. Employing 31P-NMR spectroscopy, Ambrosio et al. demonstrated that the phosphocreatine/inorganic phosphate (P-Cr/Pi) ratio, which may serve as an index of the balance between energy production and utilization, was well preserved in reperfused rabbit hearts stunned by a 20-min global ischemia [1]. This finding suggests rapid recovery of mitochondrial function and excludes impaired oxidative metabolism as a primary factor for the slow recovery of function in stunned myocardium.

Recent studies evaluating the TCA-cycle flux in reperfused canine myocardium with PET and C-11 acetate demonstrated sustained abnormalities of C-11 clearance kinetics [11]. Following a transient coronary artery occlusion of 20 min, decreased clearance rates were observed, which were largely restored within 24 hrs. Accordingly, study employing C-11 palmitate indicated altered tracer kinetics in stunned myocardium following 20 min of transient LAD occlusion [55]. It is not clear in the observed alterations of C-11 palmitate and C-11 acetate kinetics reflect an impairment of fatty acid and acetate metabolism or a decreased metabolic demand of dysfunctional (stunned) myocardium.

Following a 3 h ischemia in the chronic dog model, the metabolic abnormalities, as detected by PET, were more pronounced [54]. Delayed functional recovery was paralleld by sustained alterations of C-11 palmitate kinetics for up to 1 week. The decrease of fatty acid oxidation was matched by regionally increased FDG uptake, which suggests a shift of substrate metabolism in reperfused myocardium (Fig. 4). Correlations with histochemical staining showed that FDG uptake was increased in reversibly injured myocardium, but not in necrotic tissue [59]. Since FDG uptake

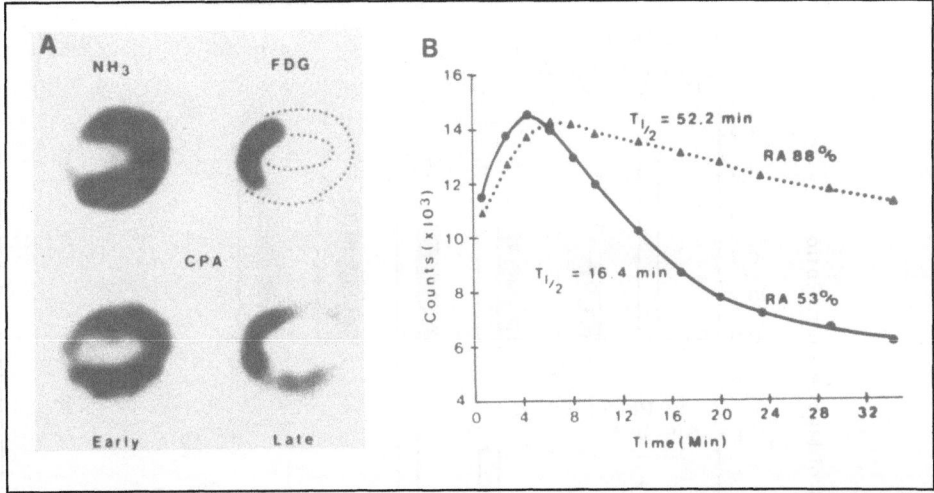

Fig. 4. A) Reconstructed cross-sectional PET images of canine myocardium obtained 24 h after reperfusion following a 3 h balloon occlusion of the LAD. The N-13 ammonia (NH_3) image (upper left), which represents the distribution of blood flow, shows reduced perfusion in the reperfused segment. In the same segment there is a "discordant" increase of FDG uptake indicating well-preserved metabolic activity (upper right). Serial images after C-11 palmitic acid (CPA) injection (lower panel) initially show a C-11 distribution which parallels blood flow and later a relatively increased activity in reperfused myocardium reflecting delayed clearance of palmitic acid. B) Time-activity curves obtained from serial images in the same dog with regions-of-interest placed over control (solid line) and reperfused myocardium (dotted line). Compared to control the C-11 clearance is markedly prolonged in the reperfused segment (half-time 52.2 vs 16.4 min) and the residual activity (RA) increased (88 vs 53%) suggesting severe impairment of fatty acid metabolism in the reperfused segment. (Reprinted with permission from J Am Coll Cardiol 1985; 6:336–347).

does not separate between oxidative or non-oxidative glucose utilization, studies employing C-14 labeled glucose have been conducted in the same model. This biochemical investigation revealed sustained non-oxidative glucose utilization at 24 h after reperfusion, as evidenced by considerable C-14 lactate production [53]. However, the regional increase of glucose utilization also resulted in increased C-14 carbon dioxide production, thus indicating enhanced oxidative glucose utilization in reperfused myocardium. These data suggest the coexistence of different degrees of cellular injury in the reperfused territory. Sustained anaerobic glycolysis may occur in the most severely injured cells, while preferential oxidative glucose utilization may reflect intermediate injury with impairment of fatty acid metabolism. Such an hypothesis is supported by the subsequent recovery of substrate metabolism in this model (Fig. 5). Four weeks after reperfusion, C-11 palmitate kinetics normalized in functionally recovered myocardium [54].

Interesting to note is the fact that glucose utilization was depressed immediately after reperfusion and increased during the first 24 h following reperfusion [51]. This unexpected finding may reflect "metabolic stunning" with sustained inhibition of the glycolytic pathway following severe ischemia. Alternatively, the ischemic injury may induce cellular processes, which lead to the expression of proteins responsible for the acceleration of glycolytic flux. Such adaptive response may require a time period of

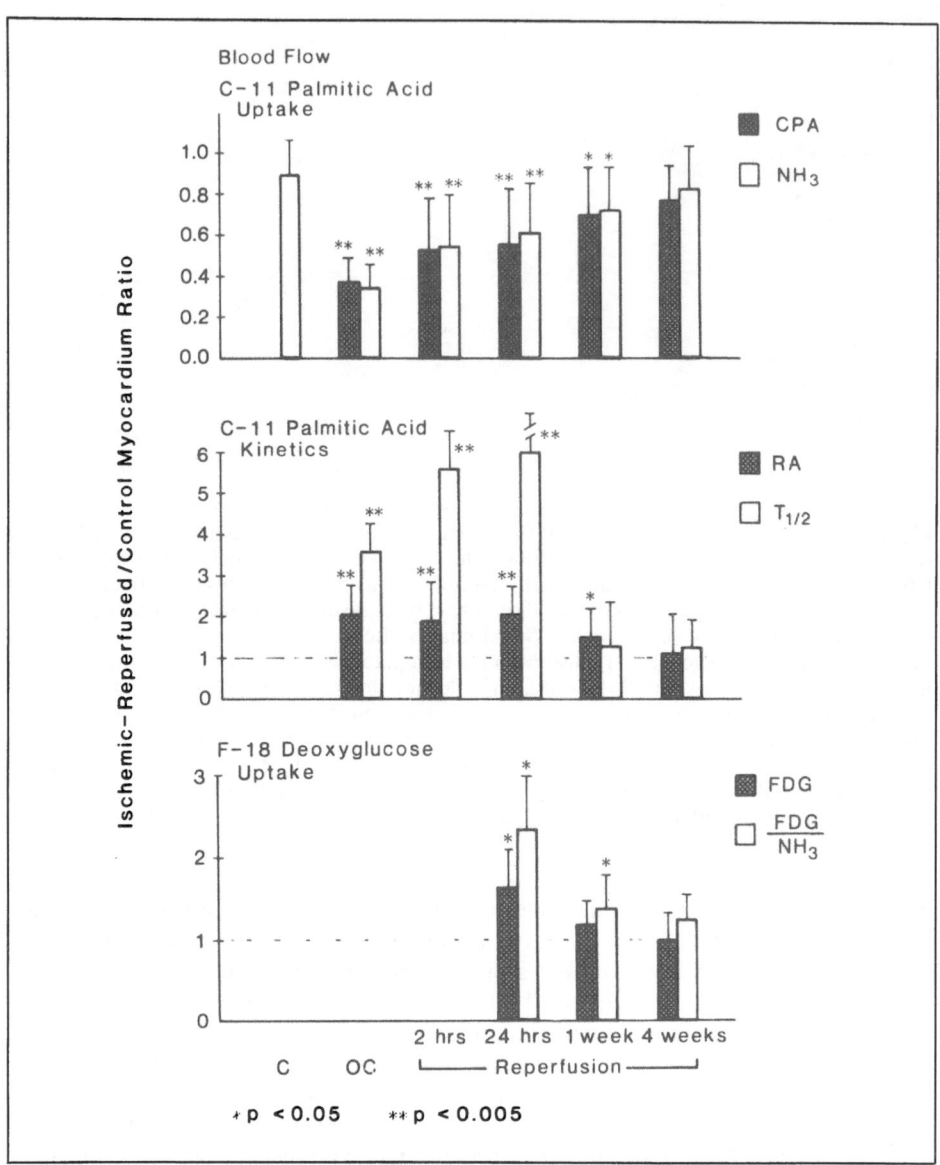

Fig. 5. Four weeks follow-up of blood flow (NH₃), C-11 palmitic acid (CPA) kinetics, and FDG uptake in a chronic dog model with 3 h balloon occlusion of the LAD followed by 4 weeks reperfusion. Values represent mean values (+ standard deviations) and are normalized to control myocardium. C-11 palmitate uptake closely parallels blood flow during the entire observation period. During the first 24 h C-11 palmitate kinetics are severely impaired as indicated by prolonged clearance half-times ($T_{1/2}$) and increased residual activity (RA). This is matched by a concomitant increase of FDG uptake and of the FDG/NH₃ ratio at 24 h in reperfused myocardium. All measured parameters approximate control values after 4 weeks of reperfusion. (Reprinted with permission from J Am Coll Cardiol 1985; 6:336–347).

several hours to develop following an ischemic injury. This hypothesis remains speculative without measurements of glycolytic enzyme levels or other markers of enzyme synthesis. However, the concept of metabolic adaptation to ischemia has been postulated in the skeletal muscle (hindlimb model) [15]. Repetitive ischemia resulted in increased glucose utilization in the affected extremity, which suggests sustained alterations in substrate metabolism.

Thus, the metabolic information obtained in vivo by metabolic tracer techniques may provide new insights into the effects of ischemia and reperfusion on substrate metabolism and into the pathophysiology of tissue recovery. However, in-vitro investigation of cellular function will be necessary to fully understand the mechanisms of the observed alterations of substrate metabolism.

Metabolic Patterns Defined by PET in Ischemic Heart Disease

The clinical application of PET in ischemic heart disease is based on the observations in animal experiments discussed above, which show distinct metabolic patterns related to different degrees of ischemic injury. In summary, a pattern of absolutely or relatively enhanced FDG uptake frequently encountered in mild to moderate ischemia has been defined as a blood flow-glucose metabolism mismatch. It documents persisting aerobic and/or anaerobic glycolysis in ischemically compromised tissue. Although this pattern is most frequently associated with abnormal left ventricular wall motion, preservation of metabolic activity is matched by histologic evidence of residual viable tissue. A second pattern with concordant decrease of blood flow and FDG uptake has been referred to as blood-flow-metabolism match. This pattern indicates the absence of metabolic activity and is related to necrosis or scarring. Metabolic imaging by the FDG method is complemented by monitoring of uptake and oxidation of radiolabeled fatty acids, which are reduced in both mild and severe ischemia. Thus, the presence of metabolic activity as a hallmark of cellular viability may be used to distinguish ischemic tissue from regions of extensive scar formation.

Demand-Induced Ischemia

Ischemic heart disease frequently manifests itself as demand-induced ischemia in patients with exertional angina. Although it has been proposed that demand-induced ischemia differs from "supply ischemia" [2] (see preceding chapter), in-vitro experiments suggest no fundamental metabolic differences between low-flow ischemia and increased oxygen demand in the presence of limited coronary reserve [34]; in both conditions, a dependence of glucose metabolism on residual perfusion for washout of metabolic end-products has been observed. However, the metabolic consequences of stress-induced ischemia in the human heart have not been well documented by conventional techniques. Transmyocardial lactate differences, although specific, did not prove to be a sensitive marker of stress-induced ischemia in man [32]. This result probably reflects the spatial metabolic heterogeneity of ischemic heart disease with the simultaneous coexistence of lactate release and uptake.

Clinical studies have demonstrated the feasibility of PET imaging with C-11 palmitate or FDG to define regional metabolic abnormalities during stress-induced ischemia [13]. When regional blood flow was measured with N-13 ammonia or Rb-82 during stress and FDG uptake in the postexercise phase, a typical mismatch pattern could be observed between postexercise FDG uptake and blood flow in the territory of the significantly stenosed coronary artery. The increased glucose utilization in the postexercise phase has been suggested to reflect either prolonged recovery of metabolic function after transient ischemic stress or sustained restoration of glycogen stores depleted during ischemia.

Other clinical studies conducted in patients with known coronary artery disease employed C-11 palmitate during control and during moderately increased workload imposed by atrial pacing [19]. Regional C-11 palmitate tissue kinetics suggested increased fatty acid oxidation in both normal and poststenotic myocardium. However, the increase was attenuated in the poststenotic segments of the left ventricle. Left ventricular segments with altered C-11 palmitate kinetics showed development of echocardiographic wall motion abnormalities in about one-half of the patients studied. The findings already discussed imply that demand-induced ischemia is characterized by a shift of substrate selection to glucose, resulting in a mismatch pattern in ischemic segments. The finding that wall motion deteriorated in only one-half of the patients, in spite of an inadaequate increase in fatty acid oxidation, may also indicate the involvement of an alternate energy-producing pathway, such as glycolysis. Theoretically, the higher ATP/O_2 ratio of carbohydrate metabolism, as compared to exclusive fatty acid oxidation, would account for an oxygen-sparing effect, which may be favorable, given the limited oxygen supply.

Metabolic Imaging in Acute Myocardial Infarction

In patients with transmural infarction, regional impairment of C-11 palmitate and FDG uptake correlated closely with the elctrocardiographic site of infarction or with regional wall motion abnormalities [58, 62] (Fig. 6). Conversely, metabolic imaging with C-11 palmitate has been used to demonstrate improved metabolic activity after successful

---

Fig. 6. Top: Cross-sectional PET images of myocardial blood flow (N-13 ammonia) and glucose utilization (F-18 deoxyglucose) in a patient with previous anterior myocardial infarction. Four continuous cross-sectional images are depicted from the base of the left ventricle to the apex. The heart is viewed from below; the septum is seen on the left, the lateral wall on the right side, and the anterior wall at the top. Note the defect of both N-13 ammonia and FDG uptake in the anteroseptal wall. Bottom: Polar maps for blood flow (left) and glucose utilization (right) in the same patient. The heart is again viewed from below and divided into 60 sectors and 12 planes. The apex of the left ventricle is depicted in the center, the septum on the left and the lateral wall on the right side, the anterior wall at the top and the inferior wall at the bottom. The concordant reduction of blood flow and FDG uptake in the anteroseptal wall is consistent with scar formation ("PET infarction"). The respective circumferential profiles for relative N-13 ammonia and FDG activities are displayed in the inset showing a matched decrease of tracer uptake from sector 1 to 17. In sectors adjacent to the matched defect FDG uptake apparently is increased relative to blood flow.

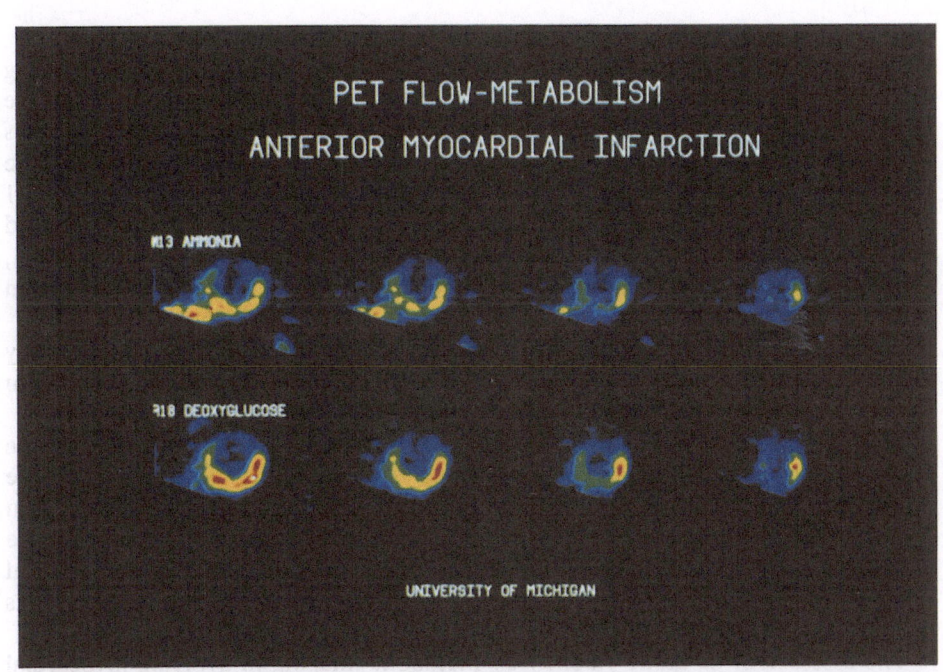

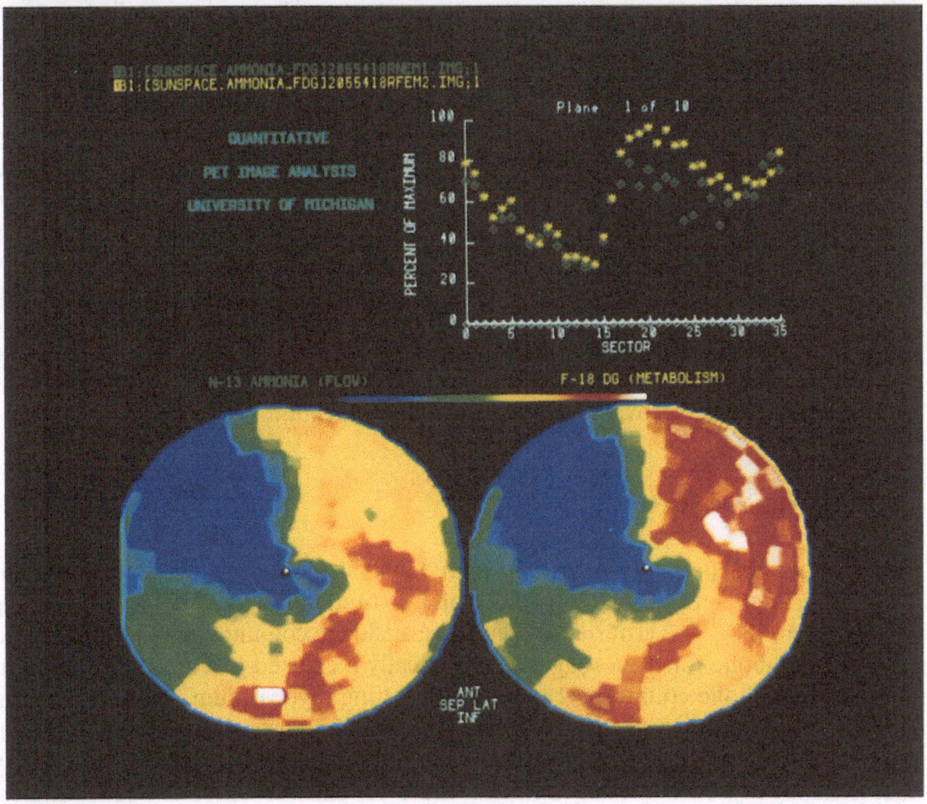

coronary thrombolysis in reperfused myocardium [57]. More recently, however, tomographic studies using C-11 palmitate or C-11 acetate have provided increasing evidence of a considerable heterogeneity of tracer uptake and clearance within the infarct territory, even in patients not subjected to acute interventions. In patients studied in the subacute stage of Q-wave infarction with C-11 acetate and PET, the impairment of oxidative metabolism ranged from a marked depression of the C-11 clearance rate in the center of the infarct zone to a mild decrease in hypoperfused peri-infarction zones, which were delineated by blood-flow imaging [64]. In parallel, studies using N-13 ammonia and FDG in patients with recent myocardial infarction showed a high incidence of residual "PET viability" [10, 33, 49]. Neither electrocardiographic changes nor the severity of associated wall motion abnormality reliably distinguished tomographically identified regions of ischemically injured myocardium from infarction.

The high incidence of viable but compromised tissue is quite compatible with the known temporal and spatial heterogeneity of ischemic injury during the subacute phase of an infarction, which has been confirmed by autopsy. Histological studies in patients who died in acute myocardial infarction have demonstrated the coexistence of myocytes with different degrees of cellular damage [30]. Due to the limited spatial resolution of PET, the images reflect metabolic activity averaged over a heterogenous population of infarcted and viable cells.

The assessment of tissue viability by metabolic imaging in evolving myocardial infarction may allow for in-vivo estimation of extent and "completeness" of myocardial necrosis and aid in selecting patients for interventional therapy. Residual metabolic activity in the infarct zone and late functional outcome have been compared in postinfarction patients [48, 49]. Ventricular segments with concordant decreases of flow and metabolism failed to recover contractile function on echocardiographic reevaluation several weeks after infarction. In contrast, functional outcome was variable in patients with evidence of tissue viability in the infarct territory. About one-half of the segments with residual glucose utilization showed improvement of function. Regional metabolic activity was significantly related to angiographic evidence of spontaneous recanalization of the infarct artery. These observations indicate the potential clinical role of PET in the selection of patients for invasive interventions such as PTCA or surgery. In addition, metabolic imaging may provide useful endpoints for early evaluation of the efficacy of therapeutic interventions designed to enhance recovery of jeopardized myocardium.

Chronic Myocardial Ischemia

When myocardial perfusion is chronically reduced, but still sufficient to maintain viability of tissue, the myocardium may exhibit impaired regional function (hibernating myocardium) [38]. The criteria of reduced perfusion in conjunction with a preserved residual metabolic function suggest a chronically increased glucose utilization pattern as a metabolic counterpart of the hibernating myocardium. In fact, positron emission tomography revealed a high prevalence of persistent tissue metabolism in patients with electrocardiographic Q-wave infarction, which supports the possibility that ischemic myocardium may enter the state of "chronic ischemia " [10]. Clinical

PET studies with FDG and N-13 ammonia showed that both blood flow and metabolism were highly heterogeneous in patients with chronic coronary artery disease. Discrete flow defects were frequently associated with either concordant or discordant metabolic defects (Fig. 6 and 7). The feature of relatively or absolutely increased myocardial glucose utilization at rest, which is common to these patients, has also been observed in patients with unstable angina [61].

Elucidation of the underlying mechanism is difficult because "chronic ischemia" in advanced human coronary artery disease cannot be reproduced easily in the animal model. It has been hypothesized that the pattern of chronically enhanced glucose utilization may represent a chronic adaptation of myocardial metabolism to a state of persistent hypoperfusion with chronic stimulation of glycolysis [14]. Alternatively, it is also possible that episodes of ischemia, caused by vasomotion or everyday activities, may repeatedly "stun" the myocardium [8, 42]. Studies with ambulatory Holter monitoring have shown that multiple episodes of painless ("silent") ischemia occur in human coronary artery disease [16]. As demonstrated in animal experiments, such alternations between ischemia and reperfusion are likely to account for the permanent impairment of contractile function in affected vascular territory [36].

Prediction of Functional Recovery and Clinical Significance

Metabolic indices defined by positron tomography can identify necrotic and reversibly injured tissue and thus detect potentially salvageable myocardium. This may be of particular significance in high-risk patients with acute myocardial infarction or with poor ventricular performance. The definition of tissue viability has gained increasing clinical importance because of the rapid development of interventional revascularization.

Tillisch et al. studies patients with impaired left ventricular function before aortocoronary bypass surgery with PET and N-13 ammonia/FDG to determine tissue viability and to assess the predictive value of metabolic imaging for tissue recovery [63]. Regional functional recovery following revascularization was compared to the preoperative PET findings. Eighty-five percent of "PET viable" segments improved following surgery, as determined by serial radionuclide ventriculography prior to and 6 weeks following surgery; 95% of segments identified as irreversibly injured did not improve regional wall motion postoperatively. These findings underscore the capability of PET to predict the extent of functional improvement after revascularization therapy and indicate the unique information obtained by metabolic imaging.

Alternative diagnostic techniques provide only limited differentiation of reversible and irreversible cell damage. Wall motion studies failed to distinguish ischemia from necrosis in both animal and clinical investigations [10, 55]. Electrocardiographic criteria, such as the presence of Q-waves, are not specific for determining the transmural extent of infarction: In 20 patients with Q-wave infarctions, a concordant decrease in flow and metabolism (PET-necrotic) was observed in only one-third of the infarct-related segments, while the remaining two-thirds were either normal or ischemic by metabolic criteria [10]. Similarly, the commonly used criterion of thallium-201 redistribution may provide only a limited definition of viable myocardium. Recent observations suggest that thallium redistribution may occur slowly and may

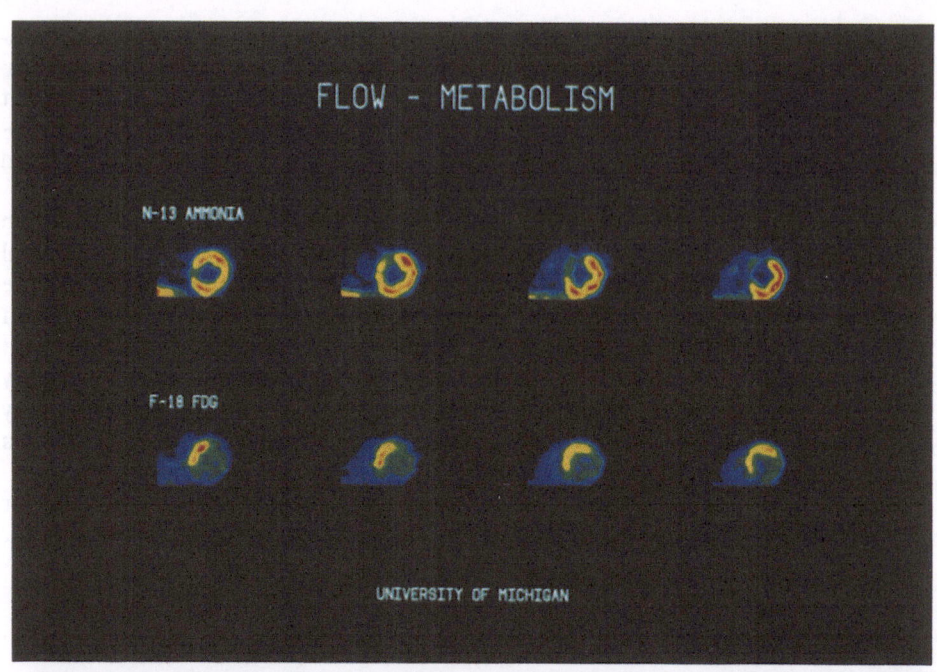

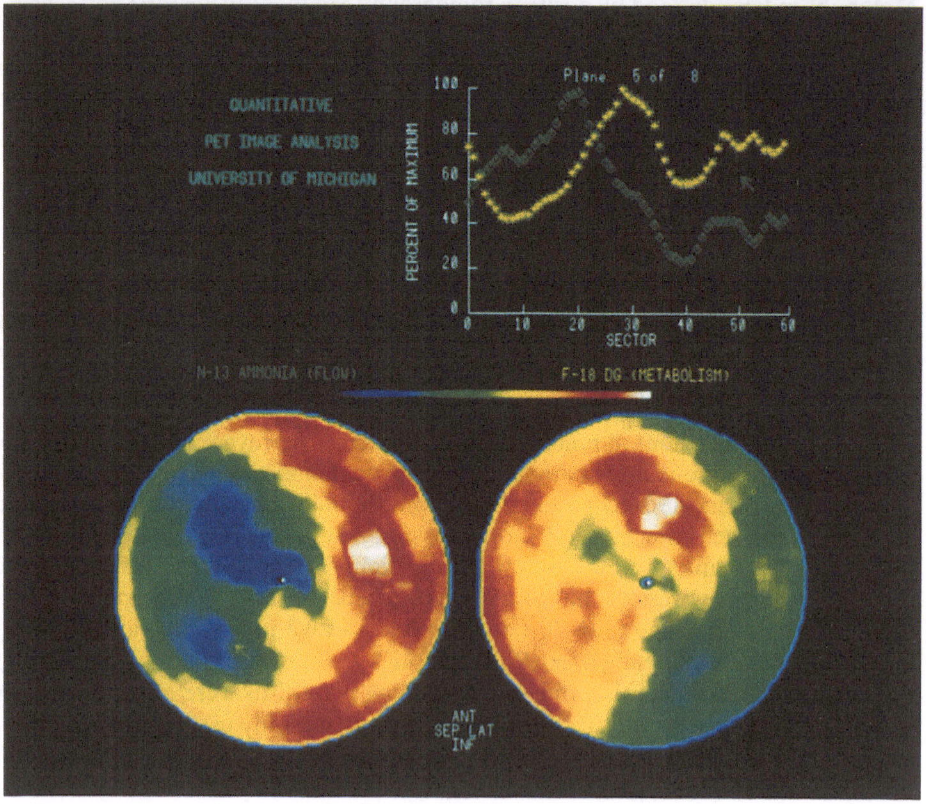

require as long as 24 h. A considerable number of fixed thallium-201 defects have been found to normalize following revascularization, thus indicating an overestimation of infarct size by this technique [9, 25]. Metabolic imaging with PET may therefore provide unique information in the diagnosis of compromised, but viable, myocardium. On the basis of such measurements, guidelines for therapeutic interventions can be developed and patients can be selected for revascularization.

Future Developments

Until recently, interpretation of PET images was based on qualitative data analysis. Recent advances in PET technology result in improved spatial and temporal resolution, which markedly improve the regional assessment of tracer concentrations. The development and validation of tracer kinetic models will enable more accurate measurements of substrate utilization rates. Future research will focus on the definition of threshold values of metabolic parameters to quantify more specifically the extent and severity of ischemic injury in the human heart. The further advance of PET as both a clinical imaging modality and a sophisticated research tool depends on the development of new radiopharmaceuticals. Besides new metabolic tracers, such as labeled amino acids for the assessment of myocardial protein synthesis, the use of radiolabeled cardiovascular drugs may permit the in-vivo definition of pharmaco-kinetics. A promising new catecholamine analogue for the evaluation of the presynaptic adrenergic nervous system of the heart has been recently introduced [23]. C-11 hydroxyephedrine traces the catecholamine uptake in adrenergic nerve terminals and may be used to assess regional neuronal function in ischemic heart disease. Clearly, PET is still in its infancy, but increased clinical application of this technique will define its diagnostic and prognostic value in a variety of cardiac diseases.

In summary, positron emission tomography (PET) allows the non-invasive quantification of regional myocardial tracer concentrations. In addition to the improved imaging technology, PET can be combined with a large number of existing short-lived radiopharmaceuticals to assess physiologic processes such as blood flow and substrate utilization. Metabolic alterations occurring during ischemia and reperfusion have been delineated with PET in animal and clinical studies. The combination of flow tracers and the glucose analogue F-18 deoxyglucose provides sensitive detection of demand-induced ischemia and has been shown to identify viable, but compromised myocardium in patients with advanced coronary disease. PET definition of

Fig. 7. Top: Cross-sectional PET images of myocardial blood flow (N-13 ammonia) and glucose utilization (F-18 FDG uptake) in a patient with coronary artery disease involving the left anterior descending artery. Images are arranged as in Fig. 6, top. The N-13 ammonia distribution shows a perfusion defect in the septum while maintained FDG uptake is increased, suggesting increased exogenous glucose utilization in compromised but viable myocardium. Bottom: Polar maps for blood flow (left) and glucose utilization (right) in the same patient. For tomographical orientation see legend of Fig. 6, bottom. The respective circumferential profiles of the relative N-13 and F-18 activity distributions (upper right) show a clear mismatch of tracer uptake from sector 24 to 60 ("PET viability").

tissue viability is clinically useful for the selection of patients considered for revascularization. Further developments of radiopharmaceuticals and instrumentation will make PET an important clinical and research tool in the non-invasive characterization of ischemic heart disease.

References

1. Ambrosio G, Jacobus W, Bergman C, Weisman H, Becker L (1987) Preserved high energy phosphate metabolic reserve in globally "stunned" hearts despite reduction of basal ATP content and contractility. J Mol Cell Cardiol 19:953–964
2. Apstein C, Grossman W (1987) Opposite initial effects of supply and demand ischemia on left ventricular compliance: the ischemia-diastolic paradox. J Mol Cell Cardiol 19:119–128
3. Armbrecht H, Buxton D, Brunken R, Phelps M, Schelbert H (1989) Regional myocardial oxygen consumption determined non-invasively in humans with [1-C-11]acetate and dynamic positron tomography. Circulation 80:863–872
4. Bergmann S, Fox K, Geltman E, Sobel B (1985) Positron emission tomography of the heart. Prog Cardiovasc Dis 28:165–194
5. Bergmann S, Fox K, Rand A, McElvany K, Welch M, Markham J, Sobel B (1984) Quantification of regional myocardial blood flow in vivo with H2 15-O. Circulation 70:724–733
6. Bergmann S, Herrero P, Markham J, Weinheimer C, Walsh M (1989) Non-invasive quantitation of myocardial blood flow in human subjects with oxygen-15 labeled water and positron emission tomography. J Am Coll Cardiol 14:639–652
7. Bolli R, Zhu W-X, Thornby J, O'Neill P, Roberts R (1988) Time course and determinants of recovery of function after reversible ischemia in conscious dogs. Am J Physiol 254:H102–H114
8. Braunwald E, Kloner R (1982) The stunned myocardium: prolonged, post-ischemic ventricular dysfunction. Circulation 66:1146–1149
9. Brunken R, Schwaiger M, Grover-McKay M, Phelps M, Tillisch J, Schelbert H (1987) Positron emission tomography detects tissue metabolic activity in myocardial segments with persistent thallium perfusion defects. J Am Coll Cardiol 10:557–567
10. Brunken R, Tillisch J, Schwaiger M, Child J, Marhsall R, Phelps M, Schelbert H (1986) Regional perfusion, glucose metabolism and wall motion in chronic electrocardiographic Q-wave infarctions: Evidence for persistence of viable tissue in some infarct regions by positron emission tomography. Circulation 73:951–963
11. Buxton D, Schwaiger M, Mody F, Krivokapich J, Nienhaber C, Armbrecht J, Luxen A, Ratib O, Phelps M, Schelbert H (1989) Regional abnormalities of oxygen consumption in reperfused myocardium assessed with [1-C-11] acetate and positron emission tomography. Am J Cardiac Imag 3:276–287
12. Buxton D, Schwaiger M, Nguyen A, Phelps M, Schelbert H (1988) Radiolabeled acetate as a tracer of myocardial tricarboxylic acid cycle flux. Circ Res 63:628–634
13. Camici P, Araujo L, Spinks T, Lammertsma A, Kaski J, Shea M, Selwyn A, Jones T, Maseri A (1986) Increased uptake of F-18-fluorodeoxyglucose in post-ischemic myocardium of patients with exercise-induced angina. Circulation 74:81–88
14. Camici P, Ferrannini E, Opie L (1989) Myocardial metabolism in ischemic heart disease: basic principles and application to imaging by positron emission tomography. Prog Cardiovasc Dis 32:217–238
15. Challiss R, Hayes D, Radda R (1986) An investigation of arterial insufficiency in the rat hind limb. Biochem J 240:395–401
16. Cohn P (1986) Total ischemic burden: Definition, mechanism and therapeutic implications. Am J Med 81 [Suppl 4A]:2–6
17. Gambhir S, Schwaiger M, Huang S, Krivokapich J, Schelbert H, Nienaber C, Phelps M (1989) Simple non-invasive quantification method for measuring myocardial glucose utilization in humans employing positron emission tomography and fluorine-18 deoxyglucose. J Nucl Med 30:359–366

18. Goldstein R, Klein M, Welch M, Sobel B (1980) External assessment of myocardial metabolism with C-11 palmitate in vivo. J Nucl Med 21:342–348
19. Griver-McKay M, Schelbert H, Schwaiger M, Sochor H, Guzy P, Krivokapich J, Child J, Phelps M (1986) Identification of impaired metabolic reserve by atrial pacing in patients with significant coronary artery stenosis. Circulation 74:281–292
20. Henes C, Bergmann S, Walsh M, Sobel B, Geltman E (1989) Assessment of myocardial oxidative metabolic reserve with positron emission tomograhy and carbon-11 acetate. J Nucl Med 30:1489–1499
21. Hoffman E, Phelps M (1986) Positron emission tomography: Principles and quantitation. In: Phelps E, Maziotta J, Schelbert H (eds) Positron emission tomography and autoradiography. New York, Raven Press, pp 237–286
22. Huang S, Phelps M (1986) Principles of tracer kinetic modeling in positron emission tomography and autoradiography. In: Phelps E, Maziotta J, Schelbert H (eds) Positron emission tomography and autoradiography. New York, Raven Press, pp 287–346
23. Hutchins G, Rothley J, Wieland D, Rosenspire K, Kuhl D, Schwaiger M (1990) Evaluation of 6-[F-18]Fluorometaraminol (FMR) kinetics in canine myocardium using positron emission tomography. J Nucl Med (in press)
24. Kalff V, Gallagher K, Nguyen N, McClanahan T, Schork A, Schwaiger M (1989) Dissociation of glucose utilization and blood flow in canine myocardial ischemia. Circulation 80 [Suppl II]:II–638 (abstr)
25. Kayden D, Sigal S, Soufer R, Zaret B, Wackers F (1989) 24 hour planar thallium-201 delayed imaging: Is reinjection necessary? Circulation 80 [Suppl II]:II–376 (abstr)
26. Khouri E, Gregg D, Rayford C (1965) Effect of exercise on cardiac output, left coronary flow and myocardial metabolism in the unanesthetized dog. Circ Res 17:427–437
27. Klein M, Goldstein R, Welch M, Sobel B (1979) External assessment of myocardial metabolism with C-11 palmitate in rabbit hearts. Am J Physiol 237:H51–H58
28. Krivokapich J, Huang S, Phelps M, Barrio J, Watanabe C, Selin C, Shine K (1982) Estimation of rabbit myocardial metabolic rate for glucose using fluorodeoxyglucose. Am J Physiol 243:H884–H895
29. Krivokapich J, Smith G, Huang S, Hoffman E, Ratib O, Phelps M, Schelbert H (1989) 13-N ammonia myocardial imaging at rest and with exercise in normal volunteers. Circulation 80:1328–1337
30. Lee J, Idekeer R, Reimer K (1981) Myocardial infarct size and location in relation to the coronary vascular bed at risk in man. Circulation 64:526–534
31. Liedtke A (1981) Alterations of carbohydrate and lipid metabolism in the acutely ischemic heart. Prog Cardiovasc Dis 23:321–336
32. Markham R, Winniford M, Firth B, Nicod P, Dehmer G, Lewis S, Hillis L (1983) Symptomatic electrocardiographic, metabolic, and hemodynamic alterations during pacing-induced myocardial ischemia. Am J Cardiol 51:1589–1594
33. Marshall R, Huang S, Nash W, Phelps M (1983) Assessment of the F-18 fluorodeoxyglucose kinetic model in calculations of myocardial glucose metabolism during ischemia. J Nucl Med 24:1060–1064
34. Marshall R, Nash W, Shine K, Phelps M, Ricchiuti N (1981) Glucose metabolism during ischemia due to excessive oxygen demand or altered coronary flow in the isolated arterially perfused rabbit septum. Circ Res 49:640–648
35. Mody-Vaghaiwalla F, Brunken R, Nienaber C, Stevenson L, Phelps M, Schelbert H (1988) Characterization of dilated and ischemic cardiomyopathy utilizing visual and circumferential profile analysis with PET. J Nucl Med 29:818 (abstr)
36. Nicklas J, Becker L, Bulkley B (1985) Effects of repeated brief coronary occlusion on regional left ventricular function and dimension in dogs. Am J Cardiol 56:473–478
37. Phelps M, Hoffman E, Selin C, Huang S, Robinson G, MacDonald N, Schelbert H, Kuhl D (1978) Investigation of [18F]2-fluoro-2-deoxyglucose for the measure of myocardial glucose metabolism. J Nucl Med 19:1311–1319
38. Rahimtoola S (1985) A perspective on the three large multicenter randomized clinical trials of coronary bypass surgery for chronic stable angina. Circulation 8:V-123–V-135
39. Ratib O, Phelps M, Huang S, Henze E, Selin C, Schelbert H (1982) Positron tomography with deoxyglucose for estimating local myocardial glucose metabolism. J Nucl Med 23:577–586

40. Rothlin M, Bing R (1961) Extraction and release of individual free fatty acids by the heart and fat depots. J Clin Invest 40:1380–1384
41. Russell M, Coleman E, Chu A, Cobb F (1989) Relation of fluorodeoxyglucose uptake in ischemic myocardium to myocardial blood flow. Circulation 80 [Suppl II]:II–638 (abstr)
42. Schelbert H, Buxton D (1988) Insights into coronary artery disease gained from metabolic imaging. Circulation 78:496–505
43. Schelbert H, Henze E, Schon H, Hansen H, Selin C, Huang S, Barrio J, Phelps M (1983) C-11 palmitate for the non-invasive evaluation of regional myocardial fatty acid metabolism with positron computed tomography. III. In vivo demonstration of the effects of substrate availability on myocardial metabolism. Am Heart J 105:492–504
44. Schelbert H, Henze E, Keen R, Schon H, Hansen H, Selin C, Huang S, Barrio J, Phelps M (1983) C-11 palmitate for the non-invasive evaluation of regional myocardial fatty acid-metabolism with positron computed tomography. IV. In vivo evaluation of acute demand-induced ischemia in dogs. Am Heart J 106:736–750
45. Schelbert H, Schwaiger M (1986) PET studies of the heart. In: Phelps M, Mazziotta J, Schelbert H (eds) Positron emission tomography and autoradiography: Principles and applications for the brain and heart. New York: Raven Press, pp 581–661
46. Schelbert H, Wisenberg G, Phelps M, Gould K, Henze E, Hoffman E, Gomes A, Kuhl D (1982) Non-invasive assessment of coronary stenoses by myocardial imaging during pharmacologic coronary vasodilation: VI. Detection of coronary artery disease in man with intravenous N-13 ammonia and positron computed tomography. Am J Cardiol 49:1197–1207
47. Schon H, Schelbert H, Najafi A, Hansen H, Robinson G, Huang S, Barrio J, Phelps M (1982) C-11 labeled palmitic acid for the non-invasive evaluation of regional myocardial fatty acid metabolism with positron computed tomography. II. Kinetics of C-11 palmitic acid in acutely ischemic myocardium. Am Heart J 103:548–561
48. Schwaiger M (1986) Metabolism and blood flow as new markers of myocardial viability in the evolution of myocardial infarction. Eur J Nucl Med 12:S62–S65
49. Schwaiger M, Brunken R, Grover-McKay M, Krivokapich J, Child J, Tillisch J, Phelps M, Schelbert H (1986) Regional myocardial metabolism in patients with acute myocardial infarction assessed by positron emission tomography. J Am Coll Cardiol 8:800–808
50. Schwaiger M, Fishbein M, Block M, Wijns W, Selin C, Phelps M, Schelbert H (1987) Metabolic and ultra-structural abnormalities during ischemia in canine myocardium: non-invasive assessment by positron emission tomography. J Mol Cell Cardiol 19:259–269
51. Schwaiger M, Hansen H, Sochor H, Parodi O (1984) Delayed recovery of regional glucose metabolism in reperfused canine myocardium by positron-CT. J Am Coll Cardiol 3:552A (abstr)
52. Schwaiger M, Krivokapich J, Ratib O, Huang S, Phelps M, Schelbert H (1988) Non-invasive quantification of coronary reserve by N-13 ammonia and positron emission tomography (PET). J Am Coll Cardiol 11:11A (abstr)
53. Schwaiger M, Neese R, Araujo L, Wyns W, Wisneski J, Sochor H, Swank S, Kulber D, Selin C, Phelps M, Schelbert H, Fishbein M (1989) Sustained non-oxidative glucose utilization and depletion of glycogen in reperfused canine myocardium. J Am Coll Cardiol 13:745–754
54. Schwaiger M, Schelbert H, Ellison D, Hansen H, Yeatman L, Vinten-Johansen J, Selin C, Barrio J, Phelps M (1985) Sustained regional abnormalities in cardiac metabolism after transient ischemia in the chronic dog model. J Am Coll Cardiol 5:336–347
55. Schwaiger M, Schelbert H, Keen R, Vinten-Johansen J, Hansen H, Selin C, Barrio J, Huang S, Phelps M (1985) Retention and clearance of C-11 palmitic acid in ischemic and reperfused canine myocardium. J Am Coll Cardiol 6:311–320
56. Shelton M, Green M, Mathias C, Weinheimer C, James H, Welch M, Bergmann S (1989) Measurement of regional blood flow using copper – PTSM and positron emission tomography (PET). J Nucl Med 30:807 (abstr)
57. Sobel B, Geltman E, Tiefenbrunn A, Jaffe A, Spadaro J, Ter-Pogossian M, Collen D, Ludbrook P (1984) Improvement of regional myocardial metabolism after coronary thrombolysis induced with tissue-type plasminogen activator or streptokinase. Circulation 69:983–990
58. Sobel B, Weiss E, Welch M, Siegel B, Ter-Pogossian M (1977) Detection of remote myocardial infarction in patients with positron emission transaxial tomography and intravenous C-11 palmitate. Circulation 55:853–857

59. Sochor H, Schwaiger M, Schelbert H, Huang S, Ellison D, Hansen H, Selin C, Parodi O, Phelps M (1987) Relationship between Tl-201, Tc-99m (Sn) pyrophosphate and F-18-2-deoxyglucose uptake in ischemically injured dog myocardium. Am Heart J 114:1066–1077
60. Spener F, Borchers T, Mukherjea M (1989) On the role of fatty acid binding proteins in fatty acid transport and metabolism. FEBS Letter 244:1–5
61. Spinks T, Araujo L, Camici P (1988) Regional myocardial glucose metabolism in angina pectoris obtained from positron emission tomography. J Thorac Imag 3:56–63
62. Ter-Pogossian M, Klein M, Markham J (1980) Regional assessment of myocardial metabolic integrity in vivo by positron emission tomography with C-11 labeled palmitate. Circulation 61:242–255
63. Tillisch J, Brunken R, Marshall R, Schwaiger M, Mandelkern M, Phelps M, Schelbert H (1986) Reversibility of cardiac wall motion abnormalities predicted by using positron tomography. N Engl J Med 314:884–888
64. Walsh M, Geltman E, Brown M, Henes C, Weinheimer C, Sobel B, Bergmann S (1989) Noninvasive estimation of regional myocardial oxygen consumption by positron emission tomography with carbon-11 acetate in patients with myocardial infarction. J Nucl Med 30:1798–1808
65. Wyns W, Schwaiger M, Huang S-C, Buxton D, Hansen H, Selin C, Keen R, Phelps M, Schelbert H (1989) Effects of inhibition of fatty acid oxidation on myocardial kinetics of C-11 labeled palmitate. Circ Res 65:1787–1797

Authors' address:
Markus Schwaiger, M.D.
University of Michigan Medical Center
1500 E. Medical Center Drive
UH B1 G505, Box 0028
Ann Arbor, MI 48109-0028

Coagulation, Thrombosis and Fibrinolysis in Myocardial Ischemia

Thomas Hohlfeld and Karsten Schrör

Institut für Pharmakologie, Heinrich-Heine-Universität Düsseldorf, FRG

Introduction

Protection from life-threatening blood loss subsequent to vessel injury requires immediate emergency mechanisms serving two major purposes: i) Formation of a platelet-fibrin hemostatic plug at the site of vessel disruption, and ii) initiation of immediate vasoconstriction to prevent further blood loss and washout of the plug. A second chain of events starts at about the same time and merely consists of the activation of a number of inhibitory control systems to: i) limit the coagulation process to the site of injury, and ii) reestablish blood flow through the occluded vessel allowing for later repair of the injury. Both clot formation and subsequent lysis are controlled by a close interaction between the vessel wall (endothelial cells and subendothelium), plasma proteins, circulating blood cells, and the fibrin fibers of the hemostatic plug. This restricts the activation of the powerful clotting and fibrinolytic systems to the site of injury and prevents a general disturbance of hemostasis, i.e., a general clotting or bleeding tendency.

It is now well established that disturbances of the blood coagulation system play a fundamental role in the pathogenesis of myocardial ischemia and infarction. There is ample evidence that atherosclerotic alterations of the vessel wall generate a prothrombotic state involving disturbed endothelial function, stimulation of the clotting cascade and platelet hyperreactivity (see [96]). Pathologic and angiographic findings have suggested a striking coincidence between thrombus formation in atherosclerotic coronary vessels and the occurence rate and severity of myocardial infarction. It has also been shown that experimental injury to the coronary vascular endothelium results in local thrombosis with subsequent ischemia, eventually associated with regional coronary arterial vasoconstriction. Thus, both clinical and experimental data suggest a tight interaction between thrombus formation, endothelial injury and myocardial ischemia and infarction [91].

This survey reviews some physiological, pathophysiological, clinical, and pharmacological aspects of thrombosis and thrombolysis and their consequences for treatment of myocardial ischemia and infarction. First, some major physiological properties of the plasmatic coagulation, platelet activation, and fibrinolytic systems will be described. Second, the sites of action of anticoagulants, such as heparin, warfarin and antiplatelet drugs, in particular aspirin, will be discussed, followed by discussion of clot lysis by fibrin-specific lytic agents, including streptokinase and tissue plasminogen activator (t-PA). Attention is also focussed upon the contribution of antiplatelet effects to antianginal effects of cardiovascular drugs, frequently used for prevention of myocardial infarction, such as organic nitrates and β-blockers.

Physiology of Hemostasis and Fibrinolysis

The hemostatic system consists of three major components: vascular endothelium and subendothelium, blood platelets, and the plasmatic coagulation system. Although these components will be discussed separately, it is important to realize that they always act together under physiological conditions and there are a number of interconnections between them. A simplified schematic presentation (modified after [18]) is shown in Fig. 1.

Endothelium-Derived Factors that Modulate Hemostasis

Healthy endothelium is a non-thrombogenic surface and under physiological circumstances reacts neither with platelets nor blood constituents. Several factors contribute to the thromboresistance of healthy endothelium. A negative surface charge repels the negatively charged platelets [127]. The endothelial lining of blood vessels is the most important site of vascular prostacyclin (PGI_2) formation [103], a major arachidonic acid metabolite of endothelial cells. PGI_2 inhibits platelet aggregation and formation of platelet-derived growth factor(s) [169]. In addition, PGI_2 is also a weak inhibitor of platelet adhesion [164]. Many vasoactive and hemostatic factors, including thrombin [83], have been shown to stimulate endothelial PGI_2 formation.

Another important endothelium-derived inhibitory mediator is "endothelium-derived relaxing factor(s)" (EDRF) (see [6]), one of them is identified as nitric oxide (NO) biosynthesized from l-arginine [104]. Like PGI_2, nitric oxide in addition to causing vasorelaxation also inhibits platelet function [3] (see also [6]). Acting in concert both compounds may antagonize vasoconstrictor, procoagulant factors, such as thrombin and thromboxane A_2. Although the in vivo significance of platelet inhibitory activity of nitric oxide needs further confirmation, because of the rapid inactivation by oxyhemoglobin, much interest is currently focussed upon this compound because it might represent the active metabolite of nitrovasodilators (see [104]). This, and the reduced formation of endothelial EDRF in human atherosclerotic vessels [82] has stimulated considerable research on this mediator for clarifying its role in coronary heart disease (see below).

Another class of endothelium-derived compounds are specific antagonists of the procoagulant activity of thrombin. Endothelial cells generate thrombin inhibitors that remove thrombin from the circulation and in addition convert its procoagulant activity into anticoagulant activity: thrombomodulin, a surface receptor [12] and heparan sulfate, a glycosaminoglycan, that activates antithrombin III [79]. After binding to the endothelial thrombomodulin, thrombin looses its ability to activate coagulation factors but has enhanced ability to activate protein C [16] which in turn inactivates the clotting factors Va and VIIIa.

Activated protein C inactivates tissue plasminogen activator inhibitor (t-PA-I) resulting in enhanced t-PA activity and stimulation of fibrinolysis [24, 81]. Thus, the endothelium also modulates fibrinolysis by generation of fibrinolytic components in response to various stimuli as well as by providing surfaces and extracellular matrices that serve to localize fibrinolytic activity and to improve the efficiency of plasminogen activation by increasing the local concentration of the fibrinolytic components (see [57]).

84

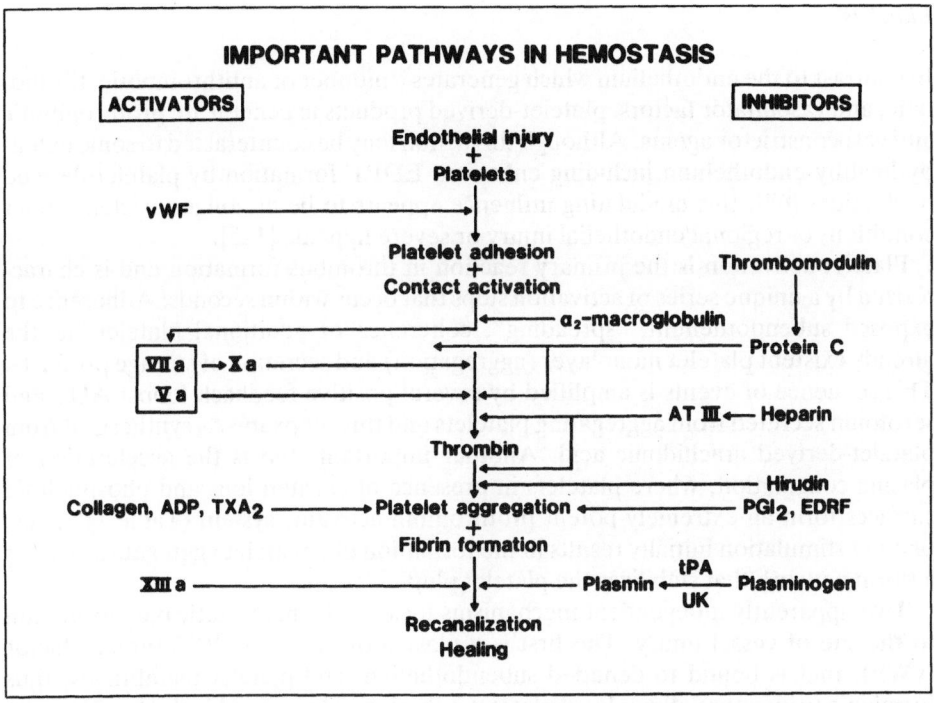

IMPORTANT PATHWAYS IN HEMOSTASIS

Fig. 1. Subsequent to endothelial injury, hemostasis is initiated by adherence of blood platelets to exposed subendothelial tissue. Von Willebrand's factor (vWF), a plasma protein, is bound to denuded subendothelium and platelet membrane receptors, allowing for platelet adhesion at the injured site, activation and development of procoagulant activity at the platelet surface. Plasma proteins interacting with the subendothelium stimulate the contact phase of coagulation. Exposed tissue releases "tissue factor" (thromboplastin) at the injured site, thereby triggering the "extrinsic" pathway of blood coagulation. Protein cofactors, such as Factor V, secreted by platelets or taken up from plasma, form enzyme/cofactor complexes at the platelet surface, thereby stimulating Factor X and subsequent conversion of the inactive plasmatic precursor enzyme (zymogen) prothrombin to thrombin. Thrombin augments its own formation by converting further plasmatic zymogens into activated cofactors (Va and VIIa). Thrombin stimulates platelet aggregation and secretion and hydrolyzes fibrinogen to fibrin monomers which then spontaneously polymerize to fibrin strands which are subsequently cross-linked by Factor XIII, resulting in stabilization of the platelet plug and improvement of its fixation to the subendothelial connective tissue. This cascade of self-amplifying events is controlled by endothelial generation of inhibitory factors, such as prostacyclin, thrombomodulin, and heparan sulfate. Inhibitors of contact activation (e.g., alpha$_2$-macroglobulin) and of serine proteases (e.g., antithrombin III) keep the hemostatic process localized by inactivating coagulation factors entering the circulating blood. At this point, activated protein C inhibits the procoagulatory action of thrombin and stimulates the fibrin-associated thrombolysis. Thrombolysis is mainly determined by conversion of the inactive plasma zymogen plasminogen into plasmin. This is realized by two classes of physiological plasminogen activators: tissue-type (t-PA) and urokinase-type plasminogen activators (UK). Similar to thrombus formation, thrombolysis is also controlled, i.e., limited to the site of thrombus formation, by specific inhibitors: Plasminogen activator-inhibitor (PA-I) and alpha$_2$-antiplasmin, inactivating circulating plasmin. Controlled recanalization of the clot reestablishes blood flow and prepares the way for the formation of fibrous tissue and wound healing. (Adapted from [18])

85

Platelets

In contrast to the endothelium which generates a number of antithrombotic, fibrinolytic, and vasodilator factors, platelet-derived products in general are prothrombotic and vasoconstrictor agents. Although this action may be counteracted to some extent by healthy endothelium including enhanced EDRF formation by platelet-derived nucleotides [62], this modulating influence appears to be no longer existent under conditions of regional endothelial injury or severe hypoxia [125].

Platelet activation is the primary reaction in thrombus formation and is characterized by a unique series of activation steps that occur within seconds: Adherence to exposed subendothelium, "spreading", coherence of additional platelets to the already existent platelet monolayer (aggregation) and secretion of storage products. This sequence of events is amplified by several positive feedback loops: ADP and serotonin secreted from aggregating platelets and thromboxane A_2 synthesized from platelet-derived arachidonic acid. Another important step is the acceleration of plasma coagulation, where platelets in presence of calcium ions and phospholipid surfaces form an extremely potent prothrombin activator system (Table 1). Thus, platelet stimulation initially results in the formation of a platelet aggregate as well as a fibrin network that stabilizes the platelet plug.

Two apparently independent mechanisms localize the hemostatic plug formation to the site of vessel injury. The first is a plasma protein, von Willebrand's factor (vWF), that is bound to denuded subendothelium and platelet membranes, thus forming a bridge that allows local platelet adhesion. The second is the formation of antiplatelet factors by the healthy adjacent endothelium, such as prostacyclin, that prevent excessive thrombus growth.

Plasmatic Coagulation Systems

Cell disruption by trauma of the vessel wall initiates the extrinsic pathway of coagulation by release of tissue factor, i.e., the membrane lipoprotein thromboplastin. In presence of (platelet) phospholipids and calcium ions it activates the plasma zymogen proconvertin (Factor VII). Factor VII then develops proteolytic activity and acti-

Table 1. Reaction rate enhancement resulting from interactions among the components of the prothrombin activation system

Component	Relative rate of thrombin formation Bovine	Human
$PT + Xa + Ca^{++}$	1	1
$PT + Xa + Ca^{++} + PL$	50–100	50
$PT + Xa + Va + Ca^{++}$	350	–
$PT + Xa + Va + PL + Ca^{++}$	19,000	20,000
$PT + Xa + Platelets + Ca^{++}$	300,000	–

PT = Prothrombin
PL = Phospholipide
from [66]

vates factor X to Xa (thrombokinase). This proteolytic enzyme converts prothrombin to thrombin. These reactions are potentiated by a number of positive feedback-loops, including stimulation of Factor VII by Factor Xa and, though less effective, by thrombin and Factor IX.

The intrinsic system starts with a contact activation of Hageman factor (Factor XII), a plasma protein which forms in the presence of a suitable surface (vascular subendothelium) a noncovalent complex with high molecular weight kininogen and prekallikrein. Within this complex, prekallikrein is converted to kallikrein, which catalyzes a proteolytic activation of Factor XII and liberates bradykinin. In a cascade of reactions, Factor XII activates factors XI and IX. The activated Factor IX, together with Factor VIII, converts Factor X into Xa (thrombokinase) and initiates the final common steps of the extrinsic and intrinsic system, i.e., thrombin formation and subsequent proteolytic cleavage of fibrinogen to fibrin. Fibrin strands are subsequently cross-linked by Factor XIII, resulting in clot stabilization, protection from fibrinolysis and fixation to subendothelial connective tissue. This may be the reason why clots become resistant to lysis when they age [34].

Because of these numerous amplification steps, minute quantities of thrombin are sufficient to activate the entire amount of fibrinogen in plasma. Therefore, effective endogenous inhibitors are necessary to prevent excessive local activation of the clotting cascade and to restrict the clotting process to the site of vessel injury. The most important regulators of contact activation are the naturally occuring plasma protease inhibitors C1-inhibitor and alpha$_2$-macroglobulin. These inhibitors differ from inhibitors of serine proteases (thrombin) in later stages of the coagulation process. The most important regulator of coagulation is antithrombin III (heparin cofactor). It inhibits thrombin and other serine proteases, including the activated clotting factors IXa and Xa [79].

Another control of the clotting process is provided by the vitamin K-dependent plasma factor protein C. This protein is bound to the endothelial surface protein thrombomodulin, undergoes proteolytic activation by thrombin and controls the coagulation factors V and VIII [16] (see above).

Fibrinolysis

The hemostatic process is terminated by formation of the platelet-fibrin-plug fixed to the injured part of the vessel wall. The control mechanisms mentioned above prevent excessive contact activation, limit thrombus formation to the site of injury, and prevent extension of platelet adhesion to adjacent parts of the vessel where the endothelium is intact. During the healing process the clot must be resolved. This dissolution is accomplished by the fibrinolytic system which, like the coagulation system, is well-controlled, site − specific, and involves the interaction of many fluid − and solid-phase components (see [57]). The principal component of the fibrinolytic system is the serine protease plasmin, which is formed from the inactive zymogen plasminogen. Therefore, control of plasminogen activation is the key event for fibrinolysis.

Two functionally different plasminogen activators have been identified: urokinase-type and tissue-type plasminogen activators (t-PA). t-PA, primarily derived from endothelial cells [81] (see above) but also from other sources (see [57]) is synthesized as a single-chain molecule and converted to a two-chain form. Fibrin enhances the affinity of t-PA for its substrate, plasminogen. This limits its fibrinolytic activity to the surface of the thrombus and explains why plasminogen in circulating blood is not activated even at high levels of circulating t-PA. Because of these properties much interest has been focussed upon the molecular biology of t-PA in order to develop fibrin-specific thrombolytic agents for rapid recanalization of occluded coronary vessels (see below).

Urokinase, another natural plasminogen activator, is synthesized by different cells including the endothelium. In plasma, urokinase is present as a single-chain zymogen (single chain urokinase-type plasminogen activator scu-PA, pro-urokinase) which is converted to high-molecular weight urokinase (HMW-UK) by hydrolysis [50]. In contrast to t-PA, the fibrinolytic activity of urokinase is not stimulated by fibrin. However scu-PA was found to lyse clots in purified systems with an efficiency and specificity equivalent to that of t-PA (see [57]). This appears to be due to the fibrin-dependent dissociation of a competitive plasmatic inhibitor [76]. Traces of plasmin seem to convert the enzyme into its active form [48] involving a circuit of positive feedback loops.

Plasmin has a low substrate specificity and when freely circulating in blood it will degrade several proteins, including fibrinogen, Factor V and VIII. Therefore, the plasminogen activation needs to be restricted to the vicinity of fibrin [17]. At several levels of activation, the fibrinolytic system is controlled by serine protease inhibitors. These include the fast-acting specific inhibitors of t-PA and urokinase, plasminogen activator inhibitor (PA-I) and alpha$_2$-antiplasmin. PA-I is released from endothelial cells and platelets and interferes with the enzymatic activity of t-PA by forming a 1 : 1 stoichiometric complex [71]. Free plasmin is rapidly inactivated by circulating alpha$_2$-antiplasmin that binds and inactivates plasmin as soon as it is liberated from the fibrin matrix of the thrombus.

Alterations of Hemostasis in Coronary Heart Disease and Myocardial Infarction

The tight interaction between thrombus formation, regional coronary artery constriction ("spasm") and endothelial injury is presented in Fig. 2 [91]. It is evident that the major consequence of disturbed interaction between vessel wall and vessel content is myocardial ischemia, resulting in reversible functional disturbances (e.g., arrhythmias) or myocardial infarction.

Disturbed Interaction Between Platelets and the Vessel Wall in Coronary Heart Disease and Myocardial Infarction

While healthy endothelium provides a non-thrombogenic surface that prevents local platelet adhesion and subsequent activation, the situation might be changed in atherosclerosis, typical for coronary heart disease. Platelets become adherent to the injured endothelium. Possible reasons for this involve reduced local generation of antiplatelet factors, such as PGI$_2$ (see [104]) and EDRF [82]. There is an enhanced

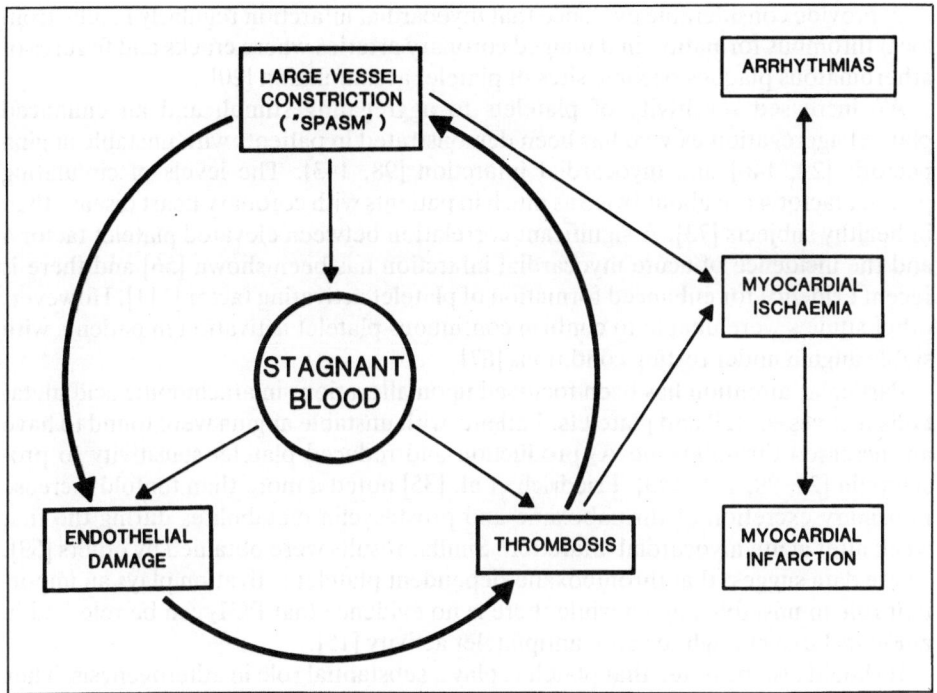

Fig. 2. Interaction between platelets, injured endothelium and vessel tone, resulting in stasis and myocardial ischemia. (For further explanation see text). (Adapted from [91])

requirement of PGI_2 to prevent platelet adhesion at the site of active atherosclerotic plaques [31]. Thus, despite a compensatory increase in total vascular prostacyclin formation in active atherosclerosis, probably mediated by stimulation of less injured endothelium by platelet products [30], this endothelial stimulation is insufficient to meet the requirements of the circulation. As a result, the balance between pro- and anticoagulant properties of the endothelium is shifted towards stimulation of pro-coagulant activity.

Endothelium-dependent release of vasoactive factors is disturbed in atherosclerosis. Thus, the relaxing activity of several stimuli, mediating endothelium-dependent dilatation (see [6, 104]) is converted into a vasoconstrictor action that can be demonstrated for example during experimental hypoxia [125].

Critical narrowing of a large coronary artery by a fixed stenosis results in cyclic reductions of blood flow, even in non-sclerotic coronary vessels [32, 157]. These flow reductions are caused by the formation of platelet aggregates at the site of coronary obstruction and can be prevented by treatment with acetylsalicylic acid [32]. These results with aspirin additionally suggest that aspirin-sensitive vasoconstrictor species, namely thromboxane A_2, are released from the platelet aggregate and are involved in platelet activation and vasoconstriction.

Arterial thrombosis is rare under normal circumstances but may develop in a stenosed atherosclerotic vessel. Morphologic [11, 26] and angiographic [11, 33, 170] data provide considerable evidence that myocardial infarction regularly results from local thrombus formation in damaged coronary arteries where cracks and fissures of atheromatous plaques become sites of platelet accumulation [20].

An increased sensitivity of platelets to aggregating stimuli and an enhanced platelet aggregation ex vivo has been demonstrated in patients with unstable angina pectoris [29, 146] and myocardial infarction [98, 143]. The levels of circulating platelet factor 4 are about twice as much in patients with coronary heart disease than in healthy subjects [73]. A significant correlation between elevated platelet factor 4 and the incidence of acute myocardial infarction has been shown [56] and there is recent evidence for enhanced formation of platelet activating factor [111]. However, other studies were unable to confirm continuous platelet activation in patients with stable angina under resting conditions [87].

Particular attention has been focussed upon alterations in arachidonic acid metabolism of vessel wall and platelets. Patients with unstable angina were found to have an increased thromboxane A_2 production and reduced platelet sensitivity to prostacyclin [29, 99, 111, 143]. Friedrich et al. [35] noted a more than tenfold increase in urinary excretion of thromboxane and prostacyclin metabolites during the first week after acute myocardial infarction. Similar results were obtained by others [59]. These data suggest that thromboxane-dependent platelet activation plays an important role in unstable angina while there is no evidence that PGI_2 can be released in amounts large enough to cause antiplatelet activity [15].

It should also be noted that platelets play a substantial role in atherogenesis. They release a potent mitogen (platelet derived growth factor, PDGF) which promotes the proliferation of fibrous and smooth vascular tissue at the site of endothelial lesions [169]. Therefore, enhanced activity of circulating platelets may represent a major detrimental factor not only for acute thrombotic complications but also the progression of coronary heart disease.

Disturbed Plasmatic Coagulation and Fibrinolysis in Coronary Heart Disease and Myocardial Infarction

In addition to disturbed platelet/vessel wall interactions, systemic changes towards an increased coagulability and/or reduced spontaneous fibrinolytic activity have been connected with the incidence and severity of myocardial ischemia. Epidemiological studies have shown that lower levels of fibrinogen, Factor VII and Factor X are found in communities with low risk of coronary heart disease (Gambia) whereas higher levels occur in areas with increased risk (England, Czechoslovakia) and high risk (Scotland, Finland) of myocardial infarction [95].

Clinical studies have confirmed that all major components of coagulation and fibrinolysis appear to be affected in more advanced stages of coronary heart disease. This includes reduced fibrinolytic activity in patients with coronary heart disease after exercise [52, 77] and an increase in plasma fibrinogen levels [77, 114], eventually resulting in elevated levels of fibrin monomers [93] and fibrinopeptide A [153]. There is also evidence for reduced levels of antithrombin III after acute myocardial infarction [115], although this could be partially explained by heparin treatment (see below). During short-term and long-term recovery from myocardial infarction, reduced plasma levels of t-PA and increased plasma levels of PA-I [44, 54, 112] have been reported.

A substantial amount of successfully reperfused coronary arteries undergoes early or late reocclusion [139]. This suggests that formation of platelet and fibrin thrombi may take place despite of antiplatelet and anticoagulant therapy and underlines the importance of improved anticoagulatory/antiplatelet/fibrinolytic therapy. Alternatively, neutrophil activation and release of deleterious mediators from these cells might significantly contribute to tissue injury [97].

Therapeutic Approaches in Myocardial Ischemia and Infarction

There are many approaches to reduce the risk of myocardial infarction and to improve the outcome of patients who suffered an acute coronary event. This includes appropriate diets, quitting cigarette smoking, changes in lifestyle, but also the use of antianginal drugs which might reduce the vascular risk. Most of these approaches, including lipid-lowering drugs, also involve changes in variables of the hemostatic and fibrinolytic systems. While these approaches might be helpful, in particular for prevention of acute coronary complications, they are beyond the scope of this survey. The following section is focussed upon agents with a primary action on hemostasis as well as some antianginal compounds. Because of the considerable interest in fibrinolytic agents, these compounds will be discussed in detail.

Antiplatelet Agents

According to their mechanism of action, antiplatelet agents can be divided into several classes of compounds: Agents that interfere with the platelet arachidonic acid metabolism; agents that elevate platelet cyclic nucleotide levels (cAMP or cGMP) by stimulation of adenylate or guanylate cyclases; specific antagonists of stimulating platelet receptors and a number of other compounds, in particular anti-anginal drugs, that in addition to their primary effects on the vasculature also inhibit platelet function.

Agents interfering with platelet arachidonate metabolism: The metabolism of arachidonic acid in human platelets results in the formation of a potent platelet activating and vasoconstrictor compound via the cyclooxygenase pathway, thromboxane A_2. There are two levels of interferences with the formation of this compound: inhibition of (platelet) cyclooxygenase and inhibition of thromboxane formation and action (Fig. 3).

Acetylsalicylic acid (ASA): ASA irreversibly blocks the platelet cyclooxygenase, i.e., the enzyme catalyzing the conversion of arachidonic acid into the prostaglandin endoperoxides PGG_2 and PGH_2. In platelets, this results in permanent inhibition of thromboxane A_2 formation, which can be antagonized only by fresh platelets. This action of ASA (and all other nonsteroidal antiinflammatory agents) is not selective for the platelet but also results in impaired formation of PGI_2 by endothelial cells.

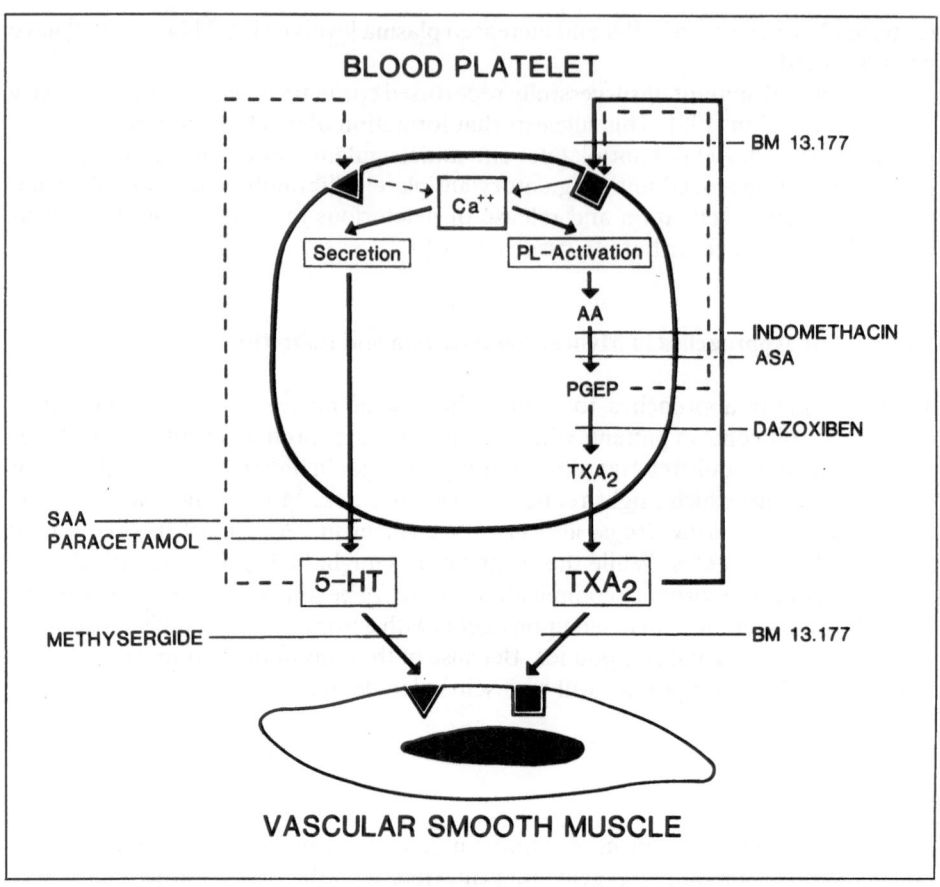

BLOOD PLATELET

BM 13.177

INDOMETHACIN
ASA

DAZOXIBEN

SAA
PARACETAMOL

METHYSERGIDE

BM 13.177

VASCULAR SMOOTH MUSCLE

Fig. 3. Release of vasoactive mediators from stimulated human platelets and their pharmacological modification. SAA: salicylic acid, BM 13.177: sulotroban, ASA: acetylsalicylic acid, 5-HT: serotonin, PL: phospholipase.

However, this effect is not permanent because endothelial cyclooxygenase may be resynthesized within less than 24 h [67]. As ASA undergoes a high first-pass metabolism in the liver, a relative selectivity for thromboxane formation in platelets may be achieved by oral administration of low doses (e.g., ≤50 mg), allowing for presystemic platelet inactivation without reducing systemic prostacyclin formation (see [113]).

The significance of blocking platelet hyperaggregability by ASA for prevention and treatment of myocardial infarction is well established (see [15, 121]). Between 1974 and 1981 seven placebo-controlled clinical trials of ASA in secondary prevention of myocardial infarction have been performed, using doses of 300 to 1500 mg/day. Whereas none of these trials showed a significant reduction in reinfarction rate by acetylsalicylic acid, this was the case if the data were pooled [121]. No dose-dependency was observed, suggesting that a low-dose therapy with acetylsalicylic acid may be as effective as high-dose treatment [121]. The usefulness of ASA for prevention of myocardial infarction in high-risk patients was confirmed by two more recent studies using 324 mg/day for 12 weeks [74] and 4 × 325 mg/day for up to 2 years [13]. Both trials demonstrated a reduction of reinfarction rate and infarct lethality by 50%. Interestingly, ASA at 30 mg/day was reported to reduce the risk of reinfarction in comparison to 1000 mg/day [60]. However, this study did not include an appropriate placebo control group. Thus, while ASA can clearly be recommended for secondary prevention in high-risk groups [65] the optimal, i.e., lowest dose of ASA providing maximum clinical benefit is not known. According to available information, doses of 160–325 mg ASA may be recommended [38].

Two large clinical trials have recently investigated ASA in primary prevention of myocardial infarction. One of them, using 325 mg every second day, was previously terminated due to a significant reduction of infarct rate by 44% at unchanged total cardiovascular mortality. Subgroup analysis further showed that the reduction in the risk of myocardial infarction was only among those who were 50 years of age or older. Whereas the other with a dose of 500 mg/day did not show a significant reduction of (non-fatal) myocardial infarctions [117]. Pooled analysis of both trials revealed a significantly lowered rate of non-lethal infarction by 33%, but no improvement of mortality [58]. Both trials have been criticized with respect to their methodology and the results as a whole appear not to suggest ASA for primary prevention of myocardial infarction. Side effects of ASA such as gastrointestinal discomfort and possible ulcer formation, and an increased bleeding tendency, eventually associated with a tendency for hemorrhagic stroke, and drug interactions require an individual estimation of the benefit/risk ratio dependent on the individual coronary risk [148]. It should also be noted that ASA does not increase the exercise tolerance of anginal patients [36].

Selective inhibitors of thromboxane formation: The advantage of thromboxane synthetase inhibitors is the selective blockade of thromboxane A_2 formation without concomitant suppression of vascular PGI_2 biosynthesis. Prostacyclin formation may rather be increased by utilization of accumulating platelet thromboxane precursors (PGG_2, PGH_2) by vascular endothelial prostacyclin synthase.

Clinical trials with thromboxane synthase inhibitors in patients with vasospastic angina [14, 171], unstable angina or acute myocardial infarction [152] did not provide encouraging results. In one study there was a significant reduction of transcardiac thromboxane gradients in pacing-induced angina and a significantly increased pacing level [63].

There is a number of possible explanations for this clinical failure of thromboxane synthetase inhibitors. First, PGG_2 and PGH_2 themselves are agonists of platelet thromboxane receptors with a molar potency comparable to TXA_2, and thus may substitute for TXA_2 [85]. Second, PGI_2 generation requires functioning endothelial prostacyclin synthase which, by definition, will not be available at the site of endothelial injury where platelets usually adhere. Third, the dominating products of redirection of the platelet prostaglandin endoperoxide pathway after inhibition of thromboxane formation are PGE_2 and PGD_2, and not PGI_2. Both compounds might interfere with leukocyte-mediated immune reactions, a subject still poorly understood. Interestingly, only a very limited number of the compounds designed to specifically inhibit thromboxane formation that have been successfully used in animal experiments have become available for clinical investigation. These compounds, including dazoxiben and dazmegrel are rather short-acting competitive inhibitors. An ASA-type blockade of TXA_2 formation, lasting for several days after a single tablet (see [113]) is not obtained. Furthermore, an about 80–90% suppression of thromboxane generating capacity is required to prevent thromboxane-dependent platelet activation.

More recently the combined use of a thromboxane synthase inhibitor and receptor antagonist has been found to exert supraadditive beneficial effects in experimental coronary thrombosis [28]. One compound (Ridogrel) has been described that contains thromboxane receptor blocker and synthase inhibitory actions in the same molecule and was found to be active in man [22]. It will be interesting to know whether this new approach will also provide clinical benefit in patients with ischemic heart disease.

Thromboxane receptor antagonists: In contrast to agents, designed to inhibit selected synthetic pathways of arachidonic acid in platelets, specific inhibitors of stimulating platelet receptors interfere with specific binding sites at the platelet membrane. Although there are a number of stimulating receptors (alpha$_2$, fibrinogen, prostaglandin endoperoxide/thromboxane A_2, 5-HT$_2$, PAF and others), this discussion will be focussed upon selective inhibitors of the prostaglandin endoperoxide/thromboxane A_2 receptor.

Thromboxane receptor antagonists differ principally from synthase inhibitors by their failure to interfere with thromboxane synthesis. Consequently, there is no accumulation of thromboxane precursors. These compounds also inhibit thromboxane A_2/PG endoperoxide-mediated platelet secretion reactions, including antagonism of serotonin release [162]. Thromboxane receptor antagonists, such as pinane thromboxane A_2 [132], sulotroban [134] and daltroban [144, 155] were found to be beneficial in experimental myocardial ischemia at antiplatelet doses. A number of other thromboxane receptor antagonists was effective as well (see [133]). Although these compounds have been shown to suppress experimental coronary thrombus formation even more efficiently than aspirin [42], it is not known whether this is the only or even the major effect to explain their cardioprotective properties. For example, daltroban was found in experimental studies to inhibit ischemia-induced neutrophil activation as well [144], probably by an initial action on platelets [155].

There is little experience with these compounds in man [124] and no study has been published so far investigating thromboxane receptor antagonists in myocardial infarction. Eventually, these compounds may be used in combination with thromboxane synthetase inhibitors (see above).

Stimulation of adenylate cyclase: PGE$_1$, prostacyclin and its analogs (e.g., iloprost) inhibit platelet function by stimulation of the platelet adenylate cyclase. This results in reduced cytosolic free calcium and a reduced response of platelets to any activating stimuli. A theoretical advantage of platelet inhibitory prostanoids is that their effect persists in the presence of high concentrations of the physiologically important stimulators collagen and thrombin. Additionally, platelet inhibitory prostaglandins may stimulate fibrinolysis [19] and counteract the inhibitory action of ASA on this parameter [7].

There is ample experimental evidence, suggesting a protective action of these compounds in experimental myocardial ischemia [129, 133]. In addition to antiplatelet effects, a number of other mechanisms has been proposed as well, contributing to the beneficial effects of PGE_1, PGI_2 and its synthetic analogs, such as iloprost. This includes inhibition of ischemia-induced catecholamine distribution within the myocardium, prevention of the loss of energy-rich nucleotides, inhibition of oxygen-centered radical formation, antineutrophil effects and others (see [129]). A unique property of these compounds is direct cytoprotection, i.e., improved myocardial tissue preservation independent of changes in coronary flow or myocardial oxygen consumption [130].

Despite these promising findings in animal experiments, the use of PGI_2 and iloprost did not result in convincing antiischemic effects in humans. The vast majority of clinical studies on PGI_2 or iloprost in vasospastic angina, unstable angina, and acute myocardial infarction [2, 45, 70, 149] did not confirm the promising animal data. Moreover, PGI_2 in stable angina instead produced acute ischemic episodes [10], probably related to the vasodilating activity of the agent.

Clinical problems with PGI_2 and iloprost may arise from the potent blood-pressure lowering activity due to significant circulating levels of the active compound. This does not occur in physiological conditions. The high doses in animal experiments (100–200-times the human dose on a weight basis) can therefore not be safely administered to man. Additionally, the number of PGI_2 receptors on platelets appears to be reduced in unstable angina and myocardial infarction [68], requiring higher doses of the compound to exert an antiplatelet effect in these patients. This might contribute to a reduced potency of PGI_2 to inhibit platelet function in myocardial infarction [105]. Additionally, the spasmolytic activity of prostacyclins in coronary vessels is also reduced in presence of platelet-derived thromboxane and serotonin [135].

PGE_1 does not reduce the arterial blood pressure at therapeutic doses in an extent comparable with PGI_2. The compound was active in unstable angina [140]. The reopening rate after intracoronary thrombolysis has been found to be significantly enhanced by intracoronary administration of PGE_1 prior to intracoronary streptokinase. This was associated with a considerable improvement of left ventricular function by PGE_1 treatment, suggesting additive effects of PGE_1 and streptokinase [138].

One recent attempt to overcome the problems connected with systemic administration of PGI_2 or iloprost is selective stimulation of endogenous PGI_2 production. Defibrotide, a single-stranded DNA fraction from mammalian lung appears to have these properties and, in addition, has fibrinolytic activities, probably associated with t-PA activation [25, 78] (Fig. 4). Defibrotide has cardioprotective actions in experimental myocardial ischemia [61, 110, 154]. However, this compound has not been tested so far for antianginal effects in man in any controlled study.

Stimulation of guanylate cyclase (Organic nitrates): The interest in agents that stimulate soluble guanylate cyclase and act as inhibitors of platelet function arose from the recent detection that organic nitrates, in particular nitroglycerin, might exert their antianginal effect via this mechanism of action (see [104]). In addition to inhibit platelet hyperreactivity by relaxing stenotic parts of coronary vessels, i.e., removing

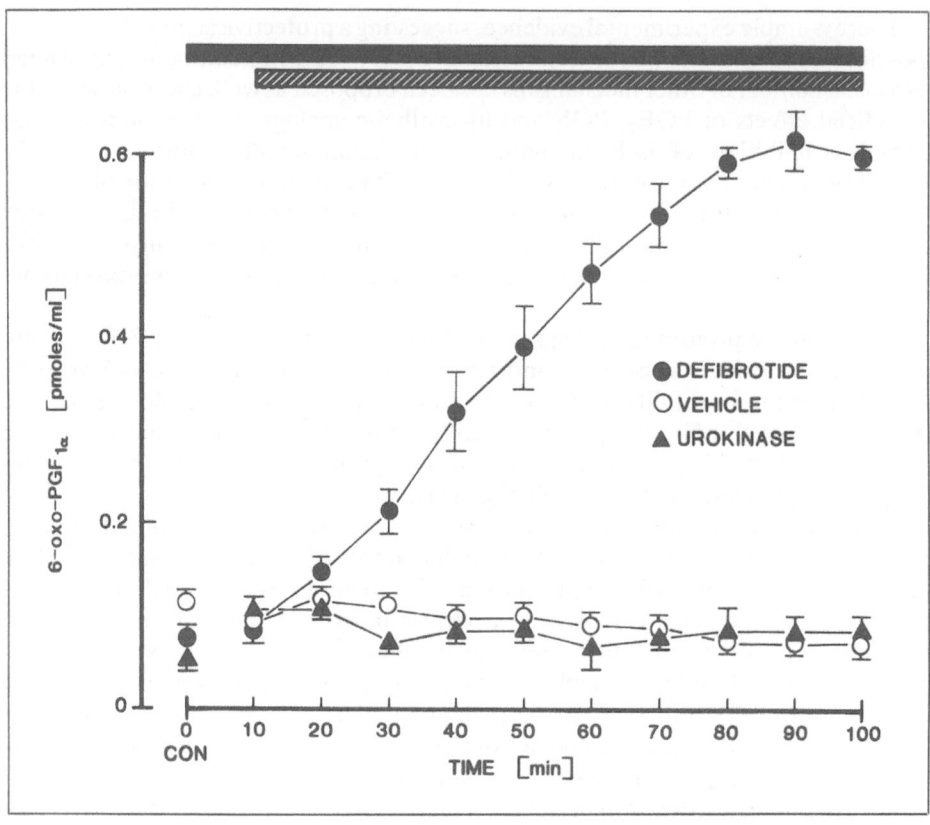

Fig. 4. Selective stimulation of PGI$_2$ (determined as 6-oxo-PGF$_{1\alpha}$) formation in the platelet-perfused guinea-pig heart by defibrotide but not by urokinase. The black line on top marks the duration of platelet infusion, the hatched line shows the infusion of defibrotide. (From [78]).

turbulent blood-flow [9], nitrates may have direct antiplatelet effects by nitric oxide release and stimulation of the platelet cyclic GMP level [3]. This and cyclic GMP-mediated vasodilatation will result in prolongation of bleeding time [21, 75]. Although the contribution of other mechanisms, e.g., stimulation of coronary vascular prostacyclin [106, 131] is not entirely clear, recent experimental data have confirmed that nitroglycerin inhibits platelet deposition in presence of deep arterial wall injury, i.e., the absence of functional endothelium [72]. A reduction of the incidence of restenosis in patients with acute myocardial infarction has been obtained by a combined treatment with streptokinase and nitroglycerin [123].

Beta-blockers: After demonstration of inhibition of platelet aggregation in vitro [53], β-blockers such as propranolol have repeatedly been shown to exert antiplatelet effects also in patients with coronary heart disease [37, 150]. Thus, platelet hyperreactivity appears to be a prerequisite for antiplatelet effects of β-blockers. The mechanism of action is not entirely clear. There appears to exist no direct relation-

96

ship to β-blockade but rather to the lipophilicity, i.e., membrane-stabilizing effects of β-blockers. Moreover, the inhibition of platelet function by propranolol, including inhibition of thromboxane formation, requires an intact platelet membrane [100]. The possible contribution of antiplatelet effects of β-blockers to their positive long-term antianginal properties is unknown but seems not to be significant [172].

Anticoagulants

Heparin: Heparin has no anticoagulant effect by its own but acts via activation of antithrombin III, the natural inhibitor of serine proteases, including thrombin and plasmin. Clinically this results in inactivation of thrombin and prevention of thrombus formation and growth.

Intravenous heparin was investigated in a placebo-controlled clinical trial in 214 patients with unstable angina. The treatment was continued for 8 weeks and was found to significantly reduce the incidence of myocardial infarction at 1 week and 8 weeks of treatment [151]. Heparin might also be useful within the second and third week following myocardial infarction. In this time, embolization of ventricular thrombi may occur and lead to stroke and renal or peripheral embolization. Preventive treatment with heparin is recommended if embolization has already occurred or if a mural thrombus appears mobile in the echocardiogram [102].

Oral anticoagulants: Oral anticoagulants decrease the bulk of circulating active procoagulants rather than truly interfering with the coagulation process, suggesting that these agents are hypocoagulants rather than true anticoagulants [96]. The compounds act as inhibitors of vitamin K-dependent synthesis of coagulation Factors II, VII, IX and X. Effective concentrations should result in a 1.5-fold prolongation of prothrombin time. However, the plasma concentrations of the antithrombotic proteins C and S are depressed as well. Since the plasma half-life of protein C is shortest, this may result in a transient prothrombotic state at the onset or termination of therapy [163].

Since the discovery of the first coumarin derivative in 1939, there have been at least 37 published reports examining the role of oral anticoagulants in patients with acute myocardial infarction (see [80]). Early studies evaluating a reduction of mortality after myocardial infarction led to controversial results (see [80]). Two out of the three controlled trials reported no significant reduction in mortality or recurrent myocardial infarction despite a significant reduction of thrombembolic complications (pulmonary embolism, stroke, and others). Furthermore, the significant reduction in mortality found in the Bronx Municipal Hospital study was restricted to women and no change was detected in men. On the other hand, there was a considerable increase in hemorrhagic complications. Thus, anticoagulation has little impact on the outcome of myocardial infarction despite significant reduction of complications from systemic embolization. Routine administration of these compounds is, therefore, considered to be neither indicated nor useful [80].

At least 26 trials have been conducted to evaluate oral anticoagulants in the secondary prevention of myocardial infarction. Four of them included a sufficient amount of patients and suitable protocols (see [80]). Only one out of these four studies, the Dutch Sixty Plus randomized double-blind trial with a 2-year follow up gave positive results: Coumarin treatment reduced the end-point mortality from 13.4% in the placebo group to 7.6% in the treated group and the reinfarction rate from 15.9% to 5.7%, significantly in both instances. Additionally, in this and all of the other trials there was a significant reduction in thromboembolic events but also a considerable increase in the rate of hemorrhagic complications. Taken together these results do not support the use of anticoagulation for secondary prevention because of an unacceptable benefit/risk ratio, in particular by comparison with ASA-type antiplatelet agents (see above).

Thrombolytic Agents

Thrombolytic therapy is based upon the use of plasminogen activators, namely streptokinase, urokinase, prourokinase, and t-PA. Similar to angioplasty and surgical revascularization, fibrinolytic therapy must be started soon after the acute event when coronary occlusion has not led to irreversible necrosis of the ischemic myocardium. Later onset of treatment results in an impaired recovery or no recovery of wall motion [92, 136, 142]. Moreover, the proteolytic action of plasminogen is not restricted to fibrin but also involves fibrinogen. Thus, fibrinogen levels in plasma may decrease as well. This certainly increases the risk of hemorrhagic events, but also may result in improved regional perfusion by reduced plasma fibrinogen levels, i.e. reduced plasma viscosity [81]. Because plasminogen bound to fibrin generally is easier to activate than in free solution, efforts have been undertaken to achieve a more localized action of these agents to the site of the thrombus, in order to minimize systemic fibrinolysis.

To critically evaluate the benefits of thrombolysis by t-PA and other thrombolytic agents, one has to consider the large variations in the control hospital mortality, varying between 13.0 [46] and 4.2% [108], suggesting different entry criteria and base-line characteristics of the patients [120], as well as the actions of the different agents on left ventricular function and infarct size [137]. Until now there has been no clear superiority of one thrombolytic agent above the others [4, 120, 141].

Streptokinase: Streptokinase forms a 1 : 1 complex with plasminogen which rapidly develops proteolytic activity and activates further molecules of plasminogen to plasmin. This plasmin, generated by the streptokinase/plasminogen complex circulates in plasma and will not only lyse fibrin but also circulating fibrinogen and Factors V and VIII. As a result, a lytic state exists, depending in its severity on the actual concentrations of physiological plasmin antagonists. Additionally, the systemic appearance of fibrin and fibrinogen degradation products may stimulate platelet activation, i.e., thromboxane generation [27] and the coagulation process. Thus, immediately following the administration of streptokinase, a paradoxic decrease of thrombolytic activity has been observed and is suspected to be responsible for failure of therapy in some patients [23].

The therapeutic efficacy of intravenous streptokinase with respect to postinfarct mortality has been convincingly demonstrated in a number of clinical trials [40, 47, 142, 156, 165]. The ISAM study demonstrated a better ejection fraction and reduced plasma creatine kinase activity in streptokinase-treated individuals. However, in this study, a higher rate of reinfarctions occured in streptokinase-treated patients. This did not affect the reduction of long-term mortality in the streptokinase-treated group [64].

It has been suspected that the favorable action of streptokinase may not be simply related to thrombolytic activity, but involve other actions on the clotting system. In fact, there is recent evidence of an enhanced formation of prostacyclin by streptokinase in smooth muscle and endothelial cells, which does not occur with urokinase and t-PA [69].

In view of the apparent systemic side effects which result from intravenous fibrinolytic therapy (see below), much interest has been focussed on intracoronary instead of intravenous administration of streptokinase. In spite of initial encouraging reports [39, 90], several carefully performed trials did not show a significant improvement of mortality in conjunction with intracoronary thrombolysis. With proper election of patients, in 0.3% major bleeding was reported and cerebrovascular events during therapy have been observed in 0.2% −1.0% [46, 64, 128]. Intracoronary infusion of streptokinase may be maintained for 1 h and reperfusion can be expected in about 75% of treated patients with acute myocardial infarction [39, 147] which, however, was not in all studies associated with a significant reduction in myocardial death despite significantly improved left ventricular function [116]. Continuous iv infusion is necessary due to the short half-life of streptokinase of about 23 min [89]. Data from the ISIS-II Study would favor the combined use of streptokinase and ASA [65], i.e., inhibition of streptokinase-induced platelet-derived thromboxane formation by ASA [27].

Due to the bacterial origin of streptokinase, allergic reactions may occur such as rash, fever, nausea, and a serum-sickness-like syndrome that rarely (0.1%) leads to anaphylactic shock [47]. Reperfusion arrhythmias are a sign of successful recanalization [64]. The antigenicity of streptokinase may be reduced by the use of fragments which retain their biologic activity but are devoid of antigenic sequences. This might be associated with a reduction of the thrombolytic activity.

Anisoylated streptokinase plasminogen complex: Acylation of plasmin at its active center removes its catalytic activity and, in addition, prevents the irreversible reaction with inhibitors such as alpha$_2$-antiplasmin. Due to the prolonged action, a single intravenous bolus injection of 30 units results in effective fibrinolysis for up to 12 h [122]. On the other hand, even acylated plasmin still binds to fibrin clots, because binding sites on the "cringle" domains are structurally apart from the acylated catalytic center [145]. Spontaneous deacylation will slowly reestablish plasmin activity localized at the thrombus, where alpha$_2$-antiplasmin is less effective.

Theoretically, this therapeutic strategy has several advantages: i) the thrombolytic agent escapes plasmatic inhibitor systems, ii) systemic plasminogen activation will be reduced, and iii) a sustained activation of plasmin will simplify the administration allowing for easier clinical therapy by single bolus injection. For clinical use, an anisoylated plasminogen-streptokinase activator complex (APSAC) with a deacylation half-life of about 40 min has been developed [145]. In fact, APSAC in comparison with heparin has been demonstrated to reduce 4-weeks mortality of patients treated within 4 h after the onset of pain from 12.6% to 5.6% [101]. Similar results were obtained by comparing APSAC with placebo in the AIMS trial: 30-days mortality was 6.4% vs 12.2%, respectively [1]. A 31%-reduction in infarct size by APSAC in comparison to heparin at 3 weeks was also observed in the APSIM trial [5].

Despite the theoretical localization of APSAC action to fibrin thrombi, a considerable systemic lytic state has been observed [122, 126]. As such sustained fibrinolytic activity cannot be readily terminated, there is the risk of severe bleeding [55]. Rethrombosis rate has been reported to be low [88], possibly due to the long half-life of the compound. The presence of streptokinase in APSAC involves an occasional allergic response.

Urokinase and pro-urokinase: Pro-urokinase (scu-PA) and urokinase catalyze the direct conversion of plasminogen to plasmin. Urokinase is not selective for fibrin. Due to the human origin, prourokinase and urokinase are not antigenic, provided that the preparations are free of foreign antigens from non-human cell lines that are used for their production.

A trial comparing intravenous urokinase with streptokinase treatment did not show a significant difference in the clinical efficacy and incidence of hemorrhagic complications [158]. Pro-urokinase has been reported to recanalize 75% of acutely occluded coronary arteries, and plasma fibrinogen declines less as compared with urokinase treatment [160] suggesting thrombus specificity of the compound in clinical conditions. A double-blind randomized trial comparing recombinant pro-urokinase with streptokinase has just been completed [118]. Patency rates after 60 min of treatment were higher in pro-urokinase treated than in streptokinase treated patients (72% vs 48%). A combination of urokinase and pro-urokinase has been reported to increase the reperfusion rate to more than 80% and to shorten lysis time of pro-urokinase alone [49]. The rather high expenses of the treatment with urokinase limit its clinical use. The intravenous treatment with urokinase has to take into account the rather short half-life of urokinase (16 min) [89].

A prolongation of plasma half-life may be achieved by coupling thrombolytic enzymes to inactivating water-soluble carriers (agarose, nylon, heparin) which slowly release the active compound. Trivalent urokinase-heparin-nitroprusside complexes have been described [86]. The specificity of urokinase for fibrin has been experimentally improved by coupling the heavy chain of plasmin, which contains the "cringle" domains with specificity for fibrin, with the low molecular weight chain of urokinase [107]. Another, very elegant approach is the combination of urokinase with a fibrin-specific [51] or platelet-specific [8] monoclonal antibody [51]. This increases selectivity considerably without reducing the enzymatic activity of urokinase.

t-PA: Tissue-type plasminogen activator (t-PA) is synthesized by endothelial cells and converts plasminogen to plasmin in vitro in the presence of fibrin 100- to 1000-times more efficiently [55]. For clinical use recombinant t-PA (rt-PA) is available. t-PA has a short half-life of 4 to 5 min [41] and therefore requires infusion in order to maintain a sufficient thrombolytic activity. After the termination of administration, the plasma level of free t-PA rapidly declines with prominence of t-PA complexes, notably with PA-I. Following 24 h after t-PA infusion, PA-I may show a reactive increase which possibly is responsible for a relatively high incidence of reocclusion, as compared with streptokinase [43, 168].

In one study a reperfusion rate of 66% was achieved and proven to be superior to streptokinase, which caused only 36% recanalization in this trial [156]. In another trial, t-PA showed only a tendency to reduced mortality in comparison to the placebo group. However, all patients in this trial received intravenous heparin [167]. Recent results from the GAUS trial demonstrated comparable reperfusion rates with t-PA and urokinase [109]. Administration of t-PA at different doses of 100, 50, and 20 mg resulted in reperfusion rates of 82%, 71%, and 50%, respectively [94], which demonstrates a dose-dependency of reperfusion with t-PA. A direct comparison of 270 patients with first myocardial infarction yielded apparently identical patency rates, i.e., 76% and 75% after intravenous streptokinase as compared to rt-PA [166]. Similar results were obtained in another recent trial, demonstrating a 51%-reduction in mortality rate at 2 weeks by comparing t-PA with placebo and even an 82% reduction in mortality if treatment was started within 3 h after onset of the symptoms [159]. Comparing t-PA with streptokinase, a significantly improved left ventricular function was noted in the t-PA group but not in the streptokinase group [84]. As already noted with urokinase, treatment with t-PA is also quite expensive in comparison with streptokinase [120].

In contrast to urokinase and streptokinase, systemic levels of fibrinogen were reported to fall only modestly (by 25%) during infusion of t-PA [161]. As fibrinogen degradation correlates with bleeding complications [119], the modest decrease of fibrinogen seen with t-PA might be expected to be associated with a reduced incidence of hemorrhage. So far this theoretical benefit has not been proven in clinical practice: the incidence of hemorrhagic events has been reported to be similar in t-PA and streptokinase treated patients [120]. Allergic complications are not likely because of the natural occurrence of the enzyme in human plasma.

Conclusions

A procoagulatory state involving enhanced platelet aggregation and coagulation has been repeatedly described in patients suffering from coronary heart disease and might be involved in its acute thrombembolic complications, i.e., myocardial infarction. One possible reason for this is a disturbed endothelial function in atherosclerotic vessels, associated with reduced formation of platelet inhibitory factors (PGI_2, EDRF), allowing local platelet adhesion and secretion, further stimulated by an insufficient production of fibrinolytic enzymes, such as plasminogen activators.

Attempts have therefore been made to specifically modify these pathophysiological alterations in order to prevent the progression of the atherosclerotic disease and the acute coronary artery occlusion as well as to improve early recanalization of the thrombus in the occluded coronary vessel. Current data clearly suggest that acetylsalicylic acid is an effective prophylactic agent in secondary prevention of myocardial infarction and might also be useful in combination with fibrinolytic agents in order to prevent platelet activation caused by these compounds. There is also clinical benefit with intravenous heparin in early stages of myocardial infarction but no proven benefit with oral anticoagulants. Similarly, neither thromboxane synthase inhibitors nor prostacyclin mimetics have been found to be beneficial in myocardial infarction in man. There might be some benefit with PGE_1, in particular in combination with fibrinolysis. However, no controlled clinical trial with any of these compounds including thromboxane receptor antagonists is available yet.

Current interest in the treatment of acute myocardial infarction is clearly focussed upon fibrinolytic agents. A number of large clinical trials have just been completed and have demonstrated clear clinical benefit for all agents used: streptokinase, anisoylated streptokinase (APSAC), prourokinase, urokinase, and t-PA. Considering all aspects of clinical treatment with these agents, including side-effects and costs, there is no clear superiority of one agent above the others.

References

1. AIMS trial study group (1988) Effect of intravenous APSAC on mortality after myocardial infarction: preliminary report of a placebo-controlled clinical trial. Lancet I:545–549
2. Armstrong PW, Langevin LM, Watts DG (1988) Randomized trial of prostacyclin infusion in acute myocardial infarction. Am J Cardiol 61:455–457
3. Azuma H, Ishikawa M, Sekizaki S (1986) Endothelium-dependent inhibition of platelet aggregation. Br J Pharmacol 88:411–415
4. Bang U, Wilhelm OG, Clayman MD (1989) Thrombolytic therapy in acute myocardial infarction. Ann Rev Pharmacol Tox 29:323–341
5. Bassand J-P, Machecourt J, Cassagne J, Anguenot T, Lusson R, Borel E, Peycelon P, Wolf E, Ducellier D (1989) Multicenter trial of intravenous anisoylated plasminogen streptokinase activator complex (APSAC) in acute myocardial infarction: Effects on infarct size and left ventricular function. J Am Coll Cardiol 13:988–997
6. Bassenge E, Busse R (1988) Endothelial modulation of coronary tone. Progr Cardiovasc Dis 30:349–360
7. Bertele G, Mussoni V, Pintucci G, del Rosso G, Romano C, de Gaetano G, Libretti A (1989) The inhibitory effect of aspirin on fibrinolysis is reversed by iloprost, a prostacyclin-analogue. Thromb Haemost 61:286–288
8. Bode C, Meinhardt G, Runge MS, Eberle T, Schuler G, Kübler W (1989) Conjugation of urokinase to an antiplatelet antibody results in a more potent fibrinolytic agent. Thromb Haemost 62:483 (abstr)
9. Brown BG, Bolson E, Peterson RE, Pierce CD, Dodge HT (1981) The mechanism of nitroglycerin action: Stenosis vasodilatation as a major component of the drug response. Circulation 64:1089–1097
10. Bugiardini R, Galvani M, Ferrini D, Gridelli C, Tollemeto D, Macri N, Puddu P, Lenzi S (1987) Myocardial ischemia during intravenous prostacyclin administration: Hemodynamic findings and precautionary measures. Am Heart J. 113:234–240
11. Buja LM, Willerson JT (1987) The role of coronary artery lesions in ischemic heart disease: Insights from recent clinicopathologic, coronary arteriographic and experimental studies. Human Path 18:451–461
12. Bush C, Owen WG (1982) Identification in vitro of an endothelial cell surface cofactor for antithrombin III. J Clin Invest 69:726–729
13. Cairns JA, Gent M, Singer J (1985) Aspirin, sulfinpyrazone or both in unstable angina. Results of a Canadian Multicenter Trial. N Engl J Med 313:1369–1375
14. Chierchia S, De Caterina R, Crea F, Patrono C, Maseri A (1982) Failure of thromboxane A_2 blockade to prevent attacks of vasospastic angina. Circulation 66:702–705
15. Chierchia S, Patrono C (1987) Role of platelets and vascular eicosanoids in the pathophysiology of ischemic heart disease. Fed Proc 46:81–88
16. Clouse LH, Comp PC (1986) The regulation of hemostasis: The protein C system. N Engl J Med 314:1298–1304
17. Collen D (1988) Fibrin-specific thrombolytic agents. Klin Wschr 66, (Suppl XII):15–23

18. Coleman RW, Marder VJ, Salzman EW, Hirsh J (1987) Overview on hemostasis. In: Coleman RW, Hirsh J, Marder VJ, Salzman EW (eds) Hemostasis and Thrombosis. Lippincott Philadelphia, pp 3–17
19. Crutchley DJ, Conanan LB, Maynard JR (1982) Stimulation of fibrinolytic activity in human skin fibroblasts by prostaglandins E_1, E_2 and I_2. J Pharmacol Exp Ther 222:544–549
20. Davies MJ, Thomas AC (1985) Plaque fissuring: the cause of acute myocardial infarction, sudden ischemic death and crescendo angina. Br Heart J 53:363–373
21. DeCaterina R, Gianessi D, Bernini W, Mazzone A (1988) Organic nitrates: direct antiplatelet effects and synergism with prostacyclin. Thromb Haemost 59:207–211
22. DeClerck F, Beetens J, van der Water A, Vercammen E, Janssen PAJ (1989) R68 070: Thromboxane A_2 synthetase inhibition and thromboxane A_2/prostaglandin endoperoxide receptor blockade combined in one molecule – II. Pharmacological effects in vivo and ex vivo. Thromb Haemost 61:43–49
23. Eisenberg PR, Sherman LA, Jaffe AS (1987) Paradoxic elevation of fibrinopeptide A after streptokinase: evidence for continued thrombosis despite intense fibrinolysis. J Am Coll Cardiol 10:527–529
24. Esmon NL (1987) Thrombomodulin. Semin Thromb Hemostasis 13:454–460
25. Fareed J, Walenga JM, Hoppensteadt DA, Kumar A, Ulutin ON, Cornelli U (1988) Pharmacological profiling of defibrotide in experimental models. Semin Thromb Hemostasis (Suppl) 14: 27–37
26. Falk E (1983) Plaque rupture with severe pre-existing stenosis precipitating coronary thrombosis: Characteristics of coronary atherosclerotic plaque underlying fatal occlusive thrombi. Br Heart J 50:127–134
27. FitzGerald DJ, Catella F, Roy L, FitzGerald GA (1988) Marked platelet activation in vivo after intravenous streptokinase in patients with acute myocardial infarction. Circulation 77:142–150
28. FitzGerald DJ, Fragetta J, FitzGerald GA (1988) Prostaglandin endoperoxides modulate the response to thromboxane synthetase inhibition during coronary thrombosis. J Clin Invest 82:1708–1713
29. FitzGerald DJ, Roy L, FitzGerald GA (1985) Enhanced prostacyclin and thromboxane A_2 synthesis in vivo in ischemic heart disease: noninvasive evidence of sporadic platelet activation in unstable angina. Circulation 72 (suppl III):III–113 (Abstract)
30. FitzGerald GA, Smith B, Pederson AK, Brash AR (1984) Prostacyclin biosynthesis is increased in patients with severe atherosclerosis and platelet activation. N Engl J Med 310:1065–1068
31. Fitscha P, Kaliman J, Sinzinger H (1985) Gamma-camera imaging after autologous human platelet labeling with ^{111}In-oxine-sulfate: A key for assessing the efficacy of prostacyclin treatment in active atherosclerosis? In: Schrör K (ed) Prostaglandins and other eicosanoids in the cardiovascular system. Karger, Basel, pp 352–357
32. Folts JD, Crowell ED, Rowe GG (1976) Platelet aggregation in partially obstructed vessels and its elimination with aspirin. Circulation 54:365–370
33. Forrester JS, Litvack F, Grunfest W, Hickey A (1987) A perspective of coronary artery disease seen through the arteries of a living man. Circulation 75:505–515
34. Francis CW, Marder VJ (1988) Increased resistance to plasmic degradation of fibrin with highly crosslinked alpha-polymer chains formed at high factor XIII concentrations. Blood 71:1361–1365
35. Friedrich T, Lichey J, Nigam S, Priesnitz M, Wegscheider K (1985) Follow-up of prostaglandin plasma levels after acute myocardial infarction. Am Heart J 109:218–222
36. Frishman WH, Christodoulou J, Weksler BB, Smithen C, Killip T, Scheidt S (1976) Aspirin therapy in angina pectoris: effects on platelet aggregation, exercise tolerance, and echocardiographic manifestations of ischemia. Am Heart J 92:3–10
37. Frishman WH, Weksler BB, Christodoulou J, Smithen C, Killip T (1974) Reversal of abnormal platelet aggregability and change in exercise tolerance in patients with angina pectoris following oral propranolol. Circulation 50:887–896
38. Fuster V, Cohen M, Halperin J (1989) Aspirin in the prevention of coronary disease. N Engl J Med 321:183–185
39. Ganz W, Buchbinder M, Marcus H, Mondkar A, Maddahi J, Charuzi Y, O'Conner L, Shell

103

W, Fishbein MC, Kass R, Miyamoto A, Swan HJC (1981) Intracoronary thrombolysis in evolving myocardial infarction. Am Heart J 101:4−19

40. Ganz W, Geft I, Shah PK, Lew AS, Rodriguez L, Weiss T, Maddahi J, Berman DS, Charuzi Y, Swan HJC (1984) Intravenous streptokinase in evolving myocardial infarction. Am J Cardiol 53:1209−1216
41. Garabedian HD, Gold HK, Leinbach RC, Johns JA, Yasuda T, Kanke M, Collen D (1987) Comparative properties of two clinical preparations of recombinant human tissue type plasminogen activator in patients with acute myocardial infarction. J Am Coll Cardiol 9:599−607
42. van der Giessen WJ, Zijlstra FJ, Berk L, Verdouw PD (1988) The effect of the thromboxane receptor antagonist BM 13.177 on experimentally induced coronary thrombosis in the pig. Eur J Pharmacol 147:241−248
43. Gold HK, Leinbach RC, Garabedian HD, Yasuda T, Johns JA, Grossbard EB, Palacios J, Collen D (1986) Acute coronary reocclusion after thrombolysis with recombinant human tissue-type plasminogen activator: Prevention by a maintainance infusion. Circulation 73:347−352
44. Gram J, Kluft C, Jespersen J (1987) Depression of tissue plasminogen activator (t-PA) actitivity and rise of t-PA inhibition and acute phase reactants in blood of patients with acute myocardial infarction (AMI). Thromb Haemost 58:817−821
45. Grose R, Greenberg M, Strain J, Mueller H (1985) Intracoronary prostacyclin in evolving acute myocardial infarction. Am J Cardiol 55:1625−1626
46. Gruppo Italiano per lo studio della streptochinasi nell' infarcto miocardico (GISSI) (1986) Effectiveness of intravenous thrombolytic treatment in acute myocardial infarction. Lancet I:397−401
47. Gruppo Italiano per lo studio della streptochinasi nell' infarcto miocardico (GISSI) (1987) Long term effects of intravenous thrombolysis in acute myocardial infarction: Final report of the GISSI study. Lancet II:871−874
48. Gurewich V, Pannell R, Louie S, Kelly P, Suddith RL, Greenlee R (1984) Effective and fibrin specific clot lysis by a zymogen precursor form of urokinase (pro-urokinase): A study in vitro and in two animal species. J Clin Invest 73:1731−1739
49. Gurewich V (1987) Experiences with pro-urokinase and potentiation of its fibrinolytic effect by urokinase and by tissue plasminogen activator. J Am Coll Cardiol 10 (5 Suppl B):16B−21B
50. Gurewich V (1989) The sequential, complementary and synergistic activation of fibrin-bound plasminogen by tissue plasminogen activator and pro-urokinase. Fibrinolysis 3:59−66
51. Haber E (1986) In vivo diagnostic and therapeutic uses of monoclonal antibodies in cariology. Ann Rev Med 37:249−261
52. Hamouratidis ND, Pertsinidis TE, Bacharoudis GP, Papazachariou GS (1988) Effects of exercise on plasma fibrinolytic activity in patients with ischaemic heart disease. Int J Cardiol 19:39−45
53. Hampton JR, Harrison MJG, Honour AJ, Mitchell JRA (1967) Platelet behaviour and drugs used in cardiovascular disease. Cardiovasc Res 1:101−107
54. Hamsten A, Wiman B, de Faire U, Blombaeck M (1985) Increased plasma levels of a rapid inhibitor of tissue plasminogen activator in young survivors of myocardial infarction. N Engl J Med 313:1557−1563
55. Handin RI, Loscalzo J (1988) Hemostasis, thrombosis, fibrinolysis, and cardiovascular disease. In: Braunwald E (ed) Heart disease. Saunders, Philadelphia, pp 1758−1781
56. Handin RI, McDonough M, Lesch M (1978) Elevation of platelet factor four in acute myocardial infarction. J Lab Clin Med 91:340−349
57. Hekman CM, Loskutoff DJ (1987) Fibrinolytic pathways and the endothelium. Semin Thromb Haemostasis 13:514−527
58. Hennekens CH, Peto R, Hutchinson GB, Doll R (1988) An overview of the British and American aspirin studies. N Engl J Med 318:923−924
59. Henriksson P, Wennmalm A, Edhag O, Vesterqvist O, Green K (1986) In vivo production of prostacyclin and thromboxane in patients with acute myocardial infarction. Br Heart J 55:543−548
60. Hoffmann W, Förster W (1987) Two years follow-up Cottbus reinfarction study with 30 and 60 mg acetylsalicylic acid. In: Sinzinger H, Schrör K (eds) Prostaglandins in clinical research. Liss, New York, pp 393−397

61. Hohlfeld T, Thiemermann C, Schrör K (1989) Protection from myocardial ischemic injury in cats and pigs by defibrotide – different mode of action? In: Schrör K, Sinzinger H (eds) Prostaglandins in Clinical Research. Liss, New York, pp 137–142

62. Houston DS, Sheperd JT, Vanhoutte PM (1986) Aggregating human platelets cause direct contraction and endothelium-dependent relaxation of isolated coronary arteries. Role of serotonin, thromboxane A_2 and adenine nucleotides. J Clin Invest 78:539–544

63. Hutton I, Tweddel AC, Rankin AC, Walker ID, Davidson JF (1983) Effects of dazoxiben on transcardial thromboxane levels and haemodynamics in coronary heart disease. Br J Clin Pharmacol 15 (Suppl 1):79S–82S

64. The ISAM study group (1986) A prospective trial of intravenous streptokinase in acute myocardial infarction (ISAM). N Engl J Med 314:1465–1471

65. ISIS-2 (Second International Study of Infarct Survival) (1988) Survival Collaborative Group: Randomized trial of intravenous streptokinase, oral aspirin, both or neither among 17187 cases of suspected myocardial infarction. Lancet II:349–360

66. Jackson CM (1987) Mechanisms of prothrombin activation. In: Coleman RW, Hirsh J, Marder VJ, Salzman EW (eds) Hemostasis and Thrombosis. Lippincott, Philadelphia, pp 135–147

67. Jaffe EA, Weksler BB (1979) Recovery of endothelial cell prostacyclin production after inhibition by low doses of aspirin. J Clin Invest 63:532–535

68. Jaschonek K, Karsch KR, Weisenberger H, Tidow S, Faul C, Renn W (1986) Platelet prostacyclin binding in coronary artery disease. J Am Coll Cardiol 8:259–266

69. Kawaguchi H, Yasuda H (1988) Effect of various plasminogen activators on prostacyclin synthesis in cultured vascular cells. Circ Res 63:1029–1035

70. Kiernan FJ, Kluger J, Regnier JC, Rutkowski M, Fieldman A (1986) Epoprostenol sodium (prostacyclin) infusion in acute myocardial infarction. Br Heart J 56:428–432

71. Kruithof EK, Tran-Thang C, Ransijn A, Bachmann F (1984) Demonstration of a fast acting inhibitor of plasminogen activators in human plasma. Blood: 64:907–913

72. Lam JYT, Chesebro JH, Fuster V (1988) Platelets, vasoconstriction, and nitroglycerin during arterial wall injury. Circulation 78:712–716

73. Levine SP, Lindenfeld J, Ellis JB, Raymond NM, Krentz LS (1981) Increased plasma concentrations of platelet factor 4 in coronary artery disease. Circulation 64:626–632

74. Lewis HD, Davis JW, Archibald DG, Steinke WE, Smitherman TC, Doherty JE III, Schnaper JW, LeWinter MM, Linares E, Pouget JM, Sabharwal SC, Chesler E, DeMots H (1983) Protective effects of aspirin against acute myocardial infarction and death in men with unstable angina. Results of a Veterans Administration Cooperative study. N Engl J Med 309:396–403

75. Lichtenthal PR, Rossi E, Louis G, Rehnberg KA, Wade LD, Michaelis LL, Fung H-L, Patrignani P (1985) Dose-related prolongation of the bleeding time by intravenous nitroglycerin. Anesth Analg 64:30–33

76. Lijnen HR, Zamarron C, Blaber M, Winkler ME, Collen D (1986) Activation of plasminogen by pro-urokinase. I. Mechanism. J Biol Chem 261:1253–1258

77. Lipinska I, Gurewich V, Meriam CM, Kosowsky BD, Ramaswamy K, Philbin E, Losordo D (1987) Lipids, lipoproteins, fibrinogen and fibrinolytic activity in angiographically assessed coronary heart disease. Artery 15:44–60

78. Löbel P, Schrör K (1985) Selective stimulation of coronary vascular PGI_2 but not platelet thromboxane formation by defibrotide in the platelet-perfused heart. Naunyn-Schmiedeberg's Arch Pharmacol 331:125–130

79. Lollar P, Owen WG (1980) Clearance of thrombin from circulation in rabbits by high affinity binding sites on endothelium: Possible role in the inactivation of thrombin by antithrombin III. J Clin Invest 66:1222–1230

80. Lopez LM, Mehta JL (1987) Anticoagulation in coronary heart disease: Heparin and warfarin trials. In: Mehta JL (ed) Thrombosis and Platelets in Myocardial Ischemia. Davis Company, Philadelphia, pp 215–229

81. Loscalzo J, Braunwald E (1988) Tissue plasminogen activator. N Engl J Med 319:925–931

82. Ludmer PL, Selwyn AP, Shook TL, Wayne RR, Mudge GH, Alexander WR, Ganz P (1986) Paradoxical vasoconstriction induced by acetylcholine in atherosclerotic coronary arteries. N Engl J Med 315:1046–1051

83. MacIntyre DE, Pearson JD, Gordon JL (1978) Localization and stimulation of prostacyclin production in vascular cells. Nature 271:549−551
84. Magnani B (1989) Plasminogen activator Italian Multicenter Study (PAIMS): Comparison of intravenous recombinant single-chain human tissue-type plasminogen activator (rt-PA) with intravenous streptokinase in acute myocardial infarction. J Am Coll Cardiol 13:19−26
85. Mais DE, Burch RM, Saussy DL Jr, Kochel PJ, Halushka PV (1985) Binding of a thromboxane A_2/prostaglandin H_2 receptor antagonist to washed human platelets. J Pharm Exp Ther 235:729−734
86. Maksimenko AV, Torchilin VP (1985) Water-soluble urokinase derivatives of combined action. Thromb Res 38:277−288
87. Mant MJ, Kappagoda CT, Taylor RF, Quinlan JE (1988) Transcoronary platelet activation and consumption in coronary artery disease: studies at rest. Thromb Res 50:201−211
88. Marder VJ, Rothbard RL, Fitzpatrick PG, Francis CW (1986) Rapid lysis of coronary artery thrombi with anisoylated plasminogen: Streptokinase activator complex. Treatment by bolus intravenous injection. Ann Intern Med 104:304−310
89. Marder VJ, Sherry S (1988) Thrombolytic therapy: Current status (Parts I and II). N Engl J Med 318:1512−1520; 1585−1595
90. Markis JE, Malagold M, Parker JA, Silverman KJ, Barry WH, Als AV, Paulin S, Grossman W, Braunwald E (1981) Myocardial salvage after intracoronary thrombolysis with streptokinase in acute myocardial infarction. N Engl J Med 305:777−782
91. Maseri A, L'Abbate A, Baroldi G, Chierchia S, Marzilli M, Ballestra AM, Severi S, Parodi O, Biagini A, Distante A, Pesola A (1978) Coronary vasospasm as a possible cause of myocardial infarction. N Engl J Med 299:1271−1277
92. Mathey DG, Sheehan FH, Schofer J, Dodge HT (1985) Time from onset of symptoms to thrombolytic therapy: a major determinant of myocardial salvage in patients with acute transmural infarction. J Am Coll Cardiol 6:518−525
93. Mathias FR (1978) Soluble plasma fibrin and platelet prostaglandin endoperoxides following myocardial infarction. Haemostasis 7:273−281
94. McNeill AJ, Shannon JS, Cunningham SR, Flannery DJ, Campbell MP, Kahn MM, Patterson GC, Webb SW, Adgey A (1988) A randomized dose ranging study of recombinant tissue plasminogen activator in acute myocardial infarction. Br Med J (Clin Res) 296:1768−1771
95. Meade TW, Stirling Y, Thompson SG, Vickers MV, Woolf L, Ajdokiewicz AB, Stewart G, Davidson JF, Walker ID, Douglas AS, Richardson JM, Weir RD, Aromaa A, Impivaara O, Maatela J, Hladovec J (1986) An international and interregional comparison of hemostatic variables in the study of ischaemic heart disease. Report of a working group. Int J Epidemiol 15:331−336
96. Mehta JL, Kitchens CS (1987) Pharmacology of platelet inhibitory drugs, anticoagulants and thrombolytic agents. In: Mehta JL (ed) Thrombosis and Platelets in Myocardial Ischemia. Davis Company, Philadelphia, pp 163−179
97. Mehta JL, Nichols WW, Mehta P (1988) Neutrophils as potential participants in acute myocardial ischemia: relevance to reperfusion. J Am Coll Cardiol 11:1309−1316
98. Mehta P, Mehta J (1985) Platelet function studies in coronary disease. V. Evidence for enhanced platelet aggregate formation activity in acute myocardial infarction. Am J Cardiol 43:757−760
99. Mehta JL, Mehta P, Conti CR (1980) Platelet function studies in coronary heart disease. IX. Increased platelet prostaglandin generation and abnormal platelet sensitivity to prostacyclin and endoperoxide analog in angina pectoris. Am J Cardiol 46:943−947
100. Mehta JL, Mehta P, Ostrowski N (1983) Influence of propranolol and 4-hydroxypropranolol on platelet aggregation and thromboxane A_2 generation. Clin Pharmacol Ther 34:559−564
101. Meinertz T, Kasper W, Schumacher M, Just H (1988) The German Multicenter Trial of anisoylated plasminogen streptokinase activator complex versus heparin for acute myocardial infarction. Am J Cardiol 62:347−351
102. Meltzer RS, Visser CA, Fuster V (1986) Intracardiac thrombi and systemic embolisation. Ann Intern Med 104:689−698
103. Moncada S, Herman AG, Higgs EA, Vane JR (1977) Differential formation of prostacyclin (PGX or PGI_2) by layers of the arterial wall: An explanation of the antithrombotic properties of vascular endothelium. Thromb Res 11:323−344

106

104. Moncada S, Palmer RMJ, Higgs EA (1987) Prostacyclin and endothelium-derived relaxing factor: biological interactions and significance. In: Verstraete M, Vermylen J, Lijnen HR, Arnout J (eds) Thrombosis and Hemostasis. Leuven University Press, pp 597–618
105. Mueller HS, Rao PS, Greenberg MA, Buttrick PM, Sussman II, Levite HA, Grose RM, Perez-Davila V, Strain JE, Spaet TH (1985) Systemic and transcardiac platelet activity in acute myocardial infarction in man: resistance to prostacyclin. Circulation 72:1336–1345
106. Nakabayashi S, Uyama O, Nagatsuka K, Uehara A, Wanaka A, Yoneda S, Kimura K, Kamata T (1985) The effect of isosorbide dinitrate and isosorbide 5-mononitrate on prostacyclin (PGI_2) and thromboxane A_2 (TXA_2) generation in rat and human arteries. Res Commun Chem Pathol Pharmacol 47:323–332
107. Nakayama Y, Shinohara M, Tani T, Kawaguchi T, Furuta T, Izawa T, Kaise H, Miyazaki W, Nakano Y (1986) The plasmin heavy chain urokinase conjugate: a specific thrombolytic agent. Thromb Haemost 56:969–970
108. National Heart Foundation of Australia Coronary Thrombolysis Group (1988) Coronary thrombolysis and myocardial salvage by tissue plasminogen activator given up to 4 hours after onset of myocardial infarction. Lancet I:203–208
109. Neuhaus KL, Tebbe U, Gottwik M, Weber MA, Feuerer W, Niederer W, Haerer W, Praetorius F, Grosser KD, Huhmann W, Hoepp HW, Alber G, Sheikhzadeh A, Schneider B (1988) Intravenous recombinant tissue plasminogen activator (rt-PA) and urokinase in acute myocardial infarction: Results of the German Activator Urokinase Study (GAUS). J Am Coll Cardiol 12:581–587
110. Niada R, Porta R, Pescador R, Mantovani M, Prino G (1985) Cardioprotective effects of defibrotide in acute myocardial ischemia in the cat. Thromb Res 38:71–81
111. Nidorf SM, Sturm M, Strophair J, Kendrew PJ, Taylor RR (1989) Whole blood aggregation, thromboxane release and the lyso-derivative of platelet activating factor in myocardial infarction and unstable angina. Cardiovasc Res 23:273–278
112. Nilsson TK, Johnson O (1987) The extrinsic fibrinolytic system in survivors of myocardial infarction. Thromb Res 48:621–630
113. Oates JA, FitzGerald GA, Branch RA, Jackson EK, Knapp HR, Roberts LJ (1988) Clinical implications of prostaglandin and thromboxane A_2 formation. N Engl J Med 319:689–698
114. O'Connor NT, Cederholm-Williams S, Copper S, Cotter L (1984) Hypercoagulability and coronary artery disease. Br Heart J 52:614–616
115. Oehler G, Büdinger M, Heinrich D, Schöndorf T (1988) Alterations of antithrombin III after acute myocardial infarction. Folia Haematol 115:319–323
116. Patel B, Kloner RA (1987) Analysis of reported randomized trials of streptokinase therapy for acute myocardial infarction in the 1980's. Am J Cardiol 59:501–504
117. Peto R, Gray R, Collins R, Wheatley K, Hennekens C, Jamrozik K, Warlow C, Hafner B, Thompson E, Norton S, Gilliand J, Doll R (1988) Randomized trial of prophylactic daily aspirin in British male doctors. Br Med J 296:313–331
118. PRIMI Trial Study Group (1989) Randomized double-blind trial of recombinant pro-urokinase against streptokinase in acute myocardial infarction. Lancet I:863–868
119. Rao AK, Pratt C, Berke A, Jaffe A, Ockene I, Schreiber TL, Bell WR, Knatterud G, Robertson TL, Terrin ML (1988) Thrombolysis in Myocardial Infarction (TIMI) trial – phase I: Hemorrhagic manifestations and changes in plasma fibrinogen and the fibrinolytic system in patients treated with recombinant tissue plasminogen activator and streptokinase. J Am Coll Cardiol 11:1–11
120. Rapaport E (1989) Thrombolytic agents in acute myocardial infarction. N Engl J Med 320:861–864
121. Reilly IAG, FitzGerald GA (1988) Aspirin in cardiovascular disease. Drugs 35:154–176
122. Renkin J, Beys CC, Lavenne-Pardonge E, Pintens H, Col J (1987) Analysis of coagulation and fibrinolysis after intravenous anisoylated plasminogen streptokinase activator complex or heparin in patients with acute myocardial infarction. A Belgian multicentre study. Drugs 33 (Suppl 3):253–260
123. Rentrop KP, Feit F, Blanke H, Steey P, Schneider R, Rey M, Horowitz S, Goldman H, Karsch K, Meilmann H, Cohen M, Siegel S, Sanger J, Slater J, Gorlin R, Fox A, Fagerstrom R, Calhoun WF (1984) Effects of intracoronary streptokinase and intracoronary nitroglycerin infusion on coronary angiographic patterns and mortality in patients with acute myocardial infarction. N Engl J Med 311:1457–1463

124. Riess H, Hiller B, Reinhardt B, Bränning C (1984) Effects of BM 13.177, a new antiplatelet drug, in patients with atherosclerotic disease. Thromb Res 35:371–378
125. Rubanyi GM, Vanhoutte PM (1985) Hypoxia releases a vasoconstrictor substance from the canine vascular endothelium. J Physiol 304:45–56
126. Samama M, Conard J, Verdy E, van Dreden P, Nguyen G, Combrisson A, Acar J (1987) Biological study of intravenous anisoylated plasminogen streptokinase activator complex in acute myocardial infarction. Drugs 33 (Suppl 3):268–274
127. Sawyer PN, Srinivasan S (1972) The role of electrochemical surface properties in thrombosis at vascular interfaces: Cumulative experience of studies in animals and man. Bull NY Acad Med 48:235–256
128. Schröder R, Neuhaus KL, Leizorovicz A, Linderer T, Tebbe U (1987) for the ISAM study group: A prospective placebo controlled double blind multicenter trial of intravenous strep-tokinase in acute myocardial infarction (ISAM): Long term mortality and morbidity. J Am Coll Cardiol 9:197–203
129. Schrör K (1987) Actions of prostaglandins on the heart. In: Gryglewski RJ, Stock G (eds) Prostacyclin and its stable analogue iloprost. Springer, Berlin Heidelberg New York Tokyo, pp 159–178
130. Schrör K (1989) Cytoprotection by prostacyclins in myocardial ischemia. In: Zor U, Naor Z, Danon A (eds) Leukotrienes and Prostanoids in Health and Disease. Karger, Basel, pp 246–251
131. Schrör K, Grodzinska L, Darius H (1981) Stimulation of coronary vascular prostacyclin and inhibition of human platelet thromboxane A_2 after low-dose nitroglycerin. Thromb Res 23:59–65
132. Schrör K, Smith EF III, Bickerton M, Smith JB, Nicolaou KC, Magolda R, Lefer AM (1980) Preservation of the ischemic myocardium by pinane thromboxane A_2. Am J Physiol 238:H87–H92
133. Schrör K, Smith EF III, Lefer AM (1988) The cat as an in vivo model for myocardial ischemia and infarction. Progr Pharmacol 6:31–91
134. Schrör K, Thiemermann C (1986) Treatment of acute myocardial ischemia with a selective antagonist of thromboxane receptors (BM 13.177). Br J Pharmacol 87:631–637
135. Schrör K, Verheggen R (1986) Prostacyclins are only weak antagonists of coronary vaso-constriction induced by authentic thromboxane A_2 and serotonin. J Cardiovasc Pharmacol 8:483–490
136. Schwartz F, Schuler G, Katus H, Hofmann M, Manthey J, Tillmanns H, Mehmel HC, Kübler W (1982) Intracoronary thrombolysis in acute myocardial infarction: Duration of ischemia as major determinant of late results after recanalization. Am J Cardiol 50:933–937
137. Selzer A (1989) Does thrombolytic therapy reduce infarct size? J Am Coll Cardiol 13:1431–1434
138. Sharma B, Wyeth RP, Gimenez HJ, Franciosa JA (1986) Intracoronary prostaglandin E_1 plus streptokinase in acute myocardial infarction. Am J Cardiol 58:1161–1166
139. Sherry S (1987) Appraisal of various thrombolytic agents in the treatment of acute myocardial infarction. Am J Med 83:31–46
140. Siegel RJ, Shah FK, Nathan M, Rodriguez L (1984) Prostaglandin E_1 infusion in unstable angina: Effects on anginal frequency and cardiac function. Am Heart J 108:863–868
141. Simoons ML (1989) Thrombolytic therapy in acute myocardial infarction. Ann Rev Med 40:181–200
142. Simoons ML, Serruys PW, van den Brand M, Res J, Verheugt FW, Krauss XH, Remme WJ, Bär F, de Zwaan C, van der Laarse A, Vermeer F, Lubsen J (1986) Early thrombolysis in acute myocardial infarction: Limitation of infarct size and improved survival. J Am Coll Car-diol 7:717–728
143. Sinzinger H, Kaliman J, Widhalm K, Pachinger O, Probst P (1981) Value of platelet sensitiv-ity to antiaggregatory prostaglandins (PGI_2, PGE_1, PGD_2) in 50 patients with myocardial infarction at young age. Prostaglandins Med 7:125–132
144. Smith EF III, Griswold DE, Egan JW, Hillegass LM, DiMartino MJ (1989) Reduction of myocardial damage and polymorphonuclear leukocyte accumulation following coronary artery occlusion and reperfusion by the thromboxane receptor antagonist BM 13.505. J Car-diovasc Pharmacol 13:715–722

145. Smith RAG, Dupe RJ, English PD, Green J (1981) Fibrinolysis with acyl-enzymes: a new approach to thrombolytic therapy. Nature 290:505−508
146. Sobel M, Salzman EW, Davies GC, Handin RI, Sweeney J, Ploetz J, Kurland G (1981) Circulating platelet products in unstable angina pectoris. Circulation 63:300−306
147. Spann JF, Sherry S (1984) Coronary thrombolysis for evolving myocardial infarction. Drugs 28:465−483
148. Steering Committee of the physician's health study research group (1989) Final report on the aspirin component of the ongoing physician's health study. N Engl J Med 321:129−135
149. Swedberg K, Held P, Wadenvik H, Kutti J (1987) Central haemodynamic and antiplatelet effects of iloprost − a new prostacyclin analogue − in acute myocardial infarction in man. Eur Heart J 8:362−368
150. Tai E, Berenzow J, Weksler BB, Klein N, LeJemtel T, Frishman W (1982) Comparative effects of oral verapamil and propranolol on platelet activation in angina pectoris: a placebo-controlled, double-blind crossover study. Circulation 66:II−323 (Abstract)
151. Telford AM, Wilson C (1981) Trial of heparin versus atenolol in prevention of myocardial infarction in intermediate coronary syndrome. Lancet I:1225−1228
152. Thaulow E, Dale J, Myhre E (1984) Effects of selective thromboxane synthetase inhibitor, dazoxiben and of acetyl salicylic acid on myocardial ischemia in patients with coronary artery disease. Am J Cardiol 53:1255−1258
153. Theroux P, Latour JG, Leger-Gauthier C, De Lara J (1987) Fibrinopeptide A and platelet factor levels in unstable angina pectoris. Circulation 75:156−162
154. Thiemermann C, Löbel P, Schrör K (1985) Usefulness of defibrotide in protecting ischemic myocardium from early reperfusion damage. Am J Cardiol 56:978−982
155. Thiemermann C, Ney P, Schrör K (1988) The thromboxane receptor antagonist, daltroban, protects the myocardium from ischaemic injury resulting in suppression of leukocytosis. Eur J Pharmacol 155:57−67
156. The TIMI study group (1985) The thrombolysis in myocardial infarction trial. Phase I findings. N Engl J Med 312:932−936
157. Uchida Y, Murao S (1974) Cyclic changes in peripheral blood pressure of partially constricted coronary artery. Jap Coll Angiol 14:383−393
158. Urokinase-streptokinase pulmonary embolism trial study group (1974) Urokinase-streptokinase pulmonary embolism trial. Phase II results. A national cooperative study. J Am Med Ass 229:1606−1613
159. Van de Werf F (1988) Lessons from the European cooperative recombinant tissue-type plasminogen activator (rt-PA) versus placebo trial. J Am Coll Cardiol 12 (Suppl A): 14A−19A
160. van de Werf F (1986) Coronary thrombolysis with recombinant single chain urokinase type plasminogen activation in patients with acute myocardial infarction. Circulation 74:1066−1070
161. van de Werf F, Arnold AER, and the European Study Group for Recombinant Tissue − Type Plasminogen Activator (rt-PA) (1988) Intravenous tissue plasminogen activator and size of infarct, left ventricular function, and survival in acute myocardial infarction. Br Med J 297:1374−1379
162. Verheggen R, Schrör K (1986) The modification of platelet-induced vasoconstriction by a thromboxane receptor antagonist. J Cardiovasc Pharmacol 8:483−490
163. Wessler S, Gitel SN (1986) Pharmacology of heparin and warfarin. J Am Coll Cardiol 6 (Suppl B):10B−20B
164. Weiss HJ, Turitto VT (1979) Prostacyclin inhibits platelet adhesion and thrombus formation on subendothelium. Blood 53:244−248
165. White HD, Norris RM, Brown MA, Takayama M, Maslowski A, Bass NM, Ormiston JA, Whitlock T (1987) Effect of intravenous streptokinase on left ventricular function and early survival after acute myocardial infarction. N Engl J Med 317:850−855
166. White HD, Rivers JT, Maslowski AH, Ormisten JA, Takayama M, Hart HH, Sharp DN, Whitlock RML, Norris RM (1989) Effect of intravenous streptokinase as compared with that of tissue plasminogen activator on left ventricular function after first myocardial infarction. N Engl J Med 320:817−821
167. Wilcox RG, von der Lippe G, Olsson CG, Jensen G, Skene AM, Hampton JR (1988) Trial of tissue plasminogen activator for mortality reduction in acute myocardial infarction. Anglo-Scandinavian study of early thrombolysis (ASSET). Lancet II:525−530

168. Williams DO, Borer J, Braunwald E, Chesebro JH, Cohen J, Dodge HT, Francis CK, Knatterud G, Ludbrook P (1986) Intravenous recombinant tissue-type plasminogen activator in acute myocardial infarction: A report from the NHLBI thrombolysis in myocardial infarction trial. Circulation 73:338–346
169. Willis AL, Smith DL (1989) Therapeutic impact of eicosanoids in atherosclerotic disease. Eicosanoids 2:69–99
170. Wilson RF, Holida MD, White CE (1986) Quantitative angiographic morphology of coronary stenosis leading to myocardial infarction or unstable angina. Circulation 73:286–293
171. Yui Y, Hattori R, Takatsu Y, Kawai C (1986) Selective thromboxane A_2 synthetase inhibition in vasospastic angina pectoris. J Am Coll Cardiol 7:25–29
172. Yusuf S, Peto R, Lewis J, Collins R, Sleight P (1985) β-blockade during and after myocardial infarction: an overview of the randomized trials. Progr Cardiovasc Dis 27:335–371

Authors' address:
Prof. Dr. K. Schrör
Institut für Pharmakologie
Heinrich-Heine-Universität
Moorenstraße 5
4000 Düsseldorf, FRG

Regional Myocardial Flow-Function Relationship in Ischemia

Kim P. Gallagher

Thoracic Surgery Research Laboratory, Departments of Surgery and Physiology, University of Michigan Medical School, Ann Arbor, Michigan, USA

The main objective of this review is to summarize recent findings on the functional consequences of acute regional ischemia. In addition to summarizing what we know about regional function during ischemia, a second objective of this paper is to emphasize what we do not know and to suggest potential avenues of additional investigation that may be worth pursuing.

Regional myocardial function has been measured experimentally with a variety of techniques when coronary inflow was restricted or eliminated by total occlusion. Tennant and Wiggers [80] began it all by using an optical myograph to measure epicardial motion in open-chest dogs anesthetized with morphine and sodium barbital. In their classic experiment, occlusion of the left anterior descending artery produced rapid changes in the myographic recordings. Within 60 s after coronary occlusion, shortening was eliminated during the ejection phase and replaced by lengthening during the isovolumic contraction phase. Thus, the basic features of regional waveforms during ischemia were identified and discussed by Tennant and Wiggers over 50 years ago. The rest of us have been reproducing their findings with one technique or another ever since. Accurately measuring regional function, however, remains an important issue since many clinical studies utilize regional function as a critical "end-point" parameter and many of the questions posed by Tennant and Wigger's findings remain to be answered.

We have concentrated on sonomicrometric measurements of regional myocardial dimensions. Ultrasonic crystals have been arrayed to measure myocardial segment lengths or wall thickness since 1969 in Europe [7] and the United States since 1973 [12, 81]. To evaluate regional flow-function relations, we have mainly used wall thickness measured with sonomicrometers, which was first described by Bugge-Asperheim et al. [7] but popularized by Sasayama et al. [71] some years later. Wall thickening provides an integrated measurement of mechanical function across all layers of the myocardium and it avoids a potential limitation of myocardial segment length measurements which may be influenced by their alignment relative to local fiber orientation [13, 23, 29]. Wall thickening also has the advantage of being measureable with many clinical techniques such as echocardiography, digital ventriculography, cine-computed tomography, or magnetic resonance imaging, facilitating extrapolation of experimental results to the clinical setting.

This chapter is organized around three specific issues that come under the purview of flow-function relations: 1) Transmural function within an ischemic area, 2) subepicardial function within an ischemic area, and 3) transmural function at the lateral margins of ischemic myocardium.

Transmural Function within an Ischemic Area

The first general issue we addressed was the effect of coronary inflow reduction on regional function within an ischemic area. The circumflex artery was equipped with an occluder, enabling production of different levels of flow restriction in the posterior wall. Sonomicrometers were used to measure wall thickness [7, 71]. They were arranged with one crystal of the pair inserted tangentially through the myocardium to the subendocardium. The second crystal, attached to a dacron patch, was sewn to the epicardial surface over the position of the subendocardial crystal. Arrayed in this manner, the sonomicrometers enabled continuous measurements of wall thickness from which estimates of thickening or percentage thickening could be derived to use as measures of systolic function. This approach has been used in both anesthetized, open-chest and conscious, chronically instrumented animals to study how regional function is influenced by restriction of coronary inflow.

With this kind of preparation, analog tracings like those shown in Fig. 1 were obtained. The recordings are from an anesthetized, open-chest dog in which five levels of flow restricting coronary stenosis were produced. The usual pattern of wall thickening during control conditions is shown in the upper left panel and different degrees of regional dysfunction are evident in panels 1−5 (labeled ST1−ST5). The waveforms in each panel represent beat averaged data, i.e., digitized averages of 20

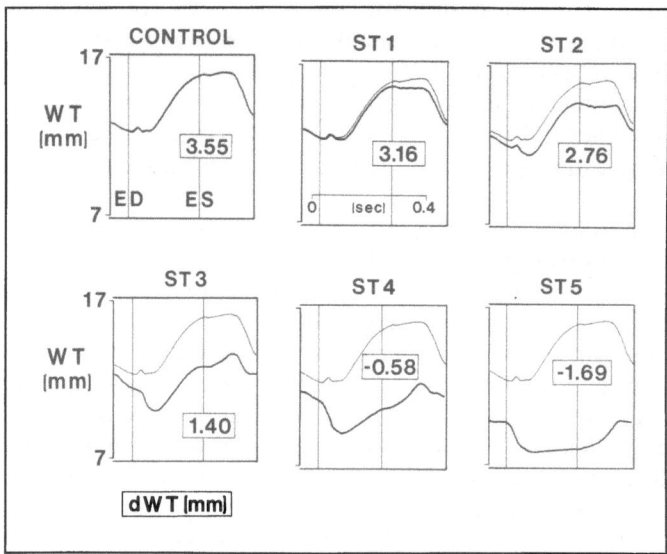

Fig. 1. Examples of beat averaged waveforms from an experiment performed in an anesthetized, open-chest dog. The CONTROL waveform (upper left) demonstrates the normal pattern of thickening that is observed from end-diastole (ED) to end-systole (ES) without coronary flow restriction. Superimposed on each panel is wall thickening excursion (dWT, in mm). In this experiment, five different levels of coronary inflow restriction or stenosis (ST) were produced that resulted in mild to quite severe alterations in regional contractile function. Microspheres were used to measure blood flow in the experiment, enabling us to correlate changes in thickening with changes in myocardial perfusion.

cardiac cycles during steady-state conditions. The numbers superimposed on each panel correspond to the amount of wall thickening excursion (in mm) measured at each condition and the control wall thickness waveform is included in each panel to emphasize the degree of change with progressively more severe stenoses.

Myocardial blood flow was measured with tracer labeled microspheres [44] when steady conditions were established at each level of coronary narrowing. Blood flow and functional data were correlated to determine regional flow-function relations such as those shown in Fig. 2. Data from four individual experiments performed in open-chest, anesthetized dogs are shown to emphasize both the linearity and reproduceability of the relationship between changes in mean transmural blood flow and transmural systolic wall thickening. Data from simultaneously measured subendo-cardial segment lengths are also shown in each panel. With one exception, the

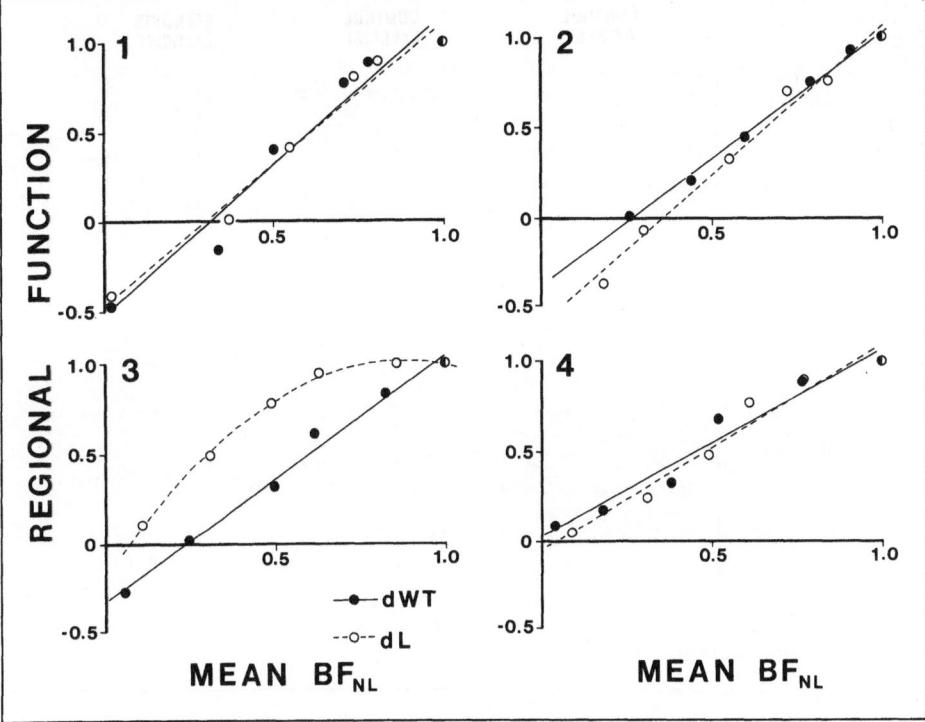

Fig. 2. Four examples of transmural blood flow-transmural wall thickening relationships derived from studies in anesthetized, open-chest dogs. Wall thickening (dWT) and subendocardial segment length shortening (dL) were measured in each experiment with sonomicrometers during different levels of coronary inflow restriction. At each level, microspheres were injected to measure regional blood flow, enabling construction of flow-function graphs like these. Mean transmural blood flow (BF_{NL}) was normalized by expressing it as a decimal fraction of control zone perfusion; regional function is presented as a decimal fraction of baseline values. Transmural wall thickening correlated closely and in linear fashion with reductions in mean transmural blood flow. Subendocardial segment shortening displayed a similar strong dependence on transmural blood flow, with the exception of the experiment shown in panel 3.

relationship between transmural perfusion and subendocardial segment shortening was very similar to the wall thickening relationship. Similar linear relationships between mean transmural blood flow and regional function (measured as wall thickening or segment shortening with sonomicrometry) have been documented in several reports involving anesthetized [15, 22, 78] or conscious animals [8, 21, 72, 82], emphasizing the close coupling that exists between blood supply and regional contractile function.

It has also been possible to determine the effects of different hemodynamic and behavioral conditions on regional flow-function relations. In Fig. 3 are shown analog tracings from an experiment in which a chronically instrumented dog ran on a treadmill while coronary inflow was restricted by adjustment of a hydraulic occluder around the circumflex artery. The degree of coronary narrowing corresponded to a

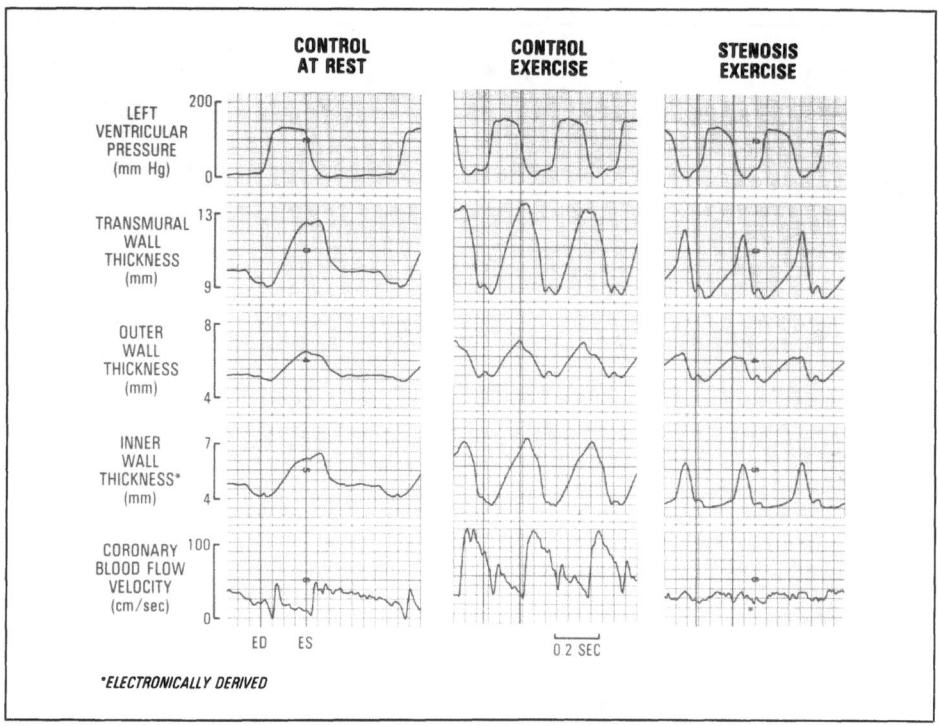

Fig. 3. Examples of analog tracing from an experiment involving a chronically instrumented dog trained to run on a treadmill. Normal patterns of wall thickness are evident during control conditions at rest (left panel). Augmented wall thickening (measured with sonomicrometers) during control exercise (middle panel) was associated with elevated heart rate, left ventricular pressure, and coronary blood flow velocity. During exercise with coronary stenosis (right panel), coronary blood flow velocity was restricted to the resting level, approximating a critical stenosis. Transmural thickening was reduced by approximately 75%. Inner wall thickening (derived by substracting outer from transmural thickness) was completely eliminated, whereas outer wall thickening (measured directly) was only modestly reduced, a pattern which parallels the change in transmural perfusion produced by exercise with a critical level of coronary stenosis. Abbreviations: ED, end-diastole; ES, end-systole.

114

critical stenosis since blood flow was limited to the resting level. Since this amount of blood flow was inadequate to meet the elevated demands of exercise, regional dysfunction was produced.

In Fig. 4 is shown the relationship between mean transmural blood flow (in ml/min/g) and transmural systolic thickening from an entire group of experiments [20]. Note that systolic wall thickening correlates well and in linear fashion with mean transmural blood flow during exercise, just as it did during resting conditions. The relationship between flow and function is shifted to the right compared to the relationship at rest, but it converges to a similar point, when flow restriction is severe and wall motion is akinetic.

Based on findings like these, we think that transmural flow-transmural function relations operate over a family of linear curves, as shown in Fig. 5. When myocardial work increases during exercise, the relationship between flow and function is shifted to the right, an observation supported by experimental evidence [20]. We propose that the relationship shifts leftward and its slope steepens when cardiac work and myocardial oxygen demand decrease from resting levels, such as may occur during administration of negative inotropic agents (calcium antagonists, beta blockers, volatile anesthetics, etc.). We have obtained indirect evidence supporting this view with the volatile anesthetic isoflurane [79] but no direct experimental demonstration of this possibility has been made, as far as I know, which is a little surprising given the widespread use of such agents. If verified experimentally, however, it would support the

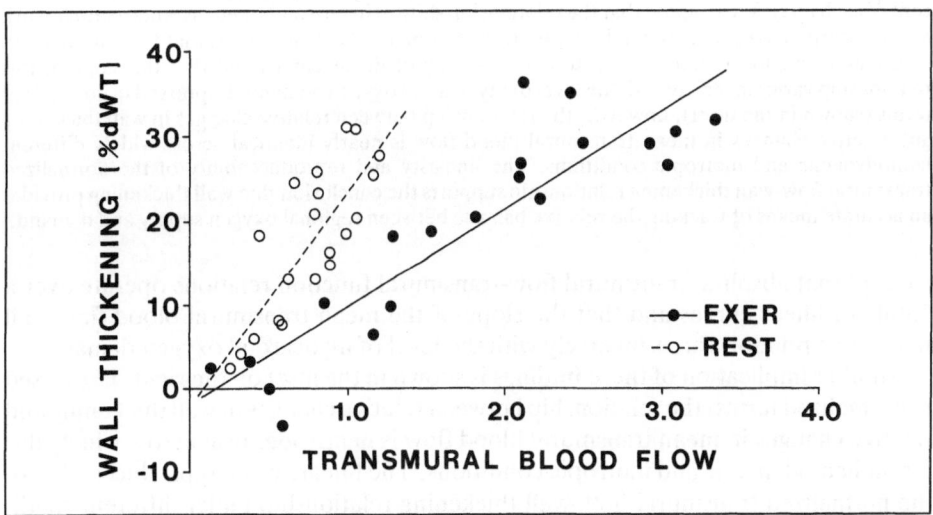

Fig. 4. Relationship between percentage wall thickening and mean transmural blood flow (ml/min/g) in chronically instrumented dogs. Data are shown at different levels of inflow restricting coronary stenosis at rest (open circles) and during treadmill exercise (solid circles) in the same dogs. Transmural wall thickening and blood flow varied linearly at rest (y = 25.3 × − 2.1, r = 0.80) and during exercise (y = 11.6 × − 1.9, r = 0.90). Exercise shifted the relationship between absolute flow and wall thickening to the right in a nonparallel manner (the slopes were significantly different but the intercepts were not). (Figure redrawn from [20] with the permission of the American Heart Association.)

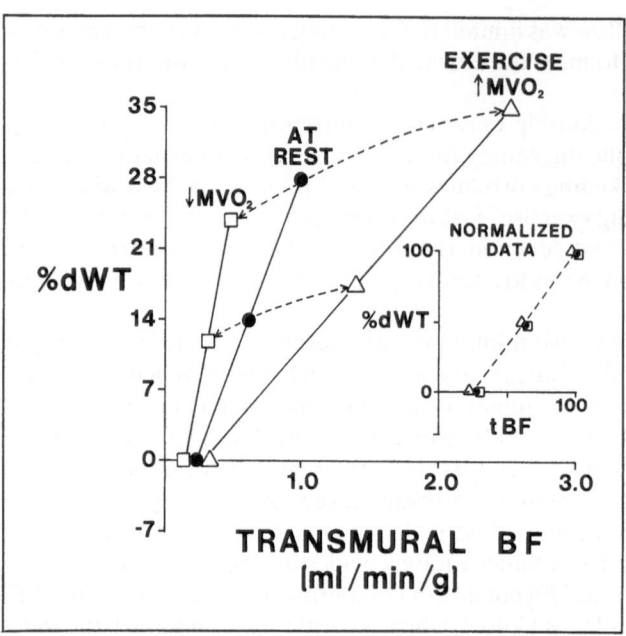

Fig. 5. Relationship between mean transmural blood flow and transmural wall thickening. When myocardial work increases during exercise, the relationship between flow and function is shifted to the right compared to the relation at rest but it converges to a similar point as the resting relationship, when flow restriction is severe and wall motion akinetic (roughly approximating the x-intercept of each line). We propose that the relationship shifts leftward and steepens when cardiac work and myocardial oxygen demand decreases from resting levels. This implies that transmural flow-transmural function relations operate over a family of linear curves and that the slope of this relationship varies inversely with the level of myocardial oxygen demand. Expressed in normalized terms (shown in the inset), however, the relationship between relative changes in wall thickening and relative changes in mean transmural blood flow is nearly identical across widely different hemodynamic and inotropic conditions. The linearity and reproduceability of the normalized transmural flow-wall thickening relationship supports the conclusion that wall thickening provides an accurate means of tracking the relative balance between regional oxygen supply and demand.

concept that absolute transmural flow-transmural function relations operate over a family of linear curves and that the slope of the mean transmural blood flow-wall thickening relation varies inversely with the level of myocardial oxygen demand.

Another implication of these findings is shown in the inset on the right. Expressed in normalized terms, the relationship between relative changes in wall thickening and relative changes in mean transmural blood flow is nearly identical across widely different hemodynamic and inotropic conditions. The linearity and reproduceability of the normalized transmural flow-wall thickening relationship under different conditions has been shown at rest [8, 15, 21, 22, 58, 72], during exercise [19, 20, 35, 43, 55, 59, 60], and with isoproterenol infusion [18], supporting the conclusion that wall thickening provides an accurate means of tracking the balance between oxygen supply and demand.

A particularly striking feature of regional flow-wall thickening relations is the observation that subendocardial flow restriction appears to dominate transmural function [8, 15, 21, 50, 67]. In Fig. 6 is shown the relationship between normalized

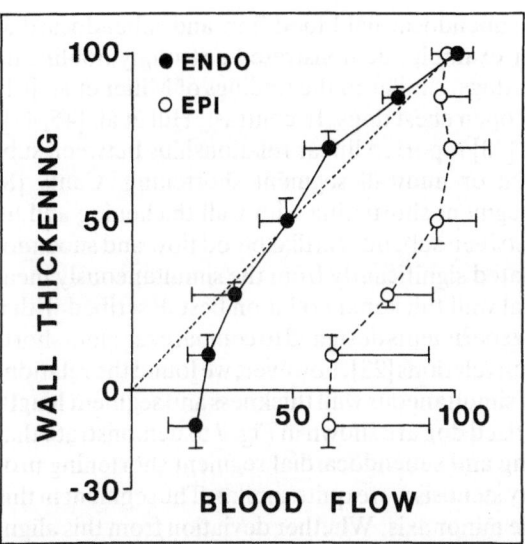

Fig. 6. Average data in six categories of regional dysfunction due to coronary inflow restriction in conscious, chronically instrumented dogs. Subendocardial (ENDO) and subepicardial (EPI) blood flow were normalized to represent percentages of blood flow in nonischemic, normally perfused myocardium. The blood flow measurements were made with tracer labeled microspheres. Wall thickening is presented as a percentage of baseline values. The dashed line is the line of identity. Note that changes in wall thickening follow changes in subendocardial perfusion quite closely. Wall thickening correlates poorly, however, with changes in subepicardial blood flow. Comparable results have been obtained in anesthetized, open-chest dogs and conscious, chronically instrumented dogs in other studies [8, 15, 20, 67], lending support to the conclusion that decrements in subendocardial perfusion dominate transmural contractile function. (Data from Gallagher et al., 1984 [21].)

wall thickening (expressed as a percentage of control values) and subendocardial or subepicardial blood flow (expressed as a percentage of control zone blood flow) derived from studies performed in conscious, chronically instrumented dogs [21]. Notice that a close, direct relation exists between subendocardial flow and transmural wall thickening. Subepicardial blood flow, however, relates poorly to changes in thickening.

Since autoregulated vasodilator reserve is exhausted in the subendocardium before the subepicardium [16, 36, 48], it is possible for subepicardial flow to be normal when subendocardial flow is markedly reduced. In these circumstances, transmural function may be reduced 50% or more because wall thickening correlates so closely with subendocardial perfusion but so poorly with subepicardial blood flow (Fig. 6). Comparable findings have been reported by Canty [8] in which wall thickening was measured with sonomicrometers in conscious dogs. We have generally found that the relationship between normalized subendocardial blood flow and wall thickening is best described with a quadratic expression but the improvement over a linear fit is small [21] lending support for the proposal by Canty (personal communication) that transmural thickening can be used effectively as a "subendocardial flowmeter".

It should also be noted that the linear or mildly nonlinear relationship between subendocardial blood flow and transmural wall thickening appears to differ from some

117

reports on the relationship between subendocardial blood flow and subendocardial segment shortening. Vatner [83], for example, demonstrated a strongly nonlinear, exponential relationship in conscious dogs, similar to the findings of Miller et al. [61] and Weintraub et al. [85] obtained in open chest dogs. In contrast, Hill et al. [46, 47], Tomoike et al. [82], and Gross et al. [33] reported linear relationships between subendocardial flow and subendocardial or midwall segment shortening. Canty [8] recently compared subendocardial segment shortening with wall thickening and he also observed a linear relationship between subendocardial blood flow and subendocardial segment shortening that deviated significantly from the simultaneously measured subendocardial flow-transmural wall thickening relation best described with a quadratic expression. In preliminary experiments designed to compare segment shortening and wall thickening flow-function relations [28], however, we found the relationships to be fairly similar. Examples of simultaneous wall thickness and segment length tracings from an open-chest, anesthetized dog are shown in Fig. 7 to demonstrate that the relative changes in wall thickening and subendocardial segment shortening produced by different degrees of coronary stenosis were quite similar. The segment in this experiment was aligned parallel to the minor axis. Whether deviation from this alignment would alter the subendocardial segment flow-function relationship, and thereby explain some of the discrepancies in the literature, remains to be determined.

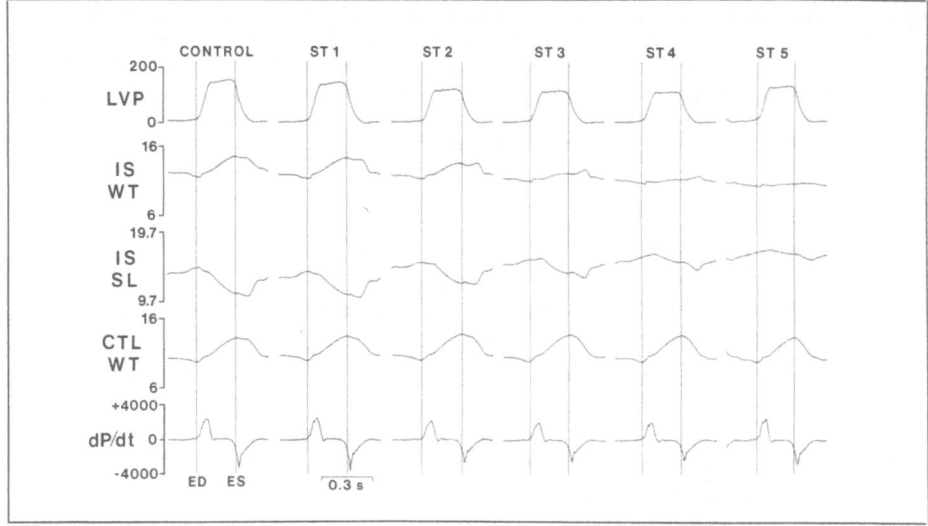

Fig. 7. Examples of analog tracings from an experiment in an anesthetized, open-chest dog in which different levels of coronary inflow restriction were produced. Left ventricular pressure (LVP) and dP/dt are shown in the upper and lower rows, respectively. Sonomicrometers were used to measure a wall thickness (WT) and a subendocardial segment length (SL) in the ischemic (IS) area supplied by the circumflex artery. Wall thickness was also measured in a control (CTL) area supplied by the left anterior descending artery. Baseline or control recordings are shown in the leftmost panel. The remaining panels (labeled ST1–ST5) demonstrate the effects of coronary stenosis. The main point of the figure is the comparison of ischemic wall thickening and subendocardial segment shortening. The changes parallel each other closely, suggesting that transmural thickening and deep segment shortening provide roughly equivalent information on the functional consequences of acute ischemia.

118

My impression is that most investigators (including the ones cited above) think that subendocardial segments and transmural wall thickness generally provide fairly equivalent information during moderate or severe ischemia (see Fig. 7). Most of the disagreement stems from the effects of mild coronary stenosis. These data points, the ones most difficult to produce experimentally, largely dictate whether a linear or nonlinear relationship is needed to describe the flow-function relation. A few years ago, we reported that we could detect a 15% reduction in thickening at rest, associated with a 20% reduction in subendocardial blood flow [21]. On the average, that remains about the best we can do to this day.

Potential differences between wall thickening and subendocardial segment shortening during reperfusion have not been explored at all, as far as I know. A superficial review of the "stunning" literature suggests to me that subendocardial segments seem to recover faster and more completely than wall thickening after a standard 15 min occlusion. It may be worth looking into this issue more closely, since it seems paradoxical that segments contained in the most severely ischemic portion of the wall should recover more rapidly than thickening measured across the entire wall. In addition, the functional characteristics of reperfused myocardium are important factors in many clinical trials, as well as experimental studies. Consequently, a thorough evaluation of regional flow-function relations in reperfused myocardium may be worthwhile. To the best of my knowledge, only Matsuzaki et al. [58] have evaluated the functional effects of prolonged episodes of *nontransmural* ischemia followed by reperfusion. Since such conditions may well exist in some human patients, this would also seem to be a good area for further investigation.

Thus far, only Bolli et al. [3] have evaluated the relationship between the intensity of ischemia and rate of functional recovery (measured as wall thickening) following release of a 15 min occlusion. Since they observed quite good correlations between ischemic blood flow and degree of recovery (at different times after reperfusion), their data suggest strongly that ischemic intensity largely determines the level of myocardial stunning. Heusch et al. [42] evaluated the relationship between blood flow during exercise with flow-restricting coronary stenosis and recovery of wall thickening after exercise. Like Bolli et al. [3], Heusch et al. [42] observed that post-ischemic dysfunction was primarily determined by the intensity of ischemia during exercise. This conclusion deviates somewhat from the widely held view that oxygen radicals, neutrophils, and/or other reperfusion related phenomena cause stunning. That these factors play a role in some conditions can not be discounted, but our current view is that ischemic intensity (time x flow deficit) determines how stunned the myocardium will be after brief periods of myocardial ischemia.

Based on the findings discussed above, five generalizations are posed in Table 1 regarding regional flow-function relations and in Table 2 are listed several questions that remain to be answered.

Subepicardial Function within the Ischemic Area

A second issue is the effect of subendocardial ischemia on contraction in overlying subepicardial muscle that is normally perfused.

119

Table 1. Generalizations on transmural flow-function relations within ischemic myocardium

Generalization 1:
Absolute or relative restriction of coronary inflow leads promptly to proportionate reductions in regional myocardial function.
Generalization 2:
The relationship between average transmural blood flow and wall thickening remains linear in different hemodynamic conditions: At rest, during exercise, and under anesthesia.
Generalization 3:
Wall thickening provides an accurate means of measuring the balance between oxygen supply and demand on a regional basis.
Generalization 4:
Transmural function is dominated by changes in subendocardial blood flow.
Generalization 5:
Changes in wall thickening can be effectively used as a "subendocardial flowmeter" that registers the relative adequacy of subendocardial oxygen delivery.

Table 2. Questions that remain to be answered regarding transmural flow-function relations within ischemic myocardium

1. What mechanism links flow and mechanical function so closely? In what way is this mechanism deranged in "stunned" myocardium?
2. Does the myocardium "down-regulate" or "hibernate" when coronary inflow is restricted? If so, is this a protective mechanism?
3. How sensitive are clinically useful techniques (such as echocardiography or cine-computed tomography) to ischemia? Do the same generalizations derived from experimental studies apply to the clinical situation?
4. Can the same generalizations about regional flow-function relations (derived from acute experimental studies in normal hearts) be applied to hypertrophied or failing myocardium?

As noted earlier, partial occlusion can result in subendocardial ischemia, whereas the subepicardium may remain normally perfused. In the late 1970s, abstracts were published [14, 31] reporting that epicardial segment shortening was reduced in these circumstances, but Weintraub et al. [85] were the first to report in a full-length paper that subepicardial segment shortening could decrease when subepicardial perfusion was normal. Combined with the observation that subepicardial flow correlated poorly with transmural function (measured as wall thickening), Weintraub's findings led to use of the term "transmural tethering" to describe the cause of nonischemic dysfunction in the subepicardium. The implication of transmural tethering is the concept that altered motion due to ischemia in the deeper layers somehow constrains motion in the overlying, normally perfused layers.

At about the same time the paper by Weintraub et al. [85] was published, we were working on the same issue in John Ross's Seaweed Canyon Laboratory (University of California, San Diego). After reporting preliminary results that supported the concept of transmural tethering [14], we realized that we had overlooked the issue of fiber orientation. We had originally aligned the epicardial sonomicrometers parallel to the minor axis, roughly parallel with the subendocardial sonomicrometers. In so doing, we had aligned the epicardial segments about 50° off the surface fiber direction in the territory perfused by the circumflex artery. When we repeated our studies with two pairs of epicardial crystals (one parallel with the fibers, the other deviated 50° from the surface fiber direction), it was observed that the parallel epicardial segment continued normal shortening until subepicardial ischemia was produced by total coronary occlusion. In contrast, the off-axis segment displayed dysfunction when ischemia was restricted to the subendocardium and subepicardial perfusion was normal [17]. These findings "muddied the waters" quite a bit. Something was going on in the subepicardium when the underlying layers were ischemic and dysfunctional, but it seemed to depend on how one measured epicardial function. Follow-up studies also demonstrated regional differences in the epicardial response to subendocardial ischemia [22, 39].

The bewildering array of data produced in studies on epicardial segment shortening is demonstrated in Fig. 8. In open-chest anesthetized dogs, sonomicrometers were arrayed to measure epicardial segment lengths in territory supplied by the circumflex or the left anterior descending artery. The ultrasonic crystals were arranged to measure three segments simultaneously, each of which was oriented differently relative to surface fiber orientation. Progressive coronary stenosis was produced with

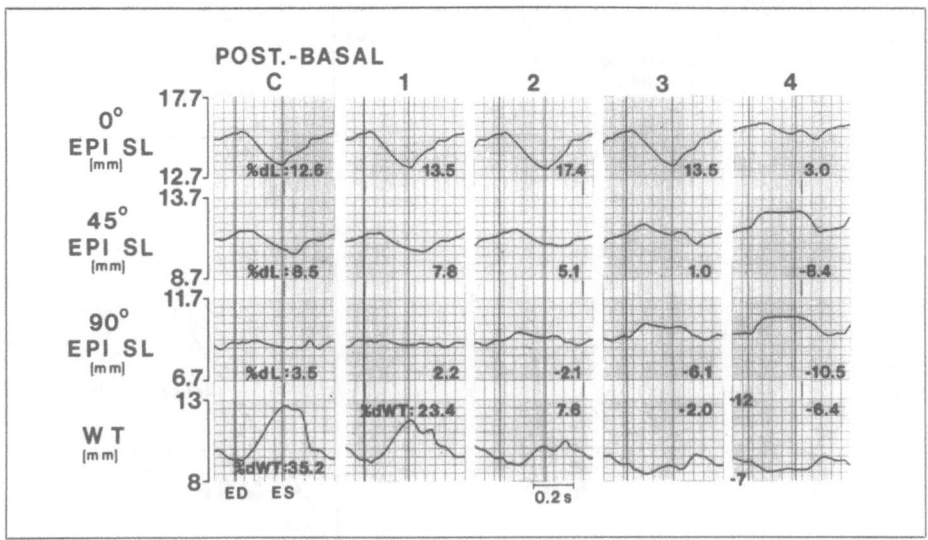

Fig. 8. Example of analog tracings from one experiment in which epicardial segment length and wall thickness sonomicrometers were implanted in the basal half of the left ventricle supplied by the left circumflex artery. Epicardial segment lengths (EPI SL) are shown in the upper three rows with average percentage shortening (%dL) superimposed on the recording. Wall thickness (WT) is presented in the bottom panel with percentage thickening (%dWT) values superimposed. The solid vertical lines represent end-diastole (ED) and end-systole (ES). Control (C) or baseline recordings are shown in the first panel. Panels 1, 2, 3 and 4 demonstrate the effects of four levels of coronary inflow restriction. Notice that the 0° epicardial segment (aligned parallel to surface fiber orientation) was characterized by normal shortening until wall thickening was eliminated, indicating that parallel shortening remains normal as long as subepicardial perfusion has not been reduced by coronary narrowing. The 45° (oblique) and 90° (perpendicular) segments, however, displayed marked decrements in shortening at levels of stenosis consistent with normal subepicardial perfusion.

121

a screw-type occluder while monitoring changes in transmural wall thickening which was used as an on-line index of transmural flow distribution and a standard of integrated transmural performance. In Fig. 8 are shown examples of analog tracings from an experiment focused on the myocardium perfused by the circumflex artery. Notice that baseline epicardial segment shortening (left panel, labeled C) depends strongly on alignment relative to surface fiber orientation. The key point of the figure, however, is the demonstration that the epicardial segmental response to ischemia varies enormously. To complicate matters further, a quite different pattern of baseline epicardial shortening and response to nontransmural ischemia exists in the anterior-apical are perfused by the left anterior descending artery [22, 29].

In an attempt to minimize the fiber orientation problems, subepicardial wall thickening instead of segment shortening has been measured, but this has also provided inconclusive results. Plotted in Fig. 9 are four examples of preliminary studies per-

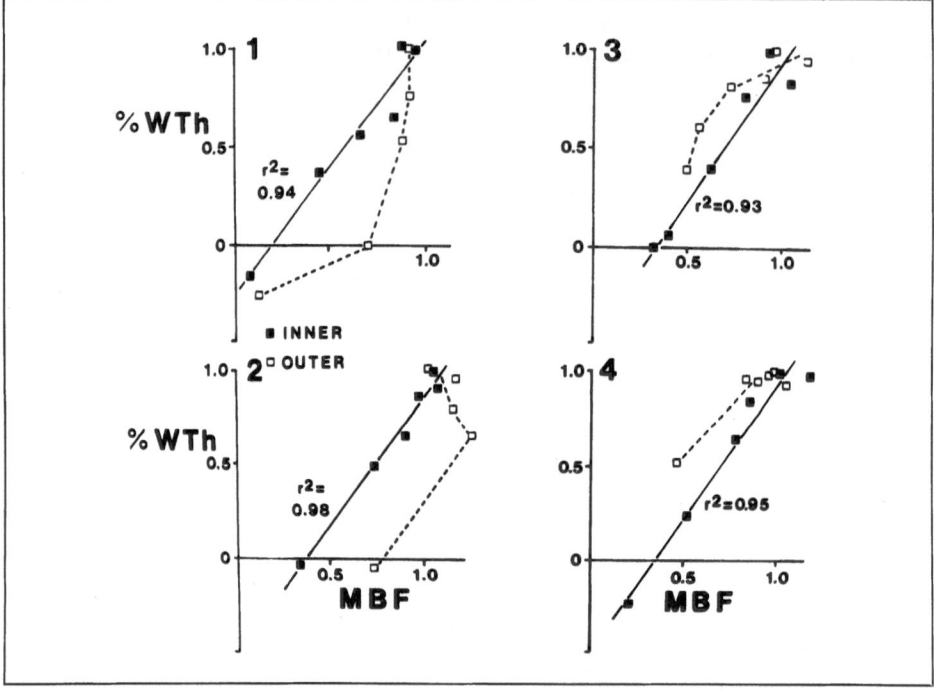

Fig. 9. Flow-function plots from four experiments in conscious, chronically instrumented dogs with different degrees of coronary stenosis used to produce inflow restriction. Wall thickening (%WTh) is plotted on the y-axis of each graph as a decimal fraction of baseline values; blood flow was normalized as a decimal fraction of control area perfusion. Inner wall thickening is plotted vs inner wall blood flow (solid squares, solid lines) and outer wall thickening is plotted vs outer wall blood flow (open squares, dashed lines). The outer wall thickness measurement was made directly and the inner wall thickness measurement was derived by subtracting the outer from a transmural wall thickness measurement. The inner wall relationships were well described by linear expressions with r^2 values of 0.93 or greater, indicating very close coupling between inner wall flow and fucntion. The outer wall relationships could not be described with sample linear equations and they varied considerably across experiments, making it difficult to reach a solid conclusion regarding transmural tethering and outer wall function.

Table 3. Questions that remain to be answered about subepicardial function during nontransmural ischemia

---

1. What is the best way to measure subepicardial mechanical function?
2. Is subepicardial function significantly influenced by "tethering"?
3. Is subepicardial function important during acute nontransmural ischemia? Does its importance increase with time in the setting of subendocardial infarction?

---

formed at the Seaweed Canyon Laboratory (University of California, San Diego) several years ago, demonstrating the relationships between inner wall blood flow and inner wall thickening and between outer wall blood flow and outer wall thickening. Inner and outer wall thickness measurements were made with sonomicrometers. Transmural wall thickness and outer wall thickness were measured directly and inner wall thickness was derived indirectly by subtracting the outer from the transmural measurement [23, 69].

The two panels on the left show that outer wall thickening decreased markedly, even when outer wall blood flow was near normal levels. In contrast, the panels on the right suggest that outer wall blood flow and thickening are more closely related. A similar conclusion is suggested by the analog tracings in Fig. 3 which illustrate the effects of exercise in the presence of critical coronary stenosis. Inner wall thickening was completely eliminated but outer wall thickening was altered minimally compared to exercise without coronary stenosis. In the study published recently by Homans et al. [50], however, it was reported that outer wall thickening declined in a manner greatly out of proportion to the outer wall flow reduction during treadmill exercise in dogs with coronary stenosis. Given contradictory evidence like this, it is difficult to come to any firm conclusions regarding outer wall function during nontransmural ischemia.

The relative role of transmural tethering (or restriction of nonischemic epicardial motion) during acute ischemia remains in dispute. Does transmural tethering of some sort occur during nontransmural ischemia? It is probably safe to answer yes, but the mechanism of tethering remains to be established. There is general agreement that the outer wall contributes a relatively small fraction to total thickening [23, 64, 69], so it could even be argued that outer wall function and how it is influenced by ischemia is not particularly important. Consequently, in Table 3 are listed only questions because for the time being, I do not think we can make any definitive generalizations about outer wall function during subendocardial ischemia.

Transmural Function at the Lateral Border of Ischemic Myocardium

The third issue concerns myocardial dysfunction at the lateral boundary of an ischemic or infarcted area. Based on observations by Kerber et al. [53, 54] in the mid-1970s that wall thickening was impaired in normally perfused myocardium adjacent to an ischemic zone, we and others [10, 32, 34, 49, 56, 66, 70] undertook studies focussed on determining the distribution of functional impairment across the perfusion boundary produced by coronary occlusion. Our starting point was the observation that the perfusion boundary between ischemic and nonischemic myocardium is sharply delineated in terms of blood flow [38, 40, 41, 63, 73, 74, 87].

To determine how far mechanical dysfunction extended into nonischemic territory we initially measured subendocardial segment shortening at or across the perfusion boundary. Three crystals were arrayed in the subendocardium in an attempt to measure two segment lengths in series, one on either side of the perfusion boundary. The idea was to use the center crystal as a transmitter for the two outside crystals (both acting as receivers). Although we were successful in measuring two segments simultaneously, positioning the center crystal precisely at the perfusion boundary was a hit-or-miss proposition. Nonetheless, we were startled to observe marked differences in shortening between the two segments, an example of which is shown in Fig. 10. This finding was reproduceable [26], suggesting that our preconceived notion of a wide zone of nonischemic dysfunction was off the mark. Furthermore, it suggested that segments tended to integrate or average function across the myocardium they subtended, a conclusion supported by the reports of Matre et al. [57] and, more recently, Nagata and Lavallee [65]. Consequently, to measure the distribution of functional impairment in a relatively narrow zone, we needed to use a different parameter of regional function. It will come as no surprise that wall thickening was our first choice as the alternative parameter.

The experimental approach we took to this problem is shown schematically in Fig. 11. Sonomicrometers arranged to measure wall thickness were placed in the myocardium on both sides of the interface between ischemic and nonischemic myocardium. As shown in Fig. 11, our initial experiments involved circumflex occlusion and we generally tried to get four to five sets of sonomicrometers into each heart: at least two in the ischemic area and two or three in the nonischemic area. The location of the perfusion boundary was determined from blood flow maps derived by using tracer-labeled microspheres and measuring blood flow in several very small tissue sections. The sections were about 3 mm wide at the endocardial surface, similar to what Schaper [73] and Murdock et al. [63] had reported using a few years earlier.

The hearts were sectioned after allowing them to fix in formalin. Since we left the sonomicrometers in place during fixation, it was relatively easy to determine their location relative to the perfusion boundary because we knew which tissue samples contained them. The location of the perfusion boundary, admittedly, was a somewhat arbitrary designation. The true perfusion boundary courses irregularly across the myocardial wall, so any position we assigned to it had an uncertainty of ± a few mm associated with it. Therefore, the position of the "perfusion boundary" served mainly as a reference point for the sonomicrometers rather than a precise identification of the ischemic-nonischemic interface.

Fig. 11 also demonstrates how we superimposed the functional results on the blood flow distribution data. The position of each bar conveys information on the location of the wall thickening measurements and the height of each bar corresponds to the relative change in thickening produced by coronary occlusion. The graph provides a relatively straightforward means of presenting how regional function is influenced in the vicinity of the perfusion boundary. The graph also demonstrates one of the problems with this approach, which is that we are restricted to "sampling" function in relatively few places. The sonomicrometers are invasive so it is not practical to put in large numbers of them. In addition, implanting sonomicrometers is a time-consum-

124

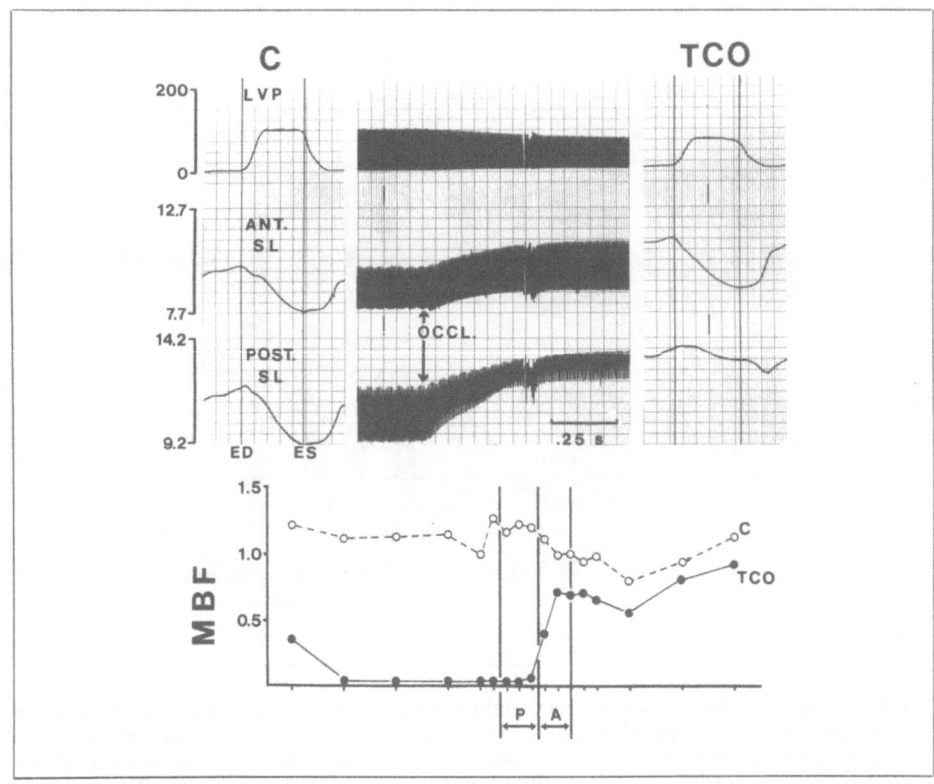

Fig. 10. Example of analog recordings and blood flow map from an anesthetized, open-chest dog in which subendocardial segment length (SL) shortening was measured with sonomicrometers on both sides of the perfusion boundary produced by total coronary occlusion (TCO) of the circumflex artery. A full circumference blood flow map, derived from microsphere measurements of blood flow (MBF) in the left ventricular cross-section containing the sonomicrometers, is shown in the lower graph. The location of the anterior (A) or nonischemic and posterior (P) or ischemic segment lengths are superimposed on the blood flow map. The salient feature of the figure is the striking disparity in shortening between the anterior (ANT.SL) and posterior (POST.SL) segment lengths after coronary occlusion. Recall that these segments shared a common ultrasonic crystal which was located at the perfusion boundary between ischemic and nonischemic myocardium (as shown in the blood flow graph). Findings such as these suggested that the functional border zone must be relatively narrow. They also suggested that segment lengths spanned too much myocardium (usually a cm or more) to effectively delineate the lateral extent of nonischemic dysfunction.

ing and exasperating process that tends to wear investigators down, placing another limit on the number of sonomicrometers that can be implanted. Given the sampling problem, we attempted to model the distribution of functional impairment by fitting curves to the wall thickening data. If a curve could be found that corresponded closely to the data, we would be in a position to extract more information from the discrete sonomicrometer measurements.

The appearance of the data suggested to us that a sigmoid curve might be appropriate and, as shown in Fig. 12, this proved to be the case (at least for the effects of circumflex occlusion). The sigmoid fits assumed that nonischemic tissue had a wall

125

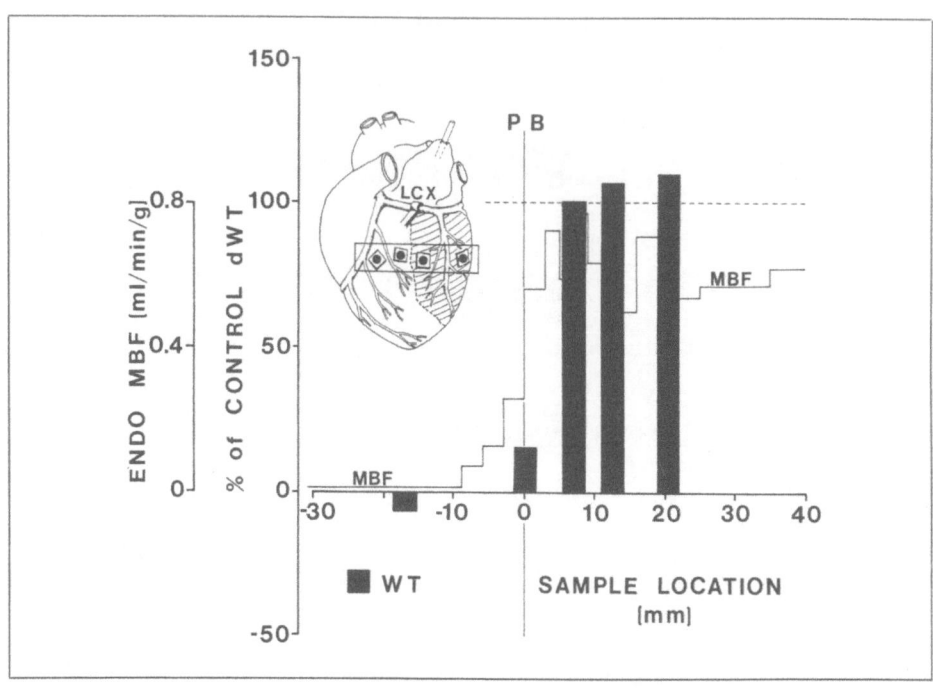

Fig. 11. Example of blood flow map and wall thickening data from an experiment performed in an anesthetized, open-chest dog subjected to circumflex coronary occlusion. Shown in the inset is the arrangement of sonomicrometers (squared solid circles) used in this study. They were situated on both sides of the perfusion boundary (PB) produced by circumflex occlusion. The circumflex supplied territory is cross-hatched. Subendocardial (ENDO) blood flow (MBF) and wall thickening (dWT) are plotted on the y-axis vs tissue sample location on the x-axis. The solid line (labeled MBF) represents blood flow in the large number of small tissue samples used to delineate the PB. The solid bars correspond to sonomicrometer location. Their height represents the percentage of baseline thickening measured after coronary occlusion. Based on findings such as these, the transition area (or functional border zone) between ischemic (dysfunctional) myocardium and non-ischemic (normally contracting) myocardium associated with circumflex occlusion appeared to be relatively narrow.

thickening asymptote, that ischemic tissue had a different wall thickening asymptote, and that wall thickening changed monotonically between the asymptotes. No doubt a sigmoid fit represents an oversimplification (especially the assumption that wall thickening remote from the perfusion boundary is adequately described by a flat line) but the sigmoid fits were generally very good, supporting their utility at least within the limited portion of the left ventricular circumference we were evaluating.

We found that the functional border zone extended approximately 30° (or 7−8 mm) of endocardial circumference from the perfusion boundary produced by circumflex occlusion. Shown in Fig. 12 are data and composite sigmoid fits from both anesthetized, open-chest [24] and conscious, chronically instrumented dogs [27]. The asymptotes differ, but the lateral extent and relative severity of nonischemic dysfunction are quite similar. Regional function in nonischemic muscle immediately adjacent

126

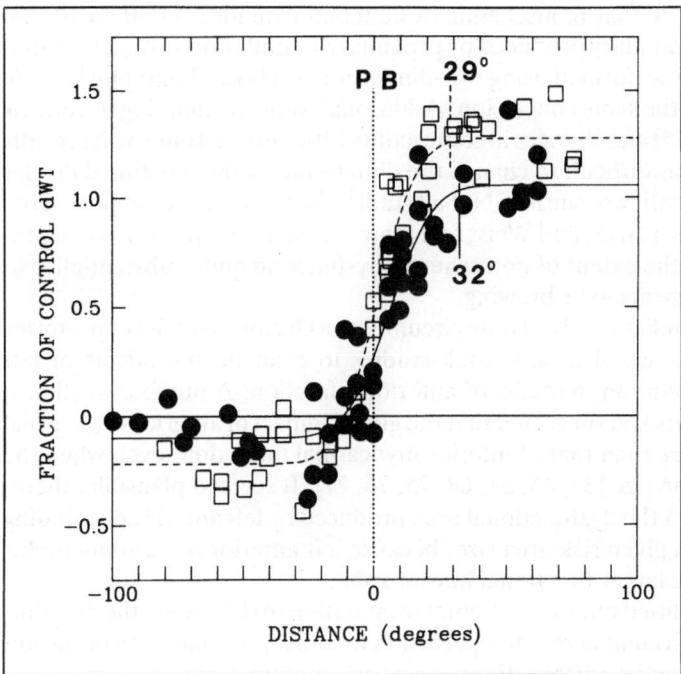

Fig. 12. Composite data sets and composite sigmoid curves from anesthetized, open-chest dogs (open squares, dashed line) and conscious, chronically-instrumented dogs (solid circles, solid line) after acute circumflex coronary occlusion. Wall thickening (dWT) is plotted as a percentage of baseline values versus location of the sonomicrometers (in degrees of circumference). The position of the perfusion boundary (PB) was used as a reference point to assign locations to the sonomicrometers. The sigmoid curve was used to model the distribution of functional impairment across the ischemic-nonischemic interface. The salient points of the figure are to illustrate 1) that a narrow zone of nonischemic dysfunction extends approximately 30° from the ischemic area into the non-ischemic zone, and 2) that the width of the functional border zone produced by circumflex occlusion is similar in open- and closed-chest dogs. (Figure from [27] with permission of the American Heart Association.)

to the perfusion boundary was significantly impaired. Though not as striking as we had originally anticipated, a narrow zone of nonischemic dysfunction surrounded ischemic myocardium.

Before we published our initial studies [24], Lima et al. [56] and Homans et al. [49] reported findings based on use of two-dimensional echocardiography to measure function around full-circumference cross-sections of the left ventricle in open-chest dogs. Lima et al. [56] reported that coronary occlusion resulted in a wide zone of nonischemic dysfunction extending beyond the perfusion boundary. Homans et al. [49], who also used subendocardial segment measurements, found that the zone was somewhat less impressive in terms of lateral extent. Sakai et al. [70] came to a similar conclusion based on sonomicrometric studies in pigs. Frankly, we were delighted that these papers came out when they did and that they appeared to be in conflict. It is much easier to publish on an issue that is perceived as modestly controversial than on an issue that no one cares about. Our results, obtained using sonomicrometers, obvi-

127

ously supported the view that nonischemic dysfunction extended less than 10 mm beyond the perfusion boundary produced by proximal circumflex occlusion. Collaborative studies were also performed using tow-dimensional echocardiography [4, 5, 6, 62] and they supported the same conclusion. Additional studies evaluating the effects of afterload elevation [25] and changes in contractility [9] confirmed our earlier results and suggested that it was difficult to change the dimensions of the functional border zone produced by circumflex occlusion. Not all studies, however, have supported the latter conclusion. Buda et al. [5] and Weiss and Marcus [86] have reported that hemodynamic changes alter the extent of nonischemic dysfunction quite substantially, so another controversy appears to be brewing.

Since our previous studies involved only circumflex occlusion, a model of posterior (or inferior) infarction, we also undertook studies to evaluate the effects of left anterior descending occlusion, a model of anterior infarction. A number of clinical and experimental reports had suggested that the global impact of anterior myocardial infarction is more severe than that of inferior myocardial infarction, even when the risk areas were the same size [37, 45, 51, 68, 75, 76, 77]. It seemed plausible, therefore, to hypothesize that the dysfunctional area produced by left anterior descending occlusion is larger for a given risk area size, because left anterior descending occlusion is associated with a larger functional border zone.

Accordingly, we modified our experimental preparation to determine the distribution of functional impairment across the perfusion boundary produced by occlusion of the left anterior descending artery. Four to six sonomicrometers were arranged in a line roughly perpendicular to the perfusion boundary which we delineated using blood flow maps (Fig. 13), just as we had done previously. This means the sonomicrometers were arrayed across the anterior wall of the left ventricle from apex to base in contrast to the anterior to posterior arrangements used to evaluate the effects of circumflex occlusion (Fig. 11). It became clear after a few experiments that the functional border zone was definitely wider with left anterior descending occlusion.

Fig. 13 is an example of a blood flow map from one of the experiments. The solid bars represent the location and relative change in wall thickening measurements. Notice that the measurements approximately 4 mm and 12 mm (marked with asterisks) to the left (or nonischemic side) of the perfusion boundary are characterized by substantially reduced wall thickening compared with baseline, despite being contained in normally perfused tissue. Compare Figs. 13 and 11. In Fig. 11, nonischemic wall thickening at comparable distances from the perfusion boundary was not different from baseline values. Data like those typified in Figs. 13 and 11 suggested that our hypothesis regarding differences in the size of the functional border zone may have been correct. Left anterior descending occlusion resulted in greater nonischemic dysfunction beyond the lateral border of the ischemic area than did circumflex occlusion. Composite data sets and composite sigmoid curves are presented in Fig. 14 to demonstrate that the differences between circumflex and left anterior descending occlusion were consistent, reproduceable, and significant.

Why is there a difference? Like so many things about regional myocardial function, the explanation remains to be determined. The shape of the ischemic bed, the radius of curvature, wall thickness, and distance from the valvular structures all constitute differences between circumflex and anterior descending territories. That they contribute to disparities in the extent of nonischemic dysfunction seems a reasonable

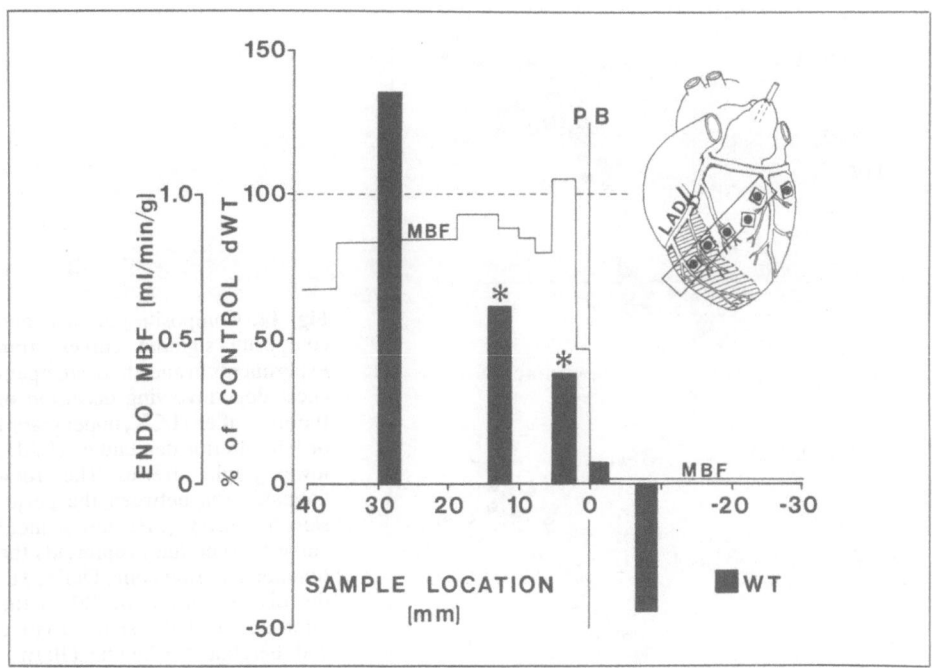

Fig. 13. Example of blood flow map and wall thickening data from an experiment performed in an anesthetized, open-chest dog subjected to occlusion of the left anterior descending (LAD) artery. The arrangement of sonomicrometers across the anterior wall of the left ventricle is shown in the inset. The sonomicrometers were implanted in a line that was roughly perpendicular to the perfusion boundary (PB) produced by LAD occlusion. Subendocardial blood flow (ENDO MBF) and wall thickening (dWT) are plotted vs sample location, similar to Fig. 11. Note that nonischemic wall thickening within 15 mm of the perfusion boundary (the solid bars highlighted with asterisks) is impaired markedly. Comparably located sonomicrometers in the circumflex occlusion experiment (Fig. 11) displayed normal or increased thickening, suggesting that LAD occlusion is associated with a wider zone of nonischemic dysfunction.

assumption, but *how* they contribute requires that we understand more about the interaction of ischemic and nonischemic myocardium. The term lateral tethering has been used in this context but it lacks a rigorous definition. Stress concentration [1, 2] has also been invoked to explain nonischemic dysfunction though no critical tests of this possibility have been pursued. Unfortunately, we lack a cohesive conceptual framework in which to fit the experimental data. Three-dimensional reconstruction of regional function [84] and more sophisticated modeling of left ventricular mechanics, perhaps involving finite element analysis, will probably be needed to resolve these questions.

In Table 4 are summarized a few generalizations that seem "safe" regarding the functional border zone. Questions that might be useful to pursue are presented in Table 5.

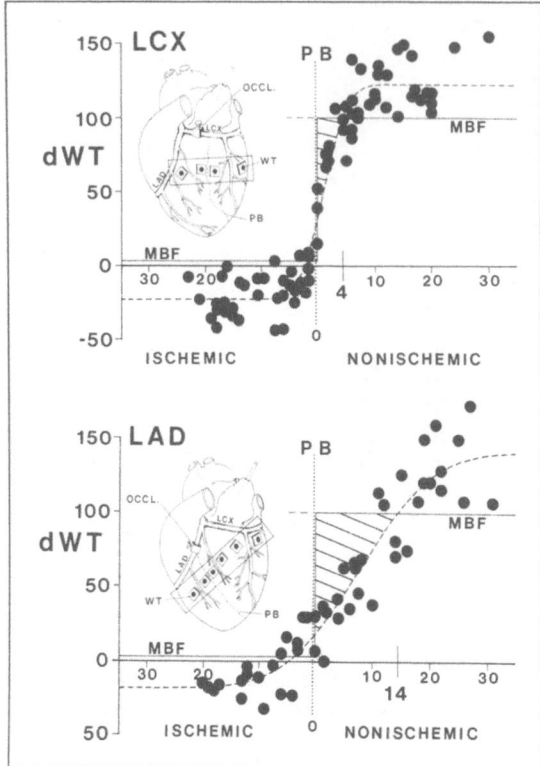

Fig. 14. Composite data sets and composite sigmoid curves from experiments in anesthetized, open-chest dogs involving occlusion of the circumflex (LCX, upper graph) or left anterior descending (LAD, lower graph) arteries. The cross-hatched area between the perfusion boundary (PB) and sigmoid curve (dashed line) represents the functional border zone. Defined as the distance from the PB to the intersection of the sigmoid curve and baseline thickening (100%), circumflex occlusion produced a functional border zone 4 mm wide, whereas LAD occlusion was associated with a functional border zone 14 mm wide.

Table 4. Generalizations on regional function adjacent to ischemic myocardium

Generalization 1:
Dysfunction extends beyond the boundaries of ischemic myocardium to form a nonischemic "functional border zone".

Generalization 2:
The size of the "functional border zone" varies depending on the location of the ischemic area.

Table 5. Questions that remain to be answered regarding regional function adjacent to ischemic or infarcted myocardium

1. How do changes in hemodynamics and replacement of infarcted tissue with scar tissue influence the functional border zone? Are the controversies related to this question due to technical, physiological, or analytical differences?
2. What is the mechanism of nonischemic dysfunction? Does it represent a type of "tethering" and can this be described in mathematically rigorous terms?

130

Conclusion

Regional function, measured as wall thickening within an ischemic area, correlates closely and directly with reductions in coronary blood flow. In so doing, wall thickening provides a reliable index of oxygen supply-demand status on a regional basis. Changes in subendocardial blood flow appear to dominate transmural function, supporting the use of wall thickening as a relative "subendocardial flowmeter". Nonischemic myocardial dysfunction may occur in normally perfused subepicardial layers overlying an ischemic subendocardial zone but the details of this phenomenon need to be clarified. More clearly defined is nonischemic dysfunction at the lateral margins of ischemic or infarcted areas which results in a functional border zone that extends into the normally perfused territory. A large number of technical, as well as physiologic questions remain to be answered about flow-function relations in ischemic (and reperfused) myocardium. Resolution of these questions is essential if regional function is to be used effectively as a key end-point parameter in clinical studies.

Acknowledgement: Supported in part by National Institutes of Health grant RO1-HL32043 and NIH Research Career Development Award KO4-HL04120.

References

1. Bogen DK, Rabinowitz SA, Needleman A, McMahon TA, Abelmann WH (1980) An analysis of the mechanical disadvantage of myocardial infarction in the canine left ventricle. Circ Res 47:728−741
2. Bogen DK, Needleman A, McMahon TA (1984) An analysis of myocardial infarction. The effect of regional changes in contractility. Circ Res 55:805−815
3. Bolli R, Zhu W-X, Thornby JI, O'Neill PG, Roberts R (1988) Time course and determinants of recovery of function after reversible ischemia in conscious dogs. Am J Physiol 254 (Heart Circ Physiol 23):H102−H114
4. Buda AJ, Zotz RJ, Gallagher KP (1986) Characterization of the functional border zone around regionally ischemic myocardium using circumferential flow-function maps. J Am Coll Cardiol 8:150−158
5. Buda AJ, Zotz RJ, Gallagher KP (1987) The effect of inotropic stimulation on normal and ischemic myocardium following coronary occlusion. Circulation 76:163−172
6. Buda AJ, Shlafer M, Gallagher KP (1988) Spatial and temporal characteristics of circumferential flow-function relations during acute ischemia in the conscious dog. Am Heart J 116:1514−1523
7. Bugge-Asperheim B, Leraand S, Kiil F (1969) Local dimensional changes of the myocardium measured by ultrasound technique. Scand J Clin Lab Invest 24:361−371
8. Canty JM Jr (1988) Coronary pressure-function and steady-state pressure-flow relations during autoregulation in the unanesthetized dog. Circ Res 63:821−836
9. Drake DH, McClanahan TB, Ning X-H, Gerren RA, Dunham WR, Gallagher KP (1987) Changes in contractility fail to alter the size of the functional border zone in anesthetized dogs. Circ Res 61:166−180
10. Force T, Kemper A, Perkins L, Gilfoil M, Cohen C, Parisi AF (1986) Overestimation of infarct size by quantitative two-dimensional echocardiography: the role of tethering and of analytic procedures. Circulation 73:1360−1368
11. Forman R, Kirk ES, Downey JM, Sonnenblick EH (1973) Nitroglycerin and heterogeneity of myocardial blood flow: reduced subendocardial blood flow and ventricular contractile force. J Clin Invest 52:905−911
12. Franklin DL, Kemper W, Patrick T, McKown D (1973) Technique for continuous measure-

ment of regional myocardial segment dimensions in chronic animal preparations (abstract). Fed Proc 32:343

13. Freeman GL, LeWinter MM, Engler RL, Covell JW (1985) Relationship between myocardial fiber direction and segment shortening in the midwall of the canine left ventricle. Circ Res 56:31–39

14. Gallagher KP, Kumada T, Reese JB, McKown D, Ross J Jr (1978) Correlation of regional myocardial blood flow and function during limited coronary inflow in dogs (abstract). Circulation 58:II–56

15. Gallagher KP, Kumada T, Koziol JA, McKown M, Kemper WS, Ross J Jr (1980) Significance of regional wall thickening abnormalities relative to transmural perfusion. Circulation 62:1266–1274

16. Gallagher KP, Folts JD, Shebuski RJ, Rankin JHG, Rowe GG (1980) Subepicardial vasodilator reserve in the presence of critical coronary stenosis in dogs. Am J Cardiol 46:67–73

17. Gallagher KP, Osakada G, Hess OM, Koziol JA, Kemper WS, Ross J Jr (1982) Subepicardial segment function during coronary stenosis and the role of myocardial fiber orientation. Circ Res 50:352–359

18. Gallagher KP, Kumada T, Battler A, Kemper WS, Ross J Jr (1982) Isoproterenol induced myocardial dysfunction in dogs with coronary stenosis. Am J Physiol 242 (Heart Circ Physiol 11):H260–H267

19. Gallagher KP, Osakada G, Matsuzaki M, Kemper WS, Ross J Jr (1982) Myocardial blood flow and function with critical coronary stenosis in exercising dogs. Am J Physiol 243 (Heart Circ Physiol 12)H698–H707

20. Gallagher KP, Matsuzaki M, Osakada G, Kemper WS, Ross J Jr (1983) Effect of exercise on the relationship between myocardial blood flow and systolic wall thickening in dogs with acute coronary stenosis. Circ Res 52:716–729

21. Gallagher KP, Matsuzaki M, Koziol JA, Kemper WS, Ross J Jr (1984) Regional myocardial perfusion and wall thickening during ischemia in conscious dogs. Am J Physiol 247 (Heart Circ Physiol 16):H727–H738

22. Gallagher KP, Stirling MC, Choy M, Szpunar CA, Gerren RA, Botham MJ, Lemmer JH (1985) Dissociation between epicardial and transmural function during acute myocardial ischemia. Circulation 71:1279–1291

23. Gallagher KP, Osakada G, Matsuzaki M, Miller M, Kemper WS, Ross J Jr (1985) Nonuniformity of inner and outer systolic wall thickening. Am J Physiol 249 (Heart Circ Physiol 18):H241–H248

24. Gallagher KP, Gerren RA, Stirling MC, Choy M, Dysko RC, McManimon SP, Dunham WR (1986) The distribution of functional impairment across the lateral border of acutely ischemic myocardium. Circ Res 58:570–583

25. Gallagher KP, Ning X-H, Gerren RA, Drake DH, Dunham WR (1987) Effect of aortic constriction on the functional border zone. Am J Physiol 252 (Heart Circ Physiol 22):H826–H835

26. Gallagher KP, Gerren RA, Choy M, Stirling MC, Dysko RC (1987) Subendocardial segment length shortening at the lateral margins of ischemic myocardium in dogs. Am J Physiol 253 (Heart Circ Physiol 22):H826–H837

27. Gallagher KP, Gerren RA, Ning X-H, McManimon SP, Stirling MC, Shlafer M, Buda AJ (1987) The functional border zone in conscious dogs. Circulation 76:929–942

28. Gallagher KP, McClanahan TB, Lynch MJ, Deeb GM (1987) Systolic wall thickening and subendocardial segment length shortening are equally sensitive to acute reductions in myocardial perfusion (abstract). Circulation 76:IV–377

29. Gallagher KP, Stirling MC, Choy M, Gerren RA (1985) The relationship between fiber direction and segment shortening in the epicardium varies by location in the left ventricle (abstract). Fed Proc 44:1380

30. Gascho JA, Lesnefsky EJ, Mahanes MS, Kaiser DL, Beller GA (1984) Effects of acute left anterior descending occlusion on regional myocardial blood flow and wall thickening in the presence of a circumflex stenosis in dogs. Am J Cardiol 54:399–406

31. Genain C, Theroux P, Thuillez C, Bourassa MG, Waters DD (1979) The interrelationships between function and flow in the subendocardial and subepicardial regions of the left ventricle (abstract). Circulation 60:II–28

32. Gibbons EF, Hogan RD, Franklin TD, Nolting M, Weyman AE (1985) The natural history of regional dysfunction in a canine preparation of chronic infarction. Circulation 71:394–402

132

33. Gross GJ, Lamping K, Warltier DC, Hardman HF (1984) Effects of three bradycardiac drugs on regional myocardial blood flow and function in areas distal to a total or partial occlusion in dogs. Circulation 69:391–399

34. Guth BD, White FC, Gallagher KP, Bloor CM (1984) Decreased systolic wall thickening in myocardium adjacent to ischemic zones in conscious swine during brief coronary artery occlusion. Am Heart J 107:458–464

35. Guth BD, Heusch G, Seitelberger R, Ross J Jr (1987) Elimination of exercise-induced regional myocardial dysfunction by a bradycardiac agent in dogs with chronic coronary artery stenosis. Circulation 75:661–669

36. Guyton RA, McClenahan JH, Newman GE, Michaelis LL (1977) Significance of subendocardial S-T segment elevation caused by coronary stenosis in the dog. Am J Cardiol 40:373–380

37. Hamby RI, Hoffman I, Hilsenrath J, Aintablian A, Shanies S, Padmanabhan VS (1974) Clinical, hemodynamic and angiographic aspects of inferior and anterior myocardial infarction in patients with angina pectoris. Am J Cardiol 34:513–519

38. Harken AH, Barlow CH, Harden WR III, Chance B (1978) Two and three dimensional display of myocardial ischemic "border zone" in dogs. Am J Cardiol 42:954–959

39. Hattori S, Weintraub WS, Agarwal JB, Bodenheimer MM, Banka VS, Helfant RH (1982) Contrasting ischemic contraction pattern by zone and layer in canine myocardium. Am J Physiol 243 (Heart Circ Physiol 12):H852–H855

40. Hearse DJ, Opie LH, Katseff IE, Lubbe WF, VanderWerff TJ, Peisach M, Boulle G (1977) Characterization of the "border zone" in acute regional ischemia in the dog. Am J Cardiol 40:716–726

41. Hearse DJ, Muller CA, Fukanami M, Kudoh Y, Opie LH, Yellon DM (1986) Regional myocardial ischemia: characterization of temporal, transmural and lateral flow interfaces in the porcine heart. Can J Cardiol 2:48–61

42. Heusch G, Guth BD, Gilpin E, Oudiz R, Matsuzaki M, Ross J Jr (1987) Determinants of recovery of regional contractile function after exercise-induced myocardial ischemia in conscious dogs (abstract). Fed Proc 46:834

43. Heusch G, Guth BD, Seitelberger R, Ross J Jr (1987) Attenuation of exercise-induced myocardial ischemia in dogs with recruitment of coronary vasodilator reserve by nifedipine. Circulation 75:482–490

44. Heymann MA, Payne BD, Hoffman JIE, Rudolph AM (1977) Blood flow measurements with radionuclide-labeled particles. Prog Cardiovasc Dis 20:55–79

45. Heyndrickx GR, Millard RW, McRitchie RJ, Maroko PR, Vatner SF (1975) Regional myocardial functional and electrophysiological alterations after brief coronary artery occlusion in the conscious dog. J Clin Invest 56:978–985

46. Hill RC, Pellom GL, Chitwood WR Jr, Sink JD, Wechsler AS (1980) The relationship of transmural myocardial blood flow to midwall function. J Surg Res 28:306–313

47. Hill RC, Kleinman LH, Tiller WH Jr, Chitwood WR, Rembert JC, Greenfield JC Jr, Wechsler AS (1983) Myocardial blood flow and function during gradual coronary occlusion in awake dogs. Am J Physiol 244 (Heart Circ Physiol 13):H60–H67

48. Hoffman JIE (1987) A critical view of coronary reserve. Circulation 75 (Suppl I):I6–I11

49. Homans DC, Asinger R, Elsperger KJ, Erlien D, Sublett E, Mikell F, Bache RJ (1985) Regional function and perfusion at the lateral border of ischemic myocardium. Circulation 71:1038–1047

50. Homans DC, Sublett E, Lindstrom P, Nesbitt T, Bache RJ (1988) Subendocardial and subepicardial wall thickening during ischemia in exercising dogs. Circulation 78:1267–1276

51. Hori M, Inoue M, Fukui S, Shimazu T, Mishima M, Ohgitani N, Minamino T, Abe H (1979) Correlation of ejection fraction and infarct size estimated from the total CK released in patients with acute myocardial infarction. Br Heart J 41:433–440

52. Kavanaugh KM, Brenner HM, Gallagher KP, Buda AJ (1988) Effect of afterload alterations on the functional border zone measured with two-dimensional echocardiography during acute coronary occlusion. Am Heart J 116:942–953

53. Kerber RE, Marcus ML, Ehrhardt J, Wilson R, Abboud FM (1975) Correlation between echocardiographically demonstrated segmental dyskinesis and regional myocardial perfusion. Circulation 52:1097–1104

54. Kerber RE, Marcus ML, Wilson R, Ehrhardt J, Abboud FM (1976) Effects of acute coronary

occlusion on the motion and perfusion of the normal and ischemic interventricular septum. Circulation 54:928−935

55. Lee J-D, Tajimi T, Guth B, Seitelberger R, Miller M, Ross J Jr (1986) Exercise-induced regional dysfunction with subcritical coronary stenosis. Circulation 73:596−605
56. Lima JAC, Becker LC, Melin JA, Lima S, Kallman CH, Weisfeldt ML, Weiss JL (1985) Impaired thickening of non-ischemic myocardium during acute regional ischemia in the dog. Circulation 71:1048−1059
57. Matre K, Hexeberg K, Lekven J (1985) Interpretation of contraction recorded from local segments. Cardiovasc Res 19:193−200
58. Matsuzaki M, Gallagher KP, Kemper WS, White F, Ross J Jr (1983) Sustained regional dysfunction during prolonged coronary stenosis: Gradual recovery of function after reperfusion. Circulation 68:170−182
59. Matsuzaki M, Gallagher KP, Patritti J, Tajimi T, Kemper WS, Ross J Jr (1984) Effects of a calcium entry blocker (diltiazem) on regional myocardial flow and function during exercise in conscious dogs. Circulation 69:801−814
60. Matsuzaki M, Guth B, Tajimi T, Kemper WS, Ross J Jr (1985) Effect of the combination of diltiazem and atenolol on exercise-induced regional myocardial ischemia in conscious dogs. Circulation 72:233−243
61. Miller MM, Thorvaldson J, Ilebekk A, Lekven J (1979) Myocardial ischemia. Relationship between local flow, function and ST-segment elevation. Eur J Cardiol 10:7−18
62. Miller RR, Olson HG, Vismara LA, Bogren HG, Amsterdam EA, Mason DT (1976) Pump dysfunction after myocardial infarction: Importance of location, extent and pattern of abnormal left ventricular segmental contraction. Am J Cardiol 37:340−344
63. Murdock RH Jr, Harlan DM, Morris JJ III, Pryor WW Jr, Cobb FR (1983) Transitional blood flow zones between ischemic and nonischemic myocardium in the awake dogs. Analysis based on distribution of the intramural vasculature. Circ Res 52:451−459
64. Myers JH, Stirling MC, Choy M, Buda AJ, Gallagher KP (1986) Direct measurement of inner and outer wall thickening dynamics with epicardial echocardiography. Circulation 74:164−172
65. Nagata M, Lavallee M (1989) Contractile function of heterogeneously perfused myocardium in conscious dogs. Am J Physiol 256 (Heart Circ Physiol 25):H352−H360
66. Pandian NG, Skorton DJ, Collins SM, Falsetti HL, Burke ER, Kerber RE (1983) Heterogeneity of left ventricular segmental wall thickening and excursion in 2-dimensional echocardiograms of normal human subjects. Am J Cardiol 51:1667−1673
67. Roan PG, Buja LM, Izquierdo C, Hashimi H, Saffer S, Willerson JT (1981) Interrelationships between regional left ventricular function, coronary blood flow and myocellular necrosis during the initial 24 hours and 1 week after experimental coronary occlusion in awake, unsedated dogs. Circ Res 49:31−40
68. Russell RO, Hunt D, Rackley CE (1973) Left ventricular hemodynamics in anterior and inferior myocardial infarction. Am J Cardiol 32:8−16
69. Sabbah HN, Marzilli M, Stein PD (1981) The relative role of subendocardium and subepicardium in left ventricular mechanics. Am J Physiol 240 (Heart Circ Physiol 9):H290−H296
70. Sakai K, Watanabe K, Millard RW (1985) Defining the mechanical border zone: A study in the pig heart. Am J Physiol 249 (Heart Circ Physiol 18):H88−H94
71. Sasayama S, Franklin D, Ross J Jr, Kemper WS, McKown D (1976) Dynamic changes in left ventricular wall thickness and their use in analyzing cardiac function in the conscious dog. Am J Cardiol 38:870−879
72. Savage RM, Guth B, White FC, Hagan AD, Bloor CM (1981) Correlation of regional myocardial blood flow and function with myocardial infarct size during acute myocardial ischemia in the conscious pig. Circulation 64:699−707
73. Schaper W (1979) Residual perfusion of acutely ischemic heart muscle. In: The pathophysiology of myocardial perfusion. In: Schaper W (ed) Elsevier/North Holland Biomedical Press, Amsterdam, pp 345−378
74. Scheel KW, Ingram LA, Gordey RL (1982) Relationship of coronary flow and perfusion territory in dogs. Am J Physiol 243 (Heart Circ Physiol 12):H738−H747
75. Schneider RM, Chu A, Akaishi M, Weintraub WS, Morris KG, Cobb FR (1985) Left ventricular ejection fraction after acute coronary occlusion in conscious dogs: Relation to the extent and site of myocardial infarction. Circulation 72:623−628

134

76. Schneider RM, Morris KG, Chu A, Roberts KB, Coleman RE, Cobb FR (1987) Relation between myocardial perfusion and left ventricular function following acute coronary occlusion: Disproportionate effects of anterior vs. inferior ischemia. Circ Res 60:60–71
77. Schneider RM, Roberts KB, Morris KG, Stanfield JA, Cobb FR (1984) Relation between radionuclide angiographic regional ejection fraction and left ventricular regional ischemia in awake dogs. Am J Cardiol 53:294–301
78. Stowe DF, Mathey DG, Moores WY, Glantz SA, Townsend RM, Kabra P, Chatterjee K, Parmley WW, Tyberg JV (1978) Segment stroke work and metabolism depend on coronary blood flow in the pig. Am J Physiol 234 (Heart Circ Physiol 3):H597–H607
79. Tatakawa S, Traber KB, Hantler CB, Tait AR, Gallagher KP, Knight PR (1987) The effects of isoflurane on myocardial blood flow, function and oxygen consumption in the presence of critical coronary stenosis. Anesth Analg 66:1073–1082
80. Tennant R, Wiggers CJ (1935) The effect of coronary occlusion on myocardial contraction. Am J Physiol 112:351–361
81. Theroux P, Franklin D, Ross J Jr, Kemper WS (1974) Regional myocardial function during acute coronary occlusion and its modification by pharmacologic agents in the dog. Circ Res 35:896–908
82. Tomoike H, Inou T, Watanabe K, Mizukami M, Kikuchi Y, Nakamura M (1983) Functional significance of collaterals during ameroid-induced coronary stenosis in conscious dogs. Circulation 67:1001–1008
83. Vatner SF (1980) Correlation between acute reductions in myocardial blood flow and function in conscious dogs. Circ Res 47:201–207
84. Waldman LK, Fung YC, Covell JW (1985) Transmural myocardial deformation in the canine left ventricle. Normal in vivo three-dimensional finite strains. Circ Res 57:152–163
85. Weintraub WS, Hattori S, Agarwal JB, Bodenheimer MM, Banka VS, Helfant RH (1981) The relationship between myocardial blood flow and contraction by myocardial layer in the canine left ventricle during ischemia. Circ Res 48:430–438
86. Weiss RM, Marcus ML (1988) The extent of regional systolic dysfunction during acute ischemia is load-dependent (abstract). Circulation 78:II–484
87. Yellon DM, Hearse DJ, Come R, Grannell J, Wyse RKG (1981) Characterization of the lateral interface between normal and ischemic tissue in the canine heart during evolving myocardial infarction. Am J Cardiol 47:1233–1249

Author's address:
Kim P. Gallagher, Ph.D.
Thoracic Surgery Research Laboratory
B560 MSRBII, Box 0686
University of Michigan Medical School
Ann Arbor, Michigan 48109, USA

Electrophysiologic and Biochemical Mechanisms Underlying Malignant Ventricular Arrhythmias during Early Myocardial Ischemia

Stephen M. Pogwizd and Peter B. Corr

Cardiovascular Division, Department of Medicine and Department of Pharmacology, Washington University School of Medicine, St. Louis, Missouri, USA

Introduction

Sudden cardiac death accounts for over 300,000 deaths per year in the United States alone [111]. It most often occurs in the setting of diffuse coronary artery disease [9], and in about 25% of cases of sudden cardiac death, the event is the initial presentation of coronary disease [87]. Death is most often due to malignant ventricular arrhythmias including complex premature ventricular complexes or ventricular tachycardia which culminate in the development of ventricular fibrillation [23].

The pathophysiologic basis for the development of these malignant arrhythmias has not yet been defined. However, at autopsy, intraluminal coronary thrombi have been found in nearly 75% of victims of sudden cardiac death [41] strongly suggesting myocardial ischemia as the critical event. The fact that sudden cardiac death is associated with marked ST segment alterations immediately preceding the fatal arrhythmic event in patients with Holter monitor recordings of the electrocardiogram also supports the concept that ischemia is the precipitating event [127, 130]. Reperfusion of ischemic myocardium, whether due to release of coronary spasm [129], platelet aggregation with subsequent lysis [85], or an increase in collateral flow [118], may also lead to the development of malignant ventricular arrhythmias and sudden cardiac death [34].

There has been little progress made in the prevention of sudden cardiac death. Therapeutic agents that are used currently are, for the most part, nonspecific in their action and of limited efficacy, primarily due to limitations in our understanding of the mechanisms underlying these lethal arrhythmias. Approaches designed to develop effective therapeutic agents will require an understanding of both the biochemical as well as the electrophysiologic mechanisms underlying arrhythmogenesis during ischemia.

Acute obstruction of a coronary artery in multiple species often leads to the development of ventricular tachycardia and ventricular fibrillation [178]. There are, however, several distinct time intervals following ischemia in experimental animals when the incidence of malignant ventricular arrhythmias is pronounced. Within the first 30 min of ischemia (Phase 1), ventricular arrhythmias including ventricular tachycardia occur frequently and often lead to ventricular fibrillation [61, 79], which

Research from the authors' laboratory was supported in part by National Institutes of Health grants HL 17646, SCOR in Ischemic Heart Disease, grant HL 28995 and by grant HL-36773.

137

is analogous to the "pre-hospital" phase of malignant ventricular arrhythmias seen with acute myocardial infarction in man [14]. Beyond 30 min there is a quiescent period during which malignant arrhythmias are rare. Twelve to 24 h after coronary obstruction there is a second interval (Phase 2) in which ventricular arrhythmias occur, although ventricular fibrillation is infrequent [59]. This may be analogous to the "in-hospital" phase of early myocardial infarction [14]. A third interval begins about 5 to 7 days after coronary obstruction (Phase 3) in which spontaneous ventricular tachycardias occur infrequently, but are often inducible by programmed electrical stimulation [48]. Phase 3 is analogous to the chronic ventricular arrhythmias following myocardial infarction in man [178]. Experimental evidence obtained from animals indicates that the mechanisms of the arrhythmias during these three phases differ considerably.

In the present review, the discussion will be confined to the arrhythmias occuring within minutes of the onset of myocardial ischemia (Phase 1). The focus will be on our current understanding of not only the underlying electrophysiologic mechanisms involved but also potential biochemical alterations which contribute to arrhythmogenesis in this setting.

The General Sequence of Events

Myocardial ischemia is characterized by a rapid decrease in developed tension occurring within seconds in the affected region [62]. Within the first few minutes there is a rapid decrease in high energy phosphates, with near complete depletion of creatine phosphate by 3 min of ischemia [84], although ATP content exhibits a somewhat slower rate of decline [84, 140]. Tissue within the ischemic region becomes progressively more hypoxic and acidotic [51, 70]. The lack of blood flow to the region also leads to the accumulation of several ions and metabolites which, as discussed in detail below, contributes significantly to the development of lethal arrhythmias. In addition, as discussed below, the autonomic nervous system is also activated which appears to contribute importantly to arrhythmogenesis [36].

Electrophysiologic Alterations during Early Myocardial Ischemia

Marked electrophysiologic alterations occur within minutes of myocardial ischemia. Transmembrane action potential recordings from the epicardial surface of the ischemic heart have demonstrated a marked decrease in resting membrane potential, action potential amplitude, rate of upstroke of phase 0 (V_{max}), and action potential duration [47, 97] (Fig. 1). Studies performed in the isolated perfused ventricle preparation exposed to ischemia have revealed that similar alterations occur in the subendocardium as well [95]. Within 10 min after ischemia, the action potential demonstrates alterations in amplitude and duration (2 : 1 alternans), with subsequent development of inexcitability and conduction block within the central ischemic zone [47]. During the 15- to 30-min interval of ischemia, excitability begins to return, but after 30 min of ischemia, the ischemic zone is almost completely inexcitable [47].

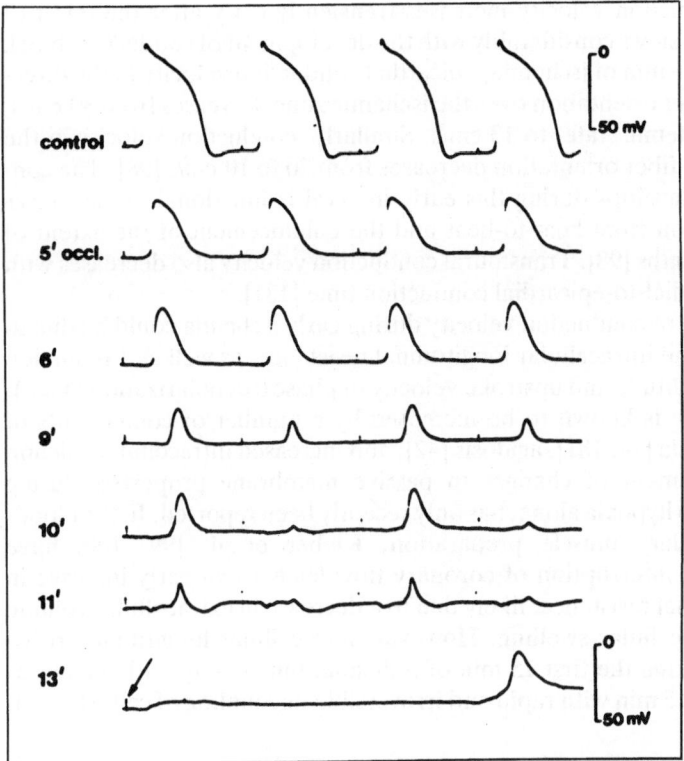

Fig. 1. Transmembrane action potentials recorded from the subepicardium of the left ventricle of an in situ pig heart before and after occlusion of the proximal left anterior descending artery. (Reproduced from [47] with permission of the American Heart Association.)

Shortly after the onset of ischemia, the refractory period in the ischemic zone decreases [108, 131, 146]. However, within several minutes, the refractory period begins to lengthen eventually to values exceeding the non-ischemic control state, with the development of a marked dispersion of the refractory periods between the ischemic and nonischemic zones [47, 146]. The refractory periods in the ischemic zone continue to increase with the development of post-repolarization refractoriness [47, 106].

The rapidity and severity of the electrophysiologic alterations during the interval early after myocardial ischemia are also reflected in recordings of extracellular electrograms from the epicardial surface of the ischemic zone. Within minutes, bipolar electrograms demonstrate a decrease in amplitude and variable increases in duration [15, 47, 150, 170]. Activation time is delayed markedly and the extracellular electrograms become more asynchronous and fractionated [15, 170]. At times, continuous electrical activity occurs, spanning the interval between a normal sinus beat and the initiation of a ventricular ectopic beat [151, 170] suggesting, albeit indirectly, that reentry is a contributing mechanism.

Paradoxically, conduction velocity increases transiently early after the onset of ischemia [73], but then slows considerably with the development of conduction block [15, 170, 176]. By 3 to 5 min of ischemia, epicardial conduction velocity in the direction longitudinal to fiber orientation over the ischemic zone decreases from 30 cm/s, in the control non-ischemic state, to 13 cm/s. Similarly, conduction velocity in the direction transverse to fiber orientation decreases from 20 to 10 cm/s [98]. The conduction block which develops during this early interval is functional in nature, as reflected by its variation from beat-to-beat and the enhancement of the extent of block at short cycle lengths [98]. Transmural conduction velocity also decreases with an increase in endocardial-to-epicardial conduction time [131].

The marked slowing of conduction velocity during early ischemia could be due to an increase in extra- and intracellular longitudinal resistance, as well as a reduction of action potential amplitude and upstroke velocity of phase 0 depolarization (V_{max}). Longitudinal resistance is known to be increased by a number of components of ischemia such as hypoxia [74, 181], acidosis [42], and increased intracellular calcium [43]. However, assessment of changes in passive membrane properties during ischemia, as opposed to hypoxia alone, has only recently been reported. In the blood-perfused rabbit papillary muscle preparation, Kleber et al. [99, 100] have demonstrated that the interruption of coronary flow leads to an early increase in extracellular longitudinal resistance, likely due to a decrease in extracellular volume secondary to osmotic cellular swelling. However, intracellular longitudinal resistance is unchanged during the first 15 min of ischemia, but subsequently increases threefold over the next 5 min with rapid and irreversible uncoupling of cells (Fig. 2).

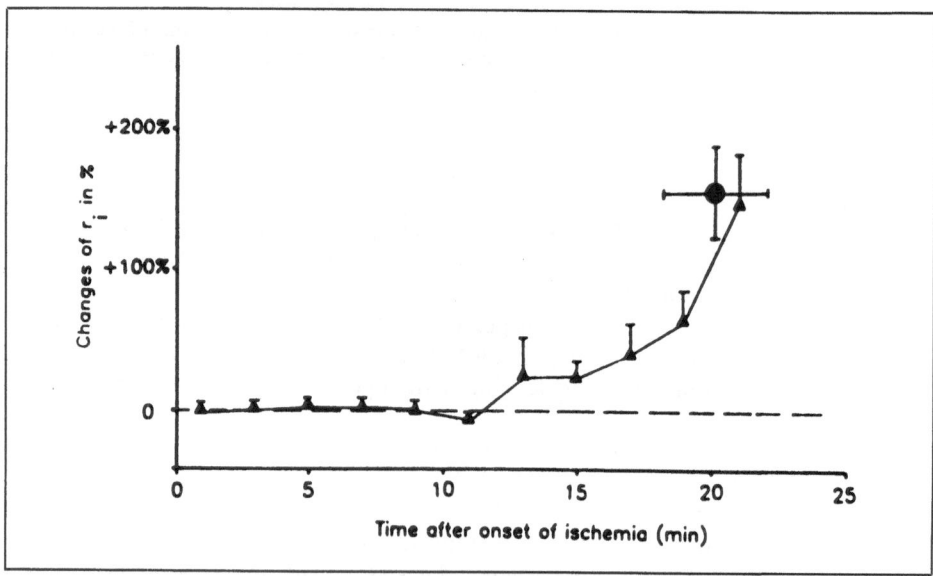

Fig. 2. Changes in intracellular longitudinal resistance in acute myocardial ischemia (mean values ± SEM) from eight isolated blood-perfused rabbit papillary muscle preparations. Values at and after 15 min are significantly different from control ($p < 0.05$). Dot indicates time and level of uncoupling at which propagation becomes discontinuous (limit for linear cable analysis). (Reproduced from [100] with permission of the American Heart Association.)

Therefore, the fact that cellular coupling is maintained during the interval early after ischemia when conduction velocity slows substantially, suggests that this mechanism is not a critical event in activation delay until much later intervals of ischemia. Beyond approximately 15 min of ischemia, cellular uncoupling may contribute importantly to slow conduction, block, and arrhythmogenesis.

It is possible that the extent of depolarization of cells within the ischemic region could inactivate the rapid Na^+ inward current, making depolarization dependent exclusively on the slow inward current (I_{si}) carried by Ca^{2+}. However, recent evidence indicates that depolarization is not dependent on the I_{si} but rather on depression of the rapid Na^+ inward current. Evidence to support this conclusion includes: 1) conduction dependent exclusively on the I_{si} is much slower than the slowest conduction velocities observed during ischemia in vivo [98], and 2) lidocaine, which at higher concentrations can decrease the rapid Na^+ inward current without affecting the I_{si} carried by Ca^{2+}, has been shown to depress action potentials recorded from ischemic cells [19].

The most striking effect of early myocardial ischemia is the marked degree of inhomogeneity of the electrophysiologic alterations. As indicated previously, there is considerable variation in the refractory periods between the ischemic and nonischemic zones [47, 146], which becomes even more pronounced as post-repolarization refractoriness develops. There is also spacial and temporal dissociation in the development of conduction delay within the ischemic zone as well as differences in the alterations in the refractory periods between the endocardium and epicardium of the ischemic zone [94]. The epicardium exhibits a greater decrease in action potential amplitude and duration and a greater conduction delay compared to the same parameters in the endocardium [94]. After 10 min of ischemia, action potential duration as well as the refractory period is shorter in the epicardium, but by 30 min of ischemia the refractory period of the epicardium exceeds that of the endocardium in the ischemic zone due to the development of post-repolarization refractoriness in the epicardium [94]. This difference between endocardium and epicardium contributes importantly to the underlying heterogeneity.

Conduction velocity through the endocardium appears to remain near normal at a time when epicardial conduction velocity slows and the wavefront becomes fractionated [15, 145, 150, 176] despite a similar degree of ischemia [145]. The reason for this discrepancy is unknown, but it has been demonstrated that the endocardium and, in particular, Purkinje fibers are more resistant to the effects of simulated ischemia in vitro (hyperkalemia, acidosis, hypoxia) than either epicardium or papillary muscle [54]. Although other studies have indicated that cell death occurs more rapidly in the endocardium compared to epicardium, independent of hemodynamic load or coronary flow, this does not appear to be the case relative to electrophysiologic endpoints. The maintenance of electrophysiologic indices in Purkinje tissue exposed to ischemia may be due to the high levels of glycogen in these cells [54] and/or the fact that intracavitary blood may prevent accumulation of noxious metabolites or extracellular ions such as K^+ in this region.

Although substantial degrees of spatial heterogeneity exist within the ischemic region, heterogeneity is most evident between the ischemic and nonischemic regions. Although earlier studies demonstrated intermediate alterations in metabolic and electrophysiologic indices in the border zone region [63], the transition between normal and ischemic tissue appears to be quite abrupt [64, 78]. In addition, a subendocardial border zone also exists with a gradient of metabolic and electrophysiologic alterations between superficial and deeper (>600 μm) cells [175].

Several of the electrophysiologic alterations which occur during myocardial ischemia and which were outlined above are consistent with the development of reentry as the mechanism underlying arrhythmias during early ischemia. In particular, slow conduction leading to unidirectional block, combined with very delayed activation and inhomogeneous recovery of excitability, could provide the necessary components for the development of reentrant circuits. The demonstration of continuous electrical activity in extracellular electrogram recordings provides additional evidence for the presence of reentry [15, 170]. Although these findings are suggestive of a reentrant mechanism, they do not constitute proof. This requires delineation of the reentrant pathway by recording simultaneously from multiple sites.

Ventricular Mapping during Early Myocardial Ischemia

Several studies have utilized ventricular mapping to delineate the mechanisms underlying arrhythmias during early ischemia. For example, Janse et al. recorded simultaneously from 60 sites (primarily epicardial but also limited intramural sites) in the isolated canine and porcine heart exposed to ischemia [81]. The initiation of ventricular tachycardia or isolated premature ventricular complexes arose in the subendocardial region in normal myocardium adjacent to the ischemic border zone. However, reentry was never demonstrated to be the initiating mechanism. In contrast, maintenance of the ventricular tachycardia was associated with epicardial reentry with slow conduction around areas of conduction block with a diameter of 1 to 2 cm. An example was also presented which suggested that reentry might involve intramural pathways. Since ventricular ectopic activity after ischemia appears to initiate in the subendocardium [89], this would further suggest that intramural processes may contribute importantly to the development of arrhythmias during early ischemia. In this study [81] the development of ventricular fibrillation was associated with fragmentation of the wavefronts and a greater degree of conduction block. Incomplete circus movements were more frequently recorded than completed circuits, which were on the order of 0.5 cm in diameter. This study, however, was limited by recording from a relatively small number of sites, primarily over a portion of the left ventricular epicardial surface. Because a large portion of the heart was not mapped, the authors could not exclude that reentry was occuring remote from the recording sites. To explain the initiation of ventricular tachycardia by nonreentrant mechanisms, they proposed that "currents of injury" flowing from ischemic to nonischemic regions were responsible. This potential mechanism is discussed in more depth below.

Due to the extensive heterogeneity in the ischemic region, the definitive assessment of the mechanisms responsible for the arrhythmias during early myocardial ischemia requires the ability to map from a very large number of sites throughout the entire heart. The mapping system designed and built in our laboratory [103, 133, 180] permits continuous and simultaneous recording from 232 intramural sites throughout the left ventricle, right ventricle and septum. Studies were performed in the feline heart in vivo. The feline model not only offers the advantage of having a coronary circulation quite similar to that of man [30], but also its small size permits enhanced resolution since mapping from 232 sites in the feline heart yields a resolution comparable to mapping from over 2700 sites in the canine heart.

Three-Dimensional Mapping during Normal Sinus Rhythm

In the following sections the results of using detailed three-dimensional mapping in the feline heart during early ischemia in vivo will be summarized [133, 134]. Normal sinus rhythm under control non-ischemic conditions was first initiated in the endocardium of the ventricular septum and conducted rapidly, both to apex and base, as well as from endocardium to epicardium (Fig. 3A). Activation was completed by 25 ± 2 ms. In marked contrast, sinus rhythm 2 to 5 min after anterior ischemia, induced by occlusion of the left anterior descending artery, was characterized by slow conduction in the ischemic region with normal conduction velocities in the posterior nonischemic zone (Fig. 3B). In addition, functional unidirectional conduction block developed in the ischemic region which could vary considerably from beat to beat. The conduction block occured both in an endocardial-to-epicardial (transverse) direction and in a side-to-side (longitudinal) direction. The most marked conduction delay was noted in the subendocardial and midmyocardial regions of the ischemic zone, with conduction delay as extensive as 220 ms.

The finding of marked conduction delay in the subendocardium and midmyocardium differs from many other studies in which the most extensive delay occured in the subepicardium after ischemia [15, 81, 150, 170]. However, our findings are not inconsistent since a moderate degree of conduction delay was noted in the epicardium (Fig. 3), and delayed subendocardial and midmyocardial activation has been noted by others as well [81]. The increased extent of conduction delay in this study is likely secondary to the enhanced resolution of the mapping procedures which permitted delineation of slow yet very small localized intramural pathways of delayed conduction within the ischemic zone. Although unidirectional block and markedly delayed activation distal to the area of block occured during sinus rhythm, this beat-to-beat heterogeneity was not evident on the surface electrocardiogram (Fig. 3B).

Mechanisms Underlying the Development of Premature Ventricular Contractions and Ventricular Tachycardia

During early myocardial ischemia in vivo, premature ventricular contractions and the initial beats of ventricular tachycardia were found to initiate by intramural reentry in 76% of cases [133]. The marked conduction delay of the preceding sinus beat in the subendocardium and midmyocardium, distal to the unidirectional block, could lead to reactivation of adjacent subendocardial tissue proximal to the block, a region in which cells had recovered their excitability (Fig. 4). The intramural location of the reentrant pathway, and the reentry initiation in the subendocardium in all but one case, explain why intramural reentry had not been noted by others using epicardial mapping alone.

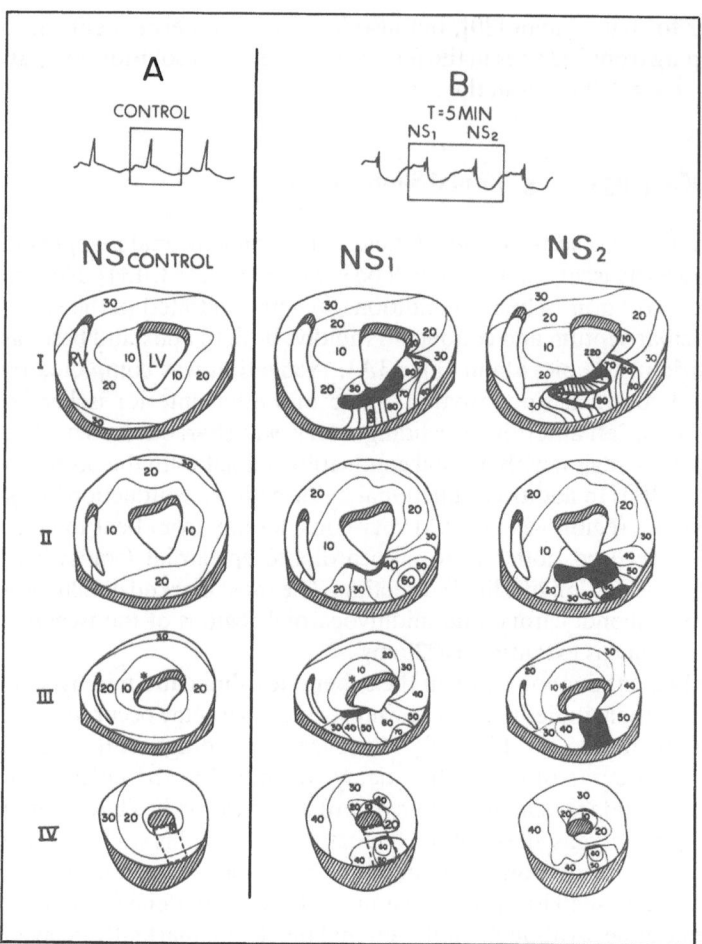

Fig. 3. Three-dimensional isochronic maps of a control sinus beat (A) and two sinus beats 5 min after occlusion of the LAD coronary artery (B) in the in vivo feline heart. Surface electrocardiograms for each beat are shown above within the box. Recordings were made from 232 intramural bipolar recording sites (eight intramural levels in the left ventricle, four intramural levels in the septum, and two intramural levels in the right ventricle). After detailed three-dimensional localization of electrode sites was performed, the hearts were sectioned transversely into 5 mm thick slices. Sections are oriented with the base on top and the most apical slice on the bottom. Right ventricular and left ventricular cavities are labeled on the most basal slice for the control nonischemic beat. Areas of conduction block are indicated by blackened areas and thickened lines. Asterisks denote sites of initial activation for each beat, and numbers within isochrones indicate time in milliseconds from initiation of each beat. (Reproduced from [133] with permission of the American Heart Association.)

Maintenance of ventricular tachycardia was often due to intramural reentry, and was dependent on the continued presence of activation which was delayed sufficiently to permit adjacent tissue to recover excitability and complete the intramural reentrant circuit (Fig. 5). Although conduction delay was necessary for continued reentry, its presence alone was not always sufficient to maintain a run of ventricular

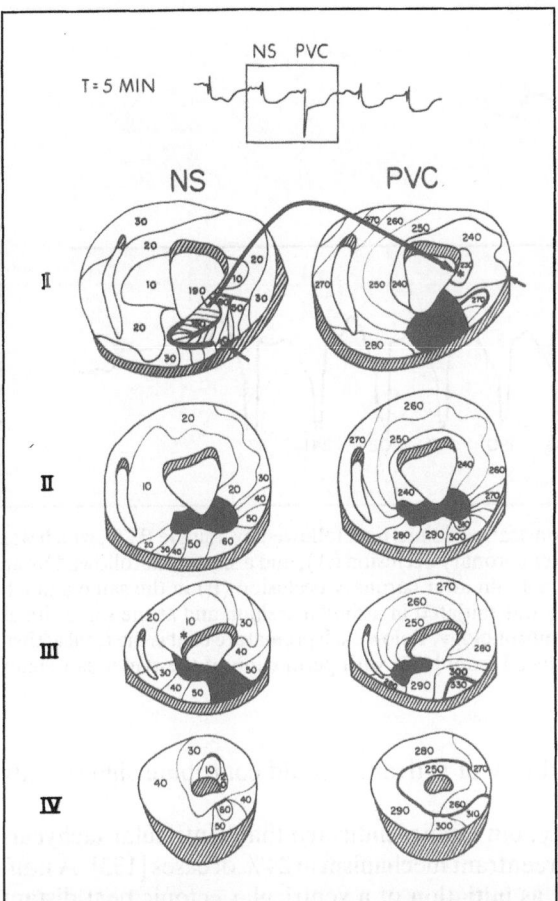

Fig. 4. Three-dimensional isochronic maps of a sinus beat (NS) followed by premature ventricular complex (PVC) 5 min after occlusion of LAD coronary artery in which initiation of the PVC occurs by intramural reentry shown by the dark arrow from NS to PVC. After initiation of the sinus beat in the septum (beat NS, level III, *), activation proceeds both apically and basally, as well as from endocardium to epicardium. By 60 ms, most of the heart had activated. There is transmural block in level III and nontransmural block evident in levels I and II. It is the very delayed activation around the nontransmural block in level I and the block in the epicardial-to-endocardial direction that results in very delayed activation of the subendocardium and midmyocardial area (190 ms isochrone), which then activates an adjacent area of the subendocardium (PVC, level I, *) to initiate the PVC by intramural reentry. Additional details are as described in Fig. 3 legend. (Reproduced from [133] with permission of the American Heart Association.)

tachycardia (Fig. 5). Ventricular tachycardia which was maintained by intramural reentry could terminate if there was not sufficient conduction delay. However, a run of ventricular tachycardia could terminate, even in the presence of continued marked conduction delay, if, for example, there was a significant change in the reentrant pathway, and delayed activation of the terminal beat now occurred in a region in which adjacent tissue had not yet recovered excitability. Thus, although conduction delay was crucial for reentry to occur, other factors such as the refractory properties

145

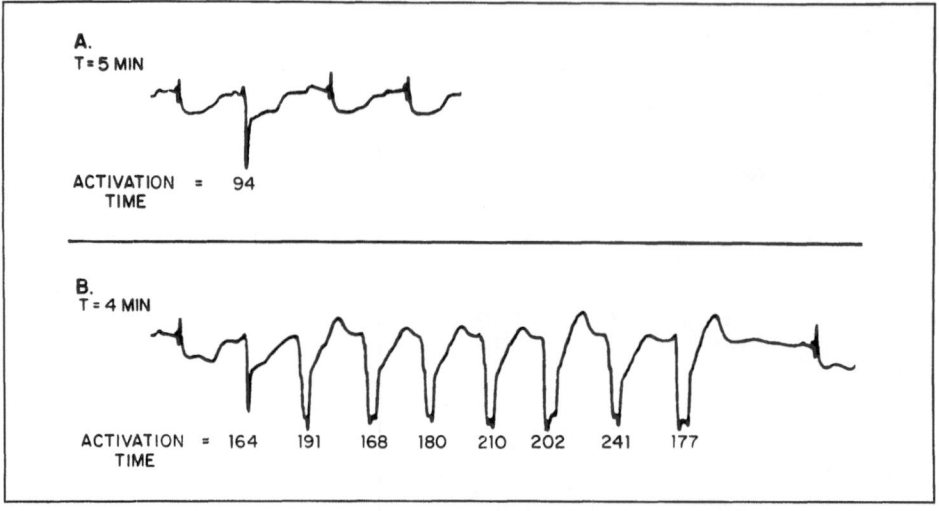

Fig. 5. Surface electrocardiograms demonstrating a sinus beat followed by a single PVC, which was initiated by intramural reentry 5 min after coronary occlusion (A), and a sinus beat followed by an eight-beat run of ventricular tachycardia 4 min after coronary occlusion (B) in the same animal. The first beat of ventricular tachycardia was initiated in a similar fashion and at the same site as PVC above and demonstrates the same morphology. Below each premature beat is the total activation time for each beat in ms. (Reproduced from [133] with permission of the American Heart Association.)

of the adjacent tissue and the pathway of activation could contribute significantly (Fig. 5).

In addition to intramural reentry, our studies indicated that ventricular tachycardia also could be initiated by a nonreentrant mechanism in 24% of cases [133]. A nonreentrant mechanism was defined as initiation of a ventricular ectopic beat distant from the site of termination of the preceding (sinus) beat with no intervening electrical activity, despite the presence of multiple intervening electrode recording sites (Fig. 6). Ventricular tachycardia could also be maintained by a nonreentrant mechanism, arising in the subendocardium or epicardium (adjacent to the ischemic border zone), which was not dependent on the degree of conduction delay of the preceding beat. The nonreentrant excitation arising in the subepicardium which could maintain ventricular tachycardia was quite interesting because it possessed several features consistent with triggered activity due to delayed after-depolarizations [37, 133, 144]: 1) it never initiated a run of ventricular tachycardia but always started after a premature ectopic beat, whether due to reentry or a nonreentrant mechanism, from another site, and 2) repetitive firing could occur at a nearly constant coupling interval before sudden cessation of activation.

The interaction between reentrant and nonreentrant mechanisms early after ischemia was quite complex. As shown in Fig. 7, a run of ventricular tachycardia could be initiated by one mechanism and maintained by another. Moreover, both reentrant and nonreentrant mechanisms could be involved in the initiation of the same beat of ventricular tachycardia (Fig. 7), demonstrating that there is a competition between the two mechanisms which, at times, can result in a "fusion beat".

146

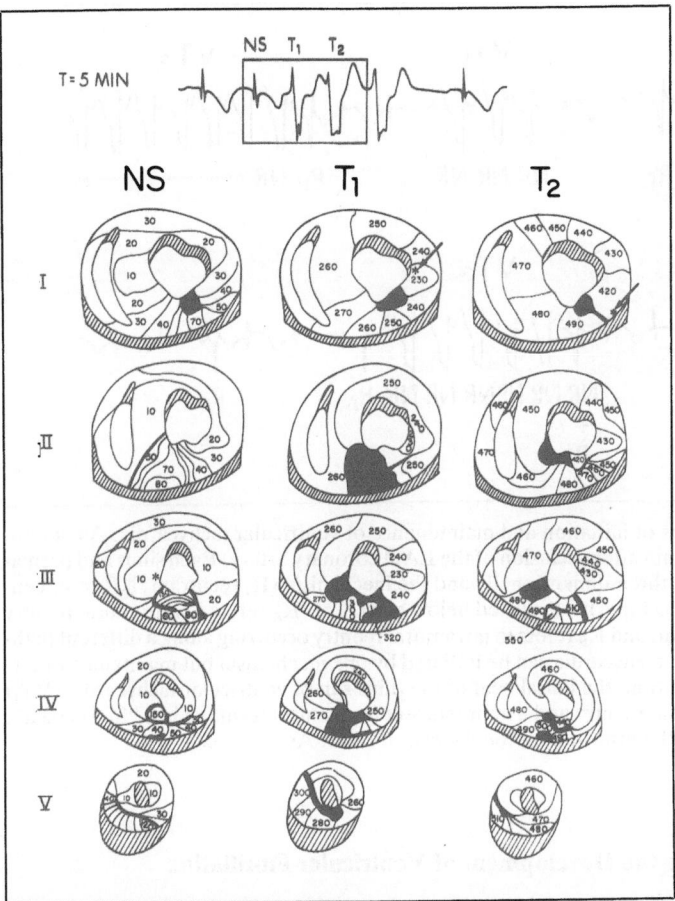

Fig. 6. Three-dimensional isochronic maps representing a sinus beat (NS) followed by the first two beats of a three-beat run of ventricular tachycardia ($T_1 - T_3$) in another animal 5 min after occlusion of the LAD coronary artery. Initiation of the sinus beat (NS) occurs in the septum (beat NS, level III, *) and spreads both apically and basally as well as from endocardium to epicardium, encountering varying degrees of conduction block. Although marked delay occurs in the subendocardial and midmyocardial regions in levels III and IV, initiation of the first beat of the tachycardia (T_1) occurs at a distance in the basal subendocardium 80 ms after termination of beat NS (beat T_1, level I, *), with no intervening activity seen despite the presence of multiple intervening electrodes. Although the first beat of the tachycardia encounters both transmural and nontransmural block in various levels with latest activity occurring in level III at the 320 ms isochrone, the second beat of the tachycardia (T_2) initiates at a distant subepicardial site in the base (beat T_2, level I, *) by a nonreentrant mechanism, again with no intervening activity seen despite multiple intervening electrode sites. See legend for Fig. 3 for additional details. (Reproduced from [133] with permission of the American Heart Association.)

147

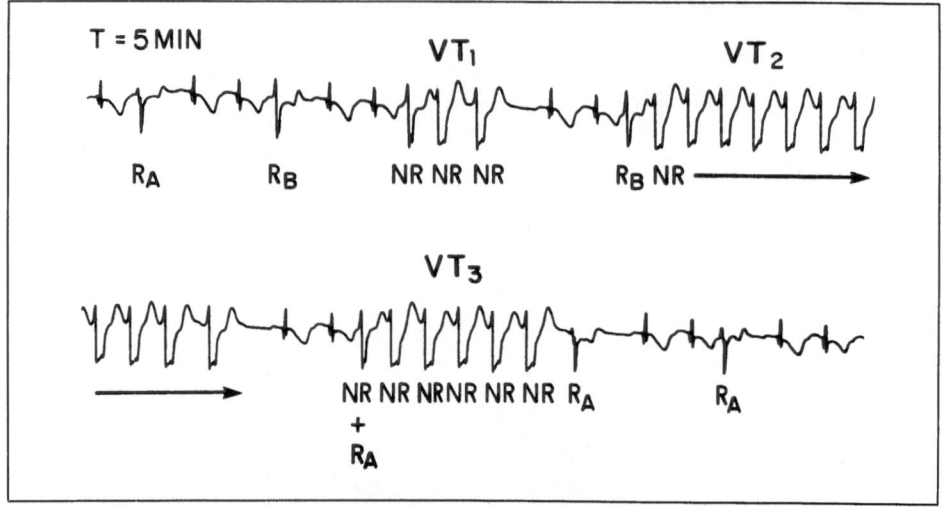

Fig. 7. Multiple mechanisms of initiation and maintenance of ventricular tachycardia. A continuous surface ECG tracing 5 min after occlusion of the LAD coronary artery. Each sinus and premature beat has been mapped three-dimensionally and the mechanism (R, reentrant; NR, nonreentrant) of initiation for each ectopic beat is listed below each beat. R_A refers to intramural reentry occurring in the apical region, and R_B refers to intramural reentry occurring along a different pathway in the base. Note that a tachycardia can be initiated by one mechanism but maintained or terminated by another. In addition, the initial beat of the third run of ventricular tachycardia (VT_3) is initiated both by intramural reentry and a nonreentrant mechanism resulting in a "fusion beat". (Reproduced from [133] with permission of the American Heart Association.)

Mechanisms Underlying the Development of Ventricular Fibrillation

In a number of cases, ventricular tachycardia did not terminate but degenerated to ventricular fibrillation. Detailed three-dimensional mapping of these complete runs of ventricular tachycardia leading to ventricular fibrillation (VT-VF) revealed that initiation occurred primarily by intramural reentry [134]. Maintenance of ventricular fibrillation was also due primarily to intramural reentry although nonreentrant mechanisms also contributed. For example, when there was insufficient activation delay for reentry to continue, successive nonreentrant beats occured, originating from endocardial or epicardial regions, which elicited further progressive conduction delay. These nonreentrant beats actually maintained the tachycardia until there was sufficient delay for the redevelopment of a reentrant circuit. The tachycardia then continued, primarily by intramural reentry.

The sequence for transition of the ventricular tachycardia to fibrillation occurred via two mechanisms. First, acceleration of the tachycardia, to a coupling interval as short as 80–90 ms due to continued intramural reentry, was associated with more extensive slowing of conduction and enhancement of the extent of conduction block. Second, very rapid and inhomogeneous recovery of excitability to values as short as 57 ms was most marked adjacent to the border zone of the ischemic region (Fig. 8). This eventually led to the development of multiple reentrant circuits with a path length as small as 10 mm. Once ventricular fibrillation had developed, chaotic activation occurred due to very slow conduction and multiple simultaneous reentrant circuits, together with rapid recovery of excitability (Fig. 9). In concert, these findings suggest that although reentry is a critical mechanism underlying the transition to ventricular fibrillation after ischemia in vivo, a nonreentrant mechanism also contributes significantly. This suggests that pharmacologic approaches should be aimed not only at interruption of the reentrant pathway but also at inhibiting this nonreentrant mechanism.

Nonreentrant Mechanism

Although the exact nature of the "nonreentrant" mechanisms occurring during ischemia is unknown, there are several possibilities. Enhancement of normal automaticity in Purkinje fibers at polarized resting membrane potentials is one possibility. However, since the idioventricular rate is either unchanged or reduced with ischemia [108, 131, 150, 151], this is unlikely. Abnormal automaticity arising from less negative resting membrane potentials is also unlikely since there is substantial data to indicate that elevation of extracellular potassium [91], which occurs rapidly within the ischemic zone [69, 70, 96], would suppress the development of abnormal automaticity.

Based on several lines of evidence, triggered activity arising from either early (EADs) or delayed (DADs) after-depolarizations may be the primary mechanism. EADs, or oscillations in the membrane potential occurring during phase 2 or 3 of the action potential, have been shown in vitro to be induced by a number of factors present in ischemic myocardium, including catecholamines, hypoxia, and acidosis [179]. Furthermore, indirect evidence has been reported that EADs may arise in ischemic myocardium [163]. DADs, or oscillations in the membrane potential occurring after repolarization is complete, are induced by a variety of pathologic states associated with an increase in intracellular calcium, such as digitalis toxicity and high concentrations of catecholamines [179]. Recently, substantial evidence has been reported supporting the concept that intracellular calcium increases during early myocardial ischemia [107] and could be the basis for the development of DADs. Additional, albeit indirect, evidence implicating the development of DADs during early ischemia includes: 1) ryanodine and caffeine, agents which prevent the release of calcium from the sarcoplasmic reticulum, both prevent ventricular arrhythmias during ischemia [164]; 2) ventricular arrhythmias occurring during early myocardial ischemia are exacerbated by rapid atrial pacing rather than overdrive suppression [131], a finding consistent with triggered activity arising from DADs [167]; and 3) monophasic action potential recordings during early ischemia indicate that DADs may arise from the subendocardium [46]. Thus, there is a considerable amount of indirect evidence to suggest that triggered rhythms arising from either EADs or DADs may contribute to arrhythmogenesis during early ischemia and may in fact be the mechanism underlying the nonreentrant activity found during our three-dimensional mapping studies of early ischemia [133].

An additional mechanism potentially responsible for nonreentrant activity in ischemic myocardium are "injury" currents flowing between the ischemic zone and

149

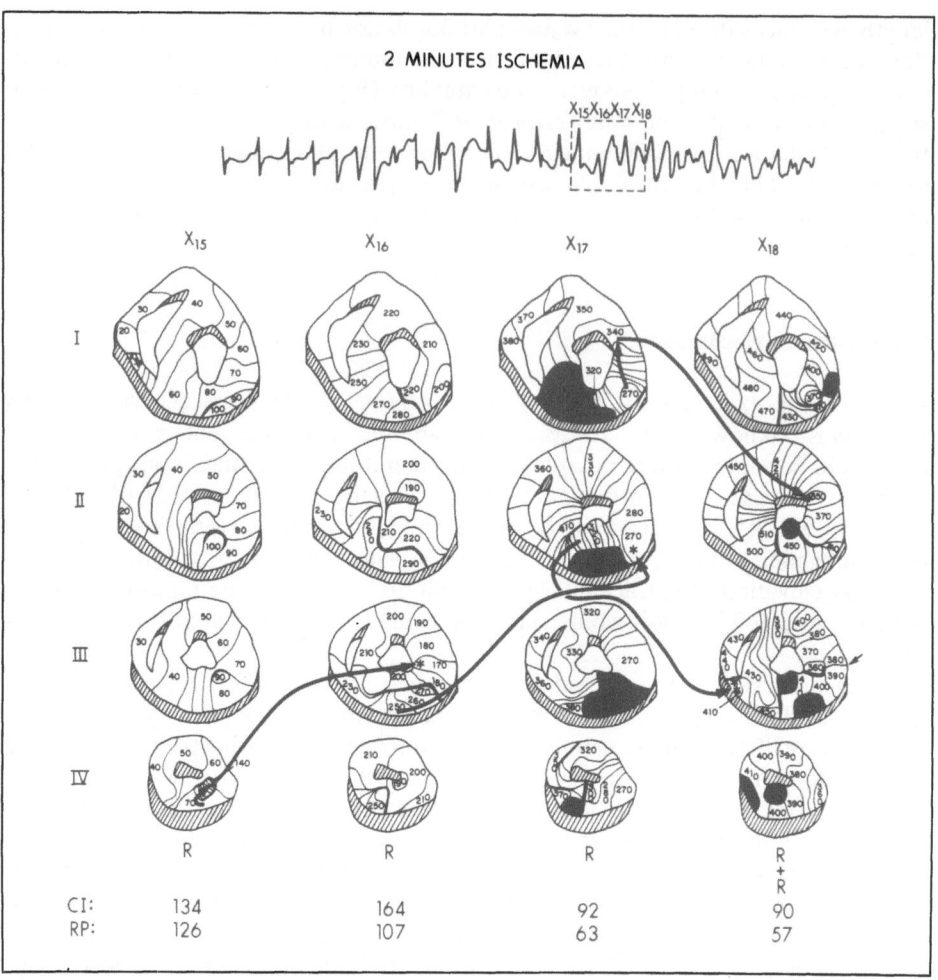

Fig. 8. Three-dimensional isochronic maps of four beats (X_{15}–X_{18}) during the transition to VF 2 min after ischemia, with acceleration of the tachycardia occurring by intramural reentry (arrows). Below the map of each beat is the mechanism of initiation (R = reentry), and the coupling interval (CI) and recovery period (RP) of the initiating site for that beat. The recovery period was an indirect measure of refractoriness and was defined as the interval between two consecutive activations at the site of initiation. After initiation of X_{16} by intramural reentry, the tachycardia begins to accelerate (to a coupling interval of 92 ms) because of rapid recovery of excitability (recovery period for X_{17} of 63 ms) and continued intramural reentry. As a result of incomplete recovery of excitability, conduction slows considerably (as indicated by the increased density of isochrones in X_{17}) and leads to the initiation of X_{18} by multiple simultaneous reentrant circuits on the order of 1 cm in path length. See legend for Fig. 3 for additional details. (Reproduced from [134] with permission of the American Heart Association.)

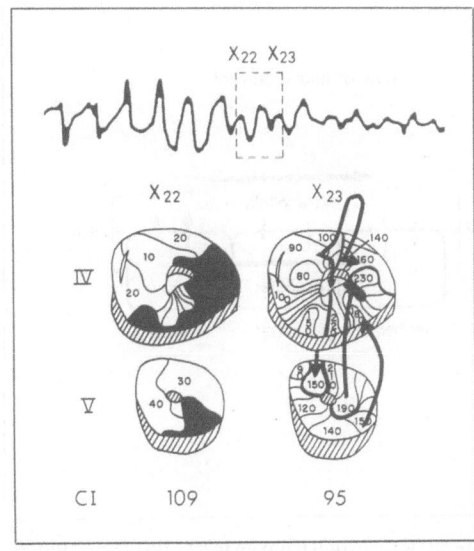

Fig. 9. Activation sequences for the apical two sections for the 22nd and 23rd beats of a run of VT leading to VF, demonstrating slow conduction and block during VF, with multiple chaotic reentrant pathways (arrows). (Reproduced from [134] with permission of the American Heart Association.)

nonischemic zone. This mechanism for initiating ectopic beats through reexcitation of cells adjacent to the ischemic border zone was originally implicated nearly three decades ago [138, 148]. More recently, Janse et al. [81, 97], utilizing isovoltaic mapping, calculated the maximal current sources at the ischemic border zone to be on the order of 2 $\mu A/mm^3$, about half as large as current sources during the propagation of the normal activation wavefront (Fig. 10). These injury currents appear to be sufficient to reexcite adjacent normal myocardium which has recovered excitability, possibly through focal reexcitation [7]. In this instance, ischemic myocardium in the border region repolarizes more rapidly but because there is a delay in activation of this tissue, ischemic tissue can be in a depolarized state while the adjacent nonischemic region has completely repolarized. If a sufficient potential difference exists at a given point in time between the depolarized ischemic tissue and the fully repolarized adjacent nonischemic tissue, current will flow, dependent on tissue resistivity, from the ischemic to nonischemic cells resulting in initial excitation of cells in the nonischemic border zone. A study by Anderson et al. [6] suggested that injury currents of the same order of magnitude may occur within 10 to 30 s after ischemia. Using a computer model, Janse and van Capelle [80] could evoke reflection across an inexcitable segment at the ischemic border zone, but required the presence of elements with suppressed pacemaker activity. They suggested that the injury current may provide the trigger for the initiation of an extrasystole, possibly due to either DADs or EADs attaining threshold. Although this mechanism is plausible, additional data will be required.

Alterations in Potassium during Early Myocardial Ischemia

Myocardial ischemia is characterized by a number of ionic and biochemical alterations which may contribute significantly to arrhythmogenesis. Extracellular potas-

151

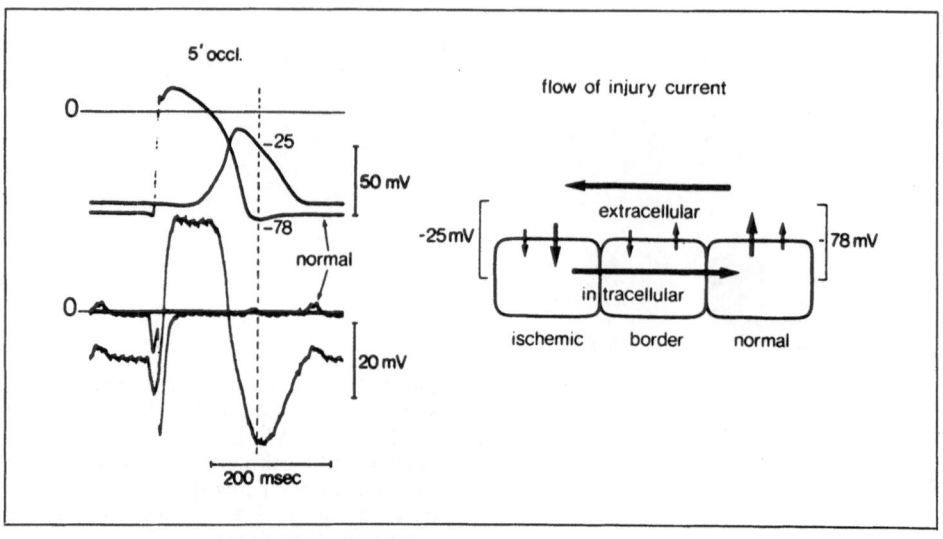

Fig. 10. Transmembrane and DC extracellular potentials recorded from an in situ dog heart. Potentials before and 5 min after LAD occlusion are superimposed, using as time reference the stimulus artifact on the atrium. Note extreme delay in activation of the epicardial ischemic cell, which is depolarized at the time normal cells are already repolarized. The negative "T-wave" in the local electrogram recorded from the ischemic zone represents in fact the intrinsic deflection caused by the delayed activity. In the diagram, the flow of injury current at the moment indicated by the dotted line is schematically depicted. At the moment of the cardiac cycle, the injury current produces current sources at the nonischemic side of the border and current sinks on the ischemic side. (Reproduced from [79] with permission of the American Heart Association.)

sium increases in the ischemic zone as early as 30 s after coronary occlusion and demonstrates a characteristic triphasic response [69, 171]. The initial rise occurs over the first 4 to 8 min and is reversible. This rise is followed by a second plateau phase. The third phase occurs after 30 min of ischemia with an increase which is heterogeneous and irreversible and concomitant with the onset of myocardial necrosis [69, 171]. Extracellular potassium can increase to as high as 15 mM in the central ischemic zone within the first phase [69, 96, 173]. Inhomogeneity is most marked in the ischemic border zone [25], but can also be considerable between endocardium and epicardium [69, 175]. An increase in extracellular potassium can elicit a number of electrophysiologic alterations including: 1) a decrease in resting membrane potential; 2) a decrease in membrane responsiveness with resultant decreases in V_{max} of phase 0 depolarization and amplitude, leading to a decreased conduction velocity [183]; and 3) a decrease in action potential duration due to a paradoxical increase in potassium conductance [72]. However, internal longitudinal resistance is not increased [45]. Although the increase in extracellular K^+ appears to relate directly to the extent of cellular depolarization, as reflected by the changes in the TQ segment of the unipolar electrogram [175], the rise in extracellular potassium does not account completely for the profound electrophysiologic alterations characteristic of early ischemia [47].

The mechanisms responsible for the early increase in extracellular potassium in reversibly injured cells has not yet been clarified. Although inhibition of Na^+/K^+-

ATPase has been postulated as one mechanism, this is unlikely to account for the rise in extracellular K^+ since intracellular Na^+ remains low during 15 min of ischemia [96]. Enhanced K^+ efflux may be secondary to efflux of lactate and inorganic phosphate, which are also generated intracellularly during early ischemia, in an attempt to maintain charge neutrality [96, 101]. It is also possible that ATP-dependent K^+-channels may contribute [128, 172], since glibenclamide, a sulfonylurea which blocks the ATP-dependent K^+-channel, has antiarrhythmic activity during early ischemia independent of direct effects on glycolysis [88]. However, these channels are usually activated when ATP falls to very low levels (<1 mM) [128], much lower than that noted during early ischemia or hypoxia [166]. Subcellular sites near the membrane ATP-dependent K^+-channel may exhibit a more rapid decline in ATP sufficient to activate this channel and thereby increase extracellular K^+. Another possibility is that other concomitants of ischemia such as lysophosphoglycerides or long-chain acylcarnitines, amphiphiles which accumulate in membranes in ischemic myocardium (see below), may interact with the membrane bound ATP-sensitive K^+-channels, although no direct data regarding this mechanism have been published.

Alterations in Calcium during Early Myocardial Ischemia

It has been hypothesized over several years that ischemia leads to an increase in intracellular calcium. However, until recently, measurements of intracellular calcium transients during ischemia have been hampered by technical limitations. Marban et al. [117], utilizing NMR spectroscopy to detect the 5,5'-difluoro derivative of the calcium chelator BAPTA, have demonstrated a threefold increase in cytosolic calcium after 10 to 20 min of ischemia, but no change after 5 min of ischemia. The slow time-course of the increase in intracellular calcium may be secondary to the extensive buffering capacity of the indicator BAPTA. This conclusion is supported by the recent report from Lee et al. [107], using indo-1 which does not buffer calcium to the same degree as BAPTA, wherein calcium transients were recorded from the epicardial surface of the isolated perfused rabbit heart exposed to ischemia. A marked increase in both systolic and end-diastolic cystolic Ca^{2+} was reported, with the most rapid increase following only 30 s of ischemia, with a subsequent plateau at 90 s. Defects in sarcoplasmic reticulum (SR) calcium transport have also been observed as early as 7 min of ischemia [104, 139]. The inhibition of calcium uptake into the SR appears to be due to inactivation of Ca^{2+}, Mg^{2+}-ATPase, followed by uncoupling of calcium transport from ATP hydrolysis [104, 139].

The increased cytosolic calcium may mediate several of the electrophysiologic alterations during early ischemia. This is supported by the antifibrillatory effect of calcium channel blocking agents in this setting [16, 22, 52, 92], although independent salutary effects on coronary flow or myocardial MVO_2 cannot be excluded. The enhanced cytosolic Ca^{2+}, in concert with Ca^{2+}-mediated Ca^{2+} release from the sarcoplasmic reticulum [49], may activate a transient inward current and potentially initiate delayed after-depolarizations and triggered activity [179] as described above. Ryanodine, which prevents release of Ca^{2+} from SR at micromolar concentrations, also is antiarrhythmic during early ischemia [165], further supporting this conclusion. Whether the transient inward current is due to a nonspecific Ca^{2+}-activated cation channel [24] or electrogenic Na^+-Ca^{2+} exchange [90], or possibly both [82], remains to be clarified. In addition, increased intracellular Ca^{2+} has been found to increase internal longitudinal resistance [43] and thereby may contribute to slow conduction and block and the development of reentry.

Contribution of Hypoxia during Early Myocardial Ischemia

Myocardial ischemia leads to a rapid reduction in the pO_2 within the ischemic zone [149]. The most striking electrophysiologic alteration induced by hypoxia is a shortening of the action potential duration along with a decrease in resting membrane potential and V_{max} of phase 0 depolarization [121]. Voltage clamp studies have indicated that the hypoxia-induced decrease in repolarization time is due to an increase in the time-independent (background) potassium current with no effect on the slow inward calcium current [168]. Hypoxia also increases internal longitudinal resistance [74, 181] and decreases the space constant [74] without altering the time constant or membrane capacitance. The above electrophysiologic alterations are exaggerated in the presence of low glucose and attenuated by high extracellular glucose concentrations, suggesting that the effects of hypoxia may be related to decreased ATP-production via glycolysis [121]. However, in isolated cell or tissue preparations superfused with artificial solutions, as used in the above studies, it is very difficult to achieve the low levels of pO_2 observed during ischemia in vivo. If the pO_2 is not sufficiently low ($pO_2 < 15$ mmHg), inhibition of β-oxidation, which occurs in vivo, will not be replicated in vitro. Inhibition of β-oxidation is a critical event during ischemia and would be required to elicit an increase in long-chain acylcarnitines (see below).

Contribution of Acidosis during Early Myocardial Ischemia

Ischemia leads to a decrease in both intracellular and extracellular pH. Extracellular pH continues to fall to a plateau between 5.5 and 6.0 [70]. Acidosis per se leads to a decrease in V_{max} of phase 0 depolarization and conduction velocity [119, 126] as well as an increase in internal longitudinal resistance [42]. Acidosis will also dramatically modify the activity of several intracellular enzymes critical for the catabolism of metabolites known to accumulate in ischemic myocardium (see below).

Contributions of the Adrenergic Nervous System

Myocardial ischemia leads to activation of myocardial sensory receptors and thereby elicits an increase in afferent nerve activity, as well as a reflex increase in efferent sympathetic nerve activity [114]. Neural activity during ischemia may be nonuniform, with increased activity in some cardiac sympathetic nerves and decreased activity in others [105]. Such asymmetric sympathetic input to the heart could be potentially arrhythmogenic with inhomogeneous alterations in conduction, repolarization, and automaticity [146] providing the milieu for not only reentrant excitation but also nonreentrant mechanisms including EADs and DADs. Addi-

154

tional evidence suggesting the deleterious influence of asymmetric efferent sympathetic input during ischemia is derived from studies using stellate ganglion stimulation and ablation. Stimulation of the left stellate ganglion during ischemia leads to a decrease in the ventricular fibrillation threshold, while stimulation of the right stellate ganglion actually increases VF threshold [154]. Chronic ablation of the left stellate ganglion decreases malignant arrhythmias during ischemia, whereas right stellectomy enhances arrhythmogenesis [154]. However, acute (as opposed to chronic) bilateral stellectomy does not attenuate ventricular tachycardia and fibrillation during ischemia [60]. The inconsistency in these findings may relate to the variable degrees of denervation and residual intramyocardial catecholamines after acute but not chronic denervation. These results strongly suggest that asymmetric adrenergic neural input can contribute significantly to the development of malignant arrhythmias, particularly during ischemia.

Whether norepinephrine is released from nerve terminals within the ischemic zone independent of nerve activation has been hypothesized but unresolved. Overall, the release of norepinephrine into the venous effluent during ischemia is not very prominent [36]. However, since norepinephrine reuptake remains intact during ischemia [137], enhanced release could occur without an increase in the concentration in the venous effluent. In addition, the venous effluent reflects predominantly flow from the nonischemic zone, with little or no flow being contributed from the ischemic zone. Thus, the lack of release into the venous effluent cannot be interpreted as a lack of enhanced release of norepinephrine from sympathetic nerve terminals. This is supported by data indicating that reperfusion of previously ischemic myocardium results in a marked release of norepinephrine into the venous effluent [142], consistent with washout of catecholamines which had accumulated during the ischemic period. Therefore, it is likely that release of intramyocardial catecholamines occurs during early ischemia and may contribute significantly to arrhythmogenesis. In the following section, we will present a synopsis of the currently available literature pertaining to the influence of catecholamine stimulation of β- or α-adrenergic receptors in the ischemic myocardium and their influence on arrhythmogenesis.

Beta-Adrenergic Receptor Stimulation

Under normal physiologic conditions, adrenergic control of the myocardium is mediated primarily through β-adrenergic receptors. Intracellular mediation of the effects of β-adrenergic receptor stimulation involves the production of 3′,5′-cyclic adenosine monophosphate (cAMP) through stimulation of adenylate cyclase and the subsequent activation of cAMP-dependent protein kinases [109]. This interaction between the β-adrenergic receptor and activation of adenylate cyclase is functionally coupled by a guanine-nucleotide binding protein (G-protein). Cyclic-AMP dependent protein kinase has been shown to phosphorylate a component of the sarcolemmal calcium channels leading to an increase in the magnitude and a decrease in threshold for activation of the slow inward current carried by Ca^{2+}. Alteration of calcium flux by β-adrenergic stimulation also occurs through cAMP-dependent phosphorylation of phospholamban [109]. Beta-adrenergic receptor stimulation also appears to regulate intracellular Ca^{2+} through a cAMP-independent modulation of

155

calcium channels, which may contribute to the electrophysiologic effects of β-adrenergic stimulation [75].

In the setting of early myocardial ischemia, β-adrenergic stimulation could be potentially arrhythmogenic through several mechanisms including: 1) an increase in sinus rate which has been shown to increase the extent of conduction delay and functional conduction block [98]; 2) an increase in the slow inward current carried by Ca^{2+} leading to an enhancement of the degree of slow conduction and block; 3) an inhomogeneous reduction in the refractory period between the ischemic and non-ischemic cells [146]; and 4) an increase in the amplitude of delayed after-depolarizations resulting in an increase in triggered rhythms [179]. In addition, the increase in heart rate would increase myocardial oxygen demand and may directly enhance the extent of myocardial ischemia.

Although β-adrenergic receptor blockade during early myocardial ischemia has been shown to reduce the occurrence of both ventricular tachycardia and ventricular fibrillation [93, 132], the results have been inconsistent [5] which may be due to intrinsic differences in the properties of the various β-adrenergic blocking agents. In contrast, a number of large multicenter clinical trials have clearly demonstrated that β-adrenergic blockade leads to a significant reduction in sudden death in patients in the chronic recovery phase after myocardial infarction [120].

The arrhythmogenic influence of β-adrenergic stimulation during ischemia may be enhanced by an increase in myocyte β-adrenergic receptors. Beta-adrenergic receptor density has been shown to increase after 60 min of ischemia in vivo associated with increased coupling, as evidenced by an increased production of cAMP and phosphorylase kinase activity in response to isoproterenol [124, 125]. Maisel et al. [112] have shown recently that the increase in β-adrenergic receptors after 60 min of ischemia was associated with a redistribution from intracellular vesicles to the cardiac cell surface. However, the increase in β-adrenergic receptors occurs quite late after initiation of ischemia, suggesting that the alterations in receptor density may be related to irreversible cell damage, which occurs initially by 40 min of ischemia in the canine heart [83, 141]. In studies from our laboratory, increases in β-adrenergic receptor density in response to hypoxia in vitro are not apparent until evidence of irreversible cellular damage is present and are likely to be only secondary to breakdown of the sarcolemmal barrier rather than a true increase in β-adrenergic receptor density [67]. Therefore, it is unclear whether an increase in β-adrenergic receptor density or functional coupling contributes to or enhances the arrhythmogenic effects of catecholamines during early myocardial ischemia.

Alpha-Adrenergic Receptor Stimulation

Although β-adrenergic receptors are the primary mediators of the effects of catecholamines in the myocardium under normal physiologic conditions, α-adrenergic receptors are present on cardiac myocytes. Alpha-adrenergic stimulation of the normal heart leads to an increase in inotropy [152, 153] as well as an increase in repolarization time and refractory periods and a decrease in automaticity in isolated Purkinje fibers in vitro [143]. These electrophysiologic effects would be expected to be antiarrhythmic rather than arrhythmogenic.

In contrast, under pathologic conditions, and in particular ischemia, the effects of α-adrenergic stimulation of the myocardium are enhanced and several lines of evidence suggest that the result is arrhythmogenic. For example, in rabbit papillary muscle which has been depolarized with high extracellular K^+, α_1-adrenergic stimulation leads to an increase in the magnitude of the slow inward current (I_{si}) carried by Ca^{2+} [122]. With reperfusion after ischemia in the cat in vivo, the increase in intracellular calcium is abolished by α_1-adrenergic blockade administered even 2 min prior to reperfusion [158]. We and others [40, 162, 165, 177] have demonstrated that α_1-adrenergic blockade significantly reduces the incidence of ventricular tachycardia and ventricular fibrillation during early myocardial ischemia or subsequent reperfusion in a variety of species [10, 39, 40, 86, 155, 158, 162, 165, 177], an effect which is independent of changes in hemodynamics or coronary flow [158]. More recently, prazosin has been shown to attenuate markedly the incidence of arrhythmic death in conscious dogs exposed to posterior-lateral ischemia in the presence of a previous anterior myocardial infarction [174].

The enhanced α_1-adrenergic responsiveness characteristic of early myocardial ischemia may, in part, be mediated by an increase in the density of α_1-adrenergic receptors and/or their intracellular coupling. Within 30 min after ischemia, α_1-adrenergic receptor number increases twofold and persists during early reperfusion, before returning to baseline values after 15 min of sustained reperfusion [31], a time interval which correlates with the time-course of enhanced α_1-adrenergic responsiveness [158]. Analogous increases in myocardial α_1-adrenergic receptors during ischemia have been reported by others [44, 113]. The increase does not appear to originate from an intracellular "light vesicle" fraction as reported for the β-adrenergic receptor during ischemia [113] but may be due to exposure of membrane-bound α_1-adrenergic receptors. To evaluate the mechanisms underlying the alterations in α_1-adrenergic receptors and their intracellular coupling during early myocardial ischemia, we utilized calcium-tolerant, isolated adult canine myocytes exposed to hypoxia as an in vitro system. In this system, 10 min of severe hypoxia at 25 or 37°C leads to a two- to threefold reversible increase in α_1-adrenergic receptors [68] analogous to the changes seen during ischemia in vivo [31]. This increase in α_1-adrenergic receptors occurs prior to any evidence of irreversible cellular injury [68].

The enhanced α_1-adrenergic responsiveness during ischemia may be due to either the increase in α_1-adrenergic receptors and/or increased coupling to intracellular events. In many types of tissues, α_1-adrenergic receptor stimulation leads to hydrolysis of phosphatidylinositol (4,5)-bisphosphate (PIP_2) to inositol (1,4,5)-trisphosphate (IP_3) and diacylglycerol (DAG) by a specific phospholipase C [11, 13]. Intracellular IP_3 leads to an increase in cytosolic calcium through interaction with the sarcoplasmic reticulum [77], whereas DAG activates protein kinase C (PKC), leading to the phosphorylation of sarcolemmal calcium channels, thereby further increasing intracellular calcium [11–13, 77]. There is evidence to support the notion that IP_3 is the intracellular second messenger mediating the effects of α_1-adrenergic stimulation in cardiac cells [17, 182]. We have recently developed a novel method to measure directly the mass of IP_3 [66] which avoids problems associated with the use of labeled precursors. In isolated canine myocytes under normoxic conditions, norepinephrine elicited a three- to fourfold increase in IP_3 30 s after stimulation [65]. After 10 min of hypoxia, a time interval which produces a two- to threefold increase in α_1-adrenergic receptors, the threshold response for an increase in IP_3 was found to occur at 100-times lower concentrations of norepinephrine. Likewise, the concentration of norepinephrine required to elicit 50% of the increase in IP_3 (EC_{50}) was eight-times lower in hypoxic compared to normoxic myocytes [65]. These findings indicate that the hypoxia-induced increase in α_1-adrenergic receptors on myocytes is associated with enhanced coupling to the production of intracellular IP_3, which could account for the increased α_1-adrenergic responsiveness during myocardial ischemia in vivo.

Since α_1-adrenergic blockade inhibits the development of delayed after-depolarization induced by hypoxia in vitro [123], this would suggest that α_1-adrenergic stimulation in vivo may also contribute to the development of the nonreentrant mechanism underlying ventricular tachycardia during early myocardial ischemia [133, 134] as well as reperfusion [135]. The coupling between the production of IP_3 and a subsequent increase in intracellular calcium could lead to activation of the transient inward current carried through a nonspecific cation channel [24] and/or Na^+/Ca^{2+}-exchange [90]. However, additional data will be required to prove this hypothesis both in vitro as well as in vivo during ischemia.

Contribution of Lysophosphatidylcholine to Arrhythmogenesis during Ischemia

The role of LPC and other lysophosphatides during ischemia has been recently reviewed in detail [29] and results will only be summarized briefly here. Lysophosphatidylcholine (LPC), an amphipathic phospholipid, is formed from phosphatidylcholine (PC), the predominant phospholipid of the cardiac sarcolemma, by cleavage of the sn-2 fatty acid by phospholipase A_2 (PLA_2) [29]. LPC is usually present in very small concentrations because of the overwhelming capacity of at least three catabolic enzymes (Fig. 11): LPC acyltransferase, which reesterifies LPC to PC; lysophospholipase, which cleaves the sn-1 fatty acid yielding glycerophosphorylcholine (GPC); and lysophospholipase transacylase, which transfers a fatty acid from one LPC molecule to another producing PC and GPC [29].

LPC accumulates in ischemic myocardium [33, 115, 116, 157] and in venular and lymphatic effluents from ischemic regions [4, 159]. Several of the metabolic alterations occuring during early ischemia likely contribute to the accumulation of LPC. First, local catecholamine release has been shown to increase the activity of PLA_2 acting through either β-adrenergic receptors [53] or α-adrenergic receptors [18, 71]; acidosis reduces markedly the activity of one of the major catabolic enzymes for LPC, membrane-bound lysophospholipase [58]; and accumulation of long-chain acylcarnitines markedly inhibits two of the other major catabolic enzymes for LPC, cytosolic lysophospholipase, as well as lysophospholipase transacylase [55, 57].

As LPC accumulates during ischemia, it becomes incorporated into the lipid bilayers. Because this amphiphile contains only one fatty acyl group, it exhibits physical characteristics distinctly different from diacyl phospholipids. The incorporation of LPC into the lipid bilayer disrupts the orderly packing of adjacent phospholipid molecules and thus leads to a marked increase in membrane fluidity [50]. This affects not only the physical properties of the membrane but also the function of integral membrane proteins, including ion channels. For example, exogenous LPC induces reversible electrophysiologic alterations in vitro, including decreases in the resting

158

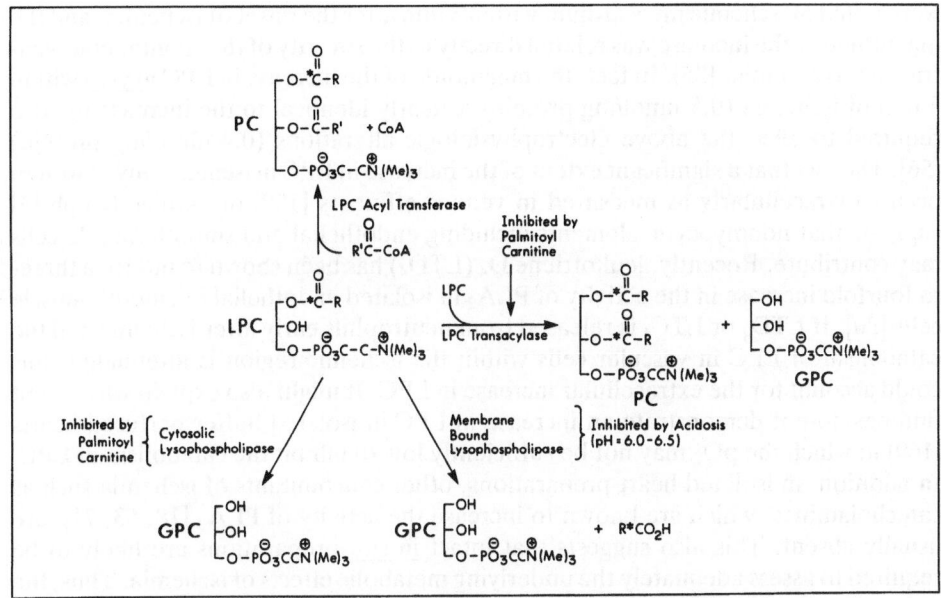

Fig. 11. Enzymatic pathways for the catabolism of lysophosphatidyl choline (LPC). Clockwise from top: LPC acyltransferase reesterifies the LPC to phosphatidylcholine (PC). LPC transacylase, which is inhibited by palmitoyl carnitine, transfers the fatty acid side chain from one molecule of LPC to another producing PC and glycerophosphoryl choline (GPC). Lysophospholipase cleaves the remaining fatty acid from LPC, leaving GPC and free fatty acid ($R*CO_2H$). The membrane-bound form of lysophospholipase is inhibited by acidosis, and the cytosolic form is inhibited by palmitoyl carnitine. (Reproduced from [28] with permission of Kluwer Academic Press.)

membrane potential, maximal rate of upstroke of phase 0 (V_{max}), and action potential duration [26, 32], alterations analogous to those which occur during early myocardial ischemia in vivo. These changes occur when as little as 1–2 mole % of LPC is incorporated into the sarcolemma based on quantitative electron microscopic autoradiography [56, 147]. The electrophysiologic effects of LPC are enhanced markedly in the presence of acidosis (pH = 6.8), comparable to that observed during early ischemia in vivo [32, 159]. LPC also leads to additional electrophysiologic effects including: 1) action potentials dependent exclusively on the slow inward calcium current [33]; 2) an increased membrane resistance and internal longitudinal resistance, and increased length and time constants [8]; 3) biphasic effects on excitability [8]; and 4) nonspecific depression of membrane channels [21]. These effects suggest that LPC may contribute to slow conduction, block, and the development of reentrant arrhythmias during early ischemia. LPC also possesses additional biochemical effects including a direct stimulation of adenylate cyclase resulting in an increase in cAMP [3] and intracellular calcium [156], suggesting that it may contribute to the development of triggered activity due to delayed after-depolarizations. We have recently demonstrated that LPC does indeed lead to DADs and triggered activity, an effect which is enhanced by catecholamines and which persists even in the presence of hyperkalemia and acidosis [136] (Fig. 12). In vivo, significant increases in LPC were noted in ischemic myocardium within 3 min after the onset of ischemia, and the magnitude of the increase was related directly to the severity of the spontaneous ventricular arrhythmias [35]. In fact, the magnitude of the increase in LPC in vivo within 3 min of ischemia (0.5 nmol/mg protein) is nearly identical to the increase in vitro required to elicit the above electrophysiologic alterations (0.4 nmol/mg protein) [56]. The fact that a significant extent of the increase in LPC in ischemic myocardium occurs extracellularly as measured in venous effluents [159] or cardiac lymph [4] suggests that nonmyocytic elements including endothelial and smooth muscle cells may contribute. Recently, leukotriene D_4 (LTD$_4$) has been shown to induce a three- to fourfold increase in the activity of PLA$_2$ in isolated endothelial or smooth muscle cells [20]. If LTD$_4$ or LTC$_4$ is released from neutrophils early after ischemia and the catabolism of LPC in vascular cells within the ischemic region is attenuated, this could account for the extracellular increase in LPC. It might also explain why recent findings do not demonstrate an increase in LPC in isolated buffer perfused hearts [169] in which the pO$_2$ may not be sufficiently low to inhibit the catabolism of LPC. In addition, in isolated heart preparations, other concomitants of ischemia such as catecholamines, which are known to increase the activity of PLA$_2$ [18, 53, 71], are usually absent. This also suggests that intact in situ preparations are likely to be required to assess adequately the underlying metabolic effects of ischemia. Thus, the accumulation of LPC may contribute significantly to arrhythmogenesis during early ischemia.

Long-Chain Acylcarnitine

Long-chain acylcarnitines are endogenous amphiphiles which are structurally similar to LPC (Fig. 13) [29]. Fatty acids metabolism (Fig. 14) begins with the uptake of free

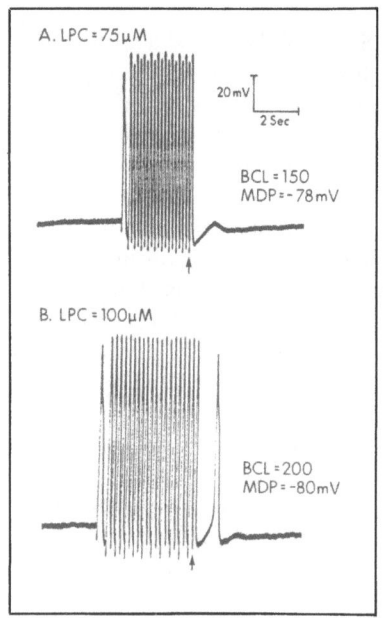

Fig. 12. Representative examples of delayed after-depolarizations and triggered activity in two canine Purkinje fiber preparations induced by infusion of lysophosphatidylcholine (LPC) at 75 μM (A) and at 100 μM (B). A) After a drive train of 20 beats at a cycle length (BCL) of 150 msec, the stimulus is terminated at the arrow and is followed by a delayed after-depolarization. B) Termination of the drive train of 21 beats at the arrow (BCL 200 msec) is followed by a triggered beat. (Reproduced from [136] with permission of the American Heart Association.)

fatty acids by the cardiac myocyte, followed by binding to an intracellular fatty acid binding protein and transport to the mitochondria [29]. In the outer mitochondrial membrane, the fatty acids are transesterified to coenzyme A to produce acyl CoA. Since acyl-CoA cannot readily cross the inner mitochondrial membrane, the fatty acid is transesterified to carnitine, producing acylcarnitine by the enzyme carnitine acyltransferase I (CAT-I). After crossing the inner mitochondrial membrane in exchange for free carnitine via carnitine-acylcarnitine translocase, the long-chain acylcarnitine is then transesterified back to CoA to form acyl-CoA on the inside or matrix of the mitochondria by the enzyme carnitine acyltransferase II (CAT-II). The acyl-CoA now undergoes β-oxidation, with catabolism of the long-chain fatty acyl-CoA to two-carbon acetyl-CoA groups producing NADH and $FADH_2$. The acetyl-CoA then enters the citric acid cycle with production of additional NADH, which enters the electron transport chain for oxidation and subsequent production of ATP [29].

Because the electron transport chain is oxygen dependent, myocardial ischemia and severe hypoxia inhibit electron transport, resulting in a large increase in NADH and $FADH_2$ which, through a negative feedback mechanism, inhibits β-oxidation. However, long-chain acylcarnitines continue to be produced and thereby accumulate in ischemic myocardium in vivo [27, 110]. Like LPC, exogenous palmitoylcarnitine induces reversible electrophysiologic alterations in normoxic myocardium in vitro analogous to that seen during ischemia, in vivo, including a marked reduction in resting membrane potential, V_{max} of phase 0, and action potential amplitude and duration [29, 32]. Whether these electrophysiologic effects are due to inhibition of Na^+/K^+-ATPase [2] or sarcoplasmic reticulum Ca^{2+}-ATPase [1], or direct activation of calcium channels leading to an increase in intracellular calcium [76, 160, 161] remains to be clarified.

Although studies performed in vitro with exogenous extracellular long-chain acylcarnitines provide evidence suggesting that this metabolite may contribute to the electrophysiologic alterations during ischemia, long-chain acylcarnitines accumulate intracellularly during ischemia or hypoxia and are not released into the extracellular space until irreversible cell damage ensues. To evaluate the effects of endogenous

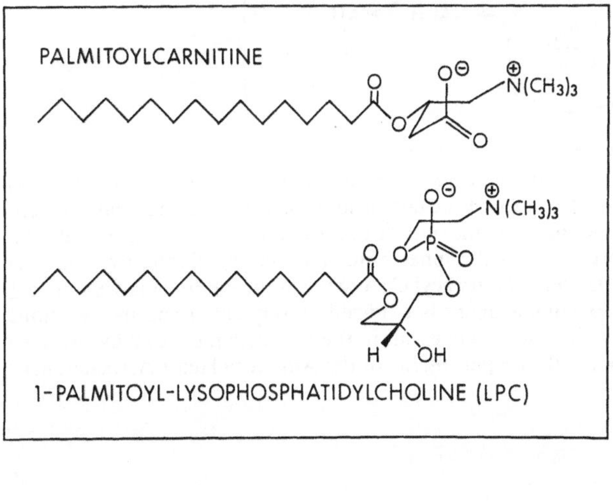

PALMITOYLCARNITINE

1-PALMITOYL-LYSOPHOSPHATIDYLCHOLINE (LPC)

Fig. 13. A comparison of the structures of long-chain acylcarnitine and lysophosphatidylcholine.

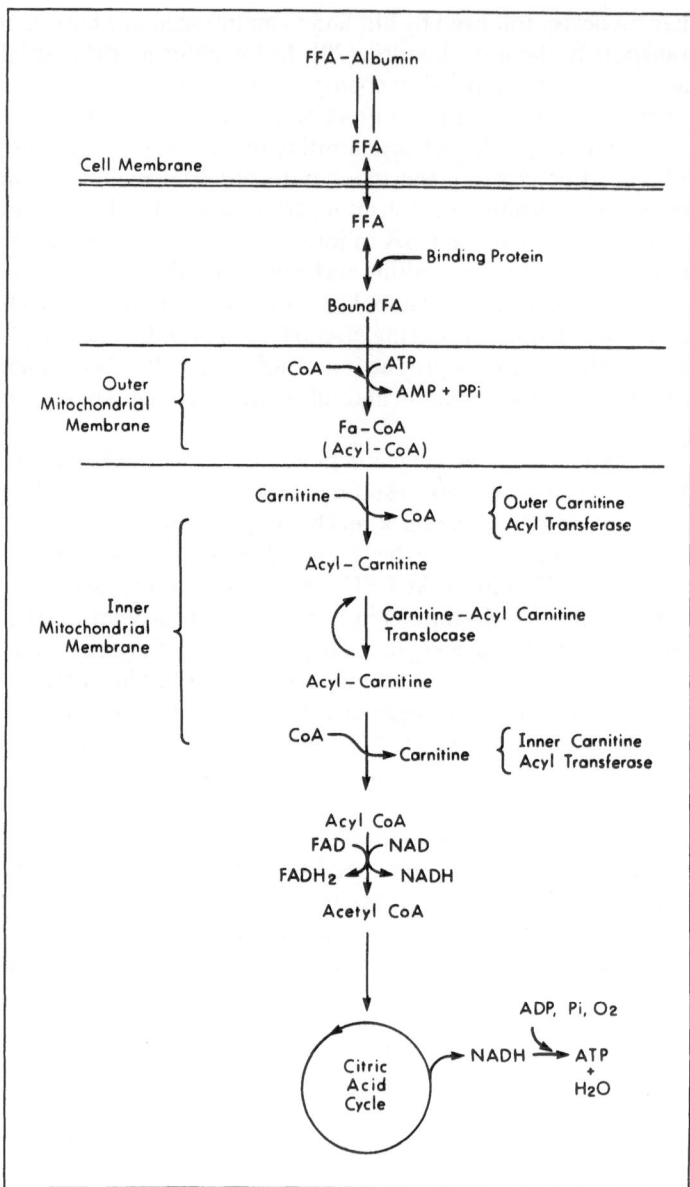

Fig. 14. Diagrammatic representation of the esterification of free fatty acids and subsequent transport into, and β-oxidation within, mitochondria. Free fatty acids transverse the sarcolemmal membrane and subsequently activate CoA lipases on the outside of the outer mitochondrial membrane. Long-chain thioesters are transported across the inner mitochondrial membrane by sequential transesterification, resulting in the delivery of fatty acyl CoA to the mitochondrial matrix compartment. Fatty acyl CoA in the matrix compartment is oxidized by sequential enzyme reactions, resulting in reducing equivalents, which ultimately result in the production of ATP by oxidative phosphorylation. (Reproduced from [29] with permission of the American Heart Association.)

intracellular accumulation of long-chain acylcarnitines, isolated myocytes were exposed to hypoxia in vitro and the subcellular localization of long-chain acylcarnitines was delineated using electron microscopic autoradiography procedures in prelabeled cells [102].

In isolated myocytes, hypoxia for 10 min induced a marked increase in total cellular long-chain acylcarnitines analogous to that found during ischemia in vivo [102]. Although the total cellular long-chain acylcarnitines increased over threefold, the content in sarcolemma increased 70-fold with marked increases also in the mitochondrial and cytoplasmic membrane structures, particularly SR (Fig. 15). Pretreatment of cells with sodium-2-[5-(4-chlorophenyl)-pentyl]-oxirane-2-carboxylate (POCA), an inhibitor of carnitine acyltransferase I (CAT I), inhibited the accumulation of long-chain acylcarnitines (Fig. 15) and attenuated markedly the electrophysiologic alterations induced by hypoxia [102]. Thus, the sarcolemmal accumulation of long-chain acylcarnitines contributes significantly to the electrophysiologic alterations induced by hypoxia and suggests that inhibition of accumulation of this amphiphile by inhibition of carnitine acyltransferase I may be particularly antiarrhythmic during myocardial ischemia in vivo.

Based on in vitro studies outlined above, the accumulation of LPC during ischemia in vivo may be due, in part, to inhibition by long-chain acylcarnitines of two major catabolic enzymes, cytosolic lysophospholipase and lysophospholipase transacylase (see Fig. 11). Therefore, inhibition of the accumulation of long-chain acylcarnitine by POCA may secondarily prevent the accumulation of LPC as well. In a series of recent experiments in vivo, occlusion of the proximal LAD coronary artery resulted in a threefold increase in long-chain acylcarnitines and a twofold increase in LPC

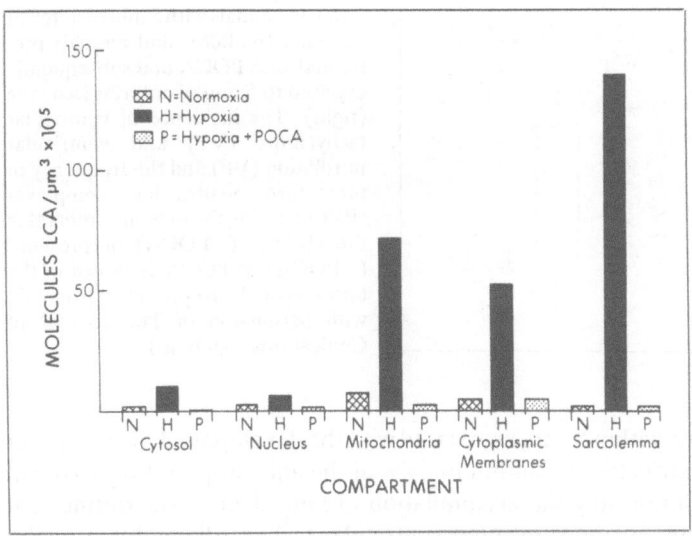

Fig. 15. Long-chain acylcarnitine (LCA) molecules/μm^3 of each source structure following normoxic (N), hypoxic (H), or hypoxic plus POCA (P) perfusion. Hypoxia resulted in an increased number of LCA molecules/μm^3 of each source compartment. POCA (10 μM), an inhibitor of carnitine acyltransferase I (CAT I) prevented accumulation of LCA. (Data derived from [102] with permission of the American Heart Association.)

163

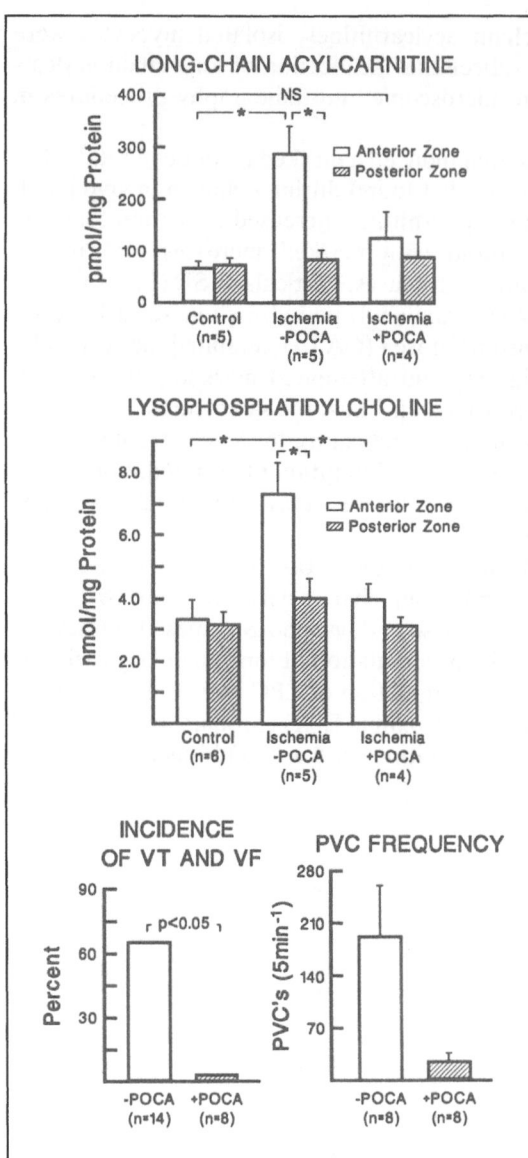

Fig. 16. Tissue content of long-chain acylcarnitine (top panel) and lyso-phosphatidylcholine (middle panel) in the anterior or ischemic zone and posterior or nonischemic zone of control nonischemic feline hearts in vivo (left), in animals with 5 min of anterior ischemia (middle), and animals pretreated with POCA and subsequently exposed to 5 min of anterior ischemia (right). The incidence of ventricular tachycardia (VT) and ventricular fibrillation (VF) and the frequency of premature ventricular complexes (PVCs) during the ischemic interval in the absence (−POCA) or presence (+POCA) of POCA is shown in the bottom panel. (Reproduced from [27] with permission of The Journal of Clinical Investigation.)

within the ischemic zone within 5 min, associated with the development of ventricular tachycardia and/or ventricular fibrillation in 64% of the animals [27]. Pretreatment with POCA prevented not only the accumulation of long-chain acylcarnitines but also LPC in ischemic tissue and prevented ventricular tachycardia and ventricular fibrillation (Fig. 16). The protective influence of POCA was independent of changes in regional myocardial blood flow or hemodynamics [27].

Because long-chain acylcarnitine accumulation in the sarcolemma during hypoxia would be expected to alter membrane fluidity, we assessed whether accumulation of this amphiphile contributed to the concomitant increase in α_1-adrenergic receptors discussed above. Pretreatment of isolated myocytes with POCA prior to hypoxia prevented not only the accumulation of long-chain acylcarnitines but also the twofold to threefold increase in α_1-adrenergic receptors [68]. Analogous increases in α_1-adrenergic receptors were induced by addition of exogenous long-chain acylcarnitines to normoxic cells [68].

In concert, these findings indicate that inhibition of CAT-I induces a potent antiarrhythmic effect during early myocardial ischemia. This effect is likely mediated by not only preventing the direct electrophysiologic effects of long-chain acylcarnitines and LPC, but also preventing the increased exposure and coupling of myocytic α_1-adrenergic receptors. Although POCA per se may not be useful clinically because of its ability to prevent utilization of fatty acids, the development of other inhibitors of CAT-I which may be specific for ischemic tissue is likely to be an important pharmacologic approach to attenuate the incidence of sudden cardiac death in man.

Conclusion

Myocardial ischemia is characterized by a number of ionic and biochemical alterations which contribute to the genesis of malignant ventricular arrhythmias. Detailed three-dimensional mapping during early myocardial ischemia has demonstrated that both intramural reentry and nonreentrant mechanisms contribute to the initiation and maintenance of ventricular tachycardia. The complex interactions between these two different mechanisms can lead to acceleration of the tachycardia by intramural reentry, with rapid inhomogeneous recovery of excitability leading to the multiple reentrant circuits and multiple simultaneous activations characteristic of ventricular fibrillation. The nature of the nonreentrant mechanism remains to be clarified but may be due to delayed after-depolarization secondary to an increase in intracellular calcium. The development of more effective therapeutic agents to attenuate the incidence of sudden cardiac death is likely to depend on a more thorough understanding of the biochemical alterations occuring in the sarcolemma in response to ischemia. Currently available evidence suggests that the development of unique inhibitors of carnitine acyltransferase I may be one such approach.

References

1. Adams RJ, Cohen DW, Gupte S, Johnson JD, Wallick ET, Wang T, Schwartz A (1979) In vitro effects of palmitylcarnitine on cardiac plasma membrane Na,K-ATPase, and sarcoplasmic reticulum Ca^{2+}-ATPase and Ca^{2+} transport. J Biol Chem 254:12 404−12 410
2. Adams RJ, Pitts BJ, McMillin-Wood J, Gende OA, Wallnick ET, Schwartz A (1979) Effect of palmitylcarnitine on ouabain binding to Na,K-ATPase. J Mol Cell Cardiol 11:941−959
3. Ahumada GG, Bergmann SR, Carlson E, Corr PB, Sobel BE (1979) Augmentation of cyclic AMP content induced by lysophosphatidyl choline in rabbit hearts. Cardiovasc Res 13:377−382
4. Akita H, Creer MH, Yamada KA, Sobel BE, Corr PB (1986) Electrophysiologic effects of intracellular lysophosphoglycerides and their accumulation in cardiac lymph with myocardial ischemia in dogs. J Clin Invest 78:271−280
5. Allen JD, Shanks RG, Zaida SA (1971) Effects of lignocaine and propranolol on experimental cardiac arrhythmias. Br J Pharmacol 42:1−12

6. Anderson GJ, Reiser J, Gough WB, Nydegger CC (1983) Intramyocardial current flow in acute coronary occlusion in the canine heart. J Am Coll Cardiol 1:436–443
7. Antzelevitch C, Jalife J, Moe GK (1980) Characteristicts of reflection as a mechanism of reentrant arrhythmias and its relationship to parasystole. Circulation 61:182–191
8. Arnsdorf MF, Sawicki GJ (1981) The effects of lysophosphatidylcholine, a toxic metabolite of ischemia, on the components of cardiac excitability in sheep Purkinje fibers. Circ Res 49:16–30
9. Bashe WJ Jr, Baba N, Keller MD, Geer JC, Anthony JR (1975) Pathology of atherosclerotic heart disease in sudden death. II. The significance of myocardial infarction. Circulation 52 (Suppl III):III-63–III-69
10. Benfey BG, Elfellah MS, Ogilvie RI, Varma DR (1984) Anti-arrhythmic effects of prazosin and propranolol during coronary artery occlusion and reperfusion in dogs and pigs. Br J Pharmacol 82:717–725
11. Berridge MJ (1984) Inositol trisphosphate and diacylglycerol as second messengers. Biochem J 220:345–360
12. Berridge MJ (1987) Inositol trisphosphate and diacylglycerol: two interacting second messengers. Ann Rev Biochem 56:159–193
13. Berridge MJ, Irvine RF (1984) Inositol trisphosphate, a novel second messenger in cellular signal transduction. Nature 312:315–321
14. Bigger JT Jr, Dresdale RJ, Heissenbuttel RH, Weld FM, Wit AL (1977) Ventricular arrhythmias in ischemic heart disease: Mechanisms, prevalence, significance and management. Progr Cardiovasc Dis 19:255–300
15. Boineau JP, Cox JL (1973) Slow ventricular activation in acute myocardial infarction: A source of reentrant premature ventricular contraction. Circulation 48:702–713
16. Brooks WW, Verrier RL, Lown B (1980) Protective effect of verapamil on vulnerability to ventricular fibrillation during myocardial ischaemia and reperfusion. Cardiovasc Res 14:295–302
17. Brown JH, Buxton IL, Brunton LL (1985) α_1-Adrenergic and muscarinic cholinergic stimulation of phosphoinositide hydrolysis in adult rat cardiomyocytes. Circ Res 57:532–537
18. Burch RM, Luini A, Axelrod J (1986) Phospholipase A_2 and phospholipase C are activated by distinct GTP-binding proteins in response to α_1-adrenergic stimulation in FRTL5 thyroid cells. Proc Natl Acad Sci 83:7201–7205
19. Cardinal R, Janse MJ, van Eeden I, Werner G, d'Alnoncourt CN, Durrer D (1981) The effects of lidocaine on intracellular and extracellular potentials, activation, and ventricular arrhythmias during acute regional ischemia in the isolated porcine heart. Circ Res 49:792–806
20. Clark MA, Littlejohn D, Conway TM, Mong S, Steiner S, Crooke ST (1986) Leukotriene D_4 treatment of bovine aortic endothelial cells and murine smooth muscle cells in culture results in an increase in phospholipase A_2 activity. J Biol Chem 261:10 713–10 718
21. Clarkson CW, Ten Eick RE (1983) On the mechanism of lysophosphatidylcholine-induced depolarization of cat ventricular myocardium. Circ Res 52:543–556
22. Clusin WT, Bristow MR, Baim DS, Schroeder JS, Jaillon P, Brett P, Harrison DC (1982) The effects of diltiazem and reduced serum ionized calcium on ischemic ventricular fibrillation in the dog. Circ Res 50:518–526
23. Cobb LA, Werner JA, Trobaugh GB (1980) Sudden cardiac death. I. A decade's experience with out-of-hospital resuscitation. Mod Concepts Cardiovasc Dis 49:31–36
24. Colquhoun D, Neher E, Reuter H, Stevens CF (1981) Inward current channels activated by intracellular Ca in cultured cardiac cells. Nature 294:752–754
25. Coronel R, Fiolet JWT, Wilms-Schopman FJG, Schaapherder AFM, Johnson TA, Gettes LS, Janse MJ (1988) Distribution of extracellular potassium and its relation to electrophysiologic changes during acute myocardial ischemia in the isolated perfused porcine heart. Circulation 77:1125–1138
26. Corr PB, Cain ME, Witkowski FX, Price DA, Sobel BE (1979) Potential arrhythmogenic electrophysiological derangements in canine Purkinje fibers induced by lysophosphoglycerides. Circ Res 44:822–832
27. Corr PB, Creer MH, Yamada KA, Saffitz JE, Sobel BE (1989) Prophylaxis of early ventricular fibrillation by inhibition of acylcarnitine accumulation. J Clin Invest 83:927–936

28. Corr PB, Dobmeyer DJ (1988) Amphipathic lipid metabolites and arrhythmogenesis: A perspective. In: Rosen M, Palti Y (eds) Lethal Arrhythmias Resulting from Myocardial Ischemia and Infarction. Kluwer Academic Publisher, pp. 91–104

29. Corr PB, Gross RW, Sobel BE (1984) Amphipathic metabolites and membrane dysfunction in ischemic myocardium. Circ Res 55:135–154

30. Corr PB, Pearle DL, Hinton JR, Roberts WC, Gillis RA (1976) Site of myocardial infarction. A determinant of the cardiovascular changes induced in the cat by coronary occlusion. Circ Res 39:840–847

31. Corr PB, Shayman JA, Kramer JB, Kipnis RJ (1981) Increased α-adrenergic receptors in ischemic cat myocardium. J Clin Invest 67:1232–1236

32. Corr PB, Snyder DW, Cain ME, Crafford WA Jr, Gross RW, Sobel BE (1981) Electrophysiological effects of amphiphiles on canine Purkinje fibers. Circ Res 49:354–363

33. Corr PB, Snyder DW, Lee BI, Gross RW, Keim CR, Sobel BE (1982) Pathophysiological concentrations of lysophosphatides and the slow response. Am J Physiol 243 (Heart and Circ 12):H187–H195

34. Corr PB, Witkowski FX (1983) Potential electrophysiologic mechanisms responsible for dysrhythmias associated with reperfusion of ischemic myocardium. Circulation 68 suppl I:I-16–I-24

35. Corr PB, Yamada KA, Creer MH, Sharma AD, Sobel BE (1987) Lysophosphoglycerides and ventricular fibrillation early after onset of ischemia. J Mol Cell Cardiol 19:34–53

36. Corr PB, Yamada KA, Witkowski FX (1986) Mechanisms controlling cardiac autonomic function and their relation to arrhythmogenesis. In: Fozzard HA et al. (eds). The Heart and Cardiovascular System. Scientific Foundations. Raven Press, New York, pp 1343–1403

37. Cranefield PF (1977) Action potentials, afterpotentials, and arrhythmias. Circ Res 41:415–423

38. Creer MH, Dobmeyer DJ, Corr PB (1989) Amphipathic lipid metabolites and arrhythmias during myocardial ischemia. In: Zipes DP, Jalife J (eds) Cardiac Electrophysiology and Arrhythmias. Grune and Stratton (in press)

39. Culling W, Penny WJ, Cunliffe G, Flores NA, Sheridan DJ (1987) Arrhythmogenic and electrophysiological effects of alpha adrenoceptor stimulation during myocardial ischaemia and reperfusion. J Mol Cell Cardiol 19:251–258

40. Davey MJ (1980) Relevant features of the pharmacology of prazosin. J Cardiovasc Pharmacol 2:S287–S301

41. Davies MJ, Thomas A (1984) Thrombosis and acute coronary artery lesions in sudden cardiac ischemic death. N Engl J Med 310:1137–1141

42. DeMello WC (1980) Influence of intracellular injection of H^+ on the electrical coupling in cardiac Purkinje fibres. Cell Biol Int Rep 4: 51–58

43. DeMello WC (1975) Effects of intracellular injection of calcium and strontium on cell communication in the heart. J Physiol 250:231–245

44. Dillon JS, Gu ZH, Nayler WG (1988) Alpha$_1$-adrenoceptors in the ischemic and reperfused myocardium. J Mol Cell Cardiol 20:725–735

45. Dominguez G, Fozzard H (1970) Influence of extracellular K^+ concentration on cable properties and excitability of sheep cardiac Purkinje fibers. Circ Res 26:565–574

46. Donaldson RM, Taggart P, Nashat F, Abed J, Richards AF, Noble D (1983) Study of the electrophysiological effects of early or subendocardial ischaemia with intracavitary electrodes in the dog. Clin Sci 65:579–588

47. Downar E, Janse MJ, Durrer D (1977) The effect of acute coronary occlusion on subepicardial transmembrane potentials in the intact porcine heart. Circulation 56:217–224

48. El-Sherif N, Hope RR, Scherlag BJ, Lazzara R (1977) Re-entrant ventricular arrhythmias in the late myocardial infarction period: I. Conduction characteristics in the infarction zone. Circulation 55:686–701

49. Fabiato A (1983) Calcium-induced release of calcium from the cardiac sarcoplasmic reticulum. Am J Physiol 245:C1–C14

50. Fink KL, Gross RW (1984) Modulation of canine myocardial sarcolemmal membrane fluidity by amphiphilic compounds. Circ Res 55:585–594

51. Fleet WF, Johnson TA, Graebner CA, Gettes LS (1985) Effect of serial brief ischemic episodes on extracellular K^+, pH, and activation in the pig. Circulation 72:922–932

167

52. Fondacaro JD, Han J, Yoon MS (1978) Effects of verapamil on ventricular rhythm during acute coronary occlusion. Am Heart J 96:81–86
53. Franson RC, Pang DC, Weglicki WB (1979) Modulation of lipolytic activity in isolated canine cardiac sarcolemma by isoproterenol and propranolol. Biochem Biophys Res Commun 90:956–962
54. Gilmour RF, Zipes DP (1980) Different electrophysiologic responses of canine endocardium and epicardium to combined hyperkalemia, hypoxia and acidosis. Circ Res 46:814–825
55. Gross RW (1983) Purification of rabbit myocardial cytosolic acyl CoA hydrolase, identity with lysophospholipase and modulation of enzymatic activity by endogenous cardiac amphiphiles. Biochemistry 22:5641–5646
56. Gross RW, Corr PB, Lee BE, Saffitz JE, Crafford WA Jr, Sobel BE (1982) Incorporation of radiolabeled lysophosphatidylcholine into canine Purkinje fibers and ventricular muscle: Electrophysiological, biochemical and autoradiographic correlations. Circ Res 51:27–36
57. Gross RW, Drisdel RC, Sobel BE (1983) Rabbit myocardial lysophospholipase-transacylase: purification, characterization, and inhibition by endogenous cardiac amphiphiles. J Biol Chem 258:15 165–15 172
58. Gross RW, Sobel BE (1982) Lysophosphatidylcholine metabolism in the rabbit heart. Characterization of metabolic pathways and partial purification of myocardial lysophospholipase-transacylase. J Biol Chem 257:6702–6708
59. Harris AS (1950) Delayed development of ventricular ectopic rhythms following experimental coronary occlusion. Circulation 1:1318–1328
60. Harris AS, Estandia A, Tillotson RF (1951) Ventricular ectopic rhythms and ventricular fibrillation following cardiac sympathectomy and coronary occlusion. Am J Physiol 65:505–512
61. Harris AS, Rojas GA (1943) Initiation of ventricular fibrillation due to coronary occlusion. Exp Med Surg 1:105–122
62. Hearse DJ (1979) Oxygen deprivation and early myocardial contractile failure: a reassessment of the possible role of adenosine triphosphate. Am J Cardiol 44:1115–1121
63. Hearse DJ, Opie LH, Katzeff IE, Lubbe WF, van der Werff TJ, Peisach M, Boulle G (1977) Characterization of the "border zone" in acute regional ischemia in the dog. Am J Cardiol 40:716–726
64. Hearse DJ, Yellon DM (1981) The "border zone" in evolving myocardial infarction: controversy or confusion. Am J Cardiol 47:1321–1334
65. Heathers GP, Evers AS, Corr PB (1989) Enhanced inositol trisphosphate response to α_1-adrenergic stimulation in hypoxic cardiac myocytes. J Clin Invest 83:1409–1413
66. Heathers GP, Juehne T, Rubin LJ, Corr PB, Evers AS (1989) Anion exchange chromatographic separation of inositol phosphates and their quantification by gas chromatography. Anal Biochem 176:109–116
67. Heathers GP, Lee PC, Yamada KA, Corr PB (1990) The influence of hypoxia on beta-adrenergic receptors and their intracellular coupling in isolated adult canine myocytes. submitted
68. Heathers GP, Yamada KA, Kanter EM, Corr PB (1987) Long-chain acylcarnitines mediate the hypoxia-induced increase in α_1-adrenergic receptors on adult canine myocytes. Circ Res 61:735–746
69. Hill JL, Gettes LS (1980) Effect of acute coronary occlusion on local myocardial extracellular K^+ activity in swine. Circulation 61:768–778
70. Hirche HJ, Franz C, Box L, Bissig R, Lang R, Schramm M (1980) Myocardial extracellular K^+ and H^+ increase and noradrenaline release as possible cause of early arrhythmias following acute coronary artery occlusion in pigs. J Mol Cell Cardiol 12:579–593
71. Ho AK, Klein DC (1987) Activation of α_1-adrenoceptors, protein kinase C or treatment with intracellular free Ca^{++} elevating agents increases pineal phospholipase A_2 activity. J Biol Chem 262:11764–11770
72. Hoffman BF, Cranefield PF (1960) Electrophysiology of the Heart. New York, McGraw-Hill Book Co.
73. Holland RP, Brooks H (1976) The QRS complex during myocardial ischemia. J Clin Invest 57:541–550
74. Ikeda K, Hiraoka M (1982) Effects of hypoxia on passive electrical properties of canine ventricular muscle. Pflügers Arch 393:45–50

75. Ingebretsen CG (1980) Interaction between alpha- and beta-adrenergic receptors and cholinergic receptors in isolated perfused rat heart: effects on cAMP-protein kinase and phosphorylase. J Cyclic Nuc Res 6:121–132
76. Inoue D, Pappano AJ (1983) L-Palmitylcarnitine and calcium ions act similarly on excitatory ionic currents in avian ventricular muscle. Circ Res 52:625–634
77. Irvine RF, Brown KD, Berridge MJ (1984) Specificity of inositol trisphosphate-induced calcium release from permeabilized Swiss-mouse 3T3 cell. Biochem J 222:269–272
78. Janse MJ, Cinca J, Morena H, Fiolet JWT, Kleber AG, de Vries GP, Becker AE, Durrer D (1979) The "border zone" in myocardial ischemia: an electrophysiological, metabolic, and histochemical correlation in the pig heart. Circ Res 44:576–588
79. Janse MJ, Kleber AG (1981) Electrophysiological changes and ventricular arrhythmias in the early phase of regional myocardial ischemia. Circ Res 49:1069–1081
80. Janse MJ, van Capelle FJL (1982) Electrotonic interactions across an inexcitable region as a cause of ectopic activity in acute regional myocardial ischemia: A study in intact porcine and canine hearts and computer models. Circ Res 50:527–537
81. Janse MJ, van Capelle FJL, Morsink H, Kleber AG, Wilms-Schopman F, Cardinal R, Naumann D' Alnoncourt C, Durrer D (1980) Flow of "injury" current and patterns of excitation during early ventricular arrhythmias in acute regional myocardial ischemia in isolated porcine and canine hearts. Evidence for two different arrhythmogenic mechanisms. Circ Res 47:151–165
82. January CT, Fozzard HA (1988) Delayed afterdepolarization in heart muscle: Mechanisms and relevance. Pharmacol Res 40:219–227
83. Jennings RB, Sommers HM, Smyth GA, Flack HA, Linn H (1960) Myocardial necrosis induced by temporary occlusion of a coronary artery in the dog. Arch Pathol 70:68–78
84. Jones CE, Thomas JX, Parker JC, Parker RE (1976) Acute changes in high energy phosphates, nucleotide derivatives, and contractile force in ischaemic and nonischaemic canine myocardium following coronary occlusion. Cardiovasc Res 10:275–282
85. Jorgensen L, Rowsell HC, Hovig T, Glynn MF, Mustard JF (1967) Adenosine diphosphate-induced platelet aggregation and myocardial infarction in swine. Lab Invest 17:616–644
86. Kane KA, Parratt JR, Williams FM (1984) An investigation into the characteristics of reperfusion-induced arrhythmias in the anaesthetized rat and their susceptibility to antiarrhythmic agents. Br J Pharmacol 82:349–357
87. Kannel WB, Doyle JT, McNamara PM, Quickenton P, Gordon T (1975) Precursors of sudden coronary death: Factors related to the incidence of sudden death. Circulation 51:606–613
88. Kantor PF, Coetzee WA, Dennis SC, Opie LH (1987) Effects of glibenclamide on ischemic arrhythmias (abstract). Circulation 76 (Suppl IV):IV–17
89. Kaplinsky E, Ogawa S, Balke W, Dreifus LS (1979) Role of endocardial activation in malignant ventricular arrhythmias associated with acute ischemia. J Electrocardiol 12:299–306
90. Kass RS, Lederer WJ, Tsien RW, Weingart R (1978) Role of calcium ions in transient inward currents and aftercontractions induced by strophanthidin in cardiac Purkinje fibres. J Physiol 281:187–208
91. Katzung BG, Hondeghem LM, Grant AO (1975) Cardiac ventricular automaticity induced by a current of injury. Pflügers Arch 360:193–197
92. Kaumann AJ, Serur JR (1975) Optical isomers of verapamil on canine heart: prevention of ventricular fibrillation induced by coronary artery occlusion, impaired atrioventricular conductance and negative inotropic and chronotropic effects. Naunyn-Schmiedeberg's Arch Pharmacol 291:347–358
93. Khan MI, Hamilton JT, Manning GW (1973) Early arrhythmias following experimental coronary occlusion in conscious dogs and their modification by β-adrenoceptor blocking drugs. Am Heart J 86:347–358
94. Kimura S, Bassett AL, Kohya T, Kozlovskis PL, Myerburg RJ (1986) Simultaneous recording of action potentials from endocardium and epicardium during ischemia in the isolated cat ventricle: relation of temporal electrophysiologic heterogeneities to arrhythmias. Circulation 74:401–409
95. Kimura S, Bassett AL, Saoudi NC, Cameron JS, Kozlovskis PL, Myerburg RJ (1986) Cellular electrophysiologic changes and "arrhythmias" during experimental ischemia and reperfusion in isolated cat ventricular myocardium. J Am Coll Cardiol 7:833–842

96. Kleber AG (1983) Resting membrane potential, extracellular potassium activity, and intracellular sodium activity during acute global ischemia in isolated perfused guinea pig hearts. Circ Res 52:442–450

97. Kleber AG, Janse MJ, van Capelle FJL, Durrer D (1978) Mechanism and time course of S-T and T-Q segment changes during acute regional myocardial ischemia in the pig heart determined by extracellular and intracellular recordings. Circ Res 42:603–613

98. Kleber AG, Janse MJ, Wilms-Schopman FJG, Wilde AAM, Coronel R (1986) Changes in conduction velocity during acute ischemia in ventricular myocardium of the isolated porcine heart. Circulation 73:189–198

99. Kleber AG, Riegger CB (1987) Electrical constants of arterially perfused rabbit papillary muscle. J Physiol 385:307–324

100. Kleber AG, Riegger CB, Janse MJ (1987) Electrical uncoupling and increase of extracellular resistance after induction of ischemia in isolated, arterially perfused rabbit papillary muscle. Circ Res 61:271–279

101. Kleber AG, Riegger CB, Janse MJ (1987) Extracellular K^+ and H^+ shifts in early ischemia: Mechanisms and relation to changes in impulse propagation. J Mol Cell Cardiol 19 (Suppl V):35–44

102. Knabb MT, Saffitz JE, Corr PB, Sobel BE (1986) The dependence of electrophysiological derangements on accumulation of endogenous long-chain acyl carnitine in hypoxic neonatal rat myocytes. Circ Res 58:230–240

103. Kramer JB, Saffitz JE, Witkowski FX, Corr PB (1985) Intramural reentry as a mechanism of ventricular tachycardia during evolving canine myocardial infarction. Circ Res 56:736–754

104. Krause S, Heses HL (1984) Characterization of cardiac sarcoplasmic reticulum dysfunction during short-term, normothermic, global ischemia. Circ Res 55:176–184

105. Lathers CM, Kelliher GJ, Robert J, Beasley AB (1978) Nonuniform cardiac sympathetic nerve discharge. Circulation 57:1058–1065

106. Lazzara R, El-Sherif N, Hope RR, Scherlag BJ (1978) Ventricular arrhythmias and electrophysiological consequences of myocardial ischemia and infarction. Circ Res 42:740–749

107. Lee H-C, Smith N, Mohabir R, Clusin WT (1987) Cytosolic calcium transients from the beating mammalian heart. Proc Natl Acad Sci USA 84:7793–7797

108. Levites R, Banka VS, Helfant RH (1975) Electrophysiologic effects of coronary occlusion and reperfusion. Observations of dispersion of refractoriness and ventricular automaticity. Circulation 52:760–765

109. Levitzki A (1986) β-Adrenergic receptors and their mode of coupling to adenylate cyclase. Physiol Rev 66:819–854

110. Liedtke AJ, Nellis S, Neely JR (1978) Effects of excess free fatty acids on mechanical and metabolic function in normal and ischemic myocardium in swine. Circ Res 43:652–661

111. Lown B (1979) Sudden cardiac death: The major challenge confronting contemporary cardiology. Am J Cardiol 43:313–328

112. Maisel AS, Motulsky HJ, Insel PA (1985) Externalization of β-adrenergic receptors promoted by myocardial ischemia. Science 230:183–186

113. Maisel AS, Motulsky HJ, Ziegler MG, Insel PA (1987) Ischemia- and agonist-induced changes in alpha$_1$- and beta-adrenergic receptor traffic in the guinea pig heart. Am J Physiol 253:H1159–1166

114. Malliani A, Lombardi F (1978) Neural reflexes associated with myocardial ischemia. In: Schwartz PJ, Brown AM, Malliani A, Zanchetti A (eds), Neural Mechanisms in Cardiac Arrhythmias. Raven Press, New York, pp 209–219

115. Man RYK (1988) Lysophosphatidylcholine-induced arrhythmias and its accumulation in the rat perfused heart. Br J Pharmacol 93:412–416

116. Man RYK, Slater TL, Pelletier MPJ, Choy PC (1983) Alterations of phospholipids in ischemic canine myocardium during acute arrhythmia. Lipids 18:677–681

117. Marban E, Kitakaze M, Kusuoka H, Porterfield JK, Yue DT, Chacko VP (1987) Intracellular free calcium concentration measured with ^{19}F NMR spectroscopy in intact ferret hearts. Proc Natl Acad Sci USA 84:6005–6009

118. Marcus ML, Kerber RE, Ehrhardt J, Abboud FM (1976) Effects of time on volume and distribution of coronary collateral flow. Am J Physiol 230:279–285

119. Marrannes R, de Hemtinne A, Leusen I (1979) Influence of lactate and other organic ions on

170

conduction velocity in mammalian heart fibers depressed by "metabolic" acidosis. J Mol Cell Cardiol 11:359-374

120. May GS, Eberlein KA, Furberg CD, Passamani ER, DeMets DL (1982) Secondary prevention after myocardial infarction: A review of long-term trials. Prog Cardiovasc Dis 24:331-352

121. McDonald TF, MacLeod DP (1973) Metabolism and the electrical activity of anoxic ventricular muscle. J Physiol 229:559-582

122. Miura Y, Inui J, Imamura H (1978) Alpha-adrenoceptor-mediated restoration of calcium dependent potentials in the partially depolarized rabbit papillary muscle. Naunyn-Schmiedebergs Arch Pharmacol 301:201-205

123. Mugelli A, Cerbia E, Amerini S, Giotti A (1985) Altered responsiveness of cardiac alpha- and beta-adrenoceptors during hypoxia and aging: relevance for arrhythmias. New Trends in Arrhythmias 1:115-123

124. Mukherjee A, Bush LR, McCoy KE, Duke RJ, Hagler H, Buja LM, Willerson JT (1982) Relationship between β-adrenergic receptor numbers and physiological responses during experimental canine myocardial ischemia. Circ Res 50:735-741

125. Mukherjee A, Wong TM, Buja LM, Lefkowitz RJ, Willerson JT (1987) Beta-adrenergic and muscarinic cholinergic receptors in canine myocardium. J Clin Invest 64:1423-1428

126. Nattel S, Elharrar V, Zipes DP, Bailey JC (1981) pH-dependent electrophysiological effects of quinidine and lidocaine on canine cardiac Purkinje fibers. Circ Res 48:55-61

127. Nikolic G, Bishop RL, Singh JB (1982) Sudden death recorded during Holter monitoring. Circulation 66:218-225

128. Noma A (1983) ATP-regulated K$^+$ channels in cardiac muscle. Nature 305:147-148

129. Oliva PB, Breckinridge JC (1977) Arteriographic evidence of coronary arterial spasm in acute myocardial infarction. Circulation 56:366-374

130. Panidis IP, Morganroth J (1983) Sudden death in hospitalized patients: cardiac rhythm disturbances detected by ambulatory electrocardiographic monitoring. J Am Coll Cardiol 2:798-805

131. Penkoske PA, Sobel BE, Corr PB (1978) Disparate electrophysiological alterations accompanying dysrhythmias due to coronary occlusion and reperfusion in the cat. Circulation 58:1023-1035

132. Pentecost BL, Austen WG (1966) Beta adrenergic blockade in experimental myocardial infarction. Am Heart J 72:790-796

133. Pogwizd SM, Corr PB (1987) Reentrant and nonreentrant mechanisms contribute to arrhythmogenesis during early myocardial ischemia: Results using three-dimensional mapping. Circ Res 61:352-371

134. Pogwizd SM, Corr PB (1990) Mechanisms underlying the development of ventricular fibrillation during early myocardial ischemia. Circ Res 66:672-695

135. Pogwizd SM, Corr PB (1987) Electrophysiologic mechanisms underlying arrhythmias due to reperfusion of ischemic myocardium. Circulation 76:404-426

136. Pogwizd SM, Onufer JR, Kramer JB, Sobel BE, Corr PB (1986) Induction of delayed afterdepolarizations and triggered activity in canine Purkinje fibers by lysophosphoglycerides. Circ Res 59:416-426

137. Preda I, Karpati P, Endsoczi E (1975) Myocardial noradrenaline uptake after coronary occlusion in the rat. Acta Physiol Hung 46:99-106

138. Prinzmetal M, Toyoshima H, Ekmekci A, Mizuno Y, Nagaya T (1961) Myocardial ischemia. Nature of ischemic electrocardiographic patterns in the mammalian ventricles as determined by intracellular electrographic and metabolic changes. Am J Cardiol 8:493-503

139. Rapundalo ST, Briggs FN, Feher JJ (1986) Effects of ischemia on the isolation and function of canine cardiac sarcoplasmic reticulum. J Mol Cell Cardiol 18:837-851

140. Reimer KA, Hill ML, Jennings RB (1981) Prolonged depletion of ATP and of the adenine nucleotide pool due to delayed resynthesis of adenine nucleotides following reversible myocardial ischemic injury in dogs. J Mol Cell Cardiol 13:229-239

141. Reimer KA, Jennings RB (1979) The "wavefront phenomenon" of myocardial ischemic cell death. II. Transmural progression of necrosis within the framework of ischemic bed size (myocardium at risk) and collateral flow. Lab Invest 40:633-644

142. Riemersma A (1982) Myocardial catecholamine release in acute myocardial ischaemia:

Relationship to cardiac arrhythmias. In: Parratt JR (ed) Early Arrhythmias Resulting from Myocardial Ischemia. Mechanisms and Prevention by Drugs. Macmillan, London, pp 125–138

143. Rosen MR, Hordof AJ, Ilvento JP, Danilo P Jr (1977) Effects of adrenergic amines on electrophysiologic properties and automaticity of neonatal and adult canine Purkinje fibers. Evidence for α- and β-adrenergic actions. Circ Res 40:390–400

144. Rosen MR, Reder RF (1981) Does triggered activity have a role in the genesis of cardiac arrhythmias? Ann Int Med 94:794–801

145. Ruffy R, Lovelace DE, Mueller TM, Knoebel SB, Zipes DP (1979) Relationship between changes in left ventricular bipolar electrograms and regional myocardial blood flow during acute coronary occlusion in the dog. Circ Res 45:764–770

146. Russell DC, Oliver MF (1978) Ventricular refractoriness during acute myocardial ischaemia and its relationship to ventricular fibrillation. Cardiovasc Res 12:221–227

147. Saffitz JE, Corr PB, Lee BI, Gross RW, Williamson EK, Sobel BE (1984) Pathophysiological concentrations of lysophosphoglycerides quantified by electron microscopic autoradiography. Lab Invest 50:278–286

148. Samson WE, Scher AM (1960) Mechanism of S-T segment alteration during acute myocardial injury. Circ Res 8:780–787

149. Sayen JJ, Pierce G, Katcher AH, Sheldon WF (1961) Correlation of intramyocardial electrocardiograms with polarographic oxygen and contractility in the non-ischemic and regionally ischemic left ventricle. Circ Res 9:1268–1279

150. Scherlag BJ, El-Sherif N, Hope R, Lazzara R (1974) Characterization and localization of ventricular arrhythmias resulting from myocardial ischemia and infarction. Circ Res 35:372–383

151. Scherlag BJ, Helfant RH, Haft JI, Damato AN (1970) Electrophysiology underlying ventricular arrhythmias due to coronary ligation. Am J Physiol 219:1665–1671

152. Scholz H (1980) Effects of beta- and alpha-adrenoceptor activators and adrenergic transmitter releasing agents on the mechanical activity of the heart. In: Szekeres L (ed) Handbook of Experimental Pharmacology, vol 54, Springer, Berlin Heidelberg New York, pp 651–733

153. Schümann HJ, Wagner J, Knorr A, Reidemeiser JC, Sadony V, Schramm G (1978) Demonstration in human atrial preparations of α-adrenoreceptors mediating positive inotropic effects. Naunyn-Schmiedebergs Arch Pharmacol 302:333–336

154. Schwartz PJ, Stone HL, Brown AM (1976) Effects of unilateral stellate ganglion blockade on the arrhythmias associated with coronary occlusion. Am Heart J 92:589–599

155. Schwartz PJ, Vanoli E, Zaza A, Zuanetti G (1985) The effect of antiarrhythmic drugs on life-threatening arrhythmias induced by the interaction between acute myocardial ischemia and sympathetic hyperactivity. Am Heart J 109:937–948

156. Sedlis SP, Corr PB, Sobel BE, Ahumada GG (1983) Lysophosphatidylcholine potentiates Ca^{2+} accumulation in rat cardiac myocytes. Am J Physiol 244:H32–H38

157. Shaikh NA, Downar E (1981) Time course of changes in porcine myocardial phospholipid levels during ischemia: a reassessment of the lysolipid hypothesis. Circ Res 49:316–325

158. Sheridan DJ, Penkoske PA, Sobel BE, Corr PB (1980) Alpha-adrenergic contributions to dysrhythmias during myocardial ischemia and reperfusion in cats. J Clin Invest 65:161–171

159. Snyder DW, Crafford WA Jr, Glashow JL, Rankin D, Sobel BE, Corr PB (1981) Lysophosphoglycerides in ischemic myocardium effluents and potentiation of their arrhythmogenic effects. Am J Physiol 241:H700–H707

160. Spedding M (1985) Activators and inactivators of Ca^{++} channels: New perspectives. J Pharmacol (Paris) 16:319–343

161. Spedding M, Mir AK (1987) Direct activation of Ca^{++} channels by palmitoyl carnitine, a putative endogenous ligand. Br J Pharmacol 92:457–468

162. Stewart JR, Burmeister WE, Burmeister J, Lucchesi BR (1980) Electrophysiologic and antiarrhythmic effects of phentolamine in experimental coronary artery occlusion and reperfusion in the dog. J Cardiovasc Pharmacol 2:77–91

163. Ten Eick RE, Singer DH, Solberg LE (1976) Coronary occlusion. Effect on cellular electrical activity of the heart. Med Clinics NA 60:49–67

164. Thandroyen FT, McCarthy J, Burton KP, Opie LH (1988) Ryanodine and caffeine prevent ventricular arrhythmias during acute myocardial ischemia and reperfusion in rat heart. Circ Res 62:306–314

172

165. Thandroyen FT, Worthington MG, Higginson LM, Opie LH (1983) The effect of alpha- and beta-adrenoceptor antagonist agents on reperfusion ventricular fibrillation and metabolic status in the isolated perfused rat heart. J Am Coll Cardiol 14:1056—1066

166. Vary TC, Angelakos ET, Schaffer SW (1979) Relationship between adenine nucleotide metabolism and irreversible ischemic tissue damage in isolated perfused rat heart. Circ Res 45:218—225

167. Vassalle M (1985) Overdrive excitation: The onset of spontaneous activity following a fast drive. In: Zipes DP, Jalife J (eds) Cardiac Electrophysiology and Arrhythmias. Grune and Stratton, New York, pp 97—107

168. Vleugels A, Vereecke J, Carmeliet E (1980) Ionic currents during hypoxia in voltage-clamped cat ventricular muscle. Circ Res 47:501—508

169. van Bilsen M, van der Vusse GJ, Willemsen PHM, Coumans WA, Roemen THM, Reneman RS (1989) Lipid alterations in isolated, working rat hearts during ischemia and reperfusion: its relation to myocardial damage. Circ Res 64:304—314

170. Waldo AL, Kaiser GA (1973) Study of ventricular arrhythmias associated with acute myocardial infarction in the canine heart. Circulation 47:1222—1228

171. Weiss J, Shine KI (1982) Extracellular K^+ accumulation during myocardial ischemia in isolated rabbit heart. Am J Physiol 242:H619—H628

172. Weiss JN, Lamp ST (1987) Glycolysis preferentially inhibits ATP-sensitive K^+ channels in isolated guinea pig cardiac myocytes. Science 238:67—69

173. Wiegand V, Güggi M, Meesmann W, Kessler M, Greitschuss F (1979) Extracellular potassium activity changes in the canine myocardium after acute coronary occlusion and the influence of beta-blockade. Cardiovasc Res 13:297—302

174. Wilber DJ, Lynch JL, Montgomery DG, Lucchesi BR (1987) Alpha-adrenergic influences in canine ischemic sudden death. Effects of alpha$_1$-adrenoceptor blockade with prazosin. J Cardiovasc Pharmacol 10:96—106

175. Wilensky RL, Tranum-Jensen J, Coronel R, Wilde AAM, Fiolet JWT, Janse MJ (1986) The subendocardial border zone during acute ischemia of the rabbit heart: an electrophysiologic, metabolic, and morphologic correlative study. Circulation 74:1137—1146

176. Williams DO, Scherlag BJ, Hope RR, El-Sherif N, Lazzara R (1974) The pathophysiology of malignant ventricular arrhythmias during acute myocardial ischemia. Circulation 50:1163—1172

177. Williams LT, Guerrero JL, Leinbach RC (1982) Prevention of reperfusion dysrhythmia by selective coronary alpha-adrenergic blockade. Am J Cardiol 49:1046 (abstract)

178. Wit AL (1982) Electrophysiological mechanisms of ventricular tachycardia caused by myocardial ischemia and infarction in experimental animals. In: Josephson ME (ed) Ventricular Tachycardia. Mount Kisco, New York: Futura, pp 33—96

179. Wit AL, Rosen MR (1986) Afterdepolarization and triggered activity. In: Fozzard HA et al. (eds) The Heart and Cardiovascular System. Scientific Foundations. Raven Press, New York, pp 1449—1490

180. Witkowski FX, Corr PB (1984) An automated simultaneous transmural cardiac mapping system. Am J Physiol 247:H661—H668

181. Wojtczak J (1979) Contractures and increase in internal longitudinal resistance of cow ventricular muscle induced by hypoxia. Circ Res 44:88—95

182. Woodcock EA, Whilt LBS, Smith AI, McLeod JK (1987) Stimulation of phosphatidylinositol metabolism in the isolated perfused rat heart. Circ Res 61:625—631

183. Zipes DP (1977) Electrolyte derangements in the genesis of arrhythmias. In: Dreifus LS, Likoff W (eds) Cardiac Arrhythmias. Grune and Stratton, New York, p 55

Authors' address:
Dr. Peter B. Corr
Washington University
School of Medicine
Cardiovascular Division
Box 8086
660 South Euclid Avenue
St. Louis, MO 63110

Reperfused Myocardium: Stunning, Preconditioning, and Reperfusion Injury

Wolfgang Schaper, Robert J. Schott, Masao Kobayashi

Max-Planck-Institute, Department of Experimental Cardiology, Bad Nauheim, FRG

Introduction

In the late 1960s, when the deathtoll from myocardial infarction had caused a high level of public awareness, the National Institutes of Health, USA, decided to fund projects directly related to the experimental treatment of infarcts, to develop animal models best suited to study infarct size after coronary occlusion, and to develop quantitative methods in animals and in man to measure infarcts. Many drugs, natural compounds, and physical methods were tested in the early phase of pragmatism and only a few forms of therapy met expectations when the measurement of infarct size became more precise. The most effective of these interventions, which is potentially able to reduce infarct size when applied within a certain window of time, is reperfusion.

Experiments from our laboratory have shown that this time window is dependent on two factors:

a) the myocardial oxygen consumption during coronary occlusion (high MVO_2 shrinks the useable interval), and

b) the collateral blood flow (low collateral flow reduces the useable interval).

Our studies showed [69–71] that, although drugs cannot prevent or halt the progression of the "wave front" [60] of necrosis, they can delay the infarct progression when they effectively reduce MVO_2, but only if the ischemic myocardium is reperfused, i.e., the window of useable reperfusion time can be opened somewhat wider using MVO_2-reducing drugs.

The studies on infarct size, MVO_2, collateral flow, and reperfusion remained somewhat academic, until cardiologists opened freshly occluded coronary arteries first mechanically and later with thrombolytic agents [30, 61, 62]. Our studies on the determinants of the time window of reperfusion became the pathophysiological basis for clinical studies, i.e., the best results of thrombolytic interventions were obtained when applied as early as possible after the onset of symptoms, i.e., already at the patient's home or in the ambulance [18].

Sometimes, however, clinicians were disappointed with the success of thrombolytic therapy in individual patients, although the interval between onset of symptoms and onset of therapy was within prescribed limits. At this stage of development of clinical thrombolysis two concepts developed in experimental animals seemed to offer help and explanation, i.e., a) the concept of stunned myocardium, and b) the concept of "reperfusion injury".

The observation of "stunned" myocardium by Heyndrickx et al. [22] showed that regional contractile function, if measured within hours or even days after reperfusion, need not be normal even if the myocardium at risk was structurally salvaged by reperfusion.

Reperfusion injury refers to the possibility of myocardial damage by reperfusion in addition to that of ischemia, especially if the myocardium was reperfused relatively late, i.e., close to the useable "open-time" of the reperfusion window.

If reperfusion injury really exists, it may explain the sometimes unexpected failure of reperfusion in the clinical setting to salvage myocardium and strategies can be developed to improve reperfusion.

Following Heyndrickx's [22] observation of long-lasting contractile dysfunction, several laboratories [3, 25, 79] became interested in the mechanisms of stunned myocardium and it soon became apparent that multiple short coronary occlusions were not additive in their damaging effect but rather exerted, paradoxically, a protection in subsequent long occlusions. This protective effect was phrased "preconditioning" [45] and it became interesting to clinicians because PTCA often necessitated repeated occlusions of a coronary artery.

Stunned Myocardium

As said in the introduction, stunned myocardium is the long-lasting (hours to days) contractile dysfunction following short coronary occlusions (between 4 and 15 min) in the absence of any cell necrosis (Fig. 1). Considerable difference of opinion exists with regard to the precise localization of the changes in the myocyte and with regard to the precise mechanisms of damage (or adaptation). Light and electron microscopy revealed no structural alterations in reperfused myocardium after these short occlusion times [67], but the myocardial ATP concentration was significantly reduced and its replenishment was also very slow, comparable to that of the mechanical function. It was therefore concluded that ATP-shortage may cause stunning [9, 29]. The difficulty with that hypothesis, however, was that the Km for ATP in the actomyosin ATPase reaction is at least one order of magnitude lower than the (reduced) ATP concentration in stunned myocardium. The explanation by Kammermeier et al. [29], that the energy level of the myocyte plays an important role, is still valid but it is difficult to assess how important a contribution that is to stunning.

One way to test the ATP-shortage hypothesis of stunning is to rapidly restore ATP levels and watch whether function returns also more quickly. The de-novo synthesis of ATP is a very slow and energy-intensive process [84] that can be considerably activated by supplying precursors like adenosine, aica-riboside or ribose [41]. However, experiments with these precursors were unable to answer the question. Adenosine was the most potent stimulator of ATP resynthesis, but questions arose as to where and in which compartment the ATP-content increased. Part of the observed increase may have occurred in endothelium [49]. In either case the rapidly replenished ATP failed to normalize contractile function [24]. With aica-riboside, the replenishment rate was relatively slow and function became significantly more depressed, probably because aica-triphosphate was also produced which is known to depress contractility. No conclusion was possible with ribose because the rate of replenishment was too slow.

Another hypothesis stated that the transfer of energy as carried out by the phosphocreatine shuttle had become damaged during ischemia and/or reperfusion [82].

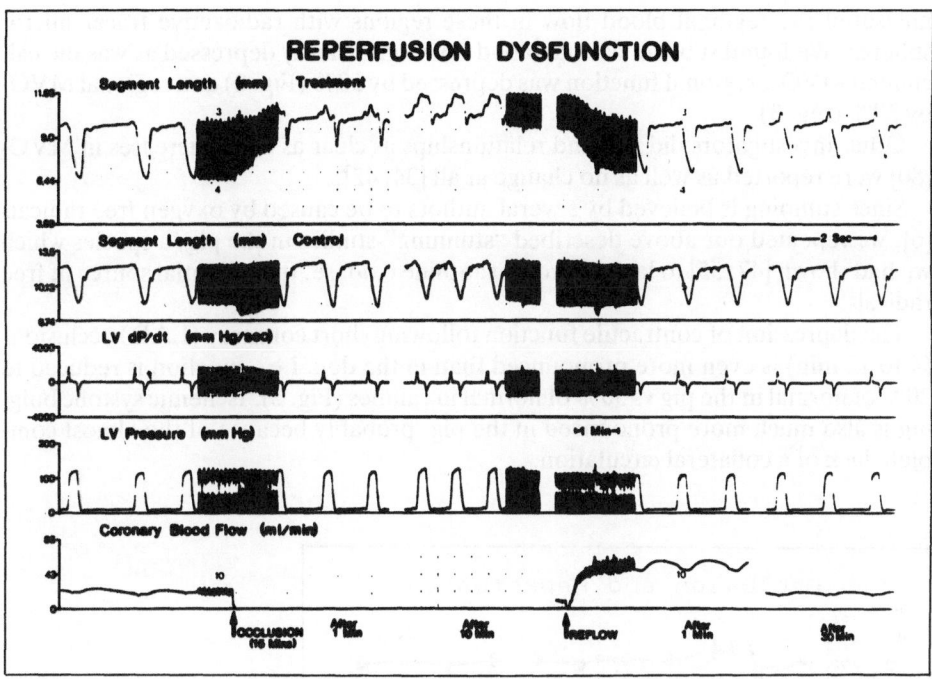

Fig. 1. Influence of LAD occlusion and reperfusion on regional contractile function in an in situ beating dog heart. From above: segment shortening (subendocardial) in an LAD-dependent area, segment shortening in a left circumflex region, left ventricular dP/dt, left ventricular pressure, and mean coronary blood flow. Note the almost complete recovery of segmental shortening immediately after reperfusion and the deterioration of function at 30 min.

This hypothesis was supported by the observation that the enzyme creatine phosphokinase may leave ischemic myocardium already after about 11 min of ischemia [44] and that myofibril-bound CK was reduced and phosphocreatine had risen above normal values in stunned myocardium. The difficulty with this hypothesis was whether the losses of CK are sufficiently large to become rate-limiting for the "shuttle". If so, a situation resembling relative energy shortage would have ensued with progressive fall of ATP and a rise in lactate production which, of course, is not the case. The observation by us and others [25], that stunned myocardium has a normal contractile reserve that can be elicited with calcium ions and catecholamines, contradicted the impaired-shuttle-hypothesis, because energy availability clearly was not a limiting factor. Neubauer et al. [50] showed with NMR-techniques that the shuttle is not impaired in stunning, and we have shown [72] that stunning is caused by limited ATP utilization rather than a relative shortage of ATP supply. The phosphocreatine overshoot is only the symptom for a slower shuttling-rate.

Our conclusions were aided by measurements of regional MVO_2 in segments of stunned myocardium. In experiments using anesthetized dogs [72], we cannulated superficial cardiac veins draining stunned myocardium and those draining normal myocardium, determined the difference of oxygen saturation between them, and

177

measured the regional blood flow in these regions with radioactive tracer microspheres. We found subendocardial blood flow significantly depressed as was the calculated MVO_2; regional function was depressed by 50% (Fig. 2) and regional MVO_2 by 30% (Fig. 3).

Other investigators did not find relationships as clear as ours; increases in MVO_2 [80] were reported as well as no change at all [34, 42].

Since stunning is believed by several authors to be caused by oxygen free radicals [6], we repeated our above described "stunning"-studies in the pig, a species which we had shown [47, 55] to lack a cardiac xanthine oxidase, the potential source of free radicals.

The depression of contractile function following short coronary (LAD) occlusions (4 to 12 min) is even more pronounced than in the dog, i.e., function is reduced to 20% of normal in the pig vs 50% of normal in canines (Fig. 2). Ischemic systolic bulging is also much more pronounced in the pig, probably because of the almost complete lack of a collateral circulation.

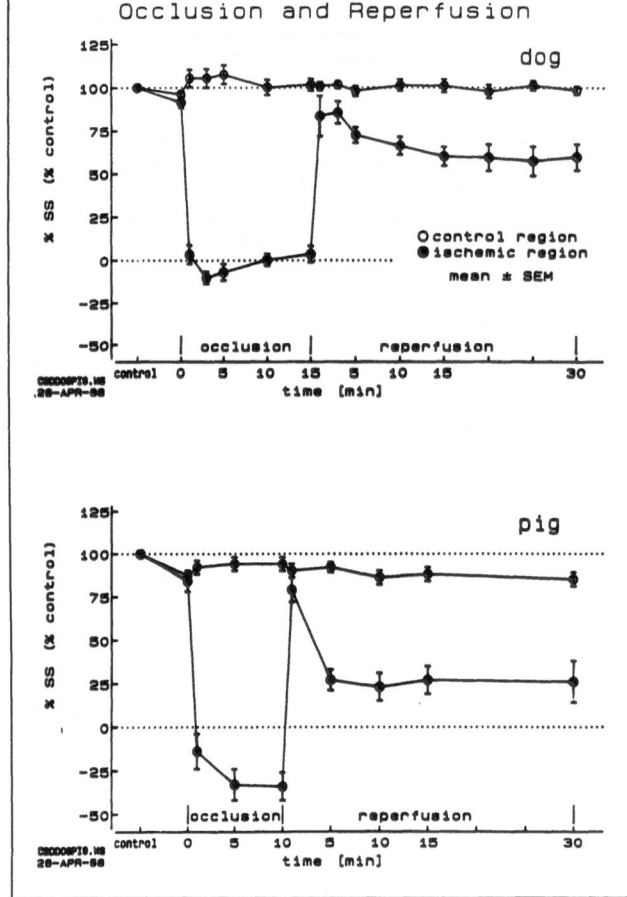

Fig. 2. Comparison of contractile impairment in the dog and pig heart following occlusion and reperfusion. Note the more marked systolic bulging in the pig heart during occlusion and the lower level of systolic segment shortening during reperfusion. Coronary occlusion lasted for 10 min in the pig and 15 min in the dog.

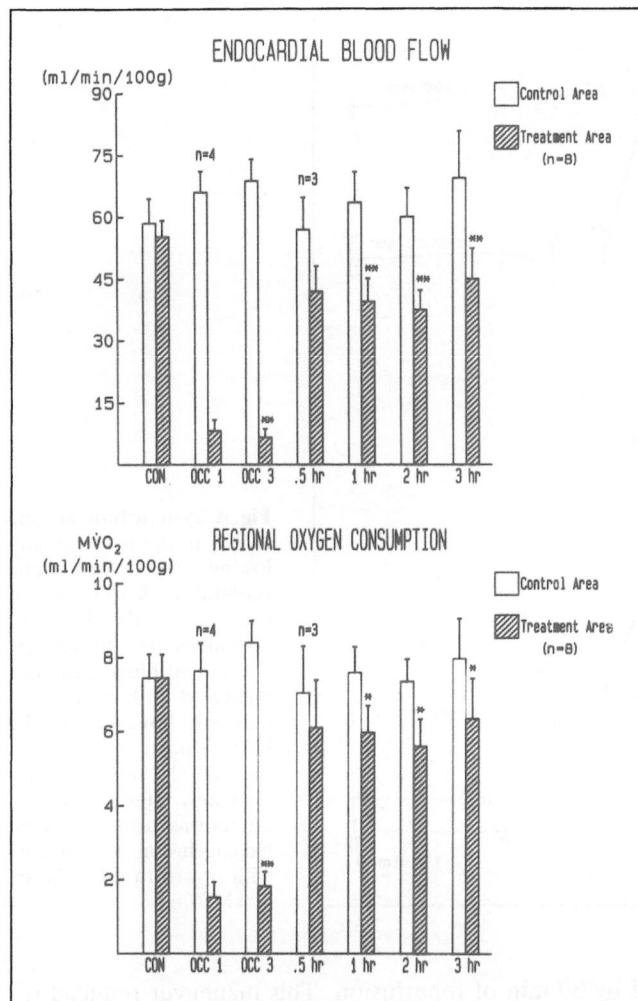

Fig. 3. Influence of coronary occlusion and reperfusion on regional coronary blood flow (subendocardium) and regional (transmural) MVO_2 in the dog. It can be assumed that subendocardial MVO_2 had decreased in proportion to subendocardial flow since venous oxygen saturation had decreased to normal levels at the time of tracer microsphere injection.

Inspite of the more pronounced depression of function in the stable stunning phase in porcines, we failed to observe a fall in regional MVO_2 (Fig. 4). Part of that failure may have been of a technical nature: In this series of pig experiments we could not use radioactive tracer microspheres because of difficulties with radioactive cadaver disposal. We therefore measured only total regional blood flow with an electromagnetic flowmeter which we felt was justified because of the extremely low admixture with collateral flow. Any endo-epi-differences that still exist may have been overlooked in these experiments. The question arising from this experiment is: which processes use as much oxygen as formerly necessary to maintain 100% contractile function in the presence of only 20% contractile function?

We repeated these pig experiments as soon as the problems with radioactive cadaver disposal were solved in the context of a stunning-preconditioning experiment with radioactive microspheres. Stunning was produced by two times 10 min of

179

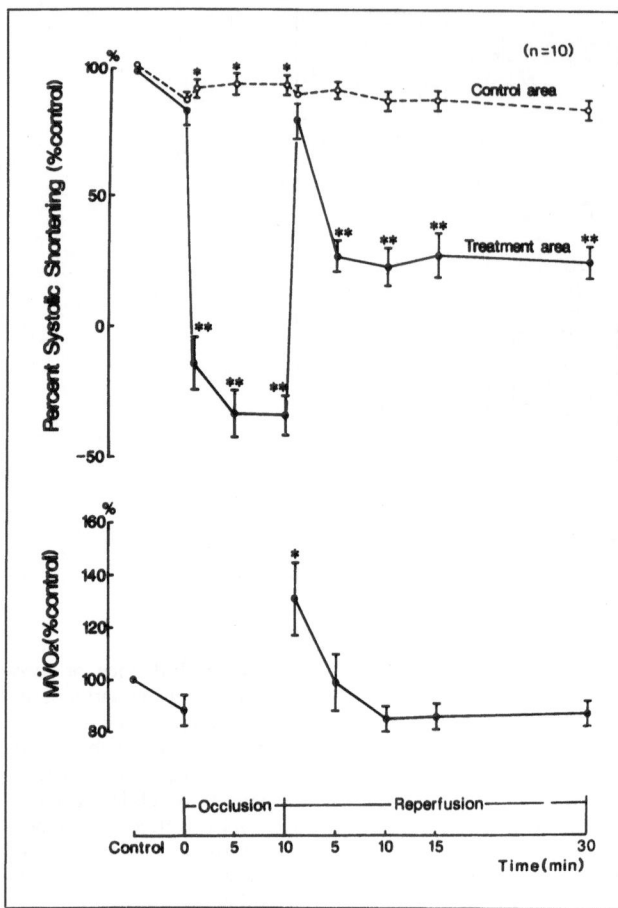

Fig. 4. Systolic function and MVO_2 in the pig heart following LAD-occlusion and reperfusion. MVO_2 was calculated on the basis of arteriovenous SO_2-differences, and blood flow was measured with an electromagnetic flowmeter in the LAD bypassline. Note that the MVO_2 increases immediately after reperfusion but returns to normal levels. No conclusions are possible with regard to subendocardial MVO_2.

LAD-occlusion interrupted by 30 min of reperfusion. This maneuver reduced regional contractile function during reperfusion to zero (i.e., more than in the previous pig experiment). In this experiment we found a 30% fall of regional MVO_2, which was, however, not statistically significant when referenced to its own initial control value but significant when referenced to the MVO_2, of a normal region of the same heart which rose during the experiment. These MVO_2 results were similar to those reported by McFalls et al. [42].

We are thus faced with the difficult question of the relationship between MVO_2 and regional function and we have two ways to advance arguments:

a) The relationship between regional contractile function and regional MVO_2 is more complex than that between global function and global MVO_2.

We should bear in mind, though, that the latter relationship is by no means simple or linear. For a long time it was believed that no relationship existed between cardiac work (pressure times stroke volume) and MVO_2 and only a not very remarkable relationship between cardiac performance (pressure times systolic flow) and MVO_2.

180

The excellent relationship between pressure development times heart rate (double product) and MVO_2 led most to believe that only pressure generation was related to MVO_2.

It took technically complex experiments and advanced statistical methods for Rooke and Feigl [65] to show that stroke volume participated in MVO_2. It should not surprise us therefore too much to see that a situation of (regional) no shortening and no lengthening in systole is costly in terms of oxygen usage. A situation of akinesia means (in terms of energy usage) that systolic lengthening (bulging) is prevented by generation of active force. This active force may be only slightly less than that needed for active shortening. The analogy for that may be the game of tug-of-war: if the two competing teams differ in strength by only 1%, the win is easy.

The relationship between regional contractile function and regional MVO_2 was recently shown [35] to be non-linear, i.e., small changes in MVO_2 causing large changes in regional function.

b) Another argument is pathophysiological: ischemic intervals causing stunning may have reversibly damaged the mitochondria, and the unexpectedly high MVO_2 is caused by uncoupling between respiration and phosphorylation. This hypothesis was investigated by Buchwald et al. [8] in mitochondria isolated from stunned pig myocardium, but no evidence for uncoupling was detected.

We conclude from our own studies that the contractile dysfunction in stunned myocardium shows species-variation: a moderate (50%) depression of regional contractile function in the canine heart was accompanied by a moderate (30%) reduction of regional MVO_2. A severe depression of regional function in the porcine heart (no systolic shortening) was only accompanied by either no reduction of MVO_2 when referenced back to its own control value or only a mild decrease (-30%) when referenced vs the normal neighbour territory. Since these different MVO_2 results originated in the same laboratory, we believe them to be true species variations. The nature of these variations remains unclear.

Impaired Ca^{2+}-Homeostasis as a Mechanism of Stunning

We [25] have shown that the intracoronary infusion of calcium chloride during reperfusion increases the systolic segment shortening of stunned canine myocardium (Fig. 5) in a dose-dependent fashion. Figure 6 shows that the level of maximal contractility before ischemia can be reached by the same Ca^{2+}-concentration after ischemia and that additional postextrasystolic potentiation has no additional effect on contractility.

The effects of Ca^{2+}-infusions are not shortlived: in our experiments infusions lasting for 30 min showed no tendency for diminished contractile responses. Stopping the infusion resulted in the same degree of stunning as before infusion. From these experiments we can conclude:
– Stunning is probably not caused by a defective energy supply because that would have limited the extent of the contractile response to calcium especially after continuous infusion.

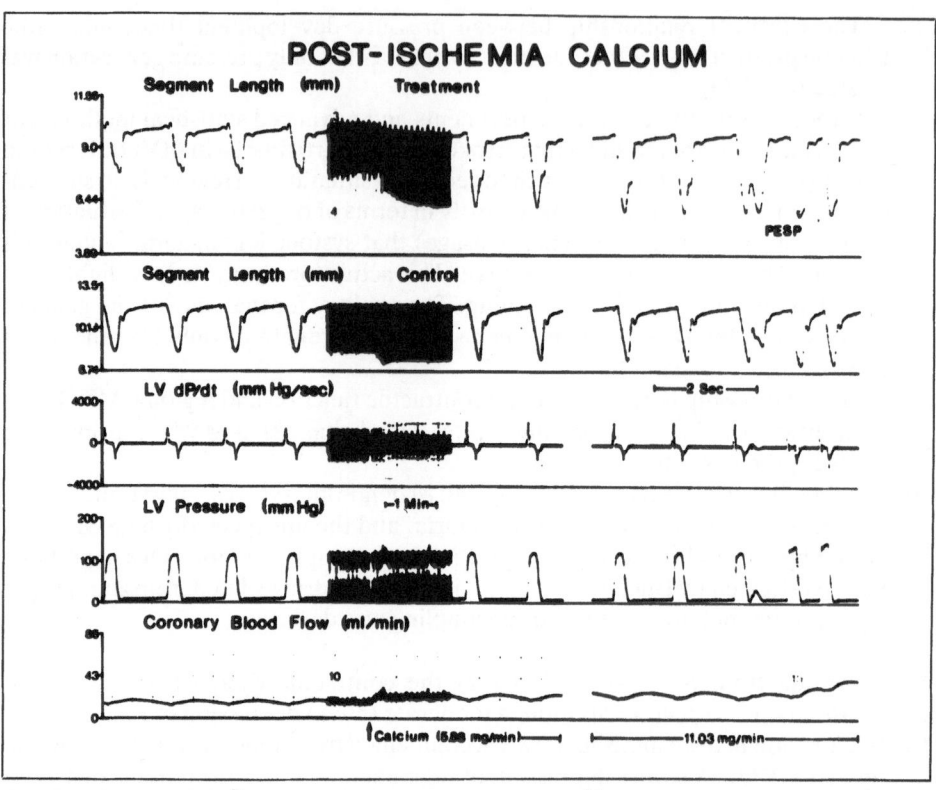

Fig. 5. Influence of Ca^{2+} on stunned dog myocardium. Ca^{2+}-infusion into the LAD increases regional contractility markedly only in the LAD-region. Postextrasystolic beats do not show further augmentation of segmental shortening. Upon switching off the Ca^{2+}-infusion, segmental shortening returns to the preinfusion level (but to no lesser level).

— Stunning is probably also not caused by a decrease in the calcium sensitivity of the myofibrils because (as Fig. 6 shows) of the unchanged relationship between Ca^{2+}-concentration and the contractile response before and after ischemia.

— Stunning is most probably caused by a change in the electromechanical coupling, in particular in the handling of cytosolic Ca^{2+}. We arrive at this diagnosis mainly by exclusion of other mechanisms. Heyndrickx et al. [22] have already shown that the topical electrocardiogram, i.e., local excitation, is normal shortly after onset of reperfusion.

Interesting experiments were reported [36] showing that the capacity of the sarcoplasmic reticulum to sequester Ca^{2+} was reduced in stunned myocardium. This would mean that the amount of Ca^{2+} released from the sarcoplasmic reticulum with electrical excitation of the cell is also decreased which would explain the reduced contractility. The ability of the sarcolemma to pump out excess Ca^{2+} is probably maintained. We hypothesize that the cell loses this faculty only with more advanced (irreversible) damage that leads to Ca^{2+}-overload. Kusuoka et al. [37] came to similar conclusions on the basis of experiments in isolated ferret hearts, i.e., cytosolic Ca^{2+} during systole

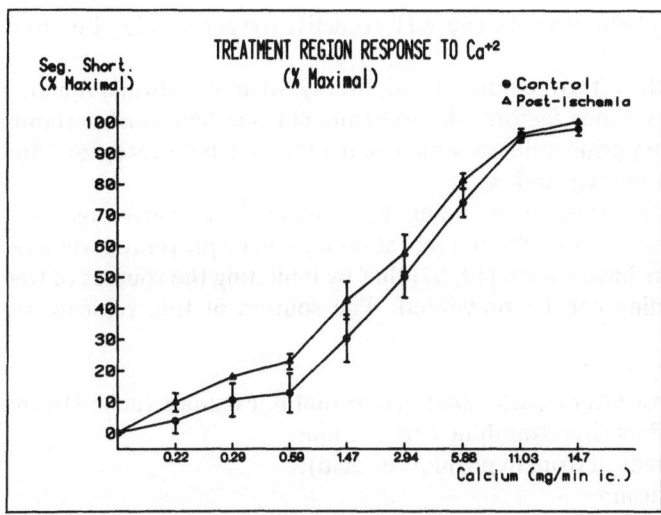

Fig. 6. Influence of Ca^{2+} on systolic segmental shortening in the dog. Control responses to Ca^{2+}-infusions are compared with responses after reperfusion. Identical dose-response relationships are obtained (corrections for flow changes) suggesting (but not proving) unchanged Ca^{2+}-sensitivity of the myofibrils.

plays an important role in stunning. Kusuoka et al. [37] do not exclude a change in myofilament sensitivity to Ca^{2+} as a contributing factor in stunning. On the basis of phosphorus NMR-studies the same authors have excluded a role of reduced ATP-concentrations in stunned myocardium.

Our experiments with intracoronary infusions of Ca^{2+} in stunned pig hearts show the same directional changes as in the dog but differ quantitatively: normal contractility is restored by Ca^{2+} but not normal contractile reserve. This may be related to the more severe depression of contractility in the pig following occlusion and reperfusion.

We have recently hypothesized [25] that stunning may belong to the mechanisms that the heart can recruit in defense of overload like the down-regulation of adrenergic receptors with chronic high ligand concentrations. Further studies are needed to support or contradict these thoughts that do not necessarily view stunning as a strictly pathological phenomenon.

Role of Oxygen Free Radicals as a Cause of Stunned Myocardium

A close look at a continuous recording of segmental shortening or wall thickening immediately before, during, and after reperfusion reveals that contractile function quickly returns to normal (often "overshoots") and then decreases over a period of 3–5 min to its new depressed stable state (Fig. 1). It is tempting to speculate that the initial return to normal function reflects the early reoxygenation before reperfusion injury had fully developed. The typical recording of regional contractile function might even be used as an argument for the existence of reperfusion injury. This reper-

183

fusion injury is commonly believed to be caused by reactive oxygen species, i.e., oxygen free radicals.

Although suggestive, the return to normal and, briefly, to above-normal function can also be explained by other factors: the ischemic episode had most certainly caused regional norepinephrine release which had occupied beta-receptors and stimulated contraction at reoxygenation.

Nevertheless, oxygen-free radicals are being discussed as the causative agents in stunning. Bolli et al. [6] have demonstrated that stunning can be prevented when radicals are trapped. Others have shown [19, 57] that by inhibiting the sources of free radical production stunning can be prevented. The sources of free radicals are believed to be:
- the mitochondria;
- xanthine dehydrogenase after transformation into xanthine oxidase fuelled by the ATP-breakdown-products hypoxanthine and xanthine;
- leukocytes (hydroxylradical from hypochlorous acid);
- catecholamine catabolism;
- prostaglandin metabolism.

Small amounts of free radicals are always generated under physiological conditions and the glutathion-glutathion peroxidase, superoxide dismutase-, and catalase systems are abundantly expressed to mop up radicals. It was shown [75] that glutathion (normally present in heart in millimolar concentrations) leaves the postischemic heart but only after long ischemic periods and late during reperfusion. If radicals were abundantly generated in mitochondria, that were unprotected by decreased glutathion levels, we would expect these organelles to be damaged because of the extreme reagibility of these species. Mitochondria, however, are neither structurally nor functionally altered in stunned myocardium [8]. Mitochondria isolated from stunned pig myocardium show normal function and normal respiratory control [8]. Although the natural antioxidants tocopherol and ascorbic acid were shown by Klein et al. [32] to reduce infarct size when given in pharmacologic amounts, these doses had no influence on the degree of stunning. Stunned myocardium in Buchwald's experiments [8] did not show increased tissue concentrations of malondialdehyde.

Measurement of Free Radicals

Reactive oxygen species are difficult to measure. Methods of detection fall into the following categories:
- indirect evidence, i.e., lipid peroxidation, malondialdehyde, tetrazolium salts, strand breaks in DNA;
- direct measurements with electron spin resonance in frozen tissue;
- direct measurements with electron spin resonance using spin traps;
- "ex iuvantibus-evidence", i.e., results of treatment with SOD and catalase.

Indirect measurements show the result of damage done to tissues by reactive oxygen species. These methods are less well accepted than the ESR techniques with the possible exception of the DNA strand breaks.

ESR of frozen and freeze-dried tissue initially showed interesting data obtained from ischemic-reperfused rat hearts. However, artifacts were introduced by grinding of the freeze-dried tissue and uncertainties existed whether the reactive species were oxygen-, carbon- or nitrogen-derived. Consequently, the direct detection of reactive oxygen species by ESR has fallen into disuse and only spin trap measurements are accepted. The problem with spin traps is that it is difficult to judge from which biological structure trapped radicals had originated. Because of their reactivity their diffusion range must be short. If the trapped radicals originate in endothelium, they may not be very important because endothelium is very resistant to damage by ischemia and reperfusion [68].

Xanthine Oxidase as a Free Radical Producing System

Ischemia and reperfusion may act in concert to produce free radicals by conversion of xanthine dehydrogenase into xanthine oxidase [66]. This is believed to be caused by the tissue-damaging effect of ischemia (no molecular mechanism offered). Conversion of xanthine dehydrogenase to oxidase occurs in part during tissue homogenization. Since ischemia leads to ATP breakdown, the substrates of this reaction, i.e., hypoxanthine and xanthine, are also present. It was therefore believed that this reaction produces free radicals and these in turn are responsible for the damage caused by stunning and reperfusion. It was, however, shown by Podzuweit [54] and Downey [11] that the human heart does not express xanthine oxidase (in contrast to hearts of rats and guinea pigs). Experiments by Muxfeldt and Schaper [47] with attempted enzyme purification showed extremely little urate production in fractions of human heart that were collected, where normally the rat enzyme appeared in gel filtration chromatography. Podzuweit [55] and Muxfeldt [47] showed also that young domestic pigs do not express a cardiac xanthine oxidase, and Downey [10] showed that rabbit hearts do not posses xanthine oxidase activity. All these species devoid of oxidase activity do develop stunning and infarction after ischemia and reperfusion and it is extremely unlikely that free radical production had participated in the damage. It would have been unlikely for yet another reason: urate was shown to be a free radical scavenger [2].

Role of Leukocytes in Stunning and Reperfusion Damage

Neutrophils have long been recognized for their role as the mobile effector cells of acute inflammation; they are equipped with membrane-bound oxidoreductases and are endowed with the capacity to produce superoxide anion [14]. Oxygen radical production, neutrophil accumulation and release of proteolytic enzymes in reperfused tissue [39], and capacity for capillary plugging in previously ischemic capillary beds (the "no reflow" phenomenon) [13] have made neutrophil contribution to reperfusion injury a subject of investigation and debate. In the early 1980s experimental data suggested that neutrophils were activated during myocardial ischemia of greater than 60 min duration, and contributed to that component of infarction which occurs with reperfusion. The evidence came from canine experiments in which neutrophil function was attenuated either by immunologic or pharmacologic means prior to an ischemic period. Following reperfusion, infarct size was found to be reduced when compared with untreated controls. In the initial experiments reported by Romson et al. [63], neutrophils were depleted with antineutrophil antibodies prior to 90 min of circumflex coronary artery occlusion, which reduced the infarct size compared with controls [63]. From the same laboratory it was subsequently shown that pre-treatment with a monoclonal antibody (clone 904, anti-Mo1), which renders neutro-phils less "sticky" by blocking adhesion promoting glycoproteins on the cell surface of neutrophils, resulted in reduction of infarct size following 90 min of circumflex occlusion and 6 h of reperfusion [76]. In other models of ischemia-reperfusion injury, anti-inflammatory agents through their neutrophil modulating effects have been shown to salvage myocardium following 60 or 90 min of ischemia [4, 63, 64, 77, 78].

Thus, there exists considerable evidence which implicates a neutrophil contribu-tion to reperfusion injury following 60 min or more of ischemia, although the major-ity of this work has been carried out in open-chest dogs. In these experiments, elimi-nation or modulation of the neutrophil "factor" reduces but does not eliminate the infarction process, suggesting that neutrophils perhaps amplify the ongoing tissue destruction in some settings, but are not the primary mechanism for myocardial death following ischemia.

With the aforementioned data incriminating neutrophils in reperfusion injury fol-lowing ischemic periods of greater than 60 min, experimental evidence was sought and subsequently found which suggested that neutrophils contributed to the contrac-tile dysfunction following brief periods of ischemia, i.e., stunning. Presumably this would be a result of intravascular free radical production, because neutrophils do not infiltrate tissue after brief periods of ischemia [17]. Several groups of investigators had demonstrated that free radical scavengers administered prior to reperfusion improved post-ischemic ventricular dysfunction in canine models [19, 57], although other investigators have not found SOD and catalase of benefit in stunned myocar-dium [46]. Since the source of oxygen radicals for which SOD/catalase confer protec-tion cannot be specified, other sources besides neutrophils must be considered. In the initial study, Engler and Covell [12] provided direct evidence that neutrophils are involved in stunning by demonstrating complete restoration of function following 15 min of ischemia in canine myocardium, when the reperfusate was filtered virtually neutrophil free. However, following a similar protocol using a neutrophil filter, Jeremy and Becker were unable to reproduce these results [28]. Schott et al. [73] using the anti-Mo1 anti-neutrophil adhesion antibody (clone 904), which limited reperfusion damage in the experiments reported by Simpson et al. [76], did not find improved stunning or reflow through 3 h of reperfusion following 15 min of LAD ischemia in a dog model. O'Neill et al. [52] eliminated 90% of neutrophils with an anti-neutrophil antibody and found no improvement in arrhythmias, vascular abnor-malities, or stunning following 15 min of regional ischemia in canine myocardium. Some of the discrepancies between reports on improvement in stunning following brief periods of ischemia may be attributable to the duration and intensity of ischemia, which Bolli et al. [7] have shown to be important determinants in the time course of recovery of stunned myocardium. That collateral flow and ischemic dura-tion are important determinants of infarct size has been known for some time [69]. The data reported by Bolli et al. [7] suggest that collateral flow during ischemia is an important predictor for recovery of stunned myocardium as well. This serves as a caveat to investigators who wish to compare interventions within and between experiments in which myocardium is subjected to less than 15 min of ischemia and then followed for recovery of stunning.

In summary, experimental evidence in acute animal models suggests that neutrophils contribute to reperfusion injury following 60 to 90 min of regional ischemia. Neutrophil participation in acute reperfusion injury outside of the acute canine model has not yet been established, and we await evidence that modulation of neutrophil function will prove a clinically useful strategy in limiting reperfusion damage in man. Neutrophil involvement in the myocardial "stunning" which follows 15 min or less of regional ischemia remains controversial at this time with the majority of reports concluding that neutrophils do not contribute to stunning. This suggests that neutrophil activation is dependent on the duration of ischemia and may be an "amplification factor" only with longer ischemic intervals, in which irreversible myocardial necrosis is already present.

Catecholamines and Free Radicals

Catecholamines were discussed within the context of free radical damage because adrenochrome, a tissue toxin, is produced from catecholamines plus free radicals [40] since norepinephrine is released in ischemia. This is not very likely for a variety of reasons:
- stunned myocardium responds to autonomic nerve stimulation which indicates sufficient norepinephrine storage in vesicles (no or little loss during the previous short period of ischemia)
- Podzuweit et al. [56] measured norepinephrine (total tissue content (HPLC) and tissue fluorescence of vesicle-stored norepinephrine) and showed that the fluorescence remained stable for up to 30 min of ischemia in pig myocardium and HPLC-measured norepinephrine also remained constant, i.e., no serious losses and hence no significant metabolism could have taken place
- as stated above: since free radical production following short occlusions and reperfusions remains doubtful, it is not very likely that adrenochrome was produced from only small amounts of norepinephrine.

As an aside it appears imaginable that a regional down-regulation of beta-receptors could have taken place due to the supposedly initially high concentrations of norepinephrine. In later stages of ischemia up-regulation of receptors was reported.

Preconditioning

Reimer's and Jennings' group [45] has shown that repeated short coronary occlusions in dogs condition the myocardium in such a way that a long occlusion, usually lethal for most myocytes at risk, does not lead to infarction, or it produces only very small ones. We have shown recently [74] that these observations can be extended to the pig heart which differs from that of the dog in the lack of a collateral circulation and by the absence of xanthine oxidase. Murry et al. [45] did these experiments in view of experiments by others [16, 23] who tried to find out whether repeated short occlusions produce cumulative damage. Zimmer et al. [84] have shown previously that the ischemic catabolism of ATP is costly because it takes such a long time to resynthesize ATP and it costs so much energy (nine molecules of ATP to produce one new molecule of ATP). We had argued that if a patient develops anginal attacks (each of which is assumed to decrease ATP) more often than can be compensated for by de-novo or salvage synthesis of ATP, the high energy stores are depleted even if no new vascular event takes place (i.e., no acute thrombosis complicating a high degree stenosis). Experiments by Swain et al. [81] and our group [23] had shown, however, that repeated occlusion not only did not produce cumulative damage to myocardial ultrastructure [21] and high energy stores, they in fact had a protective effect. Only the first few short occlusions produced a marked fall of ATP and led to measurable amounts of adenosine. In subsequent occlusions (up to 40 in one of our experiments [23]), no significant further fall of ATP occurred and even the fall of creatine phosphate was attenuated.

Since these repeated occlusions produced regional contractile dysfunction with symptoms typical for reduced ATP utilization at the myofibrillar level, the ATP-sparing effect is explained.

On the basis of our own experiments with repeated occlusions we explained Reimer's and Jennings' findings [45] as follows:

Repeated short occlusions lead to reduced electromechanical coupling and less ATP is used at the sarcomere for a less forceful contraction. Regional MVO_2 decreases, which is a more favorable condition for the myocardium when entering another longer coronary occlusion. The lower oxygen needs decrease the gap between supply (= collateral blood flow) and demand. As a result more myocardium survives longer.

Reduction of O_2-demand in the presence of a measurable collateral flow (in the canine heart up to 30% of normal flow) was believed to be the key role played in canine myocardium in preconditioning. The porcine heart has no collateral blood flow and the reduction of demand (if of the same magnitude as in the canine, i.e., 30%) should not produce similar effects because the demand-supply ratio is still infinitely high.

Much to our surprise preconditioning by short occlusions in the porcine heart did lead to significantly smaller infarcts in the absence of a significant oxygen-sparing. Obviously a different explanation for the effects of preconditioning must be found.

Our experiments went as follows:

We preconditioned swine myocardium with two 10 min episodes of ischemia, each followed by 30 min of reperfusion. We measured regional wall function with ultrasonic crystals and found that two ischemia/reperfusion cycles left the myocardium virtually akinetic in the distal LAD bed. Myocardium was then subjected to 60 min of ischemia and 90 min of reperfusion after which infarct size was measured with the nitroblue tetrazolium technique. Paradoxically, preconditioned swine myocardium which endured a total of 80 min of ischemia had about one-fifth the infarct volume of control myocardium which was subjected to only 60 min of ischemia. Tracer microspheres were used to measure blood flow to the risk area, which, consistent with our previous experience with swine, was in every case virtually nil during ischemia (<10 ml/100 g/min, transmurally). We assumed that the "energy saved" from the acontractile units of stunned myocardium was responsible for the reduction in infarct volume. To quantify this we measured myocardial oxygen consumption

(MVO$_2$) in the stunned region, which declined when referenced to the simultaneously obtained non-ischemic MVO$_2$ but not when compared with the control values for regional MVO$_2$. Thus, we were unable to explain the dramatic benefit of preconditioning in terms of reduced MVO$_2$ in stunned myocardium; more generally stated, the "energy savings" from the idling contractile machinery did not account for the greater tolerance to ischemia.

Current Hypotheses to Explain Preconditioning

Four hypotheses exist to explain preconditioning:
1) Reimer and Jennings [45] initially proposed that repeated occlusions depleted the heart's glycogen stores which they felt was beneficial for a longer subsequent occlusion because an important source for lactate was eliminated. The conversion of large molecular glycogen to small molecular lactate would cause an osmotic load which is dangerous for the ischemic cardiac myocyte. Reimer and Jennings [45] followed up on an experiment by Neely [48] who showed that anoxic perfusion of rat hearts prior to ischemia has a protective effect.
2) In a more recent experiment, Reimer's group [83] showed that the protective effect of short occlusions was neutralized by treatment with SOD and catalase, i.e., SOD and catalase produced the usual large infarcts following long occlusions, and short occlusions had lost their conditioning effect. Since they saw no effect of SOD and catalase on the degree of stunning, they concluded that stunning and preconditioning are unrelated.
3) Our group had, until recently, believed that stunning and preconditioning are indeed interrelated because of the saving of energy by the idling contractile machinery. We have shown above that this is true only for the dog heart and hence not universally applicable.
4) Our presently favored hypothesis is that short preconditioning occlusions lead to the enhanced expression of heat shock protein genes. Heat shock proteins are produced by cells and tissues following a wide variety of sublethal stresses (heat, irradiation) including ischemia [43]. These proteins confer protection against an otherwise lethal stress. Although heat shock proteins were shown to be expressed in ischemic myocardium of open-chest dogs, a protective role was so far not shown.

Reperfusion Injury

Very often problems in ischemic organs are severely aggravated at the moment of reperfusion: limbs, kidneys, and the brain swell, other organs develop hemorrhagic infarctions. Swelling and hemorrhage occur in the heart. Swelling leads to strangulation of blood vessels and no blood flow is possible after the first in-rush of reperfusion. These symptoms occur when organs are reperfused after long-lasting ischemia and it is generally agreed that they were not caused by reperfusion but by ischemia and that reperfusion had only unmasked the damage done by ischemia. To approach the problem from the other end: short coronary occlusions do not inflict any lasting structural damage when reperfused, and reperfusion completely salvages ischemic myocardium when carried out until 30 min of ischemia. Significant subendocardial necrosis had developed by 40 min of occlusion and reperfusion, and by 90 min about two-thirds of the region at risk of infarction had infarcted. These time intervals (obtained from anesthetized dogs) can be modulated by changes in MVO_2 before and during occlusion, but the final extent of infarction cannot be changed without reperfusion. The main effect of reperfusion is thus salvage of ischemic myocardium.

However, as said in the introduction, myocardium is not always salvaged by reperfusion under clinical circumstances, although the time between onset of symptoms and clot lysis may have been relatively short, and in some studies a relationship between duration of symptoms until injection of the thrombolytic agent did not show a good correlation with salvage (postlytic regional contractile function). These clinical observations plus findings in isolated buffer-perfused hearts where the re-introduction of oxygen after prolonged periods of ischemia caused accelerated cell death [20] led to the notion of reperfusion injury. It is quite obvious that people active in this field do not necessarily agree on what reperfusion injury means. These differences were recently summarized by Reimer and Jennings [59]. A good definition was offered by Buckberg et al. (cited in [59]) who stated that reperfusion injury refers to "those metabolic, functional and structural consequences of restoring coronary flow that can be avoided or reversed by modification of the conditions of reperfusion".

These modifications are relatively easy to carry out under the setting of experimental cardiac surgery (for which the definition was phrased), but in regional reperfusion of the in situ beating, blood-pumping heart reperfusion is difficult to modify. It is quite clear that experimental therapy with drugs is one of these modifications.

Another definition of reperfusion injury states that reperfusion per se may be able to injure myocytes that had only been reversibly damaged by ischemia. Modification of the conditions of reperfusion may help these cells to survive. Finally, an experimental situation should be created that shows unequivocally that more cells will survive without than with reperfusion. If such an experiment were not possible then the whole question remains more or less academic.

A number of experiments were carried out to show the benefit of treatment applied only during reperfusion. This is necessary because treatment applied also during (or even before) ischemia may have influenced the intensity of ischemia and indirectly the outcome of reperfusion.

A hypothesis favored by many states that reperfusion leads to increased Ca^{2+} uptake (overload) of the myocytes which is believed to be the initiating event in cell death [15]. Jennings et al. [27] showed recently that this hypothesis was clearly established only in buffer-perfused isolated hearts in most of which the original Krebs-Henseleit buffer was used that contains too much Ca^{2+} (2.5 mM). Jennings' experiments in blood-perfused, in situ, beating canine hearts [26] showed that only irreversibly injured cells had taken up excessive Ca^{2+} and not the surviving ischemic-reperfused myocytes. Jennings concluded that calcium overload is a consequence of cell death rather than its cause.

Klein et al. [53] infused calcium channel blockers and EGTA intracoronarily at the moment of reperfusion and found moderate decreases in infarct size but more arrhythmias. The same authors reported [32] a reduction of infarct size in pigs after a 3-day pretreatment with the antioxidants Tocopherol (vitamin E) and vitamin C.

Although the treatment was aimed at reperfusion-generated free radicals, long pre-treatment was necessary and the formal requirement of treatment at the moment and during reperfusion could not be met. Klein et al. were unable in a previous experiment to show a beneficial effect of superoxide dismutase (human, recombinant) on infarct size in porcine hearts [31], thereby contradicting reports by Lucchesi's [38] and Becker's [1] group, but confirming those of Reimer and Jennings [83].

With the possible exception of the preliminary findings of Klein et al. [53] with calcium antagonists and EGTA no convincing evidence is available to suggest that a modification of the conditions of reperfusion changes the expected consequences of restoring flow.

However, modification of the intensity of ischemia did change the consequences of reflow which supports the view that reflow only unmasks the damage done during ischemia.

Pitfalls and Sources of Errors

Salvage of ischemic myocardium by reperfusion and presence or absence of reperfusion injury are usually judged by the size of the infarct. The most difficult task when measuring infarcts is to determine the amount of necrosis after *short* (up to 3 h) periods of ischemia *without* reperfusion. The only method to distinguish a necrotic cell from a living one at ischemic intervals ranging between 30 min and 3 h without reperfusion is electron microscopy, but that method does not allow a size determination. The light microscope is completely unsuited to diagnose infarcts after short occlusions without reperfusion because light microscopy relies on vital reactions against dead cells, i.e., leukocyte invasion, cell lysis, and destruction of the nucleus. The earliest light microscopic diagnosis and size determination without reperfusion can be carried out when an occlusion had lasted for at least 12 h. When a region of myocardium was devoid of any collateral blood flow, myocytes may become "mummified", i.e., they look normal in the light microscope and may therefore be diagnosed as normal. They may even stain normally with tetrazolium salts for a time even though they are already dead. This source of error is of particular significance in leukocyte depletion or with work using antileukocyte drugs or -antibodies. It may also contribute to the notion that infarcts get larger with reperfusion time.

The fact that the effects of coronary occlusion on the viability of the myocardium cannot be quantitatively measured for at least up to 6 hours without reperfusion has certainly added to the confusion regarding the term reperfusion injury. A much better method of measuring infarcts is the use of tetrazolium salts. We systematically elucidated some years ago [33] the nature of this reaction and have described it as an enzymatic cycling. The elements of cycling are:
1) Substrate availability (i.e., lactate, malate, and pyruvate);
2) Presence of a dehydrogenase;
3) Availability of a co-substrate (NAD).

Normal and surviving tissue have sufficient substrate, co-substrate, and dehydrogenase. After incubation of fresh tissue with colorless tetrazolium solutions, the enzymatic cycling produces the water insoluble formazan dye-precipitate (depending on the type of tetrazolium used).

In tissue that had been ischemic for about 6 h and not reperfused only the dead cells are formazan-negative because the co-substrate NAD is missing. Damage to the sarcolemma in ischemia activates the enzyme glycohydrolase [51] which cleaves NAD. In older infarcts (12 h to several days) the lack of dehydrogenase as well as the lack of co-substrate become limiting. Significant errors can be made when succinate is offered as a substrate: succinate dehydrogenase is a membrane-bound (mitochondrial) enzyme that resists degradation for a long time and does not require NAD as a co-substrate. False negative infarct diagnoses will result in early and late infarction using succinate.

Although the use of tetrazolium salts reduced the time of reperfusion-free infarct sizing from 12 h to 6 h only the addition of reperfusion enabled the exact size-diagnosis of infarcts developing earlier after coronary occlusion. We were interested in the mechanisms by which reperfusion became discriminative in the diagnosis of early infarction. Studies with coronary occlusion in canine hearts showed that irreversible ultrastructural damage coincided with a fall in NAD below 50% of its normal value. Initially, in ischemia the sum of NAD plus NADH stays constant in contrast to phosphocreatine which falls precipitously, and to ATP which falls rapidly. For that initial stability NAD + NADH were used as a reference to normalized tissue ATP values from biopsies [5]. Regitz et al. [58] showed that the co-substrate NAD became displaced from its mitochondrial site after occlusion around the time of irreversible damage by ultrastructural criteria. The fact that NAD is only displaced in already irreversibly damaged cells leads to a formazan-dye precipitate on the cut surface of a heart slice, i.e., *incubation of infarcted non-reperfused myocardium in tetrazolium-buffers may lead to a false-negative diagnosis.*

Reperfusion of such tissue leads to washout of displaced NAD + NADH and only surviving tissue becomes stained.

Reperfusion hastens the decay of irreversibly injured tissue which enables also the diagnosis of early infarction by light microscopy. Infarct sizing after reperfusion by light microscopy does not differ from that using tetrazolium if equal reperfusion times are compared.

Errors Associated with the Onset of Reperfusion

Considerable disagreement exists with regard to the duration of reperfusion on final infarct size. By our definition the duration of reperfusion has little influence on infarct size. There are, however, technical problems which hinder the correct infarct sizing, and these are sometimes related to the onset of reperfusion. An example: several years ago we tried to adapt techniques to the pig heart that had proven useful in the dog heart. We occluded the LAD in the pig with a snare (as in the dog), reperfused by releasing the snare and measured infarct with tetrazolium at 45, 90, and 180 min of ischemia and at 30 min of reperfusion. Much to our surprise we observed only occasionally formazan-negative (= infarction) areas even at 180 min of occlusion and 30 min of reperfusion. The addition of tracer microsphere methodology and biochemical analysis of tissue biopsies for ATP and lactate solved the puzzle: false-formazan-positive tissue did not contain microspheres injected during reperfusion and had high lactate and low ATP levels, i.e., the hearts were not reperfused despite removal of the snare. We had fallen victim to initially unrecognized coronary artery spasm that is typical for pig coronaries, especially with snares. When different methods of occlusion were chosen (soft plastic clamps) and when indicators of reflow were used (and hearts with failed reperfusion excluded from data analysis) similar results as in canine hearts were obtained in the pig [31]. It is quite imaginable that reflow (in spite of prompt removal of the clamp) develops in a protracted way and that islands of false-positive (= false normal) myocardium present the illusion of a small infarct. With time these islands either become reperfused or the remaining NAD-NADH is destroyed and the infarcts "grow" larger, but only virtually. No argument in favor of reperfusion injury can be developed from such an experiment.

Conclusion

We have discussed stunning, preconditioning, and reperfusion damage, and we have critically examined proposed pathophysiological mechanisms. We favor changes in Ca^{2+}-homeostasis (probably situated in the sarcoplasmic reticulum and in the sarco-lemma) for stunning, expression of protective stress proteins for preconditioning, and we find relatively little evidence for the existence of reperfusion injury.

References

1. Ambrosio G, Becker LC, Hutchins GM, Weisman HF, Weisfeldt ML (1986) Reduction in experimental infarct size by recombinant human superoxide dismutase: insights into the pathophysiology of reperfusion injury. Circulation 74:1424–1433
2. Ames BN, Cather R, Schiers E, Hochstein P (1981) Uric acid provides an antioxidant defense in humans against oxidant – and radical – caused aging and cancer: a hypothesis. Proc Natl Acad Sci USA 78:6858–6862
3. Becker LC, Levine JH, DiPaula A, Guarnieri T, Aversano T (1986) Reversal of dysfunction in postischemic stunned myocardium by epinephrine and postextrasystolic potentiation. J Am Coll Cardiol 7:580–589
4. Bednar M, Smith B, Pinto A, Mullane KM (1985) Nafazatrom induced salvage of ischemic myocardium in anaesthetized dogs is mediated through inhibition of neutrophil function. Circ Res 57:131–141
5. Belle van H (1989) Personal communication.
6. Bolli R, Zhu W, Hartley CJ, Michael LH, Repine JE, Hess ML, Kukreja RC, Roberts R (1987) Attenuation of dysfunction in the post-ischemic "stunned" myocardium by dimethyl-thiourea. Circulation 76:458–468
7. Bolli R, Zhu WX, Thornby JI, O'Neill PG, Roberts R (1988) Time course and determinants of recovery of function after reversible ischemia in conscious dogs. Am J Physiol 254:H102–H114
8. Buchwald A, Klein HH, Lindert S, Oberschmidt R, Pich S, Nebendahl K, Kreuzer H (1988) Verhindert intracoronare Superoxiddismutase die Entwicklung der postischämischen myocar-dialen Kontraktionsstörung? Z Kardiol 77 [suppl I]:84 (abstr)
9. DeBoer LWV, Ingwall JS, Kloner RA, Braunwald E (1980) Prolonged derangements of canine myocardial purine metabolism after a brief coronary artery occlusion not associated with anatomic evidence of necrosis. Proc Natl Acad Sci USA 77:5471–5475
10. Downey JM, Miura T, Eddy LJ, Chambers DE, Mellert T, Hearse DJ, Yellon DM (1987) Xanthine oxidase is not a source of free radicals in the ischemic rabbit heart. J Mol Cell Cardiol 19:1053–1060
11. Eddy LJ, Stewart JR, Jones HP, Engerson TD, McCord JM, Downey JM (1987) Free radical-

producing enzyme, xanthine oxidase, is undetectable in human hearts. Am J Physiol 253:H709—H711

12. Engler R, Covell JW (1987) Granulocytes cause reperfusion ventricular dysfunction after 15 minute ischemia in the dog. Circ Res 61:20—28
13. Engler RL, Schmid-Schoenbein GW, Pavelec RS (1983) Leukocyte capillary plugging in myocardial ischemia and reperfusion in the dog. Am J Pathol 111:98—111
14. Fantone JC, Ward PA (1982) Role of oxygen-derived free radicals and metabolites in leukocyte-dependent inflammatory reactions. Am J Pathol 107:397—418
15. Fleckenstein A, Janke J, Döring HJ, Leder O (1971) Die intracelluläre Überladung mit Kalzium als entscheidender Kausalfaktor bei der Entstehung nichtkoronarogener Myokardnekrosen. Verh Dtsch Ges Kreislaufforsch 37:345—353
16. Geft IL, Fishbein MC, Ninomiya K, Hashida J, Chaux E, Yano J, Y-Rit J, Genov T, Shell W, Ganz W (1982) Intermittent brief periods of ischemia have a cumulative effect and may cause myocardial necrosis. Circulation 66:1150—1153
17. Go LO, Murry CE, Richard VJ, Weichedal GR, Jennings RB, Reimer KA (1988) Myocardial neutrophil accumulation during reperfusion after reversible and irreversible ischemic injury. Am J Physiol 255:H1188—H1198
18. Gotsman MS, Lotan C, Weiss AT, Appelbaum D, Sapoznikov D, Hasin Y, Mosseri M (1989) Early and prehospital thrombolytic therapy in acute myocardial infarction. In: Schmutzler H, Rutsch W, Dougherty FC (eds) Limitation of Infarct Size. Springer, Berlin Heidelberg New York London Paris Tokyo, pp 107—130
19. Gross GJ, Farber NE, Hardman HF, Warltier DC (1986) Beneficial actions of superoxide dismutase and catalase in stunned myocardium of dogs. Am J Physiol 250:H372—H377
20. Hearse DJ, Humphrey SM, Chain EB (1973) Abrupt reoxygenation of the anoxic potassium-arrested perfused rat heart — A study of myocardial enzyme release. J Mol Cell Cardiol 5:395—407
21. Henrichs KJ, Matsuoka H, Schaper J (1987) Influence of repetitive coronary occlusions on myocardial adenine nucleosides, high energy phosphates and ultrastructure. Basic Res Cardiol 82:557—565
22. Heyndrickx GR, Baig H, Nellens P, Leusen I, Fishbein MC, Vatner SF (1978) Depression of regional blood flow and wall thickening after brief coronary occlusions. Am J Physiol 234:H653—H659
23. Hoffmeister HM, Mauser M, Sass S, Schaper J, Schaper W (1984) Ninety minutes of coronary occlusion: prevention of infarcts by short intermittent reperfusion. Circulation 68 [Suppl II]:II—261 (abstr.)
24. Hoffmeister HM, Mauser M, Schaper W (1985) Effect of adenosine and AICAR on ATP content and regional contractile function in reperfused canine myocardium. Basic Res Cardiol 80:445—458
25. Ito B, Tate H, Kobayashi M, Schaper W (1987) Reversibly injured, post-ischemic canine myocardium retains normal contractile reserve. Circ Res 61:834—846
26. Jennings RB, Schaper J, Hill ML, Steenbergen C Jr, Reimer KA (1985) Effect of reperfusion late in the phase of reversible ischemic injury. Changes in cell volume, electrolytes, metabolites, and ultrastructure. Circ Res 56:262—278
27. Jennings RB, Reimer KA, Steenbergen C (1986) Myocardial ischemia revisited. The osmolar load, membrane damage, and reperfusion. J Mol Cell Cardiol 18:769—780
28. Jeremy R, Becker L (1988) Neutrophil depletion does not prevent myocardial stunning. Circulation 78 [Suppl II]:II—77 (abstr)
29. Kammermeier H, Schmidt P, Juengling E (1982) Free energy change of ATP-hydrolysis: a causal factor of early hypoxic failure of the myocardium. J Mol Cell Cardiol 14:267—277
30. Kennedy JW, Ritchie JL, Davis RB, Fritz JK (1983) Western Washington randomized trial of intracoronary streptokinase in acute myocardial intervention. N Engl J Med 309:1477—1482
31. Klein HH (1987) Thesis, Universität Göttingen
32. Klein HH, Pich S, Lindert S, Nebendahl K, Niedmann P, Kreuzer H (1989) Protektive Wirkung einer Kombinationsbehandlung mit den Vitaminen E und C beim experimentellen Myocardinfarkt. Z Kardiol 78 [suppl I]:136 (abstr)
33. Klein HH, Puschmann S, Schaper J, Schaper W (1981) The mechanism of the tetrazolium reaction in identifying experimental myocardial infarction. Virchows Arch (Pathol Anat) 393:287—297

34. Kobayashi M, Schmidt T, Schaper W (1987) Regional myocardial oxygen consumption and segmental function in "stunned" myocardium of the pig. Circulation 76 [suppl IV]:IV−379 (abstr)
35. Kojima S, Hori S, Fukuda K, Sato T, Kyotani S, Kusuhara M (1988) Variable regional oxygen consumption in systolic bulge. Circulation 78 [suppl II]:II−263 (abstr)
36. Krause S, Hess ML (1984) Characterization of cardiac sarcoplasmic reticulum dysfunction during short-term normothermic, global ischemia. Circ Res 55:176−184
37. Kusuoka H, Porterfield JK, Weisman HF, Weisfeldt L, Marban E (1987) Pathophysiology and pathogenesis of stunned myocardium. J Clin Invest 79:950−961
38. Lucchesi BR, Mullane KM (1986) Leukocytes and ischemia-induced myocardial injury. Ann Rev Pharmacol Toxicol 26:201−224
39. Lucchesi BR, Romson JL, Jolly SR (1984) Do leucocytes influence infarct size? In: Hearse DJ, Yellon DM (eds) Therapeutic Approaches to Myocardial Infarct Size Limitation. Raven Press, New York, pp 219−248
40. Matthews SB, Henderson AH, Campbell AK (1985) The adrenochrome pathway: the major route for adrenaline catabolism by polymorphonuclear leukocytes. J Mol Cell Cardiol 17:339−348
41. Mauser M, Hoffmeister HM, Nienaber C, Schaper W (1985) Influence of ribose, adenosine, and "AICAR" on the rate of myocardial adenosine triphosphate synthesis during reperfusion after coronary artery occlusion in the dog. Circ Res 56:220−230
42. McFalls EO, Pantely GA, Ophius TO, Anselone CG, Bristow JD (1987) Relation of lactate production to postischemic reduction in function and myocardial oxygen consumption after partial coronary occlusion in swine. Cardiovasc Res 21:856−862
43. Mehta HB, Popovich BK, Dillmann WH (1988) Ischemia induces changes in the level of mRNAs coding for stress protein 71 and creatine kinase M. Circ Res 63:512−517
44. Michael LH, Hunt JR, Weilbaecher D, Perryman MB, Roberts R, Lewis RM, Entman ML (1985) Creatine kinase and phosphorylase in cardiac lymph: coronary occlusion and reperfusion. Am J Physiol 248:H350−H359
45. Murry CE, Jennings RB, Reimer KA (1986) Preconditioning with ischemia: a delay of lethal cell injury in ischemic myocardium. Circulation 74:1124−1136
46. Murry CE, Richard VJ, Jennings RB, Reimer KA (1988) Preconditioning with ischemia: is the protective effect mediated by free radical-induced myocardial stunning? Circulation 78 [suppl II]:II−77 (abstr)
47. Muxfeldt M, Schaper W (1987) The activity of xanthine oxidase in heart of pigs, guinea pigs, rabbits, rats, and humans. Basic Res Cardiol 82:486−492
48. Neely JR, Grotyohann LW (1984) Role of glycolytic products in damage to ischemic myocardium. Dissociation of adenosine triphosphate levels and recovery of function of reperfused ischemic hearts. Circ Res 55:816−824
49. Nees S, Gerlach E (1983) Adenine nucleotide and adenosine metabolism in cultured coronary endothelial cells: formation and release of adenine compounds and possible functional implications. In: Berne RMK, Rall TW, Rubio R (eds) Regulatory Function of Adenosine. Martinus Nijhoff Publishers, The Hague Boston London, pp 347−360
50. Neubauer S, Hamman BL, Perry SB, Bittl JA, Ingwall JS (1988) Velocity of the creatine kinase reaction decreases in postischemic myocardium: a 31P-NMR magnetization transfer study of the isolated ferret heart. Circ Res 63:1−15
51. Nunez R, Calva E, Marsch M, Briones E, Lopez-Soriano F (1976) NAD glycohydrolase activity in hearts with acute experimental infarction. Am J Physiol 231:1173−1177
52. O'Neill PG, Charlat ML, Michael LH, Roberts R, Bolli R (1989) Influence of neutrophil depletion on myocardial function and flow after reversible ischemia. Am J Physiol 256:H341−H351
53. Pich S, Klein HH, Lindert S, Nebendahl K, Warneke G, Kreuzer H (1989) Therapie des Reperfusionsschadens mit Calcium-Antagonisten und verminderter freier koronarer Calcium-Konzentration beim experimentellen Myokardinfarkt. Z Kardiol 78 [suppl I]:137 (abstr)
54. Podzuweit T, Beck H, Müller A, Görlach G, Scheld HH (1988) Absence of xanthine oxidase activity in the human myocardium. J Mol Cell Cardiol 20 [suppl V]:131 (abstr)
55. Podzuweit T, Braun W, Müller A, Schaper W (1987) Arrhythmias and infarction in the ischemic pig heart are not mediated by xanthine oxidase-derived free oxygen radicals. Basic Res Cardiol 82:493−505

56. Podzuweit T, Frankenfeld-Erb C, Wagner D, Stang EV, Flaig W, Schaper W (1986) Dissozi-
    ation von Katecholaminfreisetzung und Arrhythmie-Häufigkeit im ischämischen Schweine-
    myokard. Z Kardiol 75 [suppl I]:65 (abstr)
57. Przyklenk K, Kloner RA (1986) Superoxide dismutase plus catalase improve contractile func-
    tion in the canine model of the "stunned myocardium". Circ Res 58:148−156
58. Regitz V, Paulson DJ, Hodach RJ, Little SE, Schaper W, Shug AL (1984) Mitochondrial dam-
    age during myocardial ischemia. Basic Res Cardiol 79:207−217
59. Reimer KA, Jennings RB (1988) Reperfusion injury. In: Schettler G, Jennings RB, Rapaport
    E, Wenger NK, Bernhardt R (eds) Reperfusion and Revascularization in Acute Myocardial
    Infarction. Springer, Berlin Heidelberg New York London Paris Tokyo, pp 52−55
60. Reimer KA, Jennings RB (1979) The "wavefront phenomenon" of myocardial ischemic cell
    death. II. Transmural progression of necrosis within the framework of ischemic bed size
    (myocardium at risk) and collateral flow. Lab Invest 40:633−644
61. Rentrop KP (1985) Thrombolytic therapy in patients with acute myocardial infarction. Circu-
    lation 71:627−631
62. Rentrop KP, Cohen M, Blanke H, Phillips RA (1985) Changes in collateral channel filling
    immediately after controlled coronary artery occlusion by an angioplasty balloon in human
    subjects. J Am Coll Cardiol 5:587−592
63. Romson JL, Bruce GH, Kunkel SL, Abrams GD, Schork MA, Lucchesi BR (1983) Reduction
    of the extent of ischemic myocardial injury by neutrophil depletion in the dog. Circulation
    67:1016−1023
64. Romson JL, Jolly SR, Lucchesi BR (1984) Protection of ischemic myocardium by phar-
    macologic manipulation of leukocyte function. Cardiovasc Rev Rep 5:660−709
65. Rooke GA, Feigl EO (1982) Work as a correlate of canine left ventricular oxygen consumption
    and the problem of catecholamine oxygen wasting. Circ Res 50:273−286
66. Roy RS, McCord JM (1983) Superoxide and ischemia: Conversion of xanthine dehydrogenase
    to xanthine oxidase. In: Greenwald RA, Cohen G (eds) Oxy Radicals and Their Scavenger
    Systems, vol II, Cellular and Medical Aspects. Elsevier Biomedical, New York, pp 145−153
67. Schaper J (1989) Personal communication.
68. Schaper J, Schaper W (1988) Time course of myocardial necrosis. Cardiovasc Drugs Ther
    2:17−25
69. Schaper W, Binz K, Sass S, Winkler B (1987) Influence of collateral blood flow and of varia-
    tion in MVO_2 on tissue-ATP content in ischemic and infarcted myocardium. J Mol Cell Cardiol
    19:19−37
70. Schaper W, Frenzel H, Hort W, Winkler B (1979) Experimental coronary artery occlusion. II.
    Spatial and temporal evolution of infarcts in the dog heart. Basic Res Cardiol 74:233−239
71. Schaper W, Frenzel H, Hort W (1979) Experimental coronary artery occlusion. I. Measure-
    ment of infarct size. Basic Res Cardiol 74:46−53
72. Schaper W, Ito BR (1988) The energetics of "stunned" myocardium. In: de Jong JW (ed)
    Myocardial Energy Metabolism. Martinus Nijhoff Publishers, Dordrecht Boston Lancaster,
    pp 203−213
73. Schott RJ, Nao BS, McClanahan TB, Simpson PJ, Stirling MC, Todd RF, Gallagher KP
    (1989) F(ab')₂ Fragments of Anti-Mo1 (904) monoclonal antibodies do not prevent myocardial
    stunning. Circ Res 65:1112−1124
74. Schott RJ, Rohmann S, Braun ER, Winkler B, Jürgens S, Schaper W (1989) The effect of
    ischemic preconditioning on myocardial oxygen consumption (MVO_2) and infarct size in pigs.
    J Mol Cell Cardiol 21 [Suppl II]:161 (abstr)
75. Sies H (ed) (1985) Oxidative Stress. Academic Press, London Orlando San Diego New York
    Toronto Montreal Sydney Tokyo
76. Simpson PJ, Todd RF, Fantone JC, Mickelson JK, Griffin JD, Lucchesi BR (1988) Reduction
    of experimental canine myocardial reperfusion injury by a monoclonal antibody (Anti-Mo1,
    Anti-CD 11b) that inhibits leukocyte adhesion. J Clin Invest 81:624−629
77. Simpson PJ, Mickelson J, Fantone JC, Gallagher KP, Lucchesi BR (1987) Iloprost inhibits
    neutrophil function in vitro and in vivo and limits experimental infarct size in canine heart. Circ
    Res 60:666−673
78. Simpson PJ, Mitsos SE, Ventura A, Gallagher KP, Fantone JC, Abrams GD, Schork MA,
    Lucchesi BR (1987) Prostacyclin protects ischemic reperfused myocardium in the dog by inhi-
    bition of neutrophil activation. Am Heart J 113:129−137

79. Stahl LD, Aversano TR, Becker LC (1986) Selective enhancement of function of stunned myocardium by increased flow. Circulation 74:843–851
80. Stahl LD, Weiss HR, Becker LC (1988) Myocardial oxygen consumption, oxygen supply/demand heterogeneity, and microvascular patency in regionally stunned myocardium. Circulation 77:865–872
81. Swain JL, Sabina RL, Hines JJ, Greenfield Jr JC, Holmes EW (1984) Repetitive episodes of brief ischaemia (12 min) do not produce a cumulative depletion of high energy phosphate compounds. Cardiovasc Res 18:264–269
82. Swain JL, Sabina RL, McHale PA, Greenfield JC, Holmes EW (1982) Prolonged myocardial nucleotide depletion after brief ischemia in the open-chest dog. Am J Physiol 242:H818–H826
83. Uraizee A, Reimer KA, Murry CE, Jennings RB (1987) Failure of superoxide dismutase to limit size of myocardial infarction after 40 minutes of ischemia and 4 days of reperfusion in dogs. Circulation 75:1237–1248
84. Zimmer HG, Trendelenburg C, Kammermeier H, Gerlach E (1973) De-novo-synthesis of myocardial adenine nucleotides in the rat. Acceleration during recovery from oxygen deficiency. Circ Res 32:635–642

Authors' address:
Prof. Dr. W. Schaper
Max-Planck-Institute
Department of Experimental Cardiology
Benekestrasse 2
D-6350 Bad Nauheim, FRG

Coronary Vasomotor Tone in Myocardial Ischemia

Gerd Heusch and Brian D. Guth

Abt. für Pathophysiologie, Universitätsklinikum Essen and Dept. of Medicine, University of California, San Diego, USA

Introduction

This review is focused on the pathophysiology of myocardial perfusion during regional ischemia. The importance of the extravascular component of coronary resistance, in particular for the distribution of blood flow in ischemic myocardium, should not be underestimated [6, 55, 91]. However, this contribution discusses primarily the active coronary vasomotor mechanisms during myocardial ischemia.

A knowledge of the physiology of normal coronary blood flow is a prerequisite for the understanding of the pathophysiology of myocardial perfusion in ischemia (for review see [2, 5, 26]). A schematic classification of coronary vasomotor mechanisms is proposed which provides criteria for the analysis of both physiological and pathological coronary vasomotion (Fig. 1).

Vasomotion is ultimately an active dilation or constriction of vessels as the result of relaxation or contraction of vascular smooth muscle cells. With respect to the localization within the vascular tree, a vasomotion of different coronary segments can be classified as affecting 1) the epicardial conduit arteries – with or without a stenosis –, 2) the resistive terminal vascular bed, or 3) the collateral vessels which connect a poststenotic vascular bed with its neighboring perfusion territories. The vasomotion of coronary vessels within different transmural myocardial layers can be affected homogeneously or preferentially [59]. Finally, coronary vasomotion can be classified with respect to its mediators. Myogenic vasomotion is the direct reaction of the vasomotor smooth muscle cells to changes in intraluminal or transmural pressure [80]. Metabolic vasomotion serves to match coronary blood flow to the instantaneous metabolic requirements of the myocardium [88]. More recently the importance of endothelial cells for coronary vasomotion became increasingly apparent, in particular for the dilation of epicardial coronary arteries secondary to a dilation of resistive vessels [3]. Neuronal [52, 104] and, with a certain delay, humoral mediators [15] are responsible for changes in coronary vasomotor tone during acute changes in the cardiovascular homeostasis. Finally, pharmacologic coronary vasomotion may occur by interaction with these physiologic mediators, or by independent mechanisms [9].

Compensatory Coronary Dilation during Myocardial Ischemia

Coronary blood flow is controlled by an effective autoregulation [81]. Over a range of coronary perfusion pressures from 70–130 mmHg, coronary blood flow remains

199

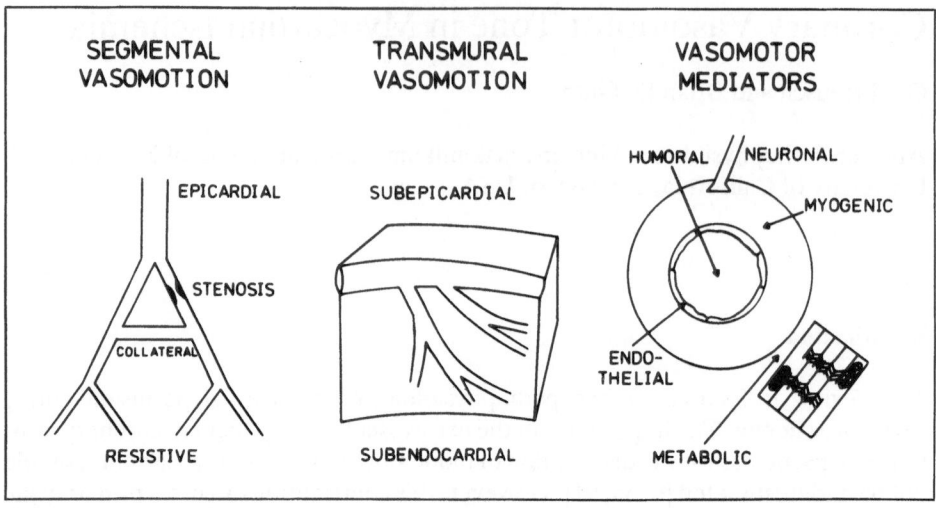

Fig. 1. Schematic classification of coronary vasomotion. For a detailed description see text.

constant by virtue of coronary vasomotion. An intact coronary vascular system is, in turn, capable of increasing coronary blood flow at an unchanged coronary perfusion pressure by about fivefold through coronary dilation [73], thereby exhibiting a marked endogenous coronary dilator reserve [66]. In the presence of a subcritical proximal coronary stenosis, this endogenous coronary dilator reserve of the post-stenotic vascular bed is recruited [37] to maintain perfusion and contractile function of the poststenotic myocardium. With increasing severity of stenosis, the endogenous coronary dilator reserve of the subendocardial myocardium is exhausted prior to that of the subepicardial myocardium since extravascular effects limiting perfusion are greatest near the endocardium [31, 42, 93]. When the endogenous coronary dilator reserve is finally exhausted such that transmural myocardial blood flow and function are already reduced, there is still a dilator reserve which can be recruited pharmacologically [1, 10, 35, 53, 89].

The mechanisms which underly coronary autoregulation and the recruitment of this endogenous coronary dilator reserve are still unclear. Myogenic mechanisms in response to a reduction in coronary perfusion pressure as well as metabolic mechanisms have been suggested [21]. Coronary autoregulation is closely coupled to coronary venous and, thus, presumably myocardial pO_2 [22]. However, the actual mediator of metabolic coronary autoregulation is still unknown. Although previously hypothesized to be this mediator, adenosine has more recently been excluded since inactivation of endogenously released adenosine by deaminase does not change poststenotic regional myocardial blood flow [33].

Epicardial Coronary Constriction

A critical reduction of coronary blood flow at the site of epicardial coronary arteries is a common feature to coronary vasospasm, dynamic coronary stenosis, as well as to

200

changes in the hemodynamic severity of a fixed coronary stenosis. However, these pathophysiological processes differ in the quantitative contribution of active coronary vasoconstriction vs fixed mechanical coronary occlusion for the initiation of myocardial ischemia. The hemodynamic severity of a coronary stenosis may change even when the stenosis is morphologically fixed. Apart from the angles of entry and exit which determine the extent of turbulences at the transition from the normal to the stenotic vascular segments [76], the intraluminal pressure determines the actual cross-sectional area of the stenotic segment. The relationship between coronary blood flow and coronary perfusion pressure is nonlinear and only an apparent coronary resistance can be calculated from the actual ratio of coronary perfusion pressure to coronary blood flow [36]. Nevertheless, changes in intraluminal pressure may change the functional hemodynamic severity of a coronary stenosis. In the presence of a fixed severe coronary stenosis, an acute dilation of the poststenotic coronary vascular bed with a resulting decrease in poststenotic perfusion pressure can markedly decrease coronary blood flow [97] due to changes in the stenotic resistance. Even though a fixed stenosis is just an obstructive mass and therefore organically inert, it can functionally be variable due to changes in the transmural pressure gradient at the obstruction. The functional changes in stenotic resistance are real and not pretended by application of Ohm's law to a nonlinear pressure-flow relationship. Rather than the observed increase in stenotic resistance, a decrease in stenotic resistance with decreasing coronary blood flow would otherwise be predicted.

An increase in the functional hemodynamic severity of a fixed coronary stenosis as the result of poststenotic coronary dilation and a consequent drop in poststenotic coronary perfusion pressure has been demonstrated in experiments after transient coronary occlusion [97], with intracoronary injection of contrast medium [97], with tachycardia by atrial pacing [57], and exercise [99]. Conversely, the hemodynamic severity of a stenosis is decreased when distal coronary perfusion pressure increases [98].

Changes in the hemodynamic severity of a coronary stenosis are much more marked if the stenotic segment also retains vasomotion. Postmortem histological examination in humans often reveals a vascular segment with an eccentric atherosclerosis together with intact vascular smooth muscle cells [28]. Thus, a dynamic coronary stenosis occurs [36, 92] when active coronary vasoconstriction is superimposed on the preexisting atherosclerotic lesion, and acute myocardial ischemia may be initiated. Acute ischemic myocardial dysfunction may be induced by isometric exercise in patients having a coronary stenosis that is compensated at rest [7]. In these patients, myocardial ischemia is induced by the reflex constriction of the stenotic segment (Fig. 2) and not by the increase in myocardial performance, since nitroglycerin prevents the constriction of a stenotic vascular segment and the resulting ischemia, despite an unchanged increase in myocardial performance. Diltiazem also prevents the reflex coronary constriction of a significant coronary stenosis and the resulting myocardial ischemia [63]. Brown et al. attribute such reflex constriction of the stenotic vascular segment to the activation of coronary α-adrenoceptors [7].

There may be only quantitative differences between the previously discussed dynamic coronary stenosis and coronary vasospasm. In fact, critical coronary constriction as well as vasospasm occur preferentially in the immediate vicinity of organic lesions [70]. With respect to this observation, McAlpin developed the theory that

201

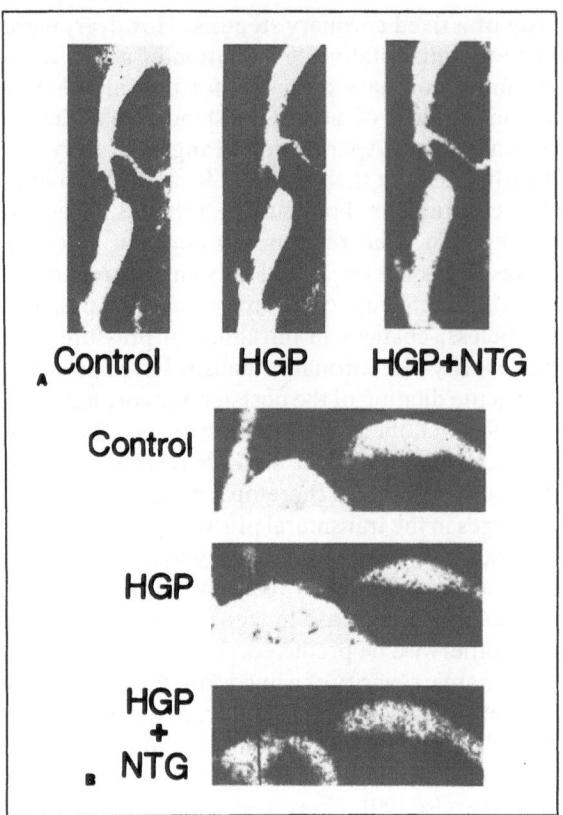

Fig. 2. Coronary arteriography demonstrates a constriction of a significant coronary stenosis during reflex sympathetic activation by isometric exercise (handgrip HGP); Nitroglycerin (NTG) prevents the reflex coronary constriction. (From [7] by permission of the American Heart Association.)

coronary vasospasm is merely the geometric result of normal vasoconstriction in a stenotic vascular segment. Thus, a given decrease in the external radius of a vascular segment with a preexisting luminal reduction results in a proportionally larger decrease in its internal radius [71]. In contrast to this purely geometric theory of coronary spasm, studies by Serruys et al. [101] emphasized the unpredictable nature of vasomotion in a stenotic epicardial coronary segment. During a provocation test with methergine, the angiographically determined luminal reduction of the stenotic segment was poorly matched to the luminal reduction predicted from the vasoconstriction of normal segments by geometric assumptions.

At present, the causes of a local hypercontractility of the coronary vascular wall remain speculative. However, a few interesting observations in this context will be discussed. Although α_1-adrenoceptor mediated constriction of epicardial coronary arteries can be induced by sympathetic activation, the reduction in vessel caliber is not sufficient to cause a measurable reduction in coronary blood flow [50]. The provocation of coronary spasm by ergonovine is considered a classic challenge for the diagnosis of spastic angina pectoris [46, 96]. The provocation of coronary spasm appears to be related to the serotonergic action of ergonovine, whereas its α-adrenergic action appears to be of minor importance [62]. With the development of choles-

202

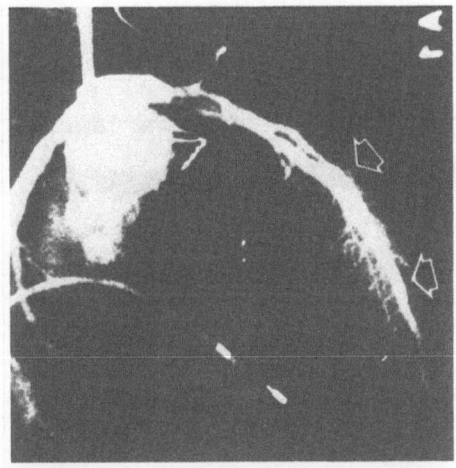

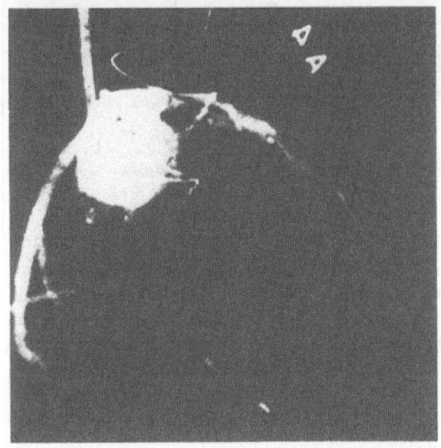

Fig. 3. Coronary angiography of a marked coronary constriction in a segment with diffuse atherosclerosis (at rest, upper panel) in response to intracoronary acetylcholine (lower panel). (From [27] by permission of Rockefeller University Press.)

terol-induced atherosclerosis, the constrictor effects of serotonin in large arterial segments are enhanced [44, 45]. Histamine has also been implicated in the initiation of coronary spasm [34].

The predisposition of epicardial coronary arteries to local hypercontractility by endothelial damage appears to be a particularly attractive hypothesis. The endothelium can release relaxant factors [29, 30]. After experimental endothelial damage an enhanced vasoconstriction of large femoral arteries during α_1-adrenoceptor activation was demonstrated [105]. In canine epicardial coronary arteries the acetylcholine-induced vasodilation is reversed to constriction after removal of the endothelium [95]; the coronary constrictive effects of serotonin are potentiated after endothelial damage [67]. In patients, acetylcholine induces predominantly a dilation of intact coronary arteries, whereas in atherosclerotic coronary segments acetylcholine regularly induces a marked constriction (Fig. 3) [27, 69].

In this context, the interaction between the coronary microcirculation and the diameter of epicardial coronary arteries (Fig. 4), which was first described by Holtz

203

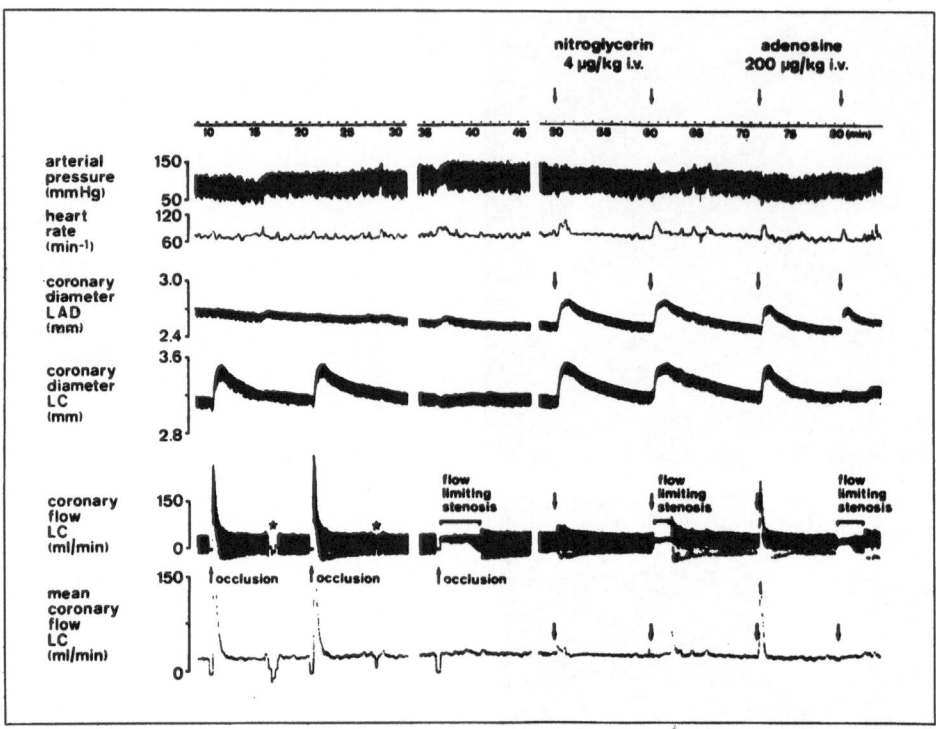

Fig. 4. Dilation of coronary resistance vessels by transient coronary occlusion or adenosine induces, with some delay, also a dilation of epicardial coronary arteries which is prevented by limitation of the increase in coronary blood flow with a stenosis. Nitroglycerin, in contrast, induces a flow-independent dilation of epicardial coronary arteries. (From [60] by permission of Raven Press.)

et al. [60, 61] and shortly thereafter by Hintze and Vatner [58], appears to be particularly important. An increase in coronary blood flow during postocclusive reactive hyperemia or during adenosine infusion induces with some delay an increase in the diameter of epicardial coronary arteries. Holtz proposed the hypothesis that a dilation of coronary arterioles increases coronary blood flow and induces an increased shear stress at the luminal surface of the endothelium which, in turn, releases a vasodilator factor. Prostanoids have been excluded as mediators of this endothelium-mediated dilation of epicardial coronary arteries, since the inhibition of cyclooxygenase by indomethacin or diclofenac does not change these responses. According to this hypothesis, the increased frequency of episodes of vasospastic angina pectoris in the night and early morning hours can be explained by the reduction in myocardial function and metabolism and a respective attenuation of the flow-dependent, endothelium-mediated coronary dilation at that time. Thus, in contrast to stress-induced myocardial ischemia in which there is a primary increase in myocardial oxygen demand, a primary reduction in myocardial demand may initiate myocardial ischemia at rest. However, it must be cautioned that 1) epicardial coronary arteries contribute only about 5% to total coronary resistance [25, 65], and 2) the experimentally observed changes in the diameter of epicardial coronary arteries in these studies have been so small that they could only contribute to reduced coronary blood flow when a pre-existing critical coronary stenosis exists. However, experimental proof for this is still lacking. Also, although there is experimental evidence for enhanced α-adrenergic or serotonergic epicardial coronary constriction, a respective clinical therapy of vasospastic angina pectoris has not been proven successful [12, 18, 103]. In the future, mediators other than the classical neurotransmitters will have to be taken into consideration such as the relatively high concentrations of neuropeptide Y found in human coronary arteries [38].

In summary, coronary blood flow can be critically reduced at the level of epicardial coronary arteries. While the hemodynamic severity of a fixed stenosis may change, epicardial vasoconstriction predominates in a dynamic stenosis with an eccentric atherosclerosis and a normal wall segment and, particularly, in coronary spasm. Despite many interesting experimental observations the causes of local hypercontractility in epicardial coronary arteries and, thus, the causes of vasospastic angina pectoris are still unclear.

Constriction of the Coronary Resistance Vessels

The classic view of the pathogenesis of myocardial ischemia assumes that coronary resistance vessels are maximally dilated during myocardial ischemia and thus are unresponsive to vasoactive stimuli. Recent studies, however, indicate that the idea of maximal coronary dilation during myocardial ischemia is not correct. A pharmacologically recruitable coronary dilator reserve exists, thereby indicating the persistence of coronary constrictive tone in spite of myocardial ischemia [1, 10, 35, 53, 89]. Even in the presence of a coronary stenosis [8] or after maximal pharmacologic dilation with adenosine [64], coronary vessels remain responsive to the α-adrenergic constrictor effects of the sympathetic transmitter norepinephrine. A persistent coronary constrictor effect in spite of myocardial ischemia was also demonstrated in experiments with leukotriene C4, thromboxane A2 [87], angiotensin [23], and vasopressin [90].

Our own studies focused on the importance of α-adrenergic coronary constrictor mechanisms in the initiation and aggravation of myocardial ischemia. In compensation for a proximal coronary stenosis the dilator reserve of the poststenotic coronary vascular bed is partially recruited. While the recruitment of coronary dilator reserve may preserve poststenotic myocardial blood flow and contractile function at rest, it also means reduced potential for further increases in coronary blood flow during conditions of increased myocardial work. Thus, electrical stimulation of cardiac sympathetic nerves induces a marked increase in myocardial work and therefore dilation of intact coronary arteries. In the presence of a moderate coronary stenosis, however, this dilation is much less since some vasodilator reserve has already been recruited (Fig. 5). In the presence of a severe coronary stenosis which actually exhausts the inherent poststenotic coronary dilator reserve, electrical stimulation of cardiac sympathetic nerves can even induce coronary vasoconstriction. This poststenotic coronary constriction, in turn, induces poststenotic myocardial ischemia as evidenced by contractile dysfunction, net lactate production and malignant

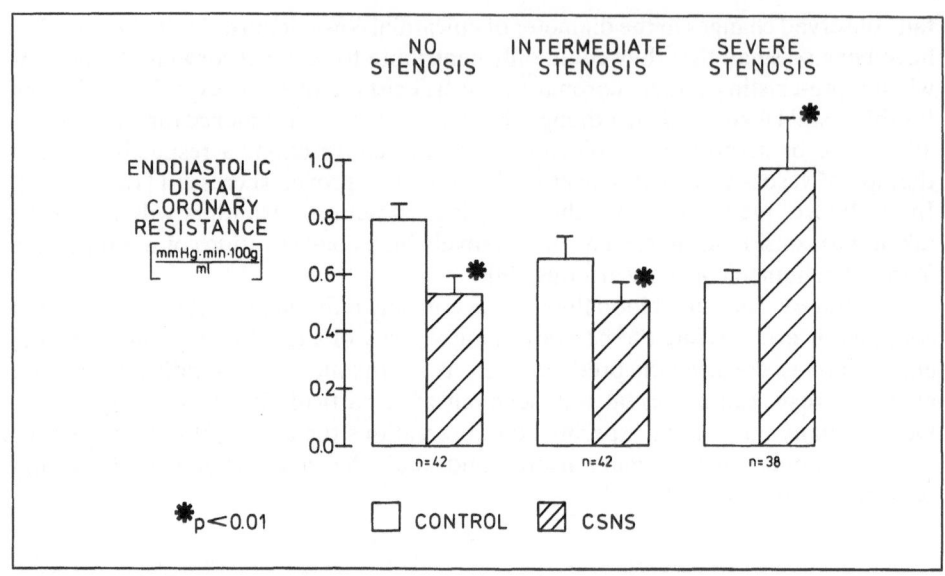

Fig. 5. Electrical stimulation of cardiac sympathetic nerves (CSNS) induces a marked decrease in coronary resistance of intact coronary arteries. With a compensatory decrease in poststenotic resistance distal to an intermediate stenosis at rest, the sympathetically induced decrease in resistance is attenuated. Distal to a severe coronary stenosis, cardiac sympathetic nerve stimulation even increases coronary resistance. (From [48] by permission of the American Heart Association.)

arrhythmias [48]. The sympathetically induced poststenotic coronary constriction is mediated by vascular α_2-adrenoceptors.

Therefore, the nonselective α-antagonist phentolamine or the selective α_2-antagonist rauwolscine can prevent the sympathetically induced poststenotic coronary constriction. The calcium antagonist nifedipine has also been shown to functionally antagonize the α_2-adrenoceptor-mediated vasoconstriction [49]. Poststenotic coronary constriction and myocardial ischemia can also be induced by reflex sympathetic activation, subsequent to carotid occlusion or pain [47, 102]. Furthermore, there is a positive feed-back between myocardial ischemia, once initiated, and the activity of cardiac sympathetic nerves which may lead to a progressive aggravation of myocardial ischemia. This vicious cycle can be interrupted, not only by α_2-blockade or nifedipine, but also by segmental epidural anesthesia of cardiac sympathetic nerves with procaine [51] and central nervous inhibition of sympathetic activity with clonidine [54]. Significant α_2-adrenoceptor mediated poststenotic coronary constriction is not only induced by electrical and reflex sympathetic activation in anesthetized dogs, but also during treadmill exercise in conscious dogs [100]. Intracoronary infusion of the selective α_2-antagonist idazoxan improves poststenotic regional myocardial blood flow and function during continued treadmill exercise (Fig. 6). The improvement in regional myocardial blood flow by idazoxan is particularly pronounced in the subendocardial layers of the ischemic myocardium [100]. Nifedipine also attenuates poststenotic coronary constriction and reduces exercise-induced myocardial ischemia. Nifedipine again particularly reduces the hypoperfusion of subendocardial and midmyocardial layers (Fig. 7) [53].

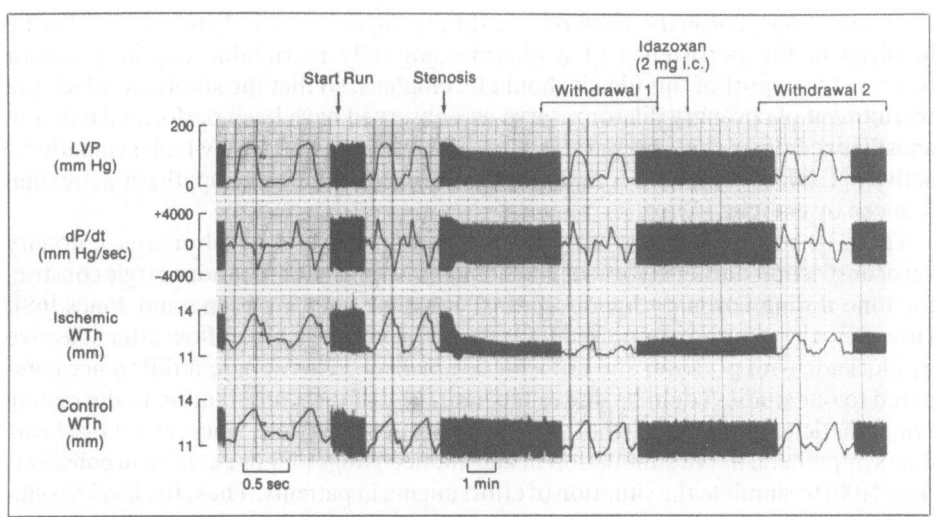

Fig. 6. Original tracing of hemodynamic reactions to treadmill exercise in a chronically instrumented dog. The poststenotic ischemic myocardial dysfunction is attenuated by intracoronary infusion of the selective α_2-antagonist idazoxan with an unchanged stenosis and an unchanged exercise level. (From [100] by permission of the American Heart Association.)

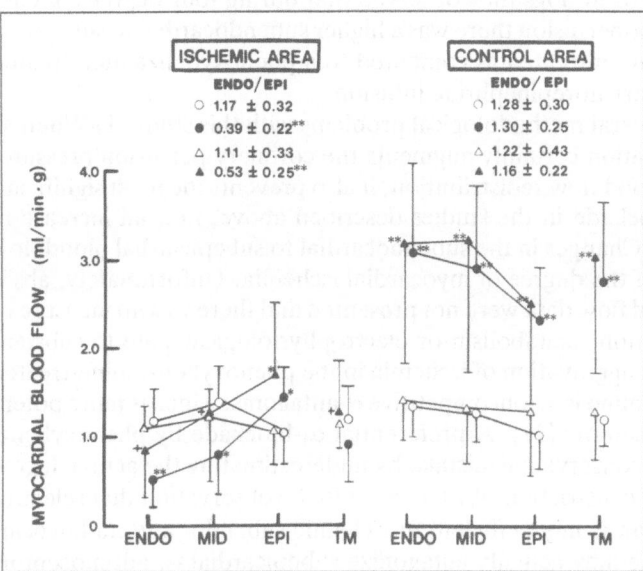

Fig. 7. Transmural distribution of regional myocardial blood flow in an ischemic and a nonischemic area. At rest (O), the distribution of blood flow in the ischemic and nonischemic area is not different. During exercise (●) blood flow in the nonischemic area is increased in all transmural layers. In contrast, in the ischemic area a transmural gradient with a decrease in subendocardial and an increase in subepicardial blood flow develops. Nifedipine does not change blood flow at rest (△), but attenuates the decrease in perfusion of the subendo- and midmyocardium during exercise (▲). (From [53] by permission of the American Heart Association.)

α_2-adrenergic coronary constriction during myocardial ischemia may also be involved in the persistence of a pharmacologically recruitable coronary dilator reserve. In support of this idea it should be emphasized that the studies in which the recruitment of coronary dilator reserve was observed have been performed either in anesthetized open-chest preparations, which are characterized by high sympathetic activity [1, 10, 35, 89], or in conscious dogs during the intense sympathetic activation induced by exercise [53].

There is, however, controversy about the importance of α_2-adrenergic coronary vasoconstriction during myocardial ischemia. The presence of α-adrenergic constrictor tone during coronary hypoperfusion was confirmed by Liang and Jones [68]. However, these authors reported an increase in coronary blood flow after selective α_1-blockade with prazosin and not with α_2-blockade. This apparent difference compared to our studies could be due to the limitation of their observation to the resting sympathetic tone of an anesthetized open-chest preparation, whereas we used cardiac sympathetic nerve stimulation in anesthetized dogs [48] or exercise in conscious dogs [100] to simulate the situation of effort angina in patients. Thus, the level of sympathetic tone may determine the observed changes in coronary vasomotor tone.

In obvious contrast to both ours [48, 100] and Liang and Jones' [68] studies which agree on a deleterious role of α-adrenoceptor mediated coronary constriction during myocardial ischemia, Nathan and Feigl [86] concluded from their study that α-adrenergic coronary constriction exerts a favorable effect on ischemic myocardium by preventing a transmural redistribution of blood flow away from the ischemic subendocardium. In anesthetized dogs they observed that during four degrees of constant inflow coronary hypoperfusion there was a higher subendocardial to subepicardial blood flow ratio in an untreated as compared to a phenoxybenzamine-treated region during intracoronary norepinephrine infusion.

However, there are several methodological problems with this study: 1) Whereas a constant inflow preparation certainly augments the coronary perfusion pressure-dependent transmural blood flow redistribution, it also prevents the most significant beneficial effect of α-blockade in the studies described above, i.e., an increase in total coronary inflow. 2) Changes in the subendocardial to subepicardial blood flow ratio are not indicative of the degree of myocardial ischemia. Unfortunately, absolute subendocardial blood flow data were not presented and there was no measure of regional contractile function, metabolism or electrophysiological state to substantiate the conclusion of an aggravation of ischemia in the phenoxybenzamine-treated region. 3) Phenoxybenzamine is a noncompetitive α-antagonist which is more potent at α_1- than at α_2-adrenoceptors [17]. A preferential α_1-blockade by phenoxybenzamine does not induce a conservative mistake by underestimating the actual degree of α-adrenergic coronary constriction. With respect to the observation that selective α_2-blockade with idazoxan is most efficient in ischemic subendo- and midmyocardium, phenoxybenzamine may actually antagonize subepicardial α_1-adrenoceptors leaving the activation of subendocardial α_2-adrenoceptors unopposed, thus inducing an artificial imbalance of transmural myocardial blood flow. There appears to be a transmurally homogenous distribution of coronary vascular α_1- and α_2-adrenoceptors [11]. Nevertheless, there is a nonhomogeneous responsiveness of coronary vascular α_1- and α_2-adrenoceptors during ischemia. With increasing severity of ischemia the α_1-adrenergic coronary constriction is progressively diminished [56]. In contrast,

α_2-adrenergic coronary constriction is unchanged during myocardial ischemia [19]. Phenoxybenzamine in the study of Nathan and Feigl [86] may have antagonized α_1-adrenoceptors in nonischemic subepicardium leaving α_2-adrenergic coronary constriction in ischemic subendocardium unopposed. Thus, rather than a protective role of the physiological transmitter norepinephrine only the pathological effect of phenoxybenzamine inducing a pharmacologic transmural steal was demonstrated.

A true protective role of cardiac sympathetic nerves innervating an ischemic myocardial region through subepicardial α-adrenergic constriction was demonstrated by Chilian and Ackell [14] in conscious dogs during exercise-induced ischemia, since subendocardial blood flow was higher in the innervated than in a phenol-denervated region. However, this protective role of subepicardial α-adrenergic coronary constriction was strictly limited to neuronally released norepinephrine, whereas, as previously shown by the same authors, the coronary vasomotor effects of norepinephrine in exercising dogs are predominantly due to humoral norepinephrine [15]. Thus, systemic α-blockade with phentolamine actually increased ischemic subendocardial blood flow in the innervated as well as in the denervated region [14].

A role for α-adrenergic coronary constriction in modulating myocardial ischemia in the clinical setting has become apparent. A significant increase in coronary resistance is induced in patients with coronary artery disease during reflex sympathetic activation [85] by the cold pressor test [72, 83, 84]. The therapeutic effectiveness of intracoronary α-blockade with the nonselective α-antagonist phentolamine during exercise-induced myocardial ischemia was recently demonstrated in two clinical studies (Fig. 8) [4, 13].

With respect to the vasomotion of resistive vessels, recent attention has been drawn to the role of nonclassical transmitters. Intracoronary infusion of neuropeptide Y in patients induced marked myocardial ischemia without angiographic evidence for a change in the diameter of epicardial coronary arteries [16]. The persistent coronary constriction in spite of myocardial ischemia induced by mediators such as

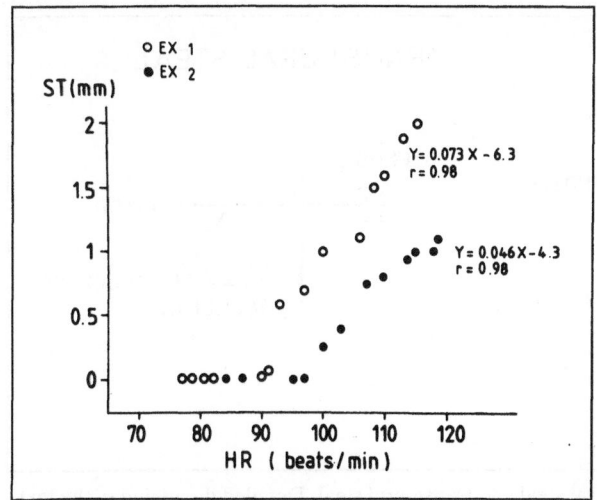

Fig. 8. With a comparable tachycardia the exercise-induced ST-depression is markedly reduced by the α-antagonist phentolamine. $\bigcirc$ = control; $\bullet$ = phentolamine. (From [4] by permission of the Dun-Donnelly Publishing Corporation.)

209

leukotriene C4, thromboxane A2 [87], angiotensin [23] and vasopressin [90] in experimental studies has already been emphasized. In conclusion, coronary resistive vessels are not maximally dilated during myocardial ischemia, but rather retain a significant constrictor tone. Sympathetic activation in experimental studies as well as in patients induces a poststenotic, α-adrenoceptor mediated coronary vasoconstriction which is capable of precipitating or aggravating myocardial ischemia. These consistent experimental and clinical findings form a solid pathophysiological basis for the therapeutic use of coronary vasodilators, in particular for the use of calcium-antagonists which can functionally antagonize α_2-adrenoceptor mediated coronary vasoconstriction [49, 53, 82].

The Distribution of Regional Myocardial Blood Flow

The regional nature of myocardial ischemia requires one to consider the interactions between the ischemic area and the adjacent nonischemic areas. Myocardial blood flow to a poststenotic area depends on two sources, the coronary arterial inflow through the stenotic vessel (F2) (Fig. 9) and collateral blood flow originating from adjacent non- or less stenotic coronary arteries (F1). During physical or mental stress, the function of the nonischemic myocardium is increased, and its increased metabolic demand is adequately met by an increase in coronary blood flow after metabolic dilation of its terminal coronary vascular bed. The dilation of the non-ischemic terminal vascular bed reduces the driving pressure gradient for collateral blood flow (P1−P2) thereby reducing blood flow to the ischemic region, a phenomenon referred to as collateral steal [94]. Collateral steal is not only the result of metabolic dilation in nonischemic myocardium, but also of enhanced extravascular coronary compression within the ischemic myocardium. Both factors contribute to the reduction in the driving pressure gradient for collateral flow [55].

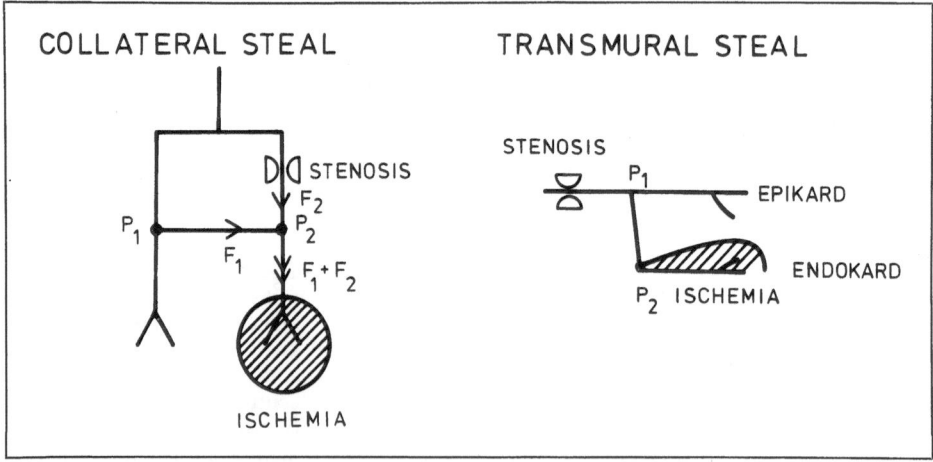

Fig. 9. Schematic diagram of collateral steal and transmural steal. Detailed discussion in the text.

210

The collateral vessels as such may undergo pharmacologic dilation [20, 24] and constrict in response to pharmacologic use of the selective α_2-agonist BHT 920 [74]. However, even chronically developed, mature collateral vessels do not respond to physiologic α-adrenoceptor activation with norepinephrine [43, 74]. Therefore, the direct vasomotion of collateral vessels appears to be of minor importance as compared to the previously discussed changes in the driving pressure gradient for collateral blood flow during myocardial ischemia.

The same consideration applies not only to laterally adjacent areas, but also to the transmural distribution of myocardial blood flow within an area supplied by the stenotic coronary artery. The dilation of subepicardial coronary vessels in concert with the greater mechanical compression of subendocardial coronary vessels by left ventricular pressure, induces a redistribution of blood flow from the subendocardium to the subepicardium, i.e., a transmural steal [32]. Such a transmural redistribution of regional myocardial blood flow during myocardial ischemia can be observed experimentally during treadmill exercise in dogs with a chronic single-vessel coronary stenosis which is fully compensated under resting conditions (Fig. 7) [39−41, 53, 77−79]. A transmural steal phenomenon can be considered as the cause of the predominantly subendocardial myocardial ischemia during angina pectoris and non-transmural myocardial infarction. With respect to the importance of increased myocardial demand for the initiation of myocardial ischemia, it must be emphasized that an increase in oxygen demand of the nonischemic, laterally or transmurally adjacent myocardium induces a redistribution of blood flow, thus causing a reduction in absolute blood flow in the ischemic myocardial region thereby precipitating myocardial ischemia.

In conclusion, the importance of the dynamic component of coronary vasomotion either in the epicardial conduit coronary arteries or in the resistive microcirculation, either in the form of a collateral or a transmural blood flow redistribution, for the initiation of myocardial ischemia must be emphasized. This dynamic component of coronary vasomotor tone may be the basis for the markedly changing susceptibility of patients with a constant coronary morphology to the precipitation of angina pectoris during situations of daily life [75].

Summary: Coronary vasomotion can be characterized according to its localization in the coronary vascular tree segmentally (epicardial, collateral, resistive), according to its localisation in the myocardium transmurally (subendocardial vs. subepicardial) or with respect to its mediators (myogenic, metabolic, endothelial, neuronal, humoral). Coronary vessels exhibit a marked dilator reserve which can be recruited to maintain regional myocardial blood flow and contractile function distal to coronary stenosis. Even in the presence of myocardial ischemia, coronary vessels retain a significant dilator reserve which can only be recruited pharmacologically. A critical reduction in blood flow at the level of epicardial coronary arteries is the principle cause for a range of pathophysiological processes from changes in the hemodynamic severity of fixed stenoses to dynamic coronary stenosis or finally to true coronary vasospasm. However, these pathophysiological processes differ in the quantitative contribution of active coronary vasoconstriction versus fixed mechanical obstruction to the initiation of myocardial ischemia. Although the mediators of epicardial coronary constriction are still largely unknown, significant α_2-adrenergic coronary constriction of the resistive vessels, predominantly in the subendocardium contributes to the initiation of poststenotic myocardial ischemia during sympathetic activation and exercise in experimental studies. Furthermore, intracoronary α-blockade with phentolamine also attenuates exercise-induced myocardial ischemia in patients with stable angina. The common feature of ischemia of all type is that the absolute regional myocardial blood flow — as a result of coronary vasomotion and blood flow redistribution — is the primary determinant of myocardial ischemia.

References

1. Aversano T, Becker LC (1985) Persistence of coronary vasodilator reserve despite functionally significant flow reduction. Am J Physiol 248:H403–H411
2. Bassenge E (1984) Physiologie der Koronardurchblutung. In: Roskamm H (Hrsg) Handbuch der Inneren Medizin. Koronarerkrankungen. Springer, Berlin Heidelberg New York Tokyo 1–48
3. Bassenge E, Busse R (1988) Endothelial modulation of coronary tone. Prog Cardiovasc Dis 30:349–380
4. Berkenboom GM, Abramowicz M, Vandermoten P, Degre SG (1986) Role of alpha-adrenergic coronary tone in exercise-induced angina pectoris. Am J Cardiol 57:195–198
5. Berne RM, Rubio R (1983) Coronary circulation. In: The cardiovascular system I. American Physiological Society, Maryland, pp 873–952
6. Bretschneider HJ (1967) Aktuelle Probleme der Koronardurchblutung und des Myokardstoffwechsels. Regensburger Jahrblatt für Ärztliche Fortbildung 15:1–27
7. Brown BG, Lee AB, Bolson EL, Dodge HT (1984) Reflex constriction of significant coronary stenosis as a mechanism contributing to ischemic left ventricular dysfunction during isometric exercise. Circulation 70:18–24
8. Buffington CW, Feigl EO (1981) Adrenergic coronary vasoconstriction in the presence of coronary stenosis in the dog. Circ Res 48:416–423
9. Busse R, Bassenge E (1984) Modulation des Koronargefäßtonus: Molekulare und zelluläre Mechanismen. Z Kardiol 73:477–491
10. Canty JM Jr, Klocke FJ (1985) Reduced regional myocardial perfusion in the presence of pharmacologic vasodilator reserve. Circulation 71:370–377
11. Chen DG, Dai X-Z, Zimmermann BG, Bache RJ (1988) Postsynaptic α1- and α2-adrenergic mechanisms in coronary vasoconstriction. J Cardiovasc Pharmacol 11:61–67
12. Chierchia S, Davies G, Berkenboom G, Crea F, Crean P, Maseri A (1984) α-adrenergic receptors and coronary spasm: an elusive link. Circulation 69:8–14
13. Chierchia S, Pratt T, DeCoster P, Maseri A (1985) Alpha-adrenergic control of collateral flow: another determinant of coronary flow reserve. Circulation 72 (Suppl 3):190
14. Chilian WM, Ackell PH (1988) Transmural differences in sympathetic coronary constriction during exercise in the presence of coronary stenosis. Circ Res 62:216–225
15. Chilian WM, Harrison DG, Haws CW, Snyder WD, Marcus ML (1986) Adrenergic coronary tone during submaximal exercise in the dog is produced by circulating catecholamines. Evidence for adrenergic denervation supersensitivity in the myocardium but not in coronary vessels. Circ Res 58:68–82
16. Clarke JG, Kerwin R, Larkin S, Lee Y, Yacoub M, Davies GJ, Hackett D, Dawbarn D, Bloom SR, Maseri A (1987) Coronary artery infusion of neuropeptide Y in patients with angina pectoris. Lancet I:1057–1059
17. Constantine JW, Lebel W (1980) Complete blockade by phenoxybenzamine of alpha 1- but not of alpha 2-vascular receptors in dogs and the effect of propranolol. Naunyn-Schmiedeberg's Arch PHarmacol 314:149–156
18. DeCaterina R, Carpeggiani C, L'Abbate A (1984) A double-blind, placebo-controlled study of ketanserin in patients with Prinzmetal's angina: Evidence against a role of serotonin in the genesis of coronary vasospasm. Circulation 69:889–894
19. Deussen A, Heusch G, Thämer V (1985) Alpha 2-adrenoceptor — mediated coronary vasoconstriction persists after exhaustion of coronary dilator reserve. Eur J Pharmacol 115:147–153

20. Diemer HP, Wichmann J, Lochner W (1977) Coronary collateral flow: effect of drugs and perfusion pressure. Basic Res Cardiol 72:332–343
21. Dole WP (1987) Autoregulation of the coronary circulation. Prog Cardiovasc Dis 29:293–323
22. Dole WP, Nuno DW (1986) Myocardial oxygen tension determines the degree and pressure range of coronary autoregulation. Circ Res 59:202–215
23. Ertl G (1987) Coronary vasoconstriction in experimental myocardial ischemia. J Cardiovasc Pharmacol 9 (Suppl 2):S9–S17
24. Ertl G, Simm F, Wichmann J, Fuchs M, Lochner W (1979) The dependence of coronary collateral blood flow on regional vascular resistances. Naunyn-Schmiedeberg's Arch Pharmacol 308:265–272
25. Fam WM, McGregor M (1968) Effect of nitroglycerin and dipyridamole on regional coronary resistance. Circ Res 22:649–659
26. Feigl EO (1983) Coronary physiology. Physiol Rev 63:1–205
27. Fish RD, Nabel EG, Selwyn AP, Ludmer PL, Mudge GH, Kirshenbaum JM, Schoen FJ, Alexander RW, Ganz P (1988) Responses of coronary arteries of cardiac transplant patients to acetylcholine. J Clin Invest 81:21–31
28. Freudenberg H, Lichtlen PR (1981) Das normale Wandsegment bei Koronarstenosen – eine postmortale Studie. Z Kardiol 70:863–869
29. Furchgott RF (1983) Role of endothelium in responses of vascular smooth muscle. Circ Res 53:557–573
30. Furchgott RF, Zawadzki JV (1980) The obligatory role of endothelial cells in the relaxation of arterial smooth muscle by acetylcholine. Nature 288:373–376
31. Gallagher KP, Folts JD, Shebuski RJ, Rankin JHG, Rowe GG (1980) Subepicardial vasodilator reserve in the presence of critical coronary stenosis in dogs. Am J Cardiol 46:67–73
32. Gallagher KP, Osakada G, Matsuzaki M, Kemper WS, Ross Jr J (1982) Myocardial blood flow and function with critical coronary stenosis in exercising dogs. Am J Physiol 243:H698–H707
33. Gewirtz H, Brautigan DL, Olsson RA, Brown P, Most AS (1983) Role of adenosine in the maintenance of coronary vasodilation distal to a severe coronary artery stenosis. Observations in conscious domestic swine. Circ Res 53:42–51
34. Ginsburg R, Bristow MR, Kantrowitz N, Baim S, Harrison DC (1981) Histamine provocation of clinical coronary artery spasm: implications concerning pathogenesis of variant angina pectoris. Am Heart J 102:819–822
35. Gorman MW, Sparks HV Jr (1982) Progressive coronary vasoconstriction during relative ischemia in canine myocardium. Circ Res 51:411–420
36. Gould KL (1980) Dynamic coronary stenosis. Am J Cardiol 45:286–292
37. Gould KL, Lipscomb K, Calvert C (1975) Compensatory changes of the distal coronary vascular bed during progressive coronary constriction. Circulation 51:1085–1094
38. Gu J, Adrian TE, Tatemoto K, Polak JM, Allen JM, Bloom SR (1983) Neuropeptide tyrosine (NPY) – a major cardiac neuropeptide. Lancet May:1008–1010
39. Guth BD, Heusch G, Seitelberger R, Ross Jr J (1987) Mechanism of beneficial effect of beta-adrenergic blockade on exercise-induced myocardial ischemia in conscious dogs. Circ Res 60:738–746
40. Guth BD, Heusch G, Seitelberger R, Ross Jr J (1987) Elimination of exercise-induced regional myocardial dysfunction by a bradycardiac agent in dogs with chronic coronary stenosis. Circulation 75:661–669
41. Guth BD, Tajimi T, Seitelberger R, Lee JD, Matsuzaki M, Ross Jr J (1986) Experimental exercise-induced ischemia: Drug therapy can eliminate regional dysfunction and oxygen supply-demand imbalance. J Am Coll Cardiol 7:1036–1046
42. Guyton RA, McClenathan JH, Newman GE, Michaelis LL (1977) Significance of subendocardial S-T segment elevation caused by coronary stenosis in the dog. Epicardial S-T segment depression, local ischemia and subsequent necrosis. Am J Cardiol 40:373–380
43. Harrison DG, Chilian WM, Marcus ML (1986) Absence of functioning alpha-adrenergic receptors in mature canine coronary collaterals. Circ Res 59:133–142

44. Heistad DD, Armstrong ML, Marcus ML, Piegors DJ, Mark AL (1984) Augmented responses to vasoconstrictor stimuli in hypercholesterolemic and atherosclerotic monkeys. Circ Res 54:711–718
45. Henry PD, Yokoyama M (1980) Supersensitivity of atherosclerotic rabbit aorta to ergonovine. J Clin Invest 66:306–313
46. Heupler FA Jr, Proudfit WL, Razavi M, Shirey EK, Greenstreet R, Sheldon WC (1978) Ergonovine maleate provocative test for coronary arterial spasm. Am J Cardiol 41:631–640
47. Heusch, G (1985) Sympathische Herznerven und Myokardischämie. Thieme, Stuttgart New York
48. Heusch G, Deussen A (1983) The effects of cardiac sympathetic nerve stimulation on the perfusion of stenotic coronary arteries in the dog. Circ Res 53:8–15
49. Heusch G, Deussen A (1984) Nifedipine prevents sympathetic vasoconstriction distal to severe coronary stenoses. J Cardiovasc Pharmacol 6:378–383
50. Heusch G, Deussen A, Schipke J, Thämer V (1984) Alpha 1- and alpha 2-adrenoceptor – mediated vasoconstriction of large and small canine coronary arteries in vivo. J Cardiovasc Pharmacol 6:961–968
51. Heusch G, Deussen A, Thämer V (1985) Cardiac sympathetic nerve activity and progressive vasoconstriction distal to coronary stenoses: feed-back aggravation of myocardial ischemia. J Auton Nerv Syst 13:311–326
52. Heusch G, Guth BD (1989) Neurogenic regulation of coronary vasomotor tone. Eur Heart J 10 suppl. F:6–14
53. Heusch G, Guth BD, Seitelberger R, Ross Jr J (1987) Attenuation of exercise-induced myocardial ischemia in dogs with recruitment of coronary vasodilator reserve by nifedipine. Circulation 75:482–490
54. Heusch G, Schipke J, Thämer V (1985) Clonidine prevents the sympathetic initiation and aggravation of poststenotic myocardial ischemia. J Cardiovasc Pharmacol 7:1176–1182
55. Heusch G, Yoshimoto N (1983) Effects of heart rate and perfusion pressure on segmental coronary resistances and collateral perfusion. Pfluegers Arch 397:284–289
56. Heusch G, Yoshimoto N, Heegemann H, Thämer V (1983) Interaction of methoxamine with compensatory vasodilation distal to coronary stenoses. Drug Res 33:1647–1650
57. Heusch G, Yoshimoto N, Müller-Ruchholtz ER (1982) Effects of heart rate on hemodynamic severity of coronary artery stenosis in the dog. Basic Res Cardiol 77:562–573
58. Hintze TH, Vatner SF (1984) Reactive dilation of large coronary arteries in conscious dogs. Circ Res 54:50–57
59. Hoffman JIE (1987) Transmural myocardial perfusion. Prog Cardiovasc Dis 29:429–464
60. Holtz J, Förstermann U, Pohl U, Giesler M, Bassenge E (1984) Flow-dependent, endothelium-mediated dilation of epicardial coronary arteries in conscious dogs: effects of cyclooxygenase inhibition. J Cardiovasc Pharmacol 6:1161–1169
61. Holtz J, Giesler M, Bassenge E (1983) Two dilatory mechanisms of anti-anginal drugs on epicardial coronary arteries in vivo: indirect, flow-dependent, endothelium-mediated dilation and direct smooth muscle relaxation. Z Kardiol 72 (Suppl 3):98–106
62. Holtz J, Held W, Sommer O, Kühne G, Bassenge E (1982) Ergonovine-induced constrictions of epicardial coronary arteries in conscious dogs: α-adrenoceptors are not involved. Basic Res Cardiol 77:278–291
63. Hossack KF, Brown BG, Stewart DK, Dodge HT (1984) Diltiazem-induced blockade of sympathetically mediated constriction of normal and diseased coronary arteries: lack of epicardial coronary dilatory effect in humans. Circulation 70:465–471
64. Johannsen UJ, Mark AL, Marcus ML (1982) Responsiveness to cardiac sympathetic nerve stimulation during maximal coronary dilation produced by adenosine. Circ Res 50:510–517
65. Kelley KO, Feigl EO (1978) Segmental alpha-receptor-mediated vasoconstriction in the canine coronary circulation. Circ Res 43:908–917
66. Klocke FJ (1987) Measurements of coronary flow reserve: defining pathophysiology versus making decisions about patient care. Circulation 76:1183–1189
67. Lamping KG, Marcus ML, Dole WP (1985) Removal of the endothelium potentiates canine large coronary artery constrictor responses to 5-hydroxytryptamine in vivo. Circ Res 57:46–54

68. Liang IYS, Jones CE (1985) Alpha 1-adrenergic blockade increases coronary blood flow during coronary hypoperfusion. Am J Physiol 249:H1070—H1077
69. Ludmer PL, Selwyn AP, Shook TL, Wayne RR, Mudge GH, Alexander RW, Ganz P (1986) Paradoxical vasoconstriction induced by acetylcholine in atherosclerotic coronary arteries. N Engl J Med 315:1046—1051
70. MacAlpin RN (1980) Relation of coronary artery spasm to sites of organic stenosis. Am J Cardiol 46:143—153
71. MacAlpin RN (1980) Contribution of dynamic vascular wall thickening to luminal narrowing during coronary arterial constriction. Circulation 60:296—301
72. Malacoff RF, Mudge GH, Holman BL, Idoine J, Bifolck L, Cohn PF (1983) Effect of the cold pressor test on regional myocardial blood flow in patients with coronary artery disease. Am Heart J 106:78—84
73. Marcus ML, Wright C, Doty D, Eastham C, Laughlin D, Krumm P, Fastenow C, Brody M (1981) Measurements of coronary velocity and reactive hyperemia in the coronary circulation of humans. Circ Res 49:877—891
74. Maruoka Y, McKirnan MD, Engler RL, Longhurst JC (1987) Functional significance of alpha-adrenergic receptors in mature coronary collateral circulation of dogs. Am J Physiol 253:H582—H590
75. Maseri A (1986) Coronary blood flow and myocardial perfusion in humans: mechanisms of acute transient myocardial ischemia. J Cardiovasc Pharmacol 8 (Suppl 3):S17—S20
76. Mates RE, Gupta RL, Bell AC, Klocke FJ (1978) Fluid dynamics of coronary artery stenosis. Circ Res 42:152—162
77. Matsuzaki M, Gallagher KP, Patritti J, Tajimi T, Kemper WS, White FC, Ross J Jr (1984) Effects of a calcium-entry blocker (diltiazem) on regional myocardial flow and function during exercise in conscious dogs. Circulation 69:801—814
78. Matsuzaki M, Guth BD, Tajimi T, Kemper WS, Ross Jr J (1985) Effects of the combination of diltiazem and atenolol on exercise-induced regional myocardial ischemia in conscious dogs. Circulation 72:233—243
79. Matsuzaki M, Patritti J, Tajimi T, Miller M, Kemper WS, Ross Jr J (1984) Effects of β-blockade on regional myocardial flow and function during exercise. Am J Physiol 247:H52—H60
80. McHale PA, Dube GP, Greenfield JC Jr (1987) Evidence for myogenic vasomotor activity in the coronary circulation. Prog Cardiovasc Dis 30:139—146
81. Mosher P, Ross Jr J, McFate PA, Shaw RF (1964) Control of coronary blood flow by an autoregulatory mechanism. Circ Res 14:250—259
82. Motulsky HJ, Snavely MD, Hughes RJ, Insel PA (1983) Interaction of verapamil and other calcium channel blockers with α_1- and α_2-adrenergic receptors. Circ Res 52:226—231
83. Mudge GH, Goldberg S, Gunther S, Mann T, Grossman W (1979) Comparison of metabolic and vasoconstrictor stimuli on coronary vascular resistance in man. Circulation 59:544—550
84. Mudge GH, Grossman W, Mills RM Jr, Lesch M, Braunwald E (1976) Reflex increase in coronary vascular resistance in patients with ischemic heart disease. New Engl J Med 295:1333—1337
85. Mueller HS, Rao PS, Rao PB, Gory DJ, Mudd JG, Ayres SM (1982) Enhanced transcardiac I-norepinephrine response during cold pressor test in obstructive coronary artery disease. Am J Cardiol 50:1223—1228
86. Nathan HJ, Feigl EO (1986) Adrenergic vasoconstriction lessens transmural steal during coronary hypoperfusion. Am J Physiol 250:H645—H653
87. Nichols WW, Mehta JL, Thompson L, Donnelly WH (1988) Synergistic effects of LTC4 and TxA2 on coronary flow and myocardial function. Am J Physiol 255:H153—H159
88. Olsson RA, Bünger R (1987) Metabolic control of coronary blood flow. Prog Cardiovasc Dis 29:369—387
89. Pantely GA, Bristow JD, Swenson LJ, Ladley HD, Johnson WB, Anselone CG (1985) Incomplete coronary vasodilation during myocardial ischemia in swine. Am J Physiol 249:H638—H647
90. Pantely GA, Ladley HD, Anselone CG, Bristow JD (1985) Vasopressin-induced coronary constriction at low perfusion pressures. Cardiovasc Res 19:433—441
91. Raff WK, Kosche F, Lochner W (1972) Extravascular coronary resistance and its relation to microcirculation. Am J Cardiol 29:598—603

92. Rafflenbeul W, Lichtlen PR (1982) Zum Konzept der "dynamischen" Koronarstenose. Z Kardiol 71:439–444
93. Rouleau J, Boerboom LE, Surjadhana A, Hoffman JIE (1979) The role of autoregulation and tissue diastolic pressures in the transmural distribution of left ventricular blood flow in anesthetized dogs. Circ Res 45:804–815
94. Rowe GG (1970) Inequalities of myocardial perfusion in coronary artery disease ("coronary steal"). Circulation 42:193–194
95. Schipke J, Heusch G, Deussen A, Thaemer V (1985) Acetylcholine induces constriction of epicardial coronary arteries in anesthetized dogs after removal of endothelium. Drug Res 35:926–929
96. Schroeder JS, Bolen JL, Quint RA, Clark DA, Hayden WG, Higgins CB, Wexler L (1977) Provocation of coronary spasm with ergonovine maleate: new test with results in 57 patients undergoing coronary arteriography. Am J Cardiol 40:487–491
97. Schwartz JS, Carlyle PF, Cohn JN (1979) Effect of dilation of the distal coronary bed on flow and resistance in severely stenotic coronary arteries in the dog. Am J Cardiol 43:219–224
98. Schwartz JS, Carlyle PF, Cohn JN (1980) Effect of coronary arterial pressure on coronary stenosis resistance. Circulation 61:70–76
99. Schwartz JS, Tockman B, Cohn JN, Bache RJ (1982) Exercise-induced decrease in flow through stenotic coronary arteries in the dog. Am J Cardiol 50:1409–1413
100. Seitelberger R, Guth BD, Heusch G, Lee JD, Katayama K, Ross J Jr (1988) Intracoronary alpha 2-adrenergic receptor blockade attenuates ischemia in conscious dogs during exercise. Circ Res 62:436–442
101. Serruys PW, Lablanche JM, Reiber JHC, Bertrand ME, Hugenholtz PG (1983) Contribution of dynamic vascular wall thickening to luminal narrowing during coronary arterial vasomotion. Z Kardiol 72 (Suppl 3):116–123
102. Tölle TR, Schipke JD, Schulz R, Thämer V, Haase J (1986) The nociceptive stimulation induced myocardial ischemia is prevented by fentanyl. Neuroscience Letter (Suppl 26):522
103. Winniford MD, Filipchuk N, Hillis LD (1983) Alpha-adrenergic blockade for variant angina: a long-term, double-blind, randomized trial. Circulation 67:1185–1188
104. Young MA, Knight DR, Vatner SF (1987) Autonomic control of large coronary arteries and resistance vessels. Prog Cardiovasc Dis 30:211–234
105. Young MA, Vatner SF (1986) Enhanced adrenergic constriction of iliac artery with removal of endothelium in conscious dogs. Am J Physiol 250:H892–H897

Authors' address:
Prof. Dr. Gerd Heusch
Abt. für Pathophysiologie
Zentrum Innere Medizin
Universitätsklinikum Essen
Hufelandstr. 55
4300 Essen
FRG

Clinical Significance of Coronary Vasomotor Tone in Myocardial Ischemia

Juan C. Kaski and Attilio Maseri

The Cardiovascular Research Unit, Royal Postgraduate Medical School, Hammersmith Hospital, London, England

Introduction

Our understanding of the pathogenic mechanisms underlying chronic stable angina, unstable angina and Prinzmetal's variant angina has increased in recent years. There is now general agreement that dynamic coronary stenoses may contribute to these ischemic syndromes [15, 24, 25]. Dynamic stenoses can be caused by increase of coronary vasomotor tone, coronary thrombosis or by a combination of the two. The thrombotic component confers to the anginal syndromes the feature of instability and the tendency to evolve towards myocardial infarction [18, 32]. It is now generally accepted that an episodic increase of the vasomotor tone of epicardial coronary arteries is not only the cause of variant angina but may also be a pathogenic component in other more common anginal syndromes [15, 18, 24, 25, 32, 50, 77]. Although the terms "coronary spasm" and "coronary vasoconstriction" are frequently used synonymous in medical literature to indicate increases in coronary vasomotor tone, a distinction between the two is necessary, as the difference between coronary spasm, as seen in variant angina, and the coronary vasoconstriction commonly observed in chronic stable angina is not only one of degree, but it seems to reflect also different underlying mechanisms.

Dynamic Stenoses: Epicardial Coronary Vasoconstriction vs Coronary Spasm. Reasons for a Distinction.

Under normal physiologic circumstances coronary blood flow matches myocardial metabolic demands due to the ability of the coronary arteries to alter their calibre dynamically. Under resting conditions large epicardial vessels are moderately compliant [45]. However, it has been documented that these vessels can change their luminal diameter in response to diverse stimuli [7, 8, 35, 39, 41, 45, 46, 55, 56, 79]. Physiologic coronary vasomotion is controlled and modulated by neural, humoral, metabolic and myogenic factors [2, 27, 40]. The role of the endothelium in mediating and modulating the vasomotion of coronary vessels has been recently documented [30, 31]. In recent years it has also become apparent that non-adrenergic non-cholinergic nervous control mechanisms involving a series of peptides that act as neurotransmitters may play a role in the regulation of coronary vasomotor tone [9, 13, 78].

Following improvement in the quality of coronary angiography it has been possible to document the ability of coronary stenotic segments to alter their calibre in response to dynamic exercise [34], handgrip [7] and intracoronary nitrates [6]. Vaso-motion at the level of critical coronary stenoses is possible due to the fact that although some stenotic segments are calcified structures with little potential for con-striction or dilation, the majority (approximately 75%) of coronary lesions have an eccentric residual lumen circumscribed by an arc of normal vascular tissue that is able to undergo changes in tone [75]. Vasomotion at the site of subintimal plaques, even if minor, may provoke dramatic changes of coronary blood flow [48]. Severely narrowed coronary arteries in which a portion of the stenotic arterial wall remains capable of vasomotion, even within the physiologic range, may lead to a variable anginal threshold in certain patients. This type of vasoconstriction, which in some patients with chronic stable angina causes transient myocardial ischemia, should not be labeled as "coronary spasm", as is often found in the medical literature. The dif-ference between coronary spasm, as seen in variant angina, and the type of vaso-constriction observed in chronic stable angina is not only one of degree but also of mechanism.

The term "coronary spasm" indicates a sudden, abnormal, exaggerated constric-tion of the coronary arteries which leads to their total or subtotal occlusion in response to stimuli that cause only mild or moderate constriction in the non-spastic segments [51] (Fig. 1). This constriction is usually focal and occurs both in arteries with atheromatous plaques and arteries which are angiographically normal [41, 51, 56]. A non-specific local coronary hyperreactivity seems to be responsible for coro-nary spasm in variant angina [41].

Several clinical features distinguish ischemic episodes triggered by coronary spasm from those depending on coronary constriction superimposed to critical coronary

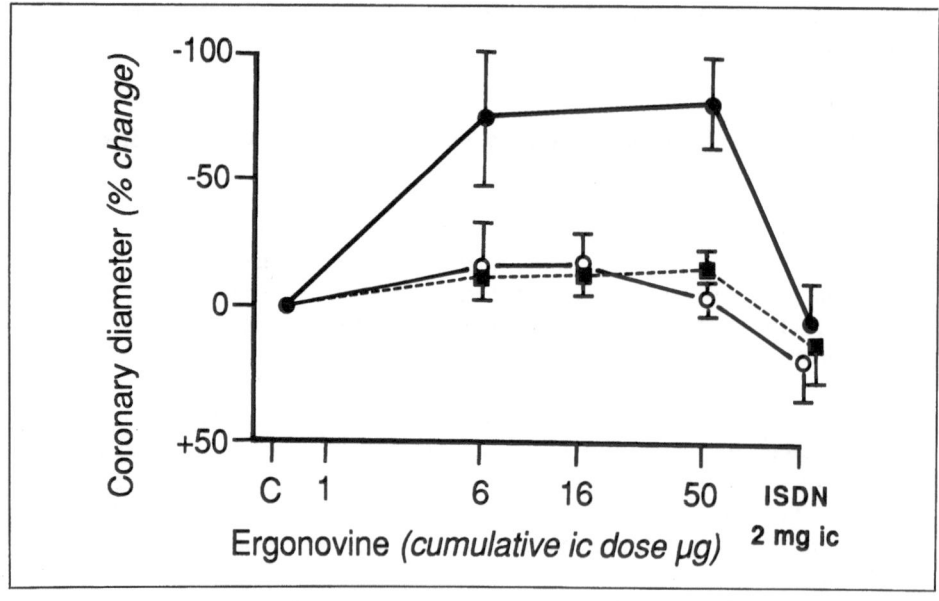

Fig. 1. Response of spastic and non-spastic segments to intracoronary ergonovine in patients with Prinzmetal's variant angina. (Used with permission from Circulation; data from [38])

218

narrowings. As seen in variant angina, transient ischemic episodes caused by spasm tend to be more frequent at night or during the early morning hours. Spontaneous anginal attacks and those provoked by ergonovine or hyperventilation are usually associated with ST segment elevation.

In patients with chronic stable angina who have the potential for dynamic stenoses, angina is mainly exertional but with a variable threshold. Anginal attacks may occur at rest or upon minimal exertion but they have a temporal distribution in contrast to those of Prinzmetal's angina. ST segment depression and not elevation is associated with spontaneous anginal attacks and also with ergonovine-induced ischemia. Hyperventilation does not induce angina or ischemic ST segment changes but the cold pressor test is often positive [15], particularly in patients with low exercise tolerance.

Although coronary thrombosis has been shown to play a major role as the underlying pathogenic mechanism in acute myocardial infarction and unstable angina, clinical observations suggest that increased coronary vasomotor tone also plays a role in acute ischemic syndromes. The prevalence of these mechanisms and their pathogenic importance however, is not yet clearly established. Platelet aggregation and endothelial dysfunction at the site of a complicated coronary lesion may result in the formation of an occlusive white thrombus and may also trigger coronary spasm or "physiologic" constriction, secondary to the release of vasoactive substances from the aggregating platelets. Coronary vasospasm, occlusive or sub-occlusive may create the conditions appropriate for a red thrombus to develop and lead to persistent occlusion of the vessel. How often a coronary artery segment becomes hyperreactive enough to cause an acute coronary spasm (and therefore without the typical clinical features of variant angina) leading to acute myocardial infarction, unstable angina or sudden death is not known. It is extremely difficult at present to identify the prevailing pathogenic mechanism in the clinical context of the acute ischemic syndromes.

Coronary Artery Spasm in Variant Angina

The widely accepted concept of coronary spasm proposed by Osler [64] to explain the "spontaneous" appearance of some anginal attacks, was gradually abandoned after the demonstration by Keefer and Resnik in 1928 [44] that the coronary arteries of patients with angina pectoris were severely fibrosed and calcified, and were therefore, those authors assumed, unable to develop spasm. Coronary artery spasm became then, "a resort of the diagnostically destitute", as Sir George Pickering wrote in The Lancet in 1951 [65]. However, the clinical observations of Prinzmetal et al. [66] suggesting the presence of "increased tonus at the site of an atherosclerotic plaque" as the mechanism underlying a "variant" form of spontaneous angina characterized by ST-segment elevation, boosted the interest in this pathogenic mechanism. But it was only less than two decades ago that the hypothesis of coronary spasm was documented as the cause of Prinzmetal's variant angina [47]. After angiographic [22, 63] and further invasive and non-invasive documentation of the phenomenon [53, 54] in patients with variant angina, coronary spasm became a "proved hypothesis" [60]. By demolishing a long-standing and widely accepted generalization, this new concept allowed the identification of "more variants of the variant" and encouraged cardiologists to study the role of dynamic stenoses in other ischemic syndromes.

Definition of Coronary Spasm

Coronary artery spasm as seen in variant angina is an exaggerated, active constriction of one or more epicardial coronary arteries. Although diffuse occasionally, it is most often focal and tends to recur in the same coronary segment. Coronary spasm is frequently associated with atheromatous plaques of different severity, but it occurs also in the absence of fixed luminal narrowings.

Clinical Presentation

The classical presentation of coronary artery spasm is Prinzmetal's variant angina. Typically, patients with coronary artery spasm experience angina at rest, associated with ST-segment elevation and they have a preserved exercise capacity [47, 66]. Coronary spasm induced by exercise, however, has been described [72, 76, 80]. Although typically myocardial ischemia triggered by coronary spasm is associated with transient ST-segment elevation, ST-segment depression or T-wave peaking can be observed during spasm, depending on whether coronary spasm is incomplete or collateral circulation is present [56, 71, 81]. Coronary spasm can be provoked by a variety of vasoconstrictor agents and manoeuvres, particularly during the active phases of the disease. Recently it has been reported that coronary spasm can be triggered by exercise in 30% to 50% of patients with active variant angina [41, 72, 76, 80].

Major Coronary Events in Variant Angina

Although episodes of coronary spasm tend to be transient and subside either spontaneously or with the administration of nitrates or calcium antagonists, serious coronary events may occasionally develop in association with coronary artery spasm. In the majority of patients with Prinzmetal's variant angina an active phase of the disease is usually followed by a long-lasting quiescent period. In others, periods of waxing and waning of the disease can be observed for weeks, months or years. Occasionally, the course of variant angina may become indistinguishable from that of unstable angina. Spontaneous coronary spasm can be the cause of unstable angina in selected patients, as demonstrated by Biagini et al. [5] in a recent study. These authors carried-out continuous Holter monitoring on 11 unstable angina patients. More than 500 episodes of transient ischemic ECG changes were recorded during 89 days of monitoring and spontaneous coronary spasm was demonstrated during episodes of ST-T changes in six of eight patients. However, the prevalence of coronary spasm as the underlying pathogenic mechanism in unselected cases of unstable angina is not known.

Major cardiac events such as myocardial infarction and sudden death develop in variant angina usually, but not necessarily, during periods of disease activity. Potentially fatal arrhythmias such as sustained ventricular tachycardia and fibrillation have been documented during coronary vasospasm with or without anginal pain [57, 61, 68]. Although in some patients these arrhythmias tend to recur during episodes of variant angina, their reproducibility is low. In patients with Prinzmetal's variant angina sudden death caused by coronary spasm in the absence myocardial infarction has been reported. In an anatomo-pathological study Roberts et al. [67] demonstrated the presence of coronary spasm, associated with fixed coronary lesions and persisting after death, in three patients with typical variant angina and documented coronary spasm who died suddenly and unexpectedly. The problem of spasm as a cause of sudden death due to reperfusion arrhythmias was addressed in an experimental study in dogs by Sheehan and Epstein [69]. However, the authors' conclusions on the role of reperfusion arrhythmias probably apply only to some animal models and extrapolation of their results to patients with variant angina should be made cautiously. As arrhythmic sudden coronary death has been documented in variant angina [57, 61, 68], coronary spasm should be considered as one of the possible causes of sudden cardiac death in the general population.

Finally, although coronary thrombosis is the mechanism responsible for the large majority of acute myocardial infarctions, in selected patients coronary spasm has been documented to be the cause of infarction, in association with a fresh coronary thrombus [52, 74]. Anatomo-pathological studies have revealed that coronary spasm can cause fatal myocardial infarction in patients with variant angina, even in the absence of fixed coronary stenoses. El-Maraghi and Sealey [23] documented coronary spasm both at postmortem coronary angiography and autopsy, where "contraction rings" were observed in the coronary arteries of a 25-year-old man who died during an acute myocardial infarction, in the absence of coronary atheromatous plaques.

Local Coronary Hyperreactivity and Susceptibility to Spasm in Other Ischemic Syndromes

There is also suggestive evidence that local coronary hyperreactivity to constrictor stimuli may play a role in patients with myocardial infarction and coronary atherosclerotic disease [3, 4, 62, 67]. Bertrand et al. have demonstrated that the coronary arteries of patients with recent transmural myocardial infarction have a high degree of reactivity to ergonovine [4], the most potent stimulus used clinically to trigger coronary spasm in variant angina. Ergonovine was positive in 20% of patients with recent acute myocardial infarction and in only 6% with old myocardial infarction. These results support the pathogenic role of a coronary hyperreactivity in acute myocardial infarction.

Bertrand et al. [3] also administered ergonovine to 1089 consecutive patients during coronary angiography. Ergonovine was negative in 99% of patients with atypical chest pain and in 98% of patients with valvular heart disease; conversely, it was positive in 38% of 203 patients with angina occuring only at rest (ergonovine was positive in 85% of patient with angina at rest and documented variant angina). The test was positive in 45% of patients with ST depression during angina and in 62% of those patients with only T-wave changes during pain. Positive ergonovine tests were documented in 14% of patients with a history of effort and rest angina and in 4% of those who had exertional angina only. Ergonovine was negative in a large group of patients with severe organic lesions. The coronary arteries of those patients, therefore, were not susceptible to constriction despite the critical reduction of luminal diameter by atheromatous plaques.

Other Types of Coronary Vasoconstriction in Diverse Ischemic Syndromes

The Coronary Constrictor Component in Chronic Stable Angina

As mentioned earlier, available evidence indicates that dynamic changes of coronary vasomotor tone superimposed on critical atheromatous coronary stenoses may interfere transiently with coronary blood supply and thus play an important role in chronic stable angina [15, 25, 50, 58, 59], although these changes of tone appear to be different from coronary spasm as seen in variant angina. Patients with stable angina usually recognize a level of exercise they cannot exceed without developing chest pain; however, angina may occur unpredictably for levels of exercise that are usually well tolerated or may even develop at rest without any detectable cause [15, 25, 50, 58, 59]. In these patients a variable anginal threshold has been documented objectively by stress testing [16] and ambulatory ECG monitoring [21]. Reduction of coronary blood supply is frequently the cause of transient myocardial ischemia during daily life in these patients [11, 12, 21]. Therefore, transient myocardial ischemia in patients with chronic stable angina results not only from an increased myocardial oxygen demand in the presence of critical atheromatous coronary obstructions, but also from a primary reduction of coronary blood supply. Reduction of coronary blood flow in chronic stable angina may be caused by vasoconstriction superimposed on critical atheromatous coronary lesions [7, 8, 15, 16, 25, 34, 43, 48, 50]. As mentioned previously, vasomotion at the level of critical coronary stenoses is possible, due to the fact that 50% to 75% of coronary lesions have an eccentric residual lumen circumscribed by an arc of normal vascular tissue that is able to undergo changes in smooth muscle tone. Minor changes of intraluminal diameter due to vasomotion at the site of subintimal plaques may provoke marked changes of coronary blood flow [48]. As the identification of patients with a potential for dynamic changes in coronary vasomotor tone at the site of critical coronary stenoses may have important clinical implications we recently attempted to identify patients with dynamic stenoses using commonly available clinical tools [43]. We observed that following the administration of sublingual isosorbide dinitrate 30% of over 200 stable angina patients markedly increased their baseline exercise capacity, which was due to an increase of coronary blood flow reserve and not only to preload reduction [43]. In those patients, but not in the nonresponders to nitrates, we were able to document the presence of dynamic changes of vasomotor tone at the site of severe coronary stenoses in response to vasoconstrictor and vasodilator stimuli (Fig. 2). Stable angina patients susceptible to dynamic changes of their coronary reserve are those exhibiting the greatest improvement in exercise performance after nitrates. Therefore, the potential for dynamic stenoses in patients with chronic stable angina can be assessed in clinical practice simply by

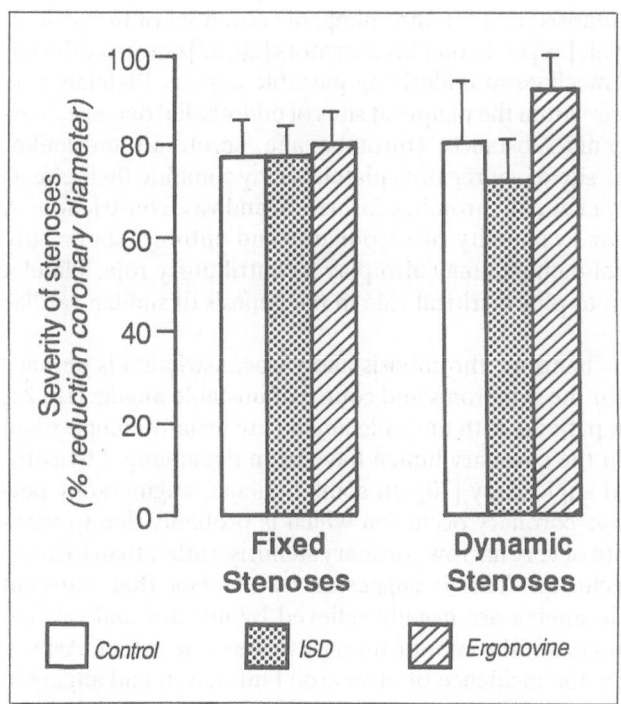

Fig. 2. Changes of coronary vasomotor tone at the site of severe coronary stenoses in patients with chronic stable angina. As mentioned in the text, those patients with a marked improvement in exercise capacity following the administration of sublingual isosorbide dinitrate, but not the nonresponders, experienced dynamic changes of coronary vasomotor tone in response to vasoconstrictor (ergonovine) and vasodilator (isosorbide dinitrate) stimuli during coronary arteriography.

means of the clinical history and repeat stress testing after the acute administration of sublingual nitrates or calcium antagonists [42].

The Coronary Constrictor Component in Unstable Angina and Myocardial Infarction

Acute ischemic syndromes are now thought to be usually initiated by injury to the atherosclerotic plaque, a common finding in the majority of acute anginal syndromes [18, 19, 24, 32, 77]. Indeed, plaque fissuring seems to be the mechanism underlying a very large proportion of cases of crescendo angina, myocardial infarction and sudden cardiac death [18, 19]. A number of plaques with a deposit of lipids undergo sudden fissuring [14]: the cap breaks, allowing blood to enter the intima. What precipitates such rupture is not known, but hemodynamic factors, mechanical fatigue of collagen, macrophage lytic activity and coronary spasm have been postulated. Following plaque injury, acute myocardial ischemia may develop as a result of the interaction of multiple mechanisms that are elicited by the damaged arterial wall. The severity of the problem will depend on the magnitude of the break of the plaque cap, the

223

severity of the pre-existing stenosis, and the thrombogenic potential of the patient. As suggested by Fitzgerald et al. [29] and other investigators [20, 33] coronary thrombosis seems to be the major mechanism underlying unstable angina. Platelets may attach to atherosclerotic debris within the plaque at sites of endothelial damage. Vasoconstrictors and thrombogenic substances (thromboxane, serotonin and leukotrienes) released locally at the site of aggregating platelets may combine their effects resulting in closure of the vessel due to thrombus formation and vasoconstriction. A decreased local production or availability of vasodilator and antithrombotic substances, related to endothelial damage, may also play a contributory role. Platelet emboli have also been shown to play a critical role in the genesis of sudden cardiac death in this setting [26].

A combination of both mechanisms, thrombosis and vasoconstriction is perhaps the most likely explanation for the symptoms and course of unstable angina [18, 24, 32, 77]. Coronary stenoses in patients with unstable angina are usually complex and eccentric [1]. Thrombi within the coronary lumen have been documented by coronary angiography [1, 10] and angioscopy [70]. In some patients, angiography performed during ischemia shows coronary occlusion which is probably due to vasoconstriction occuring at the site of very narrow coronary stenosis, rather than to intermittent total thrombotic occlusion. This is suggested by the fact that transient ischemic episodes in unstable angina are usually relieved by nitrates and calcium antagonists but do not respond to antithrombotic therapy to the same extent. Aspirin and heparin, however, reduce the incidence of myocardial infarction and angina in these patients [73]. Nitrates and calcium antagonists may exert their beneficial effects in unstable angina by reducing or preventing coronary vasoconstriction which in turn may allow thrombus washout.

Whether coronary vasoconstriction in unstable angina results from coronary hyperreactivity, as in coronary spasm, or is another type of constriction is not known. The role of vasoconstriction in the early sequence of events leading to acute myocardial infarction also remains poorly understood. Davies et al. [17] have observed that coronary occlusion during acute myocardial infarction is a dynamic process which might involve coronary constriction. The coronary vasodilator response to nitrates in the early stages of acute myocardial infarction was investigated by Hackett et al. [36] before, during and after thrombolytic therapy. Intermittent spontaneous coronary occlusion and re-opening associated to ST-segment resolution and re-elevation was observed by these authors during the early stages of acute infarction in more than 40% of their patients, both before and during thrombolysis. Intracoronary isosorbide dinitrate failed to open totally blocked infarct-related vessels, but in 50% to 86% of patients nitrates promptly re-established full coronary patency either immediately after acute re-occlusion or whenever coronary occlusion was incomplete. These findings suggest that in addition to thrombosis, other mechanisms, such as vasoconstriction, may act in the early stages of myocardial infarction to maintain the closure of the vessel. The failure of nitrates to induce coronary patency before thrombolysis does not exclude the presence of vasoconstriction but may indicate inadequate nitrate concentration at the site of occlusion, presence of a large intraluminal clot or too powerful local vasoconstrictor stimuli. After initial recanalization is achieved, nitrates are effective, probably because their local concentration becomes adequate. Vasodilatation induced by nitrates may allow improved delivery of the thrombolytic agent along

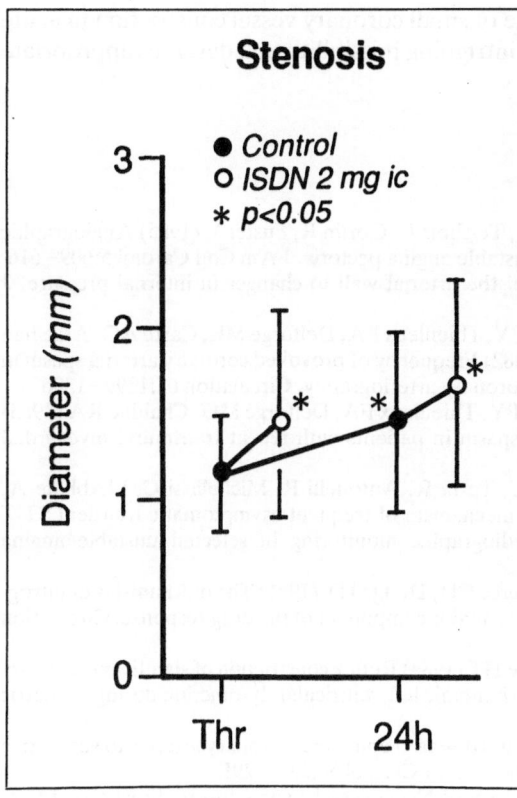

Fig. 3. Closed circles indicate minimal luminal diameter of the residual infarct-related coronary stenoses at the end of thrombolysis (Thr) and at 24 h, and open circles show changes in diameter following intracoronary nitrates. (Used with permission from the European Heart Journal; data from [37])

infarct-related arteries, thus improving patency even further. The interaction between thrombosis and vasoconstriction during early acute myocardial infarction [37] makes it necessary to consider the need for combined antithrombotic and vasodilator therapy in acute myocardial infarction to promote stable coronary recanalization (Fig. 3).

Conclusions

It is now well established that coronary vasomotor tone plays a significant role in the genesis of chronic and acute myocardial ischemia. In Prinzmetal's variant angina coronary artery spasm is caused by a local hyperreactivity of the vessel wall to constrictor stimuli that cause only mild constriction in adjacent non-spastic coronary segments and in coronary arteries of normal individuals [38]. The type of coronary vasoconstriction responsible for reductions of coronary blood flow and for the variable exercise tolerance in patients with chronic stable angina is different from that of variant angina. Various forms of epicardial coronary constriction together with coronary thrombosis play a role in unstable angina and acute myocardial infarction. The possible role of a local hyperreactivity to vasoconstrictor stimuli should be further investigated [49]. Also the role of small coronary vessel constriction in acute and chronic ischemic syndromes is an intriguing possibility that deserves appropriate investigation.

References

1. Ambrose J, Winters S, Stern A, Eng A, Teicholz L, Gorlin R, Fuster V (1985) Angiographic morphology and the pathogenesis of unstable angina pectoris. J Am Coll Cardiol 5:609–616
2. Bayliss (1902) On the local reaction of the arterial wall to changes in internal pressure. J Physiol (London) 28:220–231
3. Bertrand ME, Lablanche JM, Tilmant PY, Thienleux FA, Delforge MR, Carre AG, Asseman P, Berzin B, Libersa B, Laurent JM (1982) Frequency of provoked coronary arterial spasm in 1089 consecutive patients undergoing coronary arteriography. Circulation 65:1299–1306
4. Bertrand ME, Lablanche JM, Tilmant PY, Thieuleux FA, Delforge MG, Chahine RA (1983) The provocation of coronary arterial spasm in patients with recent transmural myocardial infarction. Eur Heart J 4:532–535
5. Biagini A, Mazzei MG, Carpeggiani C, Testa R, Antonelli R, Michelassi C, L'Abbate A, Maseri A (1982) Vasospastic ischemic mechanism of frequent asymptomatic transient ST-T changes during continuous electrocardiographic monitoring in selected unstable angina patients. Am Heart J 103:13–20
6. Brown BG, Bolson E, Petersen RB, Pierce CD, Dodge HT (1981) The mechanisms of nitroglycerin action: stenosis vasodilatation as a major component of the drug response. Circulation 69:1089–1097
7. Brown BG, Lee AB, Bolson EL, Dodge HT (1984) Reflex constriction of significant coronary stenosis as a mechanism contributing to ischemic left ventricular dysfunction during isometric exercise. Circulation 70:18–24
8. Brown BG (1985) Response of normal and diseased epicardial coronary arteries to vasoactive drugs: Quantitative arteriographic studies. Am J Cardiol 56:23E–29E
9. Burnstock G (1985) Neurohumoral control of blood vessels: some future directions. J Cardiovasc Pharmacol 7:S137–S146
10. Capone G, Wolf NM, Meyer B, Meister SG (1985) Frequency of intracoronary filling defects by angiography in angina pectoris at rest. Am J Cardiol 56:403–406
11. Chierchia S, Balasubramanian V, Muiesan L, Sonecha T, Davies A, Maseri A (1984) Transient impairment of coronary flow: a frequent cause of ischemia in chronic stable angina during normal daily life. J Am Coll Cardiol 3:579 (abstract)
12. Chierchia S, Gallino A, Smith G, Deanfield J, Morgan M, Croom M, Maseri A (1984) The role of heart rate in the pathophysiology of chronic stable angina. Lancet II:1353–1357
13. Clarke J, Davies G, Kerwin R, Hackett D, Larkin S, Dawbarn D, Lee Y, Bloom SR, Yacoub M, Maseri A (1987) Coronary artery infusion of neuropeptide Y in patients with angina pectoris. Lancet I:1057–1059
14. Constantinides P (1966) Plaque fissures in human coronary thrombosis. J Atherosclerosis Res 6:1–17
15. Crea F, Davies G, Romeo F, Chierchia S, Bugiardini R, Freedman B, Kaski JC, Maseri A (1984) Myocardial ischemia during ergonovine testing: different susceptibility to coronary vasoconstriction in patients with exertional and variant angina. Circulation 69:690–695
16. Crea F, Margonato A, Kaski JC, Rodriguez-Plaza L, Meran DO, Davies G, Chierchia S, Maseri A (1986) Variability of results during repeat exercise stress testing in patients with stable angina pectoris: role of dynamic coronary flow reserve. Am Heart J 112:249–254
17. Davies G, Chierchia S, Maseri A (1984) Prevention of myocardial infarction by very early treatment with intracoronary streptokinase. Some clinical observations. N Engl J Med 311:1488–1492
18. Davies M, Thomas A (1985) Plaque fissuring – the cause of acute myocardial infarction, sudden ischaemic death, and crescendo angina. Br Heart J 53:363–373

19. Davies M, Thomas A (1984) Thrombosis and acute coronary artery lesions in sudden cardiac ischemic death. N Engl J Med 310:1137–1140
20. De Zwaan C, Bar FW, Janssen JHA, De Swart HB, Vermeer F, Wellens HJJ (1988) Effects of thrombolytic therapy in unstable angina: clinical and angiographic results. J Am Coll Cardiol 12:301–309
21. Deanfield JE, Maseri A, Selwyn AP, Chierchia S, Ribeiro P, Krikler S, Morgan M (1983) Myocardial ischaemia during daily life in patients with stable angina: its relation to symptoms and heart rate changes. Lancet II:753–758
22. Dhurandhar RW, Watt DL, Silver MD, Trimble AS, Adelmann AG (1972) Prinzmetal's variant form of angina with arteriographic evidence of coronary arterial spasm. Am J Cardiol 30:902–905
23. El-Maraghi NRH, Sealey BJ (1980) Recurrent myocardial infarction in a young man due to coronary arterial spasm demonstrated at autopsy. Circulation 61:199–207
24. Epstein SE, Palmeri ST (1984) Mechanisms contributing to precipitation of unstable angina and acute myocardial infarction: implications regarding therapy. Am J Cardiol 54:1245–1252
25. Epstein SE, Talbot TL (1981) Dynamic coronary tone in precipitation, exacerbation and relief of angina pectoris. Am J Cardiol 48:797–803
26. Falk E (1985) Unstable angina with fatal outcome: dynamic coronary thrombosis leading to infarction and/or sudden death: autopsy evidence of recurrent mural thrombosis with peripheral embolization culminating in total vascular occlusion. Circulation 71:699–708
27. Feigl EO (1983) Coronary physiology. Physiol Rev 63:1–205
28. Feldman RL (1987) Coronary thrombosis, coronary spasm and coronary atherosclerosis and speculation on the link between unstable angina and acute myocardial infarction. Am J Cardiol 59:1187–1190
29. Fitzgerald DJ, Roy L, Catella F, Fitzgerald GA (1986) Platelet activation in unstable coronary disease. N Engl J Med 315:983–989
30. Furchgott RF, Zawadzki JV (1980) The obligatory role of endothelial cells in the relaxation of arterial smooth muscle by acetylcholine. Nature 288:373–376
31. Furchgott RF (1983) Role of endothelium in responses of vascular smooth muscle. Circ Res 53:557–573
32. Fuster V, Badimon L, Cohen M, Ambrose JA, Badimon JJ, Chesebro J (1988) Insights into the pathogenesis of acute ischemic syndromes. Circulation 77:1213–1220
33. Fuster V, Chesebro JH (1986) Mechanisms of unstable angina. N Engl J Med 315:1023–1025
34. Gage JE, Hess OM, Murakami T, Ritter M, Grimm J, Krayenbühl HP (1986) Vasoconstriction of stenotic coronary arteries during dynamic exercise in patients with classic angina pectoris: reversibility by nitroglycerin. Circulation 73:865–876
35. Ginsburg R, Bristow M, Davis K, Dibiaze A, Billingham M (1984) Quantitative pharmacologic responses of normal and atherosclerotic isolated human epicardial coronary arteries. Circulation 69:430–440
36. Hackett D, Davies G, Chierchia S, Maseri A (1987) Intermittent coronary occlusion in acute myocardial infarction. Value of combined thrombolytic and vasodilator therapy. N Engl J Med 317:1055–1059
37. Hackett D, Davies G, Maseri A (1988) Pre-existing stenoses in patients with first myocardial infarction are not necessarily severe. Eur Heart J 9:1317–1323
38. Hackett D, Larkin S, Chierchia S, Davies G, Kaski JC, Maseri A (1987) Induction of coronary artery spasm by a direct local action of ergonovine. Circulation 75:577–582
39. Horio Y, Yasue H, Rokutanda M, Nakamura N, Ogawa H, Takaoka K, Matsuyama K, Kimura T (1986) Effects of intracoronary injection of acetylcholine on coronary arterial diameter. Am J Cardiol 57:984–989
40. Johnson PC (1983) The myogenic response. In: Bohr DF, Somlyo AV, Sparks HV Jr (eds) Handbook of Physiology, Circulation 2, Vascular Smooth Muscle, pp 409–439
41. Kaski JC, Crea F, Meran D, Rodriguez L, Araujo L, Chierchia S, Davies G, Maseri A (1986) Local coronary super-sensitivity to diverse vasoconstrictive stimuli in patients with variant angina. Circulation 74:1255–1265
42. Kaski JC, Crea F (1987) Practical assessment of the role of dynamic coronary stenoses in patients with stable exertional angina. In: Maseri A, Sobel BE, Chierchia S (eds) Hammersmith Cardiology Workshop Series Vol 3. Raven Press, New York: 117

43. Kaski JC, Rodriguez-Plaza L, Meran DO, Araujo L, Chierchia S, Maseri A (1985) Improved coronary supply: the prevailing mechanism of action of nitrates in chronic stable angina. Am Heart J 110:238–245
44. Keefer CS, Resnik WH (1928) Angina pectoris. A syndrome caused by anoxemia of the myocardium. Arch Intern Med 41:769–807
45. Klassen GA, Wong AYK (1982) Coronary artery compliance in the dog. Can J Physiol Pharmacol 60:942–951
46. Ludmer PL, Selwyn AP, Shook TL, Wayne RR, Mudge GH, Alexander RW, Ganz P (1986) Paradoxical vasoconstriction induced by acetylcholine in atherosclerotic coronary arteries. N Engl J Med 315:1046–1051
47. MacAlpin RN, Kattus AA, Alvaro AB (1973) Angina pectoris at rest with preservation of exercise capacity: Prinzmetal's variant angina. Circulation 47:946–958
48. MacAlpin RN (1980) Contribution of dynamic vascular wall thickening to luminal narrowing during coronary arterial constriction. Circulation 61:296–301
49. Maseri A, Chierchia S, Davies G (1986) Pathophysiology of coronary occlusion in acute infarction. Circulation 73:233–239
50. Maseri A, Chierchia S, Kaski JC (1985) Mixed angina pectoris. Am J Cardiol 56:30E–33E
51. Maseri A, Chierchia S (1982) Coronary artery spasm: demonstration, definition, diagnosis and consequences. Prog Cardiovasc Dis 25:169–192
52. Maseri A, L'Abbate A, Baroldi G, Chierchia S, Marzilli M, Ballestra AM, Severi S, Parodi O, Biagini A, Distante A, Pesola A (1978) Coronary vasospasm as a possible cause of myocardial infarction. A conclusion derived from the study of "preinfarction" angina. N Engl J Med 299:1271–1277
53. Maseri A, Mimmo R, Chierchia S, Marchesi C, Pesola A, L'Abbate A (1975) Coronary artery spasm as a cause of acute myocardial ischemia in man. Chest 68:625–633
54. Maseri A, Parodi O, Severi S, Pesola A (1976) Transient transmural reduction of myocardial blood flow, demonstrated by thallium-201 scintigraphy, a cause of variant angina. Circulation 54:280–288
55. Maseri A, L'Abbate A, Pesola A, Ballestra AM, Marzilli M, Maltini G, Severi S, De Nes DM, Parodi O, Biagini A (1977) Coronary vasospasm in angina pectoris. Lancet I:713–717
56. Maseri A, Severi S, De Nes M, L'Abbate A, Chierchia S, Marzilli M, Ballestra AM, Parodi O, Biagini A, Distante A (1978) "Variant" angina: one aspect of a continuous spectrum of vasospastic myocardial ischemia. Am J Cardiol 42:1019–1035
57. Maseri A, Severi S, Marzullo P (1982) Role of coronary arterial spasm in sudden coronary ischemic death. Ann N Y Acad Sci 204–217
58. Maseri A (1980) Pathogenetic mechanisms of angina pectoris: expanding views. Br Heart J 43:648–660
59. Maseri A (1984) Spasm and dynamic coronary stenosis. J Cardiovasc Pharmacol 6:S683–S690
60. Meller J, Pickard A, Dack S (1976) Coronary arterial spasm in Prinzmetal's angina: a proved hypothesis. Am J Cardiol 37:938–940
61. Nakamura M, Takeshita A, Nose Y (1987) Clinical characteristics associated with myocardial infarction, arrhythmias, and sudden death in patients with vasospastic angina. Circulation 75:1110–1116
62. Oliva PB, Breckinridge JC (1977) Arteriographic evidence of coronary arterial spasm in acute myocardial infarction. Circulation 56:366–374
63. Oliva PB, Potts DE, Pluss RG (1973) Coronary arterial spasm in Prinzmetal angina. Documentation by coronary arteriography. N Engl J Med 288:745–751
64. Osler W (1910) The Lumlean lectures on angina pectoris II. Lancet I:839–844
65. Pickering GW (1951) Vascular spasm. Lancet II:845–850
66. Prinzmetal M, Kennamer R, Merliss R (1959) Angina pectoris. A variant form of angina pectoris. Am J Med 27:375–388
67. Roberts WC, Curry RC Jr, Isner JM, Waller BF, McManus BM, Mariani-Constantini R, Ross AM (1982) Sudden death in Prinzmetal's angina with coronary spasm documented by angiography. Am J Cardiol 50:203–210
68. Severi S, Davies G, Maseri A, Marzullo P, L'Abbate A (1980) Long-term prognosis of "variant" angina with medical treatment. Am J Cardiol 46:226–232

69. Sheehan FH, Epstein SE (1982) Determinants of arrhythmic death due to coronary spasm: effect of preexisting coronary artery stenosis on the incidende of reperfusion arrhythmia. Circulation 65:259–264
70. Sherman CT, Litvack F, Grundfest W, Lee M, Hickey A, Chaux A, Kass R, Blanche C, Matloff J, Morgenstern L, Ganz W, Swan HJC, Forrester J (1986) Coronary angioscopy in patients with unstable angina pectoris. N Engl J Med 315:913–919
71. Shubrooks SJ (1979) Variant angina: more variants of the variant. Am J Cardiol 43:1245–1247
72. Specchia G, De Servi S, Falcone C, Bramucci E, Angoli L, Mussini A, Marinoni GP, Montemartini C, Bobba P (1979) Coronary arterial spasm as a cause of exercise-induced ST-segment elevation in patients with variant angina. Circulation 59:948–954
73. Theroux P, Ouimet H, McCans J, Latour JG, Joly P, Levy G, Pelletier E, Juneau M, Stasiak J, DeGuise P, Pelletier GB, Rinzler D, Waters DD (1988) Aspirin, heparin, or both to treat acute unstable angina. N Engl J Med 319:1105–1111
74. Vincent MG, Anderson JL, Marshall HW (1983) Coronary spasm producing coronary thrombosis and myocardial infarction. N Engl J Med 309:220–223
75. Vlodaver Z, Edward JE (1971) Pathology of coronary atherosclerosis. Prog Cardiovasc Dis 14:256–259
76. Waters DD, Szlachcic J, Bonan R, Miller DD, Davue R, Theroux P (1983) Comparative sensitivity of exercise, cold pressor and ergonovine testing in provoking attacks of variant angina in patients with active disease. Circulation 67:310–315
77. Willerson JT, Hillis LD, Winniford M, Buja LM (1986) Speculation regarding mechanisms responsible for acute ischemic heart disease syndromes. J Am Coll Cardiol 8:245–250
78. Yanagisawa M, Kurihara H, Kimura S, Tomobe Y, Kobayashi M, Mitsui Y, Yazaki Y, Goto K, Masaki T (1988) A novel potent vasoconstrictor peptide produced by vascular endothelial cells. Nature 332:411–415
79. Yasue H, Horio Y, Nakamura N, Fujii H, Imoto N, Sonoda R, Kugiyama K, Obata K, Morikami Y, Kimura T (1986) Induction of coronary artery spasm by acetylcholine in patients with variant angina: possible role of the parasympathetic nervous system in the pathogenesis of coronary artery spasm. Circulation 74:955–963
80. Yasue H, Omote S, Takizawa A, Nagao M, Miwa K, Tanaka S (1979) Circadian variation of exercise capacity in patients with Prinzmetal's variant angina: Role of exercise-induced coronary arterial spasm. Circulation 59:938–948
81. Yasue H, Omote S, Takizawa A, Masao N, Hyon H, Nishida S, Horie M (1981) Comparison of coronary arteriographic findings during angina pectoris associated with S-T elevation or depression. Am J Cardiol 47:539–546

Authors' address:
J. C. Kaski, M.D.
The Cardiovascular Research Unit
Royal Postgraduate Medical School
Hammersmith Hospital
London, England

The Challenge of Silent Myocardial Ischemia

Alberto Malliani

Istituto Ricerche Cardiovascolari, CNR; Patologia Medica, Centro "Fidia", Ospedale "L. Sacco"; Università Milano, Italy

From a pathophysiological point of view, derived more from a clinical oversimplification than from adequate observations, there has been a traditional tendency to consider cardiac pain as the most alarming message from the jeopardized heart.

This is not to deny the general validity of the biological principle that pain has a protective, or self-preserving, value to the organism, but rather to anticipate that, where the heart is concerned and probably all viscera, "noxious" does not necessarily mean pain.

To better contrast what is new in our understanding of the peripheral mechanisms of cardiac pain, a few notes should be addressed to what can be defined as old.

The Traditional Interpretation

This view, based on a complex net of highly heterogeneous notions, is responsible for having convinced entire generations of physicians that acute myocardial ischemia, as a rule, leads to pain and that this symptom constitutes a reliable warning signal.

The Afferent Pathway

The concept that the afferent fibers running in the cardiac sympathetic nerves were the only essential pathway for the transmission of cardiac pain arose from observations on both humans and experimental animals. In man, thoracic sympathectomy or section of the higher thoracic dorsal roots was found to be a maneuver capable of relieving anginal pain [66], while thoracic sympathectomy could abolish behavioral reactions accompanying coronary occlusion in acute experiments [9, 57]. Observations that were consistent with Langley's statement that "most of the afferent fibers, which on electrical stimulation give rise to pain, pass by the sympathetic strands and not by the vagus" [24]. More in general, Langley's opinion was that the sympathetic nervous system was a pure outflow. The afferent nerve fibers contained within it were considered without reflex function, merely subserving nociception and somatic in nature as they had their cell bodies located in the dorsal ganglia together with somatic neurons [24].

From this conceptual matrix originated the holistic view according to which pain of somatic and visceral origin was due to the "direct stimulation of a common system of pain nerves" [26].

The Adequate Stimulus

When Sherrington [55] characterized the "nocuous" event that threatens the integrity of a tissue as the stimulus capable of triggering nociception, he furnished a more general concept for the interpretation of somatic pain rather than a universal key for predicting all types of nociception. For instance at about the same time it became known that "in acute endocarditis, pain is rarely present, and ulceration of valves or of the wall may proceed to a most extreme degree without any sensory disturbance" [48]. That is to say that for the heart an ambiguous relation between damage and pain has long been known.

Around the 1930s the problem of the adequate stimulus for cardiac pain was sharply focused in experimental terms. On the side of chemical stimuli the foreground was occupied by Lewis' appealing proposal of a "factor P" [25] capable of inducing nociception and accumulating in the tissue spaces when a muscle is exercising under ischemic conditions. However, at the same time, evidence was equally available on the adequacy of a pure mechanical stimulus to elicit pain from the heart [33, 41]. Although the problem was far from being solved, Lewis' prestige transformed his hypothesis into a common belief and the abnormal accumulation of chemical substances became the nub of the traditional view of how the adequate stimulus is engendered in the ischemic myocardium.

Two Contrasting Hypotheses: "Intensity" or "Specificity"

The general problem concerning the afferent code transmitting nociception has assumed, through the years, the vest of two main hypotheses proposing, respectively, "intensity" or "specificity" [33, 37].

The "intensity" mechanism, the most obvious and probably the first to be formulated assumes that pain results from an excessive stimulation of receptive structures [18]. Alternatively, pain may be conceived of as a "specific" sensation [52], that is the product of the excitation of a well defined nociceptive apparatus, the functional characteristics of which make it responsible only to a limited class of events – the "nocuous" stimuli that threaten the integrity of a tissue.

According to these premises, it should be clear that the "specificity" theory was also in the epicenter of the traditional interpretation of cardiac pain. Indeed the assignment of the afferent sympathetic path to the exclusive transmission of pain, coincided with the most committed conceptualization of a specific nociceptive channel: not only in the usual terms of a peculiar contingent of small diameter afferent fibers [10] but as the whole sensory input contained in an ensemble of nerves. Hence, the sympathetic sensory endings within the heart were considered "specific" nociceptors.

The New Experimental Findings

The Functional Characteristics of Cardiac Sympathetic Sensory Endings

The possible existence in the heart of specific nociceptors is a fundamental point that can be explored experimentally. Peripheral sensors, purely nociceptive in function, should have no background discharge [10] as a consequence of their high threshold which renders them unresponsive to normal events and excitable only with strong stimuli, likely to be noxious. Thus, the *recruitment* of their silent fibers by a peripheral stimulus could represent an unambiguous signal to the centers. It is well known, for instance, that on the somatic side there is a population of afferent fibers, innervating the skin, which have receptors that seem to fit the criteria for nociceptors [10, 52]: no background discharge and recruitment only with strong mechanical or thermal stimuli.

Recently an intense electrophysiological investigation [11, 29, 30, 36] was carried out into the properties of either the small myelinated or unmyelinated ventricular

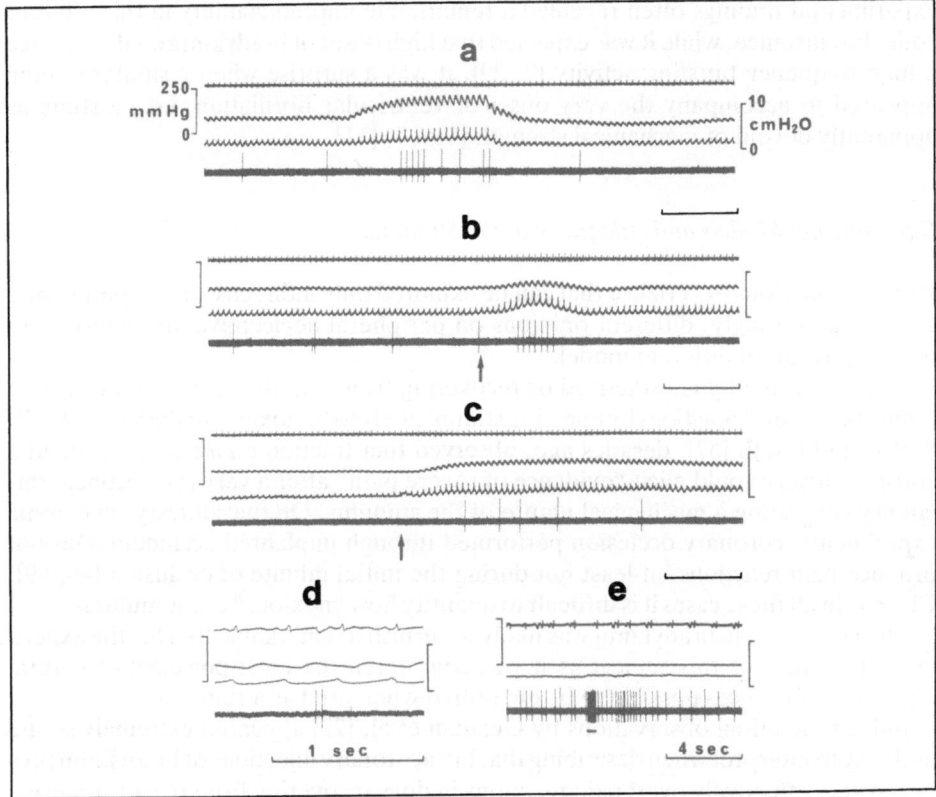

Fig. 1. Activity of an afferent sympathetic nonmyelinated nerve fiber with its receptive field located in the left ventricle. Tracings represent from top to bottom, the ECG, the systemic arterial pressure, right atrial pressure, nervous activity (all tracings are cathode-ray oscilloscope recordings). a) Occlusion of the descending thoracic aorta (indicated by the rise in arterial pressure); b) i.v. injection of 5 ml warm saline, beginning at the arrow; c) occlusion of the inferior vena cava released at the arrow; d) electrical stimulation of the left inferior cardiac nerve activating the afferent fiber; the biphasic first deflexion is the stimulus artifact and the second deflexion is the action potential of the fiber. The approximate length of the fiber was 3.8 cm. The conduction velocity calculated for this fiber was 0.92 m/s; e) mechanical probing of an area of the external surface of the left ventricle, performed on the non-beating heart, after bleeding the animal to death; notice the after-discharge which is typical of the group C fibers, (From [11].)

233

sympathetic afferent fibers, i.e., the afferent fibers which are more likely to convey cardiac nociception. It was found that these fibers possess a mechanosensitivity that makes them tonically active and responsive to normal hemodynamic events (Fig. 1). Coronary occlusion or intracoronary administration of bradykinin, i.e., possible algesic stimuli, increased markedly their tonic impulse activity, but a recruitment of silent afferent fibers could not be appreciated [2, 29].

It was concluded that the "intensity" mechanism appeared as the most likely candidate to account for the properties of the neural substrate subserving cardiac nociception [30, 33, 37]. Hence, ventricular sympathetic sensory endings are not "specific" nociceptors, but low-threshold polymodal receptors [30, 37].

Moreover, concerning the intensity of the afferent neural activity in relation to the magnitude of the hemodynamic or chemical stimuli, it should be stressed that the experimental findings often revealed a remarkable unpredictability in this afferent code. For instance, while it was expected that high doses of bradykinin could produce a high frequency bursting activity [2, 29], it was a surprise when a similar finding appeared to accompany the very onset of ventricular fibrillation, i.e., a stimulus apparently devoid of mechanical strength (Fig. 2) [11].

Experimental Models and Adequacy of the Stimulus

Pain is a conscious experience that can be explored only indirectly with experimental models: accordingly, different opinions on peripheral nociceptive mechanisms are often the result of different models.

In animals, lightly anesthetized or recovering from anesthesia, it is quite easy to obtain behavioral reactions by applying stimuli likely to be noxious to the heart [9, 57]. Sutton and Lueth [57], decades ago, observed that traction on a ligature around a coronary artery could elicit "evidence of severe pain" after a very few seconds, this latency suggesting a mechanical nature of the stimulus. On the contrary, in chronic experiments, coronary occlusion performed through implanted occluders does not produce pain reactions, at least not during the initial minute of occlusion [49, 59]. Clearly, in all these cases it is difficult to quantify how "noxious" the stimulus is.

The nonapeptide bradykinin was likely to furnish a remarkable tool for the experimental analysis of this subject as it was considered the most powerful of natural algogenic substances possible to be quantified when used as a stimulus.

Indeed the initial observations by Guzman et al. [22] appeared extremely sound and easy to interpret when describing that intracoronary injections of bradykinin produced very effectively overt pain reactions in dogs recovering from recent surgery.

However, we have recently analyzed the reflex hemodynamic effects of the chemical stimulation with bradykinin of the cardiac sensory innervation in conscious dogs after full recovery from the operation necessary for their instrumentation [50].

In these animals the injections of bradykinin into either the left anterior descending or circumflex coronary artery produced consistently a marked gradual increase in systemic arterial pressure and heart rate as well as left ventricular pressure and dP/dt (Fig. 3). It is important to point out that these changes were never accompanied by any pain reaction, as expressed by agitation and vocalization of the animals. In this study the amounts of bradykinin injected into the cannulated coronary artery ranged from 10 ng/kg (the threshold dose for the response) to 3 µg/kg. When this latter, very large dose was used, however, the direct vasodilatory effects of the drug prevailed and hypotension and tachycardia were observed, again in absence of any pain reactions. A similar pressor response, in absence of pain reactions, was also seen when bradykinin was injected into the pericardial sac [50].

As shown in Fig. 4 a clear dose-response curve was detectable starting from a threshold dose of 10 ng/kg. With increasing doses of intracoronary bradykinin, the response rose up to a maximum, which was obtained with 100 ng/kg, after which even the use of greater doses produced no further increase in the response.

The importance of recovery from anesthesia and recent surgery in explaining the apparent discrepancy with the finding by Guzman et al. [22] was demonstrated by experiments performed soon after surgery, when recovery of the animals was still incomplete. The injections of bradykinin into the coronary bed of nine animals dur-

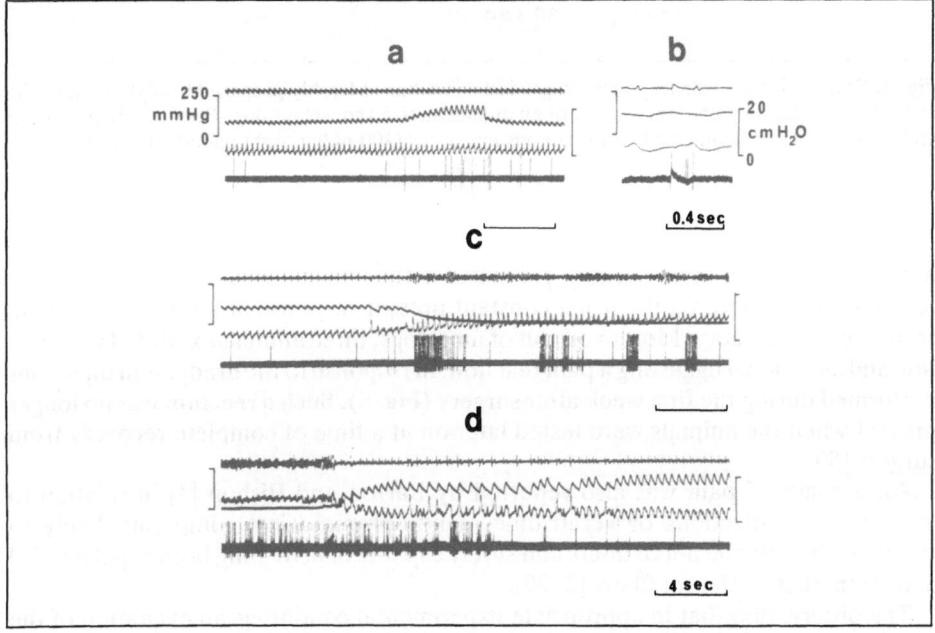

Fig. 2. Activity of a nerve filament containing impulses of two different non-myelinated fibers: the fiber yielding biphasic action potentials had its receptive field in the depth of the left ventricle, while the receptive field of the fiber with monophasic potentials was in the left atrium. Tracings represented from top to bottom, the ECG, the systemic arterial pressure, the right atrial pressure, the nervous activity (all tracings are cathode-ray oscilloscope recordings). a) Occlusion of the descending thoracic aorta; b) electrical stimulation of the left inferior cardiac nerve activating both afferent nerve fibers; the biphasic first deflection is the artifact of the stimulus, followed by a double activation of the atrial fiber and by the potential of the ventricular fiber. The approximate length of the fiber was 4.8 cm. The conduction velocity calculated for the ventricular and atrial fibers was respectively 0.32 m/s and 0.53 m/s; c) an episode of ventricular fibrillation is induced by gentle mechanical stimulation of the right ventricle, corresponding to the ectopic beat preceding the episode itself by a few cardiac cycles; d) the ventricles spontaneously return to a normal action after about 80 s. (From [11].)

235

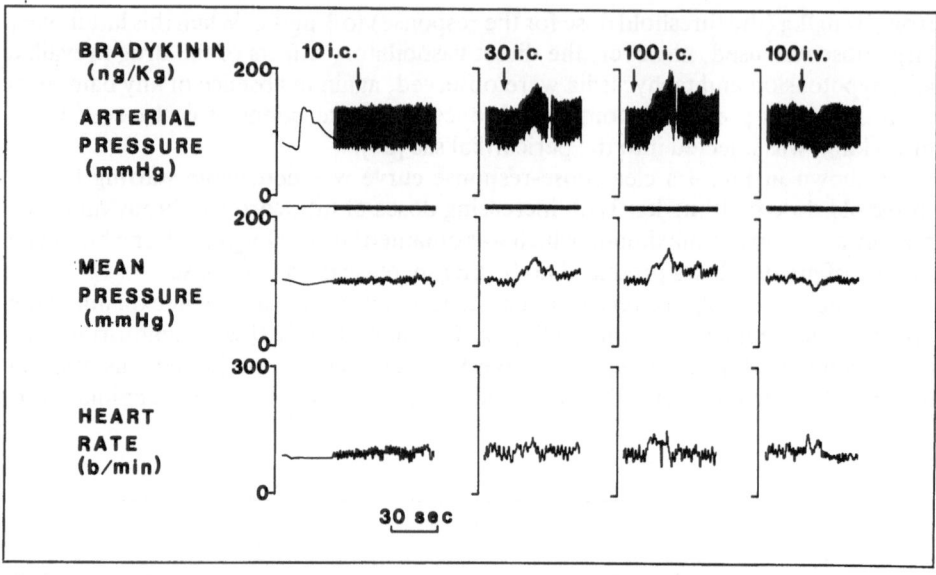

Fig. 3. Progressive pressor responses to graded injections of freshly prepared bradykinin into the cannulated circumflex coronary artery of an unanesthetized conscious dog. Note that hypotension and tachycardia are produced by intravenous injections (100 ng/kg, right panel). (from [50].)

ing the first week after surgery provoked either an early pressor or depressor response which became the usual constant pressor response at a time of complete recovery from surgery. Moreover, out of nine dogs, three animals exhibited vocalization and agitation, suggesting a pain reaction, in response to the bradykinin injections performed during the first week after surgery (Fig. 5). Such a reaction was no longer present when the animals were tested later on at a time of complete recovery from surgery [50].

An absence of pain was also reported by Barron and Bishop [4] in relation to intracoronary injections of veratridine, a non-physiological compound likely to stimulate directly the nerve fibers and surely capable of activating both vagal [60, 61] and sympathetic afferent fibers [2, 29].

The observation that in appropriate experimental conditions an excitation of the cardiac sensory supply, likely to be massive, did not elicit pain appears as a total defeat for the "specificity" theory, at least if nociceptors were postulated to be exquisitively sensitive to bradykinin. On the other hand, the "intensity" theory also does not explain in simple terms the lack of induction of pain.

As a working hypothesis we proposed [30, 31, 37] a modified version of the intensity mechanism. Cardiac pain would result from the extreme excitation of a spatially restricted population of afferent sympathetic fibers. Accordingly, coronary occlusion, performed for more than 2 min in chronic dogs, seems to be a maneuver more effective than intracoronary injection of bradykinin to elicit overt signs of pain. This difference might be due to the fact that during coronary occlusion the excitation of the afferent sympathetic fibers is probably not only quite relevant, a condition which

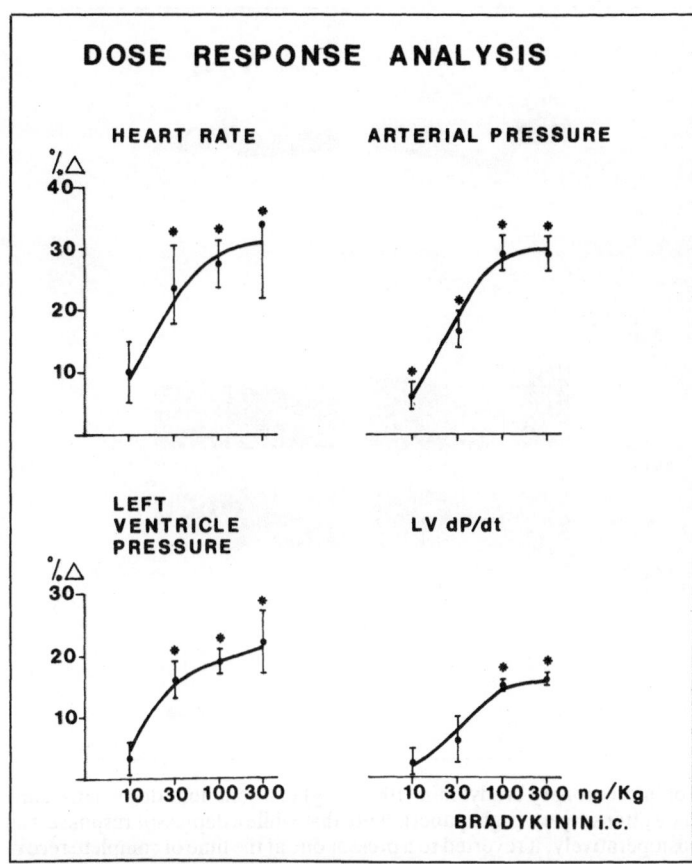

Fig. 4. Dose-response analysis of the hemodynamic effects of intracoronary injections of bradykinin in conscious dogs. Doses of bradykinin are in ng/kg, hemodynamic responses are indicated as percent change from control. (From [50].)

surely is also duplicated with bradykinin, but also characterized by a marked spatial heterogeneity, as the mechanical and chemical components of the stimulus are likely to be different at the center or at the boundaries of the ischemic region. Thus, the afferent neural code resulting from this complex event could be of an intense spatio-temporal characterization [30, 31, 37].

A fortuitous observation in a chronic dog after full recovery from surgery seemed to corroborate this point of view. In the course of one of the above-mentioned experiments [50] the coronary cannula had slipped out of the lumen and was lying, as revealed by autopsy, just outside the wall of the vessel, below the adventitia. This peculiar position maintained by the cannula determined that each minute injection of liquid distended a very limited portion of this highly innervated tissue: and constantly the animal displayed overt pain reactions, even when only saline was injected [38] (Fig. 6).

One experiment should not be the basis for a theory: but it is well known that experimental casuality has often offered new perspectives on a problem. The impli-

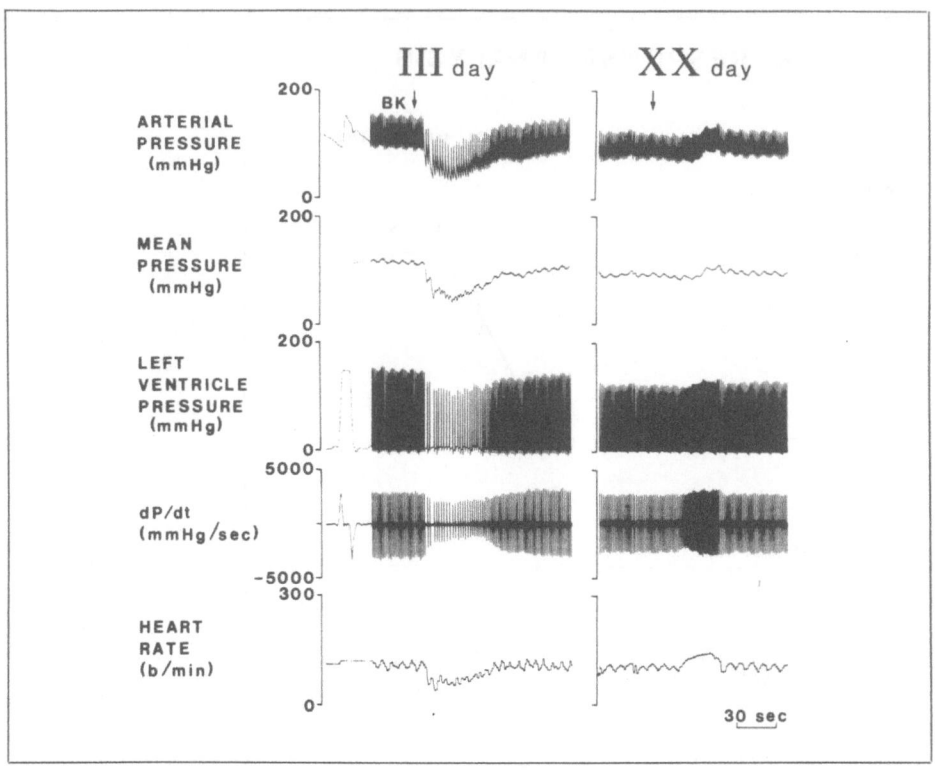

Fig. 5. Contrasting effects of intracoronary bradykinin (100 ng/kg) in a conscious dog when examined early (left panel) and late after surgery (right panel). Note that while a depressor response was obtained in the third day postoperatively, it reverted to a pressor one at the time of complete recovery from surgery (20th day postoperatively). (From [50].)

cations of this hypothesis could be relevant. A spatially restricted mechanical abnormality localized in the myocardium or in the abundantly innervated coronary arteries could lead to pain more efficiently than a more widely distributed myocardial ischemia with its chemical components.

According to this hypothesis, when the activation of the cardiac sympathetic afferent fibers is widely and homogeneously distributed, as in the case of intracoronary injections of bradykinin, or, more currently, during a marked increase in arterial pressure, central inhibitory modulations [63] would prevent the onset of pain. Conversely, recent thoracic surgery by inducing a localized somatic afferent barrage could decisively contribute to the genesis of the peculiar algogenic code through mechanisms of convergence at the spinal level [17].

Finally, the contribution by vagal afferent fibers to the mechanisms of cardiac nociception should be considered with caution and further explored. Indeed, the anginal pain referred to the jaw, head and neck, more frequent after sympathectomy (the phenomenon of "migration of pain") [27] may indicate an additional central site, besides the spinal cord, where the mechanisms for referred pain would be activated:

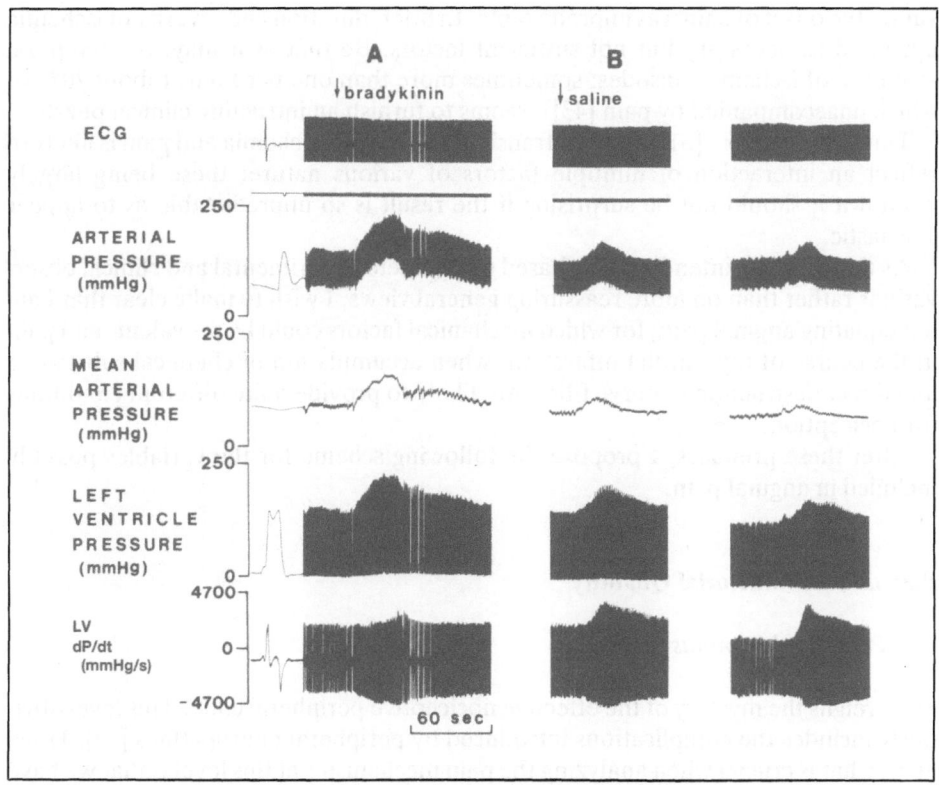

Fig. 6. Effects of mechanical distensions of the adventitia of the left circumflex coronary artery produced in a conscious dog by sub-adventitial injections of either bradykinin or saline. Similar pain reactions (vocalization and agitation) were observed even if the pressor response was blunted by alpha adrenergic receptor blockade with phentolamine (1 mg/kg between A and B and again between B and C). L.V. left ventricle. (From [38].)

in this case by cardiac vagal afferent fibers. Moreover, even when nociception is transmitted through afferent sympathetic fibers, an important modulatory role on threshold and characteristics of pain may be exerted by vagal afferents [16].

The Elusive Link

In recent years, it has been abundantly documented that in patients exhibiting spontaneous and reversible electrocardiographic changes typical of episodes of transient myocardial ischemia, the hemodynamic profile of the crises can appear substantially similar, whether or not they are accompanied by pain [12, 13, 19–21, 42, 46]. A careful analysis revealed that several factors are likely to be implicated in the genesis of pain such as duration of the episode, or severity of ischemia. For instance, ischemic crises were usually painless when shorter than 3 min and associated with increases in left ventricular filling pressure smaller than 7 mmHg [43]. However, above these values the onset of pain was unpredictable. In brief, duration and severity of ischemia appeared as necessary but not sufficient factors. Be that as it may, the temporal sequence of ischemic episodes, sometimes more than one per hour, (about 70% of which unaccompanied by pain [43]), seems to furnish an intriguing clinical puzzle.

This "elusive link" [31] between transient myocardial ischemia and pain is likely to reflect an interaction of multiple factors of various nature: these being largely unknown it should not be surprising if the result is so unpredictable as to appear stochastic.

As this article is intended to be based on concrete experimental and clinical observations rather than on more reassuring general views, I wish to make clear that I am not equating anginal pain, for which mechanical factors could be prevalent, and pain in the course of myocardial infarction, when accumulation of chemical substances and direct destruction of nerve fibers are likely to provide quite different conditions for nociception.

After these premises, I propose the following scheme for the variables possibly included in anginal pain.

Pain as a Multifactorial Quantity

The Peripheral Stimulus

Here reigns the mystery of the effective nociceptive peripheral code. This level obviously includes the complications introduced by peripheral neuropathies [47]. However, what is crucial when analyzing the pain mechanisms at this level is that we have no way yet to define the various levels of afferent sympathetic excitation on the basis of the changes which are measured in hemodynamic variables.

In other words, an ischemic episode accompanied by a greater ST tract abnormality or of longer duration does not necessarily correspond to a more marked excitation occuring in the afferent channel: the examples derived from the electrophysiological studies (Fig. 2) indicate how cautious we have to be in inferring the magnitude of an afferent code.

The Spinal Modulatory Factors

Here converge neural mechanisms capable of either accentuating [8, 17] or reducing nociception. The latter would comprise transcutaneous electrical nerve stimulation [40] or dorsal column stimulation [45] and the excitation of cardiac vagal afferents [16], the latter likely to be co-activated during the same normal or abnormal cardiac events exciting sympathetic afferents [7]. Here also are located the endogenous peptides [65], a new legion of candidates about which the problem is to discriminate between mere possibilities and sound pathophysiological mechanisms.

Some of the new trends in pain research [62] are centered upon these two levels, the peripheral input and the spinal cord integration. As no spinal pathway has been found to carry exclusively visceral sensation but rather convergent information from the soma and from the viscera, it is likely that the processing of either input could exert a reciprocal influence on the other. On the same grounds, some of the new findings on somatic nociception could be relevant to the mechanisms of visceral pain.

For instance, it has been shown that afferent non-myelinated fibers from somatic deep tissues differ from C-afferent fibers innervating the skin in that they trigger a long duration increase of excitability in spinal cells [64]. This heterosynaptic facilitation could enable neurones normally responding only to noxious stimuli to transmit information from low threshold inputs as well. This phenomenon seems to last for many minutes: and a whole spectrum of central changes in these stimulus-response relations can result from the activation of non-myelinated fibers of different origin [67].

As to the possible interpretations of these various changes, it is interesting that the chemistry of deep non-myelinated fibers appears partly different from that of C-fibers innervating the skin [44]. In addition, it has been reported quite recently that brief stimulations of non-myelinated muscle afferents can induce long-lasting increases in the cutaneous receptive fields of C-afferent fibers: moreover, in some instances, units that originally responded only to noxious cutaneous mechanical stimulation, began to respond to low threshold stimuli as well [14].

In general, modulatory mechanisms should be envisaged as occuring at times comprised between milliseconds and months [62], which suggests that no unique explanation is likely to be found.

Central Factors

Central factors at the very least comprise cognition, attention, situational context, and psychological distress [5].

It is in the complexity of this frame that one should evaluate interesting findings such as the facilitatory role of tachyarrhythmias on the occurence of pain during ischemic episodes [6]: as the arrhythmias could represent an additional mechanical stimulus or, for instance, a further element of psychological distress. On similar grounds, the subjective threshold for pain [15, 53] could depend on various mechanisms. However, while this individual characteristic gives some explanation concerning the asymptomatic patients, it gives no clue to the most frequent occuring events which are the asymptomatic episodes vs the symptomatic ones.

Again, in this complexity one should consider that angioplasty in humans – a sort of acute experimental ischemia – induces pain only in a variable manner [56].

It is in the light of this complexity that one should consider the recent report by Sylvén et al. [58], affirming that a pain similar to that of angina pectoris can be provoked by intravenous injection of adenosine in healthy volunteers. However, in another report by Barnes et al. [3] adenosine produced abdominal pain and nausea. In short, a possible similarity in the character of pain is unlikely to give insight not only into its mechanism, but also about the gross localization of the receptor endings.

The Continuum

From the heart, as probably from most of the other organs, a flow of tonic information reaches the centers. Normal and abnormal events produce reflexes. For example, neural mechanisms involved in the cardiovascular regulation are likely to be not only a link between normal and abnormal but also a solid ground for the novel certainty that health and disease are both innervated entities [32].

During episodes of transient myocardial ischemia an increased afferent activity reaches the centers through cardiac vagal [7] and sympathetic fibers [30]. Accordingly, neural reflexes can occur of a prevalent inhibitory [7] or excitatory [30, 32, 34] nature.

In the clinical setting most of the episodes of transient myocardial ischemia, as revealed with Holter monitoring, are characterized by a depression of the ST segment paralleled by a simultaneous increase in heart rate: this pattern is not influenced by the presence or absence of pain [23]. As it is known that the ischemic episodes are most frequently accompanied by an increase in arterial pressure [28] the rise in heart rate should not depend on a baroreceptive mechanism but rather upon a pressor reflex from the heart [30, 34, 35, 39].

It is quite interesting that power spectral analysis of heart rate and arterial pressure variabilities seems to offer some markers of sympathetic and vagal efferent activities and of sympatho-vagal interaction in man and in conscious dogs [51]: hence it seems possible to monitor the build up of sympathetic afferent activity as reflected on the efferent side, not only by hemodynamic changes but by these sensitive markers represented by rhythms.

Figure 7 illustrates the case of an episode of painless myocardial ischemia documented with a Holter recording (from Lombardi et al., work in progress). In the top panels the electrocardiographic differences between "control" and "ischemia" are clearly evident. In the bottom panels the power spectrum analysis of the heart rate variability for two periods of 256 beats is presented. In the "Control" state three major deflections can be noticed: proceeding from left to right, the first deflection near to 0 Hz corresponds to a DC component, with no biological significance. More to the right a component, the major one, is present at about 0.07 Hz which corresponds to what is currently called the low frequency component, a marker of sympathetic activity [51]. Finally, at about the center of the horizontal axis, a component of about 0.28 Hz, the high frequency component corresponding to the respiratory frequency, provided a marker of vagal tone. During the ischemic episode the sympatho-vagal balance is drastically modified and the sympathetic tone appears largely increased: simultaneously the marker of vagal tone almost disappears. This change is likely to reflect an excitatory reflex from the heart [31, 34], i.e., the result of an increased afferent sympathetic barrage from the heart occuring in the absence of pain.

Similar increases in the low frequency component of heart rate and arterial pressure variabilities can be measured in the conscious dog during experimental coronary occlusion [54]; in this case as well a marked excitatory reflex occurs in the absence of signs of pain.

Thus, there is now additional proof of the fact that in the conscious state excitations likely to be quite consistent can occur in the absence of pain in the afferent neural channel arising from the heart, known to convey the algogenic code when present. Yet, this afferent excitation can produce reflexes.

The proposal of this continuum may perhaps stimulate a better appreciation of how reflexes and pain can both result from the same stimuli acting on the heart: how-

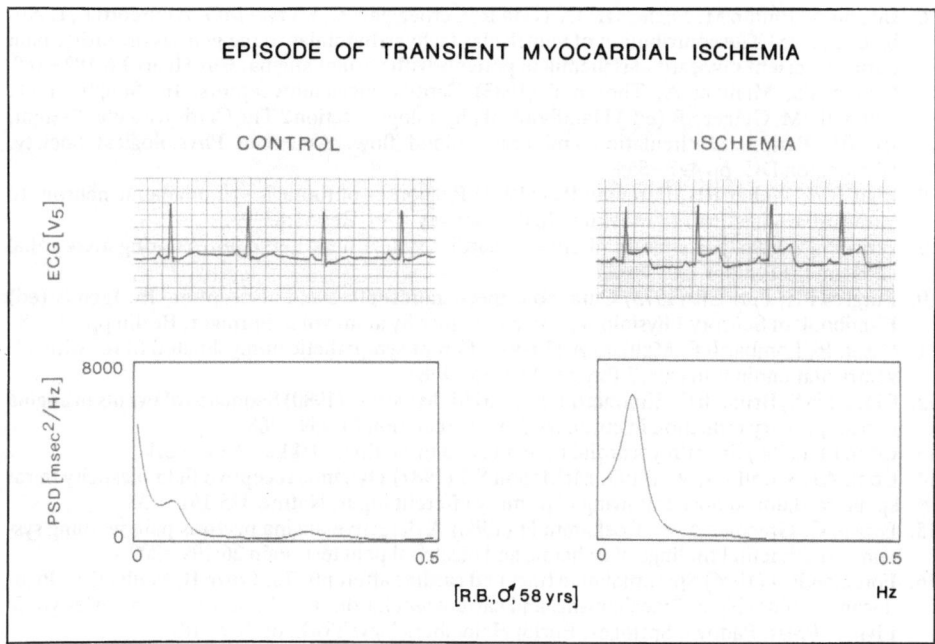

Fig. 7. Power spectral density of heart rate variability in a patient before (control) and during (ischemia) an episode of painless transient myocardial ischemia (see text). (From Lombardi et al., work in progress.)

ever, the characteristics of the codes giving rise to one of the events or to both of them are still unknown.

The analysis in the frequency domain [1, 51] surely opens the door to the possibility of studying experimental and clinical events with the same techniques. A better quantification of the reflex actions induced by myocardial ischemia seems promising. Whether this approach will also provide some light on the mysterious step existing between an afferent neural activity and the onset of pain is at the moment pure speculation.

References

1. Akselrod S, Gordon D, Übel FA, Shannon DC, Barger AC, Cohen RJ (1981) Power spectral analysis of heart rate fluctuation: A quantitative probe of beat-to-beat cardiovascular control. Science 213:220–222
2. Baker DG, Coleridge HM, Coleridge JCG, Nerdrum T (1980) Search for a cardiac nociceptor: stimulation by bradykinin of sympathetic afferent nerve endings in the heart of the cat. J Physiol 306:519–536
3. Barnes PJ, Conradson TBG, Dixon CMS, Fuller RW, Maxwell DL (1986) Effect of infused adenosine on cardiovascular and airway function in conscious man. J Physiol 374:22 (abstract)
4. Barron KW, Bishop VS (1982) Reflex cardiovascular changes with veratridine in the conscious dog. Am J Physiol 242:H810–817
5. Barsky AJ (1986) Palliation and symptomatic relief. Arch Intern Med 146:905–909

6. Biagini A, Emdin M, Michelazzi C, Testa R, Carpeggiani C, Mazzei MG, Andreotti F, L'Abbate A (1988) The contribution of ventricular tachyarrhythmias to the genesis of cardiac pain during transient myocardial ischemia in patients with variant angina. Eur Heart J 6:182–188

7. Bishop VS, Malliani A, Thoren P (1983) Cardiac mechanoreceptors. In: Shepherd JT, Abboud FM, Geiger SR (eds) Handbook of physiology. Section 2 The Cardiovascular System, vol III. Peripheral circulation and organ blood flow. American Physiological Society, Washington DC, pp 497–555

8. Blair RW, Weber RN, Foreman RD (1982) Responses of thoracic spinothalamic neurons to intracardiac injection of bradykinin in the monkey. Circ Res 51:83–94

9. Brown AM (1967) Excitation of afferent cardiac sympathetic nerve fibres during myocardial ischaemia. J Physiol 190:35–53

10. Burgess PR, Perl ER (1973) Cutaneous mechanoreceptors and nociceptors. In: Iggo A (ed) Handbook of Sensory Physiology, Somatosensory System, vol 2. Springer, Berlin pp 22–78

11. Casati R, Lombardi F, Malliani A (1979) Afferent sympathetic unmyelinated fibres with left ventricular endings in cats. J Physiol 292:135–148

12. Chierchia S, Brunelli C, Simonetti I, Lazzari M, Maseri A (1980) Sequence of events in angina at rest: primary reduction in coronary flow. Circulation 61:759–768

13. Cohn PF (1986) Silent myocardial ischemia and infarction. Dekker, New York

14. Cook AJ, Woolf CJ, Wall PD, McMahon SB (1987) Dynamic receptive field plasticity in rat spinal cord dorsal horn following C-primary afferent input. Nature 325:151–153

15. Droste C, Greenlee MW, Roskamm H (1986) A defective angina pectoris pain warning system: experimental findings of ischemic and electrical pain test. Pain 26:199–209

16. Foreman RD (1986) Spinothalamic tract and cardiac afferents. In: Lown B, Malliani A, Prosdocimi M (eds) Neural mechanisms and cardiovascular disease. Fidia Research Series vol 5, Liviana Press, Padova. Springer, Berlin Heidelberg New York, pp 169–181

17. Foreman RD, Ohata CA (1980) Effects of coronary artery occlusion on thoracic spinal neurons receiving viscerosomatic inputs. Am J Physiol 238:H667–H674

18. Goody W (1957) On the nature of pain. Brain 80:118–131

19. Guazzi M, Polese A, Fiorentini C, Magrini F, Bartorelli C (1971) Left ventricular performance and related haemodynamic changes in Prinzmetal's variant angina pectoris. Br Heart J 33:84–94

20. Guazzi M, Polese A, Fiorentini C, Magrini F, Olivari MT, Bartorelli C (1975) Left and right heart haemodynamics during spontaneous angina pectoris. Comparison between angina with ST segment depression and angina with ST segment elevation. Br Heart J 37:401–413

21. Guazzi M, Olivari MT, Polese A, Fiorentini C, Magrini F (1976) Repetitive myocardial ischemia of Prinzmetal type without angina pectoris. Am J Cardiol 37:923–927

22. Guzman F, Braun C, Lim RKS (1962) Visceral pain and the pseudoaffective response to intra-arterial injection of bradykinin and other algesic agents. Arch Int Pharmacodyn 136:353–384

23. Hoberg E (1987) Symptomatic versus asymptomatic ischemic episodes during Holter monitoring: patterns of high resolution trend recordings of ST segment and heart rate: In: von Arnim Th and Maseri A (eds) Silent Ischemia. Steinkopff, Darmstadt, pp 125–130

24. Langley JN (1903) The autonomic nervous system. Brain 26:1–26

25. Lewis T (1932) Pain in muscular ischemia – its relation to anginal pain. Arch Intern Med 49:713–727

26. Lewis T, Kellgren JH (1939) Observations relating to referred pain, viscero-motor reflexes and other associated phenomena. Clin Sci 4:47–71

27. Lindgren I, Olivecrona H (1947) Surgical treatment of angina pectoris. J Neurosurg 4:19–39

28. Littler WA, Honour J, Sleight P, Stott FD (1973) Direct arterial pressure and electrocardiogram in unrestricted patients with angina pectoris. Circulation 48:125–134

29. Lombardi F, Della Bella P, Casati R, Malliani A (1981) Effects of intracoronary administration of bradykinin on the impulse activity of afferent sympathetic unmyelinated fibers with left ventricular endings in the cat. Circ Res 48:69–75

30. Malliani A (1982) Cardiovascular sympathetic afferent fibers. Rev Physiol Biochem Pharmacol 94:11–74

31. Malliani A (1986a) The elusive link between transient myocardial ischemia and pain. Circulation 73:201–204

32. Malliani A (1986b) Homeostasis and instability: the hypothesis of tonic interaction in the car-

diovascular regulation of negative and positive feedback mechanisms. In: Lown B, Malliani A, Prosdocimi P (eds) Neural mechanisms and cardiovascular disease. Fidia Research Series vol 5. Liviana Press, Padova, Springer, Berlin – Heidelberg – New York, pp 1–9

33. Malliani A, Lombardi F (1982) Consideration of the fundamental mechanisms eliciting cardiac pain. Am Heart J 103:575–578
34. Malliani A, Schwartz PJ, Zanchetti A (1969) A sympathetic reflex elicited by experimental coronary occlusion. Am J Physiol 217:703–709
35. Malliani A, Peterson DF, Bishop VS, Brown AM (1972) Spinal sympathetic cardiocardiac reflexes. Circ Res 30:158–166
36. Malliani A, Recordati G, Schwartz PJ (1973) Nervous activity of afferent cardiac sympathetic fibres with atrial and ventricular endings. J Physiol 229:457–469
37. Malliani A, Pagani M, Lombardi F (1989) Visceral versus somatic mechanisms. In: Wall PD, Melzack R (eds). Textbook of Pain 2nd ed., Churchill Livingstone, Edinburgh, pp 128–140
38. Malliani A, Lombardi F, Pagani M (1986a) Sensory innervation of the heart. In: Cervero F, Morrison JFB (eds) Visceral sensations. Progress in Brain Research, Elsevier Science Publ, Amsterdam, vol 67, pp 39–48
39. Malliani A, Pagani M, Lombardi F (1986b) Positive feedback reflexes. In: Zanchetti A, Tarazi RC (eds) Handbook of hypertension, vol 8: Pathophysiology of hypertension. Regulatory Mechanisms. Elsevier Scientific Publications, pp 69–81
40. Mannheimer C, Carlsson CA, Vedin A, Wilhelmsson C (1986) Transcutaneous electrical nerve stimulation (TENS) in angina pectoris. Pain 26:291–300
41. Martin SJ, Gorham LW (1938) Cardiac pain. An experimental study with reference to the tension factor. Arch Intern Med 62:840–852
42. Maseri A, Severi S, De Nes M, L'Abbate A, Chierchia S, Marzilli M, Ballestra AM, Parodi O, Biagini A, Distante A (1978) ‚Variant‘ angina: one aspect of a continuous spectrum of vasospastic myocardial ischemia. Am J Cardiol 42:1019–1035
43. Maseri A, Chierchia S, Davies G, Glazier J (1985) Mechanisms of ischemic cardiac pain and of silent myocardial ischemia. Am J Med 79:7–11
44. McMahon SR, Sykova F, Wall PD, Woolf CL, Gibson SJ (1984) Neurogenic extravasation and substance P levels are low in muscle as compared to skin in the rat hindlimb. Neurosci Lett 52:235–240
45. Murphy DF, Giles KE (1987) Dorsal column stimulation for pain relief from intractable angina pectoris. Pain 28:365–368
46. Nademanee K, Intarachot V, Josephson MA, Rieders D, Mody FV, Singh BN (1987) Prognostic significance of silent myocardial ischemia in patients with unstable angina. J Am Coll Cardiol 10:1–9
47. Nesto RW, Phillips RT (1986) Asymptomatic myocardial ischemia in diabetic patients. Am J Med 80 (Suppl 4c):40–47
48. Osler W (1910) The Lumleian lectures on "angina pectoris" (lecture II). Lancet I:839–844
49. Pagani M, Baig H, Sherman A, Manders WT, Quinn P, Franklin D, Vatner SF (1978) Technique for measurement of pressure-dimension relations in the aorta and great arteries of conscious unanesthetized animals. Am J Physiol 235:H610–H617
50. Pagani M, Pizzinelli P, Furlan R, Guzzetti S, Rimoldi O, Sandrone G, Malliani A (1985) Analysis of the pressor sympathetic reflex produced by intracoronary injections of bradykinin in conscious dogs. Circ Res 56:175–183
51. Pagani M, Lombardi F, Guzzetti S, Rimoldi O, Furlan R, Pizzinelli P, Sandrone G, Malfatto G, Dell'Orto S, Piccaluga E, Turiel M, Baselli G, Cerutti S, Malliani A (1986) Power spectral analysis of heart rate and arterial pressure variabilities as a marker of sympatho-vagal interaction in man and conscious dog. Circ Res 59:178–193
52. Perl ER (1971) Is pain a specific sensation? J Psychiatr Res 8:273–287
53. Procacci P, Zoppi M, Padeletti L, Maresca M (1976) Myocardial infarction without pain. A study of the sensory functions of the upper limbs. Pain 2:309–313
54. Rimoldi O, Pierini S, Ferrai A, Cerutti S, Pagani M, Malliani A (1990) Analysis of the short term oscillations of R-R and arterial pressure in conscious dogs. Am J Physiol, in press
55. Sherrington CS (1906) The integrative action of the nervous system. Yale University Press, New Haven
56. Sigwart V, Grbic M, Payot M, Goy J, Essinger A, Fischer A (1984) Ischemic events during

coronary artery balloon obstruction. In: Rutishauser W, Roskamm H (eds) Silent myocardial ischemia. Springer, Berlin Heidelberg New York, pp 29–36

57. Sutton DC, Lueth HC (1930) Experimental production of pain on excitation of the heart and great vessels. Arch Intern Med 45:827–867

58. Sylvén C, Beermann B, Jonzon B, Brandt R (1986) Angina pectoris-like pain provoked by intravenous adenosine in healthy volunteers. Br Med J 293:227–230

59. Theroux P, Ross J, Franklin D, Kemper WS, Sasayama S (1976) Regional myocardial function in the conscious dog during acute coronary occlusion and responses to morphine, propranolol, nitroglycerin and lidocaine. Circulation 53:302–314

60. Thorén P (1979) Role of the cardiac vagal c-fibers in cardiovascular control. Rev Physiol Biochem Pharmacol 86:1–94

61. Von Bezold A, Hirt L (1867) Über die physiologischen Wirkungen des essigsauren Veratrins. Untersuch Physiol. Lab. Würzburg, 1:73–156

62. Wall PD (1985) Future trends in pain research. Phil Trans R Soc, (Lond) 308:393–401

63. Wall PD, Melzack R (eds) (1989) Textbook of pain. 2nd ed. Churchill Livingstone, Edinburgh

64. Wall PD, Woolf CJ (1984) Muscle but not cutaneous C-afferent input produces prolonged increases in the excitability of the flexion reflex in the rat. J Physiol 356:443–458

65. Weidinger F, Hammerle A, Sochor H, Smetana R, Frass M, Glogar D (1986) Role of beta-endorphins in silent myocardial ischemia. Am J Cardiol 58:428–430

66. White JC (1957) Cardiac pain. Anatomic pathways and physiologic mechanisms. Circulation 16:644–655

67. Woolf CJ, Wall PD (1986) Relative effectiveness of C primary afferent fibers of different origin of the flexor reflex in the rat. J Neurosci 6:1433–1442

Authors' address:
Alberto Malliani
Istituto Ricerche Cardiovascolari
via Bonfadini 214
20138 Milano, Italy

246

Exercise-Induced Myocardial Ischemia: The Role of Heart Rate Reduction in Therapeutic Approach

Brian D. Guth and Gerd Heusch

Seaweed Canyon Laboratory, Department of Medicine, University of California, San Diego, and Department of Pathophysiology, University of Essen

Rational therapy for exercise-induced myocardial ischemia, as manifested by effort angina pectoris, must be based on a thorough understanding of the disease process and its effects on regional myocardial blood flow and contractile function. The clinician has an ever increasing array of pharmaceuticals available with which to treat the coronary disease patient, including β-adrenergic blockers, calcium channel antagonists, and nitrates. This divergent set of agents has markedly different effects on the coronary vasculature, the myocardial contractile state, and the peripheral circulation, thereby making combined pharmacologic therapy attractive in some situations. However, such complicated therpeutic approaches further emphasize the need to fully appreciate the anatomical, physiological, and pathological mechanisms present in the patient, together with the mode of action of each pharmacologic agent utilized. The purpose of this chapter is to discuss the role of tachycardia in the precipitation of myocardial ischemia, as well as the role of heart rate reduction for the treatment of exercise-induced myocardial ischemia in the context of the underlying pathophysiology.

Pathophysiology of Exercise-Induced Ischemia

Regional Myocardial Blood Flow

Central to the existence of effort angina pectoris is the failure of myocardial blood flow to increase in proportion to increases in myocardial work and oxygen consumption. Normally myocardial blood flow can increase to four- or fivefold the resting level during strenuous exercise due to metabolic vasodilation of the coronary vasculature. This vasodilation is particularly impressive considering the tachycardia accompanying severe exercise since the markedly reduced diastolic interval at high heart rates severely limits the time available for subendocardial perfusion [2, 3]. Nevertheless, a uniform transmural blood flow is normally maintained and subendocardial blood flow increases linearly with myocardial workload [37].

The existence of coronary artery disease can interrupt this normal pattern if a functionally significant coronary stenosis develops. A drop in pressure across the stenosis results in an abnormally low coronary perfusion pressure within the distal vascular bed and as vasodilatory reserve becomes exhausted, the perfusion of the myocardium becomes pressure-dependent. Two laterally adjacent perfusion beds connected by collateral channels can thus undergo unequal vasomotion leading to the phenomenon of "lateral" or "collateral steal" [39]. This is particularly critical for perfusion of the subendocardium where the mechanical resistance to blood flow is greatest [34] and where dilatory reserve becomes limited first. The persistence of vasodilatory reserve in the outer wall at a time when subendocardial vascular reserve is exhausted causes a preferential perfusion of the outer wall, a phenomenon known as "transmural steal" [8, 10, 11, 12, 30], which further limits subendocardial perfusion. More recently, it was recognized that in arteries that perfuse areas of both the left and right ventricles (left anterior descending and right coronary arteries), a "steal" may occur from the left to the right ventricular myocardium [15] due, presumably, to the differential autoregulatory patterns of the two ventricles [29].

The common feature of any coronary "steal" phenomenon is the existence of nonuniform vasomotion resulting in pressure gradients within a perfusion territory. Exercise precipitates such blood flow maldistributions since it serves as a powerful stimulus for metabolic vasodilation, while at the same time providing a significant mechanical limitation on subendocardial perfusion. As mentioned above, tachycardia during exercise greatly shortens the diastolic duration which is crucial to subendocardial perfusion since blood flow to the inner wall of the left ventricle is confined primarily to diastole [7, 17]. Secondly, the increase in left ventricular chamber pressure during exercise can further impede subendocardial perfusion. Once ischemia is present, the increased diastolic pressures which usually result further impair subendocardial blood flow.

The above mechanisms may also be operative in the situation where a coronary artery is completely occluded (due to advanced coronary disease) but where coronary collateral circulation provides sufficient blood flow to meet the myocardial oxygen requirements at rest. During exercise, the increased myocardial oxygen demand cannot be met by the oxygen delivery through collateral channels. The vasodilatory reserve of the collateral-dependent myocardium becomes exhausted and the metabolic vasodilation within the other regions of the heart causes a decrease in the driving pressure for collateral blood flow [21].

Regional Myocardial Function

The functional consequence of regional ischemia is apparent in the close relationship between subendocardial perfusion and myocardial contractile function which occurs not only at rest but also during exercise. Gallagher et al. have shown that in dogs undergoing acute coronary artery stenosis during treadmill exercise, overall wall contractile function is linearly related to the perfusion of the inner-third of the ventricular wall [11]. Perfusion of the outer wall is less affected by the stenosis due to the transmural "steal". Nevertheless, it appears that subendocardial perfusion is most critical for the maintenance of the contractile performance of the entire ischemic wall. Consequently, evaluation of pharmacologic interventions for the treatment of exercise-induced regional myocardial ischemia may be based upon their efficacy for enhancing regional contractile function as a reflection of subendocardial perfusion.

Myocardial oxygen demand and oxygen supply are exactly matched under physiological circumstances through the control of coronary vascular resistance by local metabolic mechanisms [9, 31]. These mechanisms permit the coronary vascular bed to exhibit a marked ability to autoregulate such that coronary blood flow is relatively maintained over a wide range of coronary perfusion pressures [5, 28]. Myocardial ischemia results only when the coronary perfusion pressure falls below the range of autoregulation so that coronary blood flow then becomes directly related to the perfusion pressure and further decreases in coronary perfusion pressure result in reductions of myocardial blood flow. This can occur when there is a primary reduction in myocardial blood flow as would occur during coronary vasospasm or with acute thrombosis. Coronary perfusion pressure falls distal to the site of the lesion or spasm causing reduced myocardial blood flow. This has been referred to as a "supply type" of ischemia in that the cause of the ischemia is an acute reduction in the blood flow and consequently oxygen supply. Alternatively, ischemia can occur if the demands for oxygen of the myocardium increase but without a simultaneous increase in blood flow and oxygen supply. This would occur in a setting where an atherosclerotic lesion on an epicardial coronary artery permits the maintenance of resting blood flow but restricts any increase in coronary blood flow, i.e., a "critical stenosis". As myocardial oxygen demands increase with exercise or other cardiovascular stress, the ability of the myocardium to increase blood flow is restricted by the coronary stenosis and an imbalance between the limited oxygen supply and increased demand develops. This type of ischemia has been referred to as ischemia of the "demand type".

Exercise-induced ischemia could reasonably be classified as a "demand type" ischemia since exercise produces a powerful sympathetic stimulation of the cardiovascular system with increases in heart rate, contractility and, consequently, myocardial oxygen consumption. However, such a classification is overly simplistic since not only is oxygen demand increased, but subendocardial blood flow is also reduced due to the reduction in diastolic duration and "steal" phenomena observed as described above. The severity of the ischemic supply-demand imbalance during exercise can thus be influenced by either affecting the amount of reduction in myocardial oxygen supply or the amount of the increase in myocardial oxygen demand.

This conceptual model of exercise-induced ischemia is useful whether resting blood flow is maintained primarily through a stenotic vessel with limited collateral blood flow or the poststenotic myocardium is completely collateral-dependent. In the experimental model of canine single-vessel coronary artery stenosis described below, there is absolutely no differentiation between animals in which a true "critical stenosis" is present from those in which the epicardial artery is occluded but with adequate resting blood flow through collateral channels. In either case, exercise elicits similar hemodynamic changes (the same increases in myocardial oxygen demand) and similar changes in myocardial blood flow pattern in the post-stenotic myocardium typified by reduced subendocardial perfusion, increases in subepicardial perfusion and a marked decrease in the endo to epi blood flow ratio. Thus, the effects are similar regardless whether blood flow utilizes collateral channels, a stenotic epicardial vessel, or some combination of the two. Of primary importance is how much blood arrives at the post-stenotic myocardium while the manner in which it gets there appears irrelevant. Pharmacologic strategies for affecting these two situations, however, may be different.

Based on this description of exercise-induced ischemia, strategies for its treatment can be broadly classified into two groups: 1) attempts to reduce myocardial oxygen demand (e.g., β-adrenergic blockade, nitrates, calcium channel blockers), and 2) attempts to enhance flow to the ischemic region (e.g., calcium channel blockers). Heart-rate reduction (or restriction of tachycardia) for the treatment of exercise-induced ischemia is an attractive approach since it reduces myocardial oxygen requirements in such a way that also allows for enhanced perfusion. Thus, the reduction in the number of beats per minute has a direct effect in reducing oxygen demands (oxygen consumed per minute) while the increased diastolic time allows for enhanced perfusion of the inner ventricular wall. Additionally, as the oxygen demands of non-ischemic adjacent myocardium are reduced (fewer contractions per minute), a reversal of the "steal" effects can occur [27]. An experimental model for demonstrating the efficacy of heart rate reduction for attenuating exercise-induced ischemia will be described next.

Exercise-Induced Ischemia: Experimental Approach

Studies designed to evaluate the role of heart-rate reduction on exercise-induced ischemia have been performed in a canine model having a fixed, single vessel coronary artery stenosis (Fig. 1) [13, 14, 20, 25]. Placement of an ameroid constrictor around the proximal left circumflex coronary artery produces gradual constriction of the vessel over a two- to three-week period. Such constriction is associated with the stimulation of collateral development to the region distal to the stenosis [38]. Dogs are studied at a time when resting blood flow can be maintained through the stenosis or collateral circulation but hyperemia is severely limited. Exercise at this time produces regional ischemia and contractile abnormalities in the region distal to the stenosis (Fig. 2). The sympathetic stimulation of exercise is evident by the abrupt increase in left ventricular pressure development (dP/dt), heart rate and contractile function as indicated by the increase in systolic wall thickening measured using ultrasonic transducers. The increased wall thickening in the poststenotic region is short-lived however, as contractile dysfunction quickly ensues, but the enhanced contractile performance of the control region is maintained. After approximately 2 min of exercise, function in the ischemic region reaches a steady state in which systolic wall thickening is reduced in proportion to the reduction in subendocardial perfusion [33] where myocardial blood flow and its transmural distribution is measured using radiolabeled tracer microspheres. Paired runs performed 3 h apart have been documented to be completely reproducible [26], thereby allowing for the testing of pharmacologic agents in this model.

β-Adrenergic Blockade in Exercise-Induced Ischemia

β-adrenergic blockade has become a frontline treatment for stable angina pectoris by decreasing myocardial oxygen requirements through reductions in exercise heart

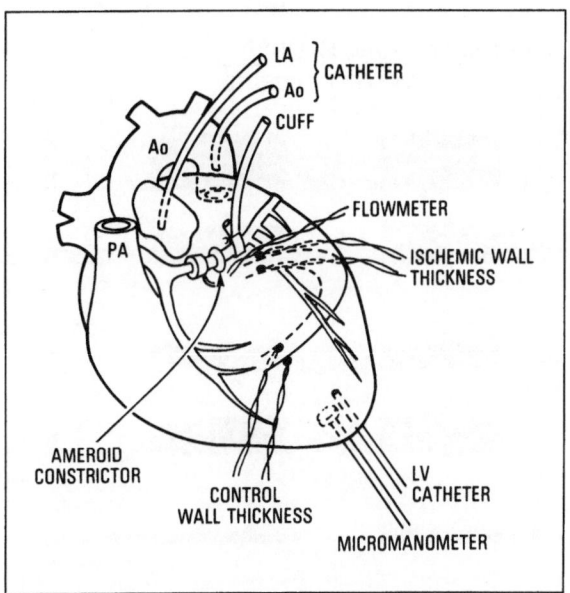

Fig. 1. Schematic diagram of the experimental canine model. Ventricular hemodynamics were assessed by measuring left ventricular pressure with a micromanometer calibrated against a catheter in the left ventricle (LV catheter). Regional contractile function was measured using wall thickness sonomicrometers in the anterior control wall and the posterolateral ischemic wall. Catheters were placed in the left atrium (LA) and descending thoracic aorta (Ao) for the microsphere technique to measure myocardial blood flow. The proximal left circumflex coronary artery was instrumented with a Doppler flow probe, an ameroid constrictor and a pneumatic occlusive cuff.

rate and contractility. The efficacy of cardioselective β-blockade was reported initially by Matsuzaki et al. in the model of experimental exercise-induced ischemia described above [27], and later by Guth et al. [13]. In a control run before β-blockade, exercise elicited an increase in heart rate from 107 ± 20 (S.D.) to 217 ± 25 beats/min, an increase in left ventricular dP/dt from 3052 ± 428 to 4907 ± 936 mmHg/s and an increase in the heart rate-left ventricular peak pressure product to 2.3-times that of control. Systolic wall thickening in the post-stenotic lateral wall during the steady state response was reduced to 27% of the resting value, whereas wall thickening in the control anterior wall increased by 28%. Subendocardial blood flow in the ischemic wall was reduced by 33% during the control exercise bout while subepicardial blood flow increased to 2.2-times that of control; the endocardial-to-epicardial blood flow ratio was reduced from 1.26 to 0.39.

Cardioselective β-adrenergic blockade (atenolol, 1.0 mg/kg) reduced the exercise-heart rate from 217 ± 25 to 166 ± 15 beats per minute and blunted the contractile response to exercise as indicated by significant reductions in left ventricular dP/dt (3051 ± 410 vs 4907 ± 936 mmHg/s, and the systolic wall thickening of the control, non-ischemic wall ($29.2 \pm 7.5\%$ during the β-blocked run compared to $41.1 \pm 12.4\%$ during the control run). However, systolic wall thickening in the ischemic region was increased from 8.3 ± 8.5 to $11.4 \pm 8.2\%$ (Fig. 3); wall thickening thus fell to only 50%

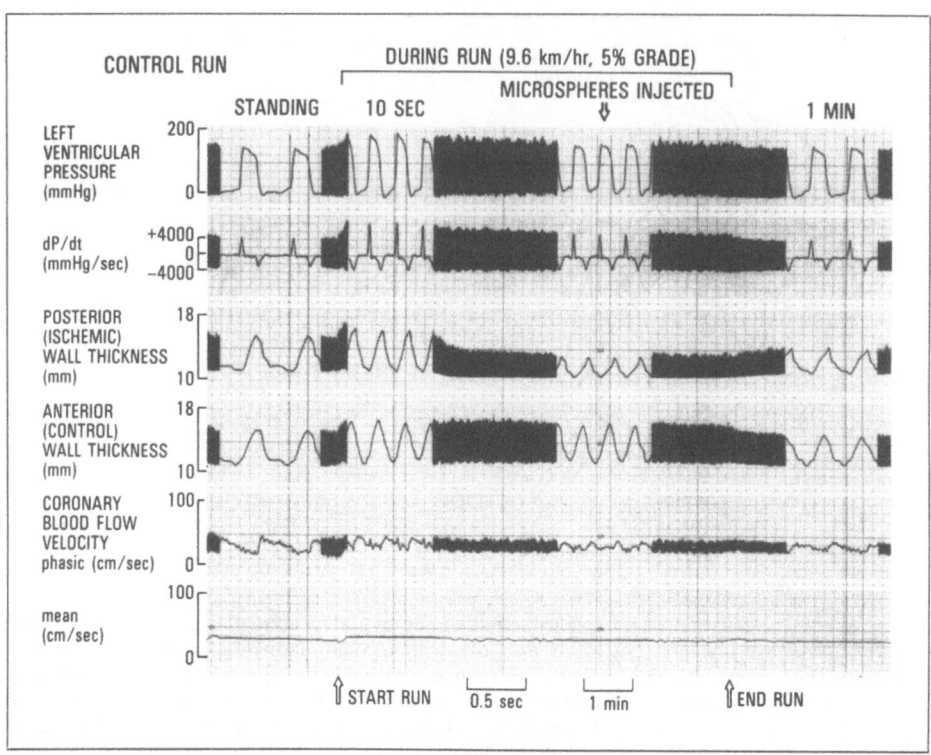

Fig. 2. An original tracing from a dog having a fixed single-vessel coronary stenosis due to the implantation of an ameroid constrictor on the proximal left circumflex coronary artery. Dogs are studied when the stenosis prevents the normal hyperemia of exercise, resulting in regional ischemia and contractile dysfunction. Microsphere measurements of myocardial blood flow are always performed in a hemodynamic steady state. Note the lack of increased coronary blood flow velocity in the left circumflex artery with the onset of running and the development of contractile dysfunction in the posterior (ischemic) wall. (From [26] by permission of the American Heart Association.)

of the resting level as compared to 27% during the control run. The study by Matsuzaki et al. demonstrated an increase in subendocardial blood flow of 50% (0.4 to 0.6 (ml/min)/g) during β-blockade together with an increase in regional contractile function [27]. These results indicated that cardioselective β-adrenergic blockade results in a favorable effect on the post-stenotic myocardium during exercise-induced ischemia. The effects of β-blockade brought about an increase in the subendocardial blood flow during exercise and, consequently, a higher level of contractile function could be supported. The precise mechanism for this beneficial action, however, was not evident.

Both a reduction in the exercise heart rate and a reduction of inotropic state by β-blockade could have resulted in an improved supply-demand ratio, hence, reduced ischemia. The relative importance of the negative inotropic and negative chronotropic effects was investigated by preventing the β-blockade-produced bradycardia by atrial pacing [13]. During the run with atenolol, after a steady-state ischemia was

252

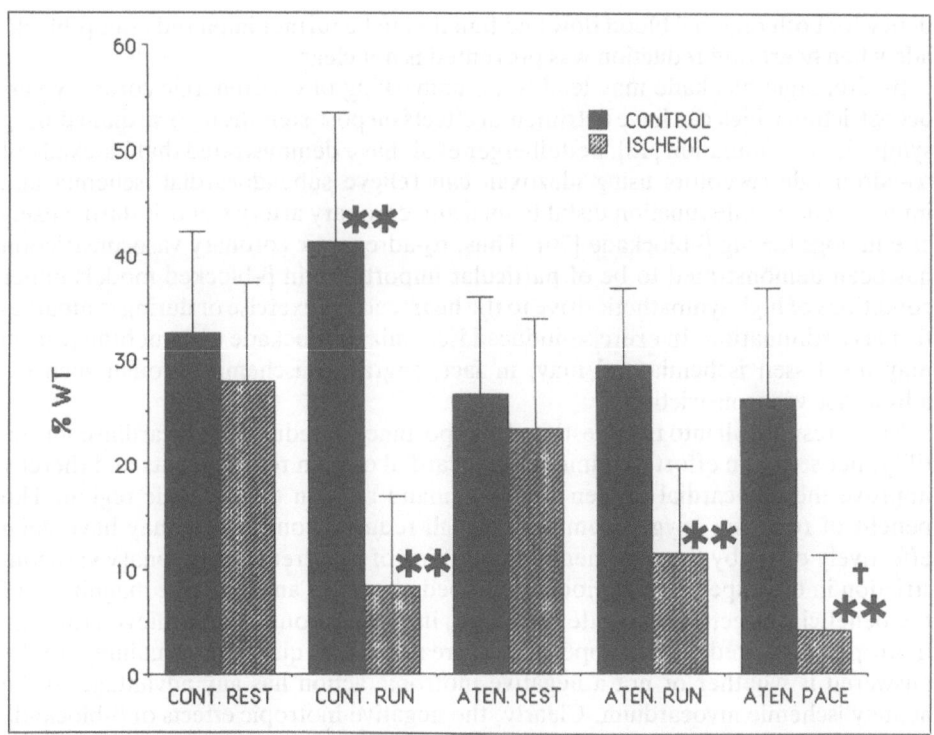

Fig. 3. Summary of regional myocardial function (assessed as systolic wall thickening, %WT) in the anterior control wall (black bars) and posterolateral ischemic wall (shaded bars) during control resting conditions (CONT.REST) and during a control run (CONT.RUN) and after the administration of cardioselective β-blockade at rest (ATEN.REST), during a subsequent run (ATEN.RUN), and with the addition of atrial pacing to prevent the bradycardia (ATEN.PACE). Ischemic wall thickening was improved after the β-blockade when heart rate was reduced but not after atrial pacing prevented the bradycardia. (** = p < 0.01 vs CONT.REST; + = p < 0.05 vs CONT.RUN). (From [13] by permission of the American Heart Association.)

achieved, the heart rate was paced (left atrial pacing) to exactly match the heart rate observed previously during the control exercise period. Pacing to prevent the heart rate reduction of β-blockade eliminated any benefical effect in regards to either regional myocardial blood flow or contractile function. Subendocardial blood flow during the run with β-blockade but at matched heart rate showed no improvement compared to the previous control run and actually tended to be reduced (0.72 ± 0.19 vs 0.56 ± 0.22 (ml/min)/g, NS). Blood flow to the middle and outer wall was unchanged under these conditions. Similarly, there was no longer any improvement in systolic wall thickening after β-blockade when heart rate reduction was prevented, being 7.2 ± 8.1% in the control run and only 4.1 ± 7.1% in the run with β-blockade and heart rate matching (Fig. 3). Thus, this study demonstrated that reduction of exercise heart rate is essential for the beneficial effect of cardioselective β-blockade in this model of exercise-induced regional myocardial ischemia. The mechanism of the tendency for both regional blood flow and function to be further impaired after β-blockade when heart rate reduction was prevented is not clear.

β-adrenergic blockade may lead to an unmasking of α-adrenergic coronary vasoconstriction which can have detrimental effects on post-stenotic myocardium during sympathetic stimulation [18]. Seitelberger et al. have demonstrated that blockade of α_2-adrenergic receptors using idazoxan can relieve subendocardial ischemia and improve contractile function distal to an acute coronary artery stenosis during exercise in dogs having β-blockade [36]. Thus, α_2-adrenergic coronary vasoconstriction has been demonstrated to be of particular importance in β-blocked models under conditions of high sympathetic drive to the heart such as exercise or during sympathetic nerve stimulation. In exercise-induced ischemia, β-blockade without bradycardia may not lessen ischemia and may, in fact, aggravate ischemia through such α_2-adrenergic vasoconstriction.

These results call into the question the importance of reducing myocardial contractility, per se, in an effort to minimize myocardial oxygen requirements and thereby improve the myocardial oxygen supply-demand ratio in the ischemic region. The benefit of reducing oxygen demand through reduced contractility may have been effectively offset by the detrimental unmasking of α_2-adrenergic coronary vasoconstriction in the experimental model described above. In any case, the magnitude of the beneficial effect of heart rate reduction, in comparison to that achieved through inotropic state reduction, appears far greater. The question remaining to be answered is whether or not a negative inotropic action has any advantage to the acutely ischemic myocardium. Clearly, the negative inotropic effects of β-blockade on the normally perfused regions of the heart are dramatic and may compromise the overall pump function of the ventricle. This may be of importance during acute regional ischemia when an ischemic portion of the ventricular wall is dysfunctional, thereby making the cardiac output and coronary perfusion pressure dependent upon the performance of the non-ischemic myocardium.

Specific Bradycardic Agents in Exercise-Induced Ischemia

The results from studies on β-adrenergic blockade suggest that the bradycardia, and not the negative inotropic effect, is most useful in minimizing the amount of ischemia observed during exercise when coronary inflow is limited. The emergence of specific bradycardic agents as a new class of cardiovascular agents will give cardiologists an additional tool to selectively prevent tachycardia while preserving the contractile state of the myocardium. The bradycardic agent UL-FS 49 (1 mg/kg) was used in this canine model of exercise-induced ischemia to test the efficacy of selective bradycardia on both regional blood flow and contractile function [14].

Figure 4 shows a control run in which exercise elicited an increase in heart rate from 114 ± 20 to 230 ± 19 beats per minute, an increase in left ventricular dP/dt from 3227 ± 762 to 5027 ± 715 mmHg/s, and an increase in the pressure-rate product to 2.5 times that of control. During the control run, systolic wall thickening decreased from the resting value of $23.3 \pm 5.2\%$ to $9.3 \pm 5.0\%$. Regional subendocardial blood flow during the run was decreased from 1.04 ± 0.30 to 0.55 ± 0.40 (ml/min)/g, whereas subepicardial blood flow increased from 0.97 ± 0.31 to 1.67 ± 0.45, indica-

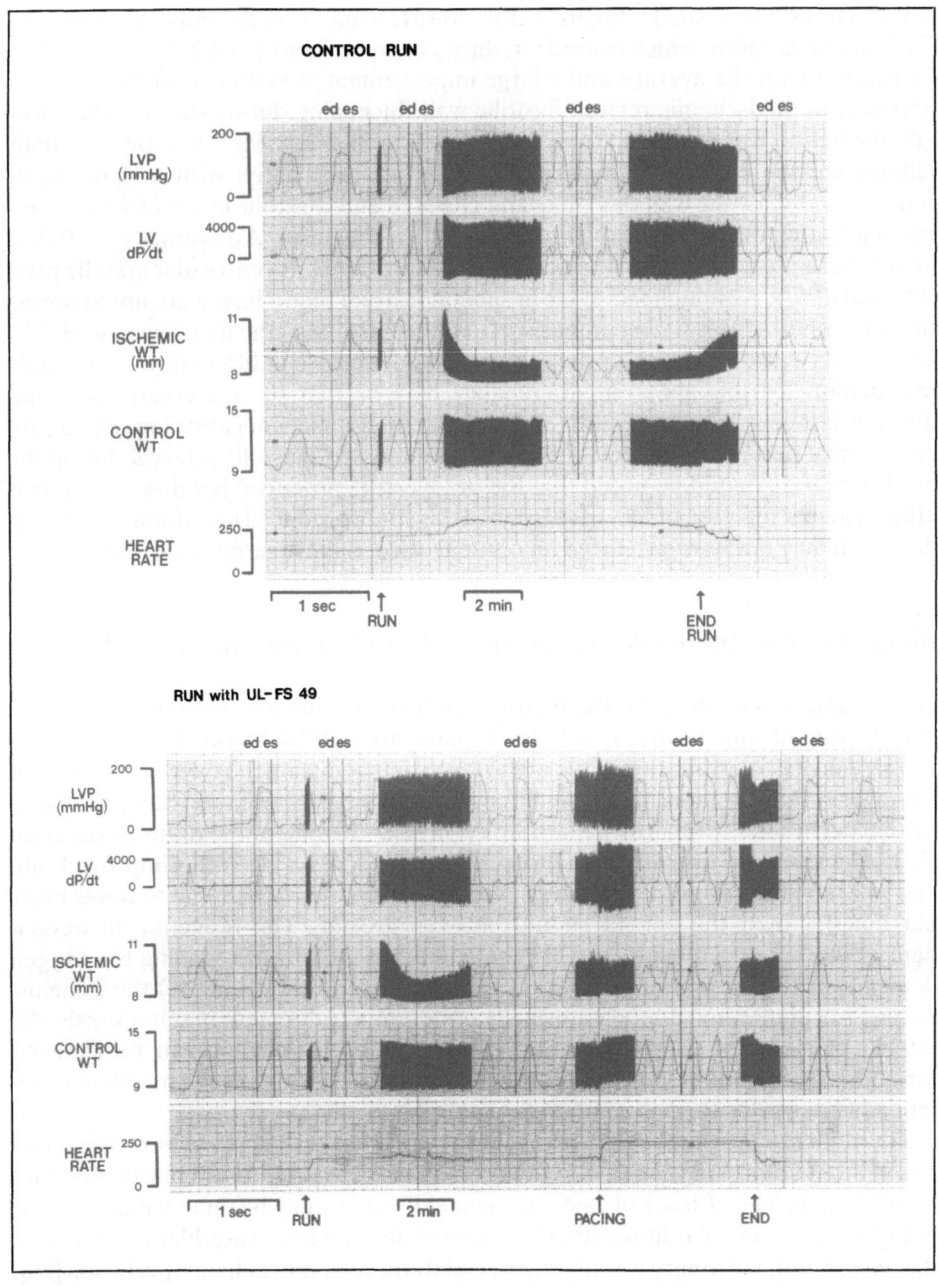

Fig. 4. Original tracing of a dog exercising before (CONTROL RUN) and after the administration of a bradycardic agent (RUN with UL-FS 49). The amount of ischemic wall dysfunction observed during the run after the administration of UL-FS 49 is significantly reduced. Timing of end-diastole (ed) and end-systole (es) is indicated. The addition of atrial pacing in the RUN with UL-FS 49 partially but incompletely reversed the beneficial effect of UL-FS 49. (From [14] by permission of the American Heart Association.)

tive of a transmural "steal". Figure 4 also shows tracings from the same dog after UL-FS 49 administration which caused a reduced exercise heart from 230 ± 19 to 139 ± 10 beats/min on the average and a large improvement in systolic wall thickening is apparent in the ischemic region. Systolic wall thickening during the run after drug administration was improved from 9.3 ± 5.0% to 21.5 ± 8.4%, thereby essentially eliminating the contractile dysfunction. This was associated with a significantly reduced end-diastolic wall thickness and elevated end-diastolic left ventricular pressure indicating utilization of the Frank-Starling mechanism. Additionally, wall thickening in the control region was not impaired nor were left ventricular systolic pressure and dP/dt. Although increases in subendocardial blood flow were noted during the run with bradycardia in comparison to the control run, the increase in wall thickening was consistently larger than the observed increase in blood flow. A possible explanation for this apparent discrepancy may be that the extremely high end-diastolic left ventricular pressure (35.7 ± 3.0 mmHg) had a negative impact on subendocardial perfusion, even with the greatly increased diastolic interval during the bradycardia. It should be noted that the bradycardia produced is a dose-dependent effect and that the dose of UL-FS 49 used in this study produced a profound bradycardia which may not have produced an optimal effect on myocardial ischemia.

Bradycardia and Relationship to Pathophysiology of Exercise-Induced Ischemia

As a therapeutic strategy for the treatment of exercise-induced ischemia, bradycardia has the advantage that it acts to increase myocardial blood flow through an increase in diastolic duration while at the same time reducing the oxygen demands of the myocardium by reducing the number of contractions per minute. Since subendocardial perfusion is primarily a diastolic event, lengthening of the diastolic duration should have a beneficial effect on blood flow to the inner wall. If this improved subendocardial perfusion is coupled with a reduced oxygen demand due to fewer beats per minute, then the benefit should be even greater. The reduction in oxygen demand with reduced heart rate has the additional benefit of reducing the oxygen demands of the myocardium adjacent to (laterally and transmurally) the ischemic subendocardium where blood is normally preferentially directed. In other words, the "steal" phenomena observed during exercise-induced ischemia can be lessened through reduction of the oxygen requirements of these neighboring areas with resultant advantages for the perfusion of the central ischemic zone.

The ability to provide bradycardia without either a direct negative inotropic effect and without unmasking α_2-adrenergic coronary vasoconstriction is a reasonable clinical objective. The efficacy of calcium channel blockers may be due, in part, to their bradycardic effects. Additionally, these agents have a direct vasodilatory action on the post-stenotic coronary vasculature or collateral vessels to enhance perfusion [20]. Calcium antagonists also avoid the problem of α-adrenergic coronary vasoconstriction by functionally blocking α_2-adrenergic receptors [19]. The importance of coronary vasodilation for the treatment of myocardial ischemia has been emphasized by the observation that ischemia does not induce a maximal coronary vasodilation and that significant vasodilator reserve can be recruited using appropriate vasodilators both at rest [1, 6, 32] and during exercise [20]. Thus, the decision of whether or not to utilize calcium blockers may be determined by secondary considerations such as the variable amounts of negative inotropism produced by these drugs and other undesirable side effects.

Heart Rate Effects on Dynamic Coronary Stenoses

The discussion thus far has focused on ischemia due to the presence of a stenosis having a fixed diameter. However, it is clear that certain stenoses are dynamic in nature and can alter their shape in response to hemodynamic changes or vasoactive agents [35]. The effective diameter of such a stenosis is a function of the luminal pressure gradient as well as the pressure gradient across the stenosis. Changes in stenotic resistance subsequent to changes in heart rate were reported by Santamore et al. [35]. They observed that increases in heart rate were associated with increased stenotic resistance and decreased blood flow distal to the stenosis. Heart rate reduction, consequently, could result in the opposite effect and provide increased blood flow. However, Heusch et al. later reported that the effect of heart rate on the resistance of a dynamic stenosis was dependent upon the severity of the stenosis [22]. A stenosis which limited, but did not prevent reactive hyperemia following a 15-s complete coronary occlusion, was shown to increase its resistance with increased heart rate as distal coronary pressure fell after metabolic dilation of the poststenotic coronary vasculature. Conversely, a critical coronary stenosis which prevented any reactive hyperemia to a 15-s coronary occlusion was found to have decreased stenotic resistance with increasing heart rate when distal coronary pressure increased after extravascular compression of the poststenotic vasculature. Thus, heart rate changes may significantly alter the resistance to coronary flow calculated for the entire coronary system [21], but the type of change observed in the stenotic segment and in the poststenotic vascular bed is a functon of the stenosis severity. Buck et al. demonstrated that β-adrenergic blockade (atenolol, metoprolol, and propranolol) reduces the severity of a dynamic stenosis in open-chest dogs [4]. The β-blockade was shown to result in a reduction in flow due to autoregulation of nonischemic regions which increased distal coronary perfusion pressure, thereby effectively reducing the stenotic resistance and improving flow to ischemic regions.

Clinical Role for Bradycardia in Treatment of Myocardial Ischemia

Although the detailed mechanisms of bradycardia in the treatment of clinical ischemia are less clear, evidence has been accumulating which suggests that heart rate reduction may be a key factor in determining the proper treatment of many coronary artery disease patients. Furthermore, it has been suggested that heart rate itself may be an important risk factor for the development of coronary artery disease. In this regard, high heart rates have been associated with elevated LDL and low HDL levels. Thus it appears that heart rate may play a role in coronary atherogenesis [40]. Once coronary disease is established, the role of heart rate becomes even more evident.

It is well accepted that acute infarction of the inferior wall is often associated with a vagally-mediated bradycardia [16]. These patients generally have a more favorable outcome than those sustaining anterior infarction and having normal or elevated heart rates [16]. From the animal studies discussed above, this enhanced prognosis in inferior infarct patients may be due to both a reduction in myocardial oxygen requirements during the period of low flow, as well as to an improvement in the subendocardial perfusion in the ischemic region.

The beneficial effects observed in clinical trials of β-adrenergic blocking drugs have, to a large extent, been attributable to the reduction in heart rate [23]. Studies in which β-blockers have been administered within 6 h after the onset of symptoms were analyzed in relationship to the resultant infarct size as estimated using serial CK analyses [24]. These studies indicate that the reduction in infarct size can be correlated to the degree of heart rate reduction observed. These findings have been supported by the observation that myocardial ischemia in patients was unchanged when heart rate was kept constant by atrial pacing [23]. However, the ability of β-blockers to reduce heart rate is limited and the associated negative inotropic effects of these agents depress the myocardium and may therefore be inappropriate in patients with severely compromised ventricular performance. Consequently, there is increasing interest in the relatively new class of cardiovascular drugs known as "specific bradycardic agents". This diverse group of heart-rate lowering drugs offers the potential for independently inducing bradycardia together with the associated beneficial effects but without the negative inotropism of the β-adrenergic blocking agents. Such drugs may have an important role in the treatment of coronary patients in the near future.

In conclusion, effort-induced ischemia is characterized by increased myocardial work in the face of reduced or constant absolute myocardial blood flow resulting in subendocardial ischemia and regional contractile dysfunction. Thus, both a reduced oxygen supply and increased oxygen demand play a role. Heart-rate reduction (or limitation of tachycardia) is efficacious since it acts to both improve subendocardial perfusion of the ischemic tissue as well as reduce the myocardial oxygen requirements. β-adrenergic blockade is effective in the treatment of exercise-induced ischemia due to its negative chronotropic effects but is limited by its negative inotropic effect. Thus, the benefit to the ischemic myocardium may well be offset by overall ventricular depression in normal regions of the heart. Specific bradycardic drugs have proven effective in experimental exercise-induced ischemia and contractile dysfunction without the negative inotropism of the β-adrenergic blockers.

Summary: The role of heart rate reduction in the treatment of exercise-induced ischemia was investigated using a conscious canine model of single vessel coronary artery stenosis. In these dogs, an ameroid constrictor on the proximal left circumflex coronary artery produces a fixed stenosis such that exercise on a treadmill elicits regional ischemia and contractile dysfunction in the posterolateral wall of the left ventricle. Cardioselective β-adrenergic blockade has been shown to reduce the amount of ischemia and improve regional myocardial function in these dogs. However, the beneficial effects of the β -blockade were lost when heart rate reduction by the β-blockade was prevented by atrial pacing. The efficacy of heart rate reduction in exercise-induced ischemia was further demonstrated by using a bradycardic agent which reduced exercise heart rate but without the negative inotropic effect of β-blockade. In models of dynamic coronary stenosis, changes in heart rate can affect the stenotic resistance due to changes in distal coronary perfusion pressure.

References

1. Aversano T, Becker LC (1985) Persistence of coronary vasodilator reserve despite functionally significant flow reduction. Am J Physiol 248:H403−H411
2. Bache RJ, Cobb FR (1977) Effect of maximal coronary vasodilation on transmural myocardial perfusion during tachycardia in the awake dog. Circ Res 41:648−653
3. Ball RM, Bache RJ, Cobb FR, Greenfield JC Jr (1975) Regional myocardial blood flow during graded treadmill exercise in the dog. J Clin Invest 55:43−49
4. Buck JD, Hardman HF, Warltier DC, Gross GJ (1981) Changes in ischemic blood flow distribution and dynamic severity of a coronary stenosis induced by β-blockade in the canine heart. Circulation 64:708−715
5. Canty JM Jr (1988) Coronary pressure-function and steady-state pressure-flow relations during autoregulation in the unanesthetized dog. Circ Res 63:821−836
6. Canty JM Jr, Klocke FJ (1985) Reduced regional myocardial perfusion in the presence of pharmacologic vasodilator reserve. Circulation 71:370−377
7. Downey JM, Kirk ES (1974) Distribution of the coronary blood flow across the canine heart wall during systole. Circ Res 34:251−257
8. Fedor JM, Rembert JC, McIntosh DM, Greenfield JC Jr (1980) Effects of exercise- and pacing-induced tachycardia on coronary collateral flow in the awake dog. Circ Res 46:214−220
9. Feigl EO (1983) Coronary Physiology. Physiol Rev 63:1−205
10. Gallagher KP, Folts JD, Shebuski RJ, Rankin JHG, Rowe GG (1980) Subepicardial vasodilator reserve in the presence of critical coronary stenosis in dogs. Am J Cardiol 46:67−73
11. Gallagher KP, Osakada G, Matsuzaki M, Kemper WS, Ross J Jr (1982) Myocardial blood flow and function with critical coronary stenosis in exercising dogs. Am J Physiol 243:H698−H707
12. Gewirtz H, Williams DO, Ohley WH, Most AS (1983) Influence of coronary vasodilation on the transmural distribution of myocardial blood flow distal to a severe fixed coronary artery stenosis. Am Heart J 106:674−680
13. Guth BD, Heusch G, Seitelberger R, Ross J Jr (1987) Mechanism of beneficial effect of β-adrenergic blockade on exercise-induced myocardial ischemia in conscious dogs. Circ Res 60:738−746
14. Guth BD, Heusch G, Seitelberger R, Ross J Jr (1987) Elimination of exercise-induced regional myocardial dysfunction by a bradycardic agent in dogs with chronic coronary stenosis. Circulation 75:661−669
15. Guth BD, Schulz R, Thaulow E, Ross J Jr (1988) Right ventricular "steal" during hypoperfusion of the left anterior descending coronary artery in swine. J Am Coll Cardiol 11:23A (Abstract)
16. Hands ME, Lloyd BL, Robinson JS, DeKlerk N, Thompson P (1986) Prognostic significance of electrocardiographic site of infarction after correction for enzymatic size of infarction. Circulation 73:885−891
17. Hess DS, Bache RJ (1976) Transmural distribution of myocardial blood flow during systole in the awake dog. Circ Res 38:5−15
18. Heusch G, Deussen A (1983) The effects of cardiac sympathetic nerve stimulation on the perfusion of stenotic coronary arteries in the dog. Circ Res 48:416−423
19. Heusch G, Deussen A (1984) Nifedipine prevents sympathetic vasoconstriction distal to severe coronary stenoses. J Cardiovasc Pharmacol 6:378−383
20. Heusch G, Guth BD, Seitelberger R, Ross J Jr (1987) Attenuation of exercise-induced myocardial ischemia in dogs with recruitment of coronary vasodilator reserve by nifedipine. Circulation 75:482−490

21. Heusch G, Yoshimoto N (1983) Effects of heart rate and perfusion pressure on segmental coronary resistances and collateral perfusion. Pfluegers Arch 397:284–289
22. Heusch G, Yoshimoto N, Müller-Ruchholtz ER (1982) Effects of heart rate on hemodynamic severity of coronary artery stenosis in the dog. Basic Res Cardiol 77:562–573
23. Kjekshus J (1987) Heart rate reduction – a mechanism of benefit? Eur Heart J 8 (Suppl L):115–122
24. Kjekshus JK (1986) Importance of heart rate in determining β-blocker efficacy in acute and long-term myocardial infarction intervention trials. Am J Cardiol 57:43F–49F
25. Kumada T, Gallagher KP, Shirato K, McKnown D, Miller M, Kemper WS, White F, Ross J Jr (1980) Reduction of exercise-induced regional myocardial dysfunction by propranolol. Studies in a canine model of chronic coronary artery stenosis. Circ Res 46:190–200
26. Matsuzaki M, Gallagher KP, Patritti J, Tajimi T, Kemper WS, White FC, Ross J Jr (1984) Effects of a calcium-entry blocker (diltiazem) on regional myocardial flow and function during exercise in conscious dogs. Circulation 69:801–814
27. Matsuzaki M, Patritti J, Tajimi T, Miller M, Kemper WS, Ross J Jr (1984) Effects of β-blockade on regional myocardial flow and function during exercise. Am J Physiol 247:H52–H60
28. Mosher P, Ross J Jr, McFate PA, Shaw RF (1964) Control of coronary blood flow by an autoregulatory mechanism. Circ Res 14:250–259
29. Murakami H, Kim SJ, Downey HF (1989) Persistent right coronary flow reserve at low perfusion pressure. Am J Physiol 256:H1176–H1184
30. Neill WA, Oxendine JM, Phelps NC, Anderson RP (1975) Subendocardial ischemia provoked by tachycardia in conscious dogs with coronary stenosis. Am J Cardiol 35:30–36
31. Olsson RA, Bünger R (1987) Metabolic control of coronary blood flow. Prog Cardiovasc Dis 29:369–387
32. Pantely GA, Bristow JD, Swenson LJ, Ladley HD, Johnson WB, Anselone CG (1985) Incomplete coronary vasodilation during myocardial ischemia in swine. Am J Physiol 249:H638–H647
33. Ross J Jr, Gallagher KP, Matsuzaki M, Lee J-D, Guth BD, Goldfarb R (1986) Regional myocardial blood flow and function in experimental myocardial ischemia. Can J Cardiol suppl. A:9A–18
34. Sabbah HN, Stein PD (1982) Effect of acute regional ischemia on pressure in the subepicardium and subendocardium. Am J Physiol 242:H240–H244
35. Santamore WP, Bove AA, Carey RA (1982) Tachycardia induced reduction in coronary blood flow distal to a stenosis. Int J Cardiol 2:23–37
36. Seitelberger R, Guth BD, Heusch G, Lee JD, Katayama K, Ross J Jr (1988) Intracoronary α₂-adrenergic receptor blockade attenuates ischemia in conscious dogs during exercise. Circ Res 62:436–442
37. Stowe DF, Mathey DG, Moores WY, Glantz SA, Townsend RM, Kabra P, Chatterjee K, Parmley WW, Tyberg JV (1978) Segment stroke work and metabolism depend on coronary blood flow in the pig. Am J Physiol 234:H597–H607
38. Tomoike H, Franklin D, Kemper WS, McKown D, Ross J Jr (1981) Functional evaluation of coronary collateral development in conscious dogs. Am J Physiol 241:H519–H524
39. Wilcken DEL, Paoloni HJ, Eikens E (1971) Evidence for intravenous dipyridamole (Persantin) producing a "coronary steal" effect in the ischemic myocardium. Aust N Z J Med 1:8–14
40. Williams PT, Haskell WL, Vranizan KM (1985) Associations of resting heart rate with concentrations of lipoprotein subfractions in sedentary men. Circulation 71:441–449

Author's address:
Brian D. Guth, Ph.D.
Seaweed Canyon Laboratory, A-011
University of California, San Diego
La Jolla, CA 92093

Treatment of Myocardial Ischemia with Beta-Blockers

Åke Hjalmarson, Johan Herlitz, and Finn Waagstein

Division of Cardiology, Department of Medicine I, University of Göteborg, Sahlgren's Hospital, Göteborg, Sweden

Introduction

Beta-blocking agents have been in clinical use for 25 years and their therapeutic effects are well established in the treatment of angina pectoris, hypertension, arrhythmias, and in other manifestations. This group of drugs was the first to show that prophylactic institution after myocardial infarction could reduce mortality and morbidity. It is well established that beta-blockers can prevent and limit pain in patients with various forms of angina pectoris which have been extensively reviewed. This chapter will therefore review the literature on the use of beta-blocking agents to prevent severe ischemic events in clinical studies of patients with threatened and definite myocardial infarction during long-term follow-up after myocardial infarction.

Chest Pain and ST-Segment Analysis

In 1975, it was first reported in a double-blind study by Waagstein and Hjalmarson [40] that intravenous injections of three selective beta$_1$-blocking agents resulted in an immediate release of pain among patients with a definite acute myocardial infarction. Two of the three beta-blockers have intrinsic stimulatory activity (practolol and H 87/ 07) and one has no such property (metoprolol). Metoprolol had the most marked effect on pain relief and it was postulated that this was due to its better reduction of heart work, e.g., heart rate and systolic blood pressure. The observation of pain relief after intravenous injection in patients with acute myocardial infarction was later confirmed in a large number of studies on propranolol, atenolol, metoprolol, and timolol, as reviewed recently [18]. In the Göteborg Metoprolol Trial [10, 19] pain score was analyzed in patients with suspected acute myocardial infarction before and 15 min after intravenous injection of metoprolol. Metoprolol significantly reduced ongoing pain and prevented later recurrent pain and use of analgesics. In the large MIAMI multicenter Trial [26, 27], including 5778 patients, early intravenous administration of metoprolol in patients with suspected acute myocardial infarction reduced the duration of early pain as well as the need of analgesics. The effect was more marked in patients with elevated heart work (rate-pressure product).

Precordial mapping of ST-segments has been utilized to study effects of beta-blockers on severity of ischemia. Several small studies from the early 1970s have documented a significant reduction of the elevated ST-segments in patients with acute myocardial infarction immediately after intravenous administration of different beta-blockers, i.e. propranolol, practolol, and metoprolol [33, 35, 40]. In

261

these studies, severity of chest pain was also reduced supporting the idea that severity of myocardial ischemia was favorably influenced. The antiischemic effect of beta-blockers in patients with acute myocardial infarction has also been demonstrated by reduced lactate production and improved glucose utilization after i.v. administration of propranolol [28]. Two large placebo-controlled trials, the Göteborg Metoprolol Trial [10, 19] and the International Collaborative Study on Early Timolol [21] have demonstrated a significant reduction in ST-segment elevation by i.v. beta-blockers compared to placebo. In both studies, beta-blocker administration also caused a more marked reduction in severity of pain.

The favorable effects of beta-blockers on chest pain and ST-segment elevation might be caused by other effects than through reduction of severity of myocardial ischemia [20, 29]. It has been found that there is a good correlation between the effects of beta-blockers on reduction of ST-segment elevation, on chest pain, and on later Q- and R-wave changes as well as clinical complications [18]. This strongly supports the idea of a cause-specific effect of beta-blockers on initial severity of myocardial ischemia among patients with threatened infarction and later complications.

Infarct Size

There are several clinical indirect methods for estimation of infarct size in man. These include analysis of Q- and R-waves from ECG, serum enzyme curves, radionuclide scintigraphy, echocardiography and positron emission tomography. There are difficulties in the interpretation of analyses in individual patients. However, for larger groups of patients there is a good correlation between the various indices of infarct size estimation and also a good prediction from these estimations of short- and long-term prognoses [15, 16].

It was first reported by Peter et al. [34] that propranolol given within 4 h after onset of pain resulteed in a lower serum enzyme rise (CK activity) compared to a control group. Since the methods applied involve indirect indices of infarct size, well randomized and controlled studies are needed. Due to very large individual variations of final infarct size in a nonselected control or placebo group, relatively large sample sizes are required to provide valid information. Table 1 shows six randomized large

Table 1. Enzymatic indices of infarct size in beta-blocker studies[a]

Drug (ref.)	No. of patients	Start of treatment (h)	Reduction (%)
Atenolol [46]	296	<12	30
Metoprolol [10]	936	≤12	17
Metoprolol [7]	391	≤ 6	14
Timolol [21]	132	≤ 4	30
Propranolol [36]	267	≤ 8	14
Metoprolol [27]	2786	≤ 7	11

[a] Studies are included in which >100 patients have been evaluated and when a beta-blocker was given i.v. within 12 h after onset symptoms.

studies in which a beta-blocker was given i.v. to patients shortly after onset of symptoms suggestive of an acute myocardial infarction [7, 10, 21, 27, 36, 46]. All studies indicate a favorable effect on infarct size estimated from serum enzyme analyses. The favorable findings were made when treatment was started within 12 h of onset of symptoms. When treatment was started later, no such effect was demonstrated. The reduction of the peak enzyme value was on the average about 20%.

In the Göteborg Metoprolol Trial [10] and the MIAMI Trial [27], a more favorable effect of the beta-blocker was seen among patients with a higher initial heart rate. In the MIAMI Trial, a more favorable effect was found in patients with a higher systolic blood pressure. This was not observed in the Göteborg Metoprolol Trial. Infarct location was not found to be of importance for the effects seen in the MIAMI Trial, while in the Göteborg Metoprolol Trial patients with an inferior infarct location responded less favorably. It can be concluded that when reviewing subgroups from various studies on infarct size estimated from serum enzymes, a rather short delay time (<12 h) is the only factor of major importance for a favorable effect of a beta-blocker.

In three well-controlled studies, Q- and R-wave changes have been analyzed for estimation of infarct size. Yusuf et al. [45, 46] reported that i.v. administration of atenolol to patients with acute myocardial infarction preserved R-waves in standard ECG better than among control patients. In the Göteborg Metoprolol Trial [10], the final total sum of R-waves was higher and the sum of Q-waves was lower in patients given metoprolol within 12 h after onset of an acute myocardial infarction compared to placebo. No difference was found when treatment was started later than 12 h. In the Göteborg Metoprolol Trial, precordial mapping was used containing 24 chest leads, and with this technique only patients with anterior wall infarction could be evaluated. In the International Collaborative Study on Early Timolol [21], the QRS vector was analyzed and the changes were much smaller in patients treated with this beta-blocker within 4 h after onset of symptoms compared to placebo. In the study from Göteborg, delay time was found to be important, but apart from this, due to the rather small number of patients, no conclusions can be drawn from subgroup analysis.

A good correlation has been found between Q- and R-wave changes and the serum peak enzyme value [17, 44]. It is known that beta-blockers will influence pressure-volume relations and might therefore influence the Q- and R-wave amplitude by means other than by the extent of ischemic myocardial damage. Other methods for estimation of infarct size, such as radionuclide ejection fraction, technetium-99m-labeled pyrophosphate, and angiocardiography, have only been used in a few studies with a small number of patients [36]. The results so far have been inconclusive.

Infarct Development

It was first reported in a rather small study by Norris et al. [30] that early propranolol administration in patients with threatened infarction at randomization caused less patients to develop a definite myocardial infarction in the propranolol group compared to placebo (55% vs 96%). Similar observations were made by Yusuf et al. [45] who reported in a study of 170 patients that 49% of those treated with atenolol developed a definite infarction compared to 66% in the control group. In a study of 104 patients with intermediate coronary syndrome, atenolol administration, on the other hand, did not protect patients from developing transmural myocardial infarction [38].

Two studies on metoprolol, the Göteborg Trial and the MIAMI Trial [10, 27], have shown a prevention of infarct development by 5% to 10% if treatment was started within 7 to 12 h. In the larger MIAMI Trial, including 5778 patients, the most marked effect was seen in patients with no significant pathological ECG changes on admission to hospital, in which the development of a definite infarction was reduced by 22% compared to the placebo group (Fig. 1). However, in a study on metoprolol from Belfast [7], and the PREMIS Study evaluating propranolol [31], in which treatments were started within 6 and 4 h, respectively, no favorable effects on infarct development were found. The large ISIS-1 Trial [22] did not appear to decrease the number of patients in whom cardiac enzymes rose to about twice the local upper limit of normal. No details were given on the number of patients developing a definite acute myocardial infarction according to WHO or other criteria.

In the two trials on metoprolol [10, 27], subgroup analyses have indicated that besides a short delay time also a high initial rate-pressure product was of importance for a favorable effect of early administration of metoprolol for infarct development. The same observation was made for the effect on indices of infarct size.

Pooled data from 24 trials on beta-blockers compared to placebo in a postinfarction follow-up (18 841 patients totally) have shown that the number of reinfarctions

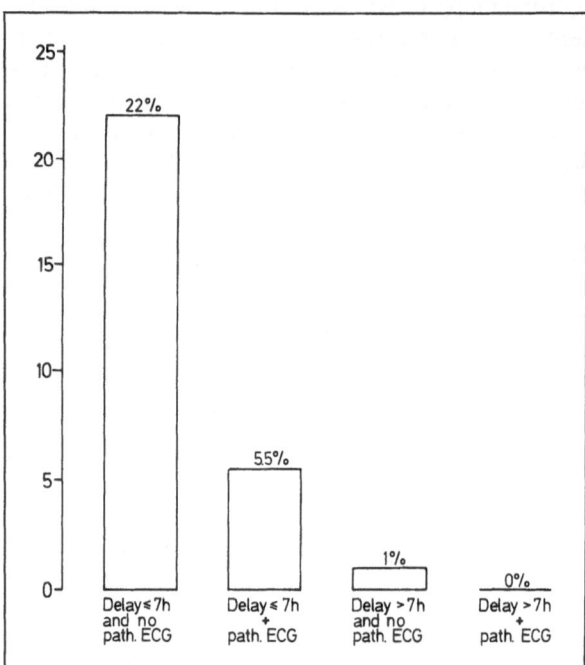

Fig. 1. Prevention of infarct development according to WHO criteria after metoprolol treatment compared to placebo in relation to delay time and ECG signs at entry.

over an average of about 1 year was reduced by 24%, from 7.5% to 5.7% [47]. In the Göteborg Metoprolol Trial there was a significant prevention of late definite infarctions (days 4–90 by 35%; p < 0.05) and to a similar extent already within 2 weeks [10]. During days 4 to 15 of the MIAMI Trial [27], 3.9% of patients in the placebo group developed a reinfarction compared to 3.0% in the metoprolol group, a difference of 30%. In the ISIS Trial [22], corresponding figures within one week were 2.8% vs 2.5%. The fact that beta-blockers can prevent a reinfarction strongly supports the idea that a beta-blocker can prevent infarct development if administered early.

Arrhythmias

Beta-blockers have been reported to reduce supraventricular as well as ventricular arrhythmias in patients with acute myocardial infarction. In several of the large randomized trials, the occurence of various arrhythmias has been reported based upon clinical observations and/or continuous ECG recordings over a longer period. In the Göteborg Metoprolol Trial [10, 19], early institution of the beta-blocker markedly reduced the occurrence of supraventricular and ventricular arrhythmias. Supraventricular and ventricular tachycardia, electrically converted or overall treated, were reduced by more than 50% (p < 0.001). Several studies have shown a reduced incidence of supraventricular tachyarrhythmias, including atrial fibrillation [10, 27, 46].

It was first reported from the Göteborg Trial [10, 19, 37] that metoprolol prevented the incidence of ventricular fibrillation during 15 days of blind treatment (Fig. 2). In this study, lidocaine was used only in patients with ventricular tachycardia causing hemodynamic disturbances or otherwise sustained for 60 s. Lidocaine was therefore used only in 5.5% of placebo-treated patients and 2.3% of metoprolol-treated patients. Similar occurrence was seen of ventricular fibrillation among patients with or without chronic beta-blockade before entry to the trial. Ventricular fibrillation has also been reported to be reduced in a previous small atenolol study [46]. In addition, Norris et al. [31] reported that during a trial period of 48 h in patients with suspected acute myocardial infarction 14 patients in the control group developed ventricular fibrillation compared to two of the propranolol-treated patients (p < 0.006). In the large MIAMI Trial [26, 27] there was no significant difference between patients randomized to placebo or metoprolol. There was, however, a marked trend towards a higher number of episodes of ventricular fibrillation during days 5 to 10 in the placebo group compared to the metoprolol group.

In the ISIS-1 Trial [22], the incidence of ventricular fibrillation has not been reported. The number of cardiac arrests did not differ between control and atenolol-treated patients in this trial (198 vs 189 patients with cardiac arrest). When all studies reporting cardiac arrest were pooled [22], beta-blockade reduced the odds of cardiac arrest by 15% (p < 0.05).

In view of the published literature, beta-blockers can reduce the incidence of supraventricular as well as ventricular tachyarrhythmias caused by acute ischemia. Several studies have demonstrated a favorable effect on ventricular fibrillation or cardiac arrest (nonspecified). The fact that sudden death is most markedly reduced by beta-blockade in postinfarction patients (about 25% in pooled materials) also supports a protective property of beta-blockers against ventricular fibrillation.

265

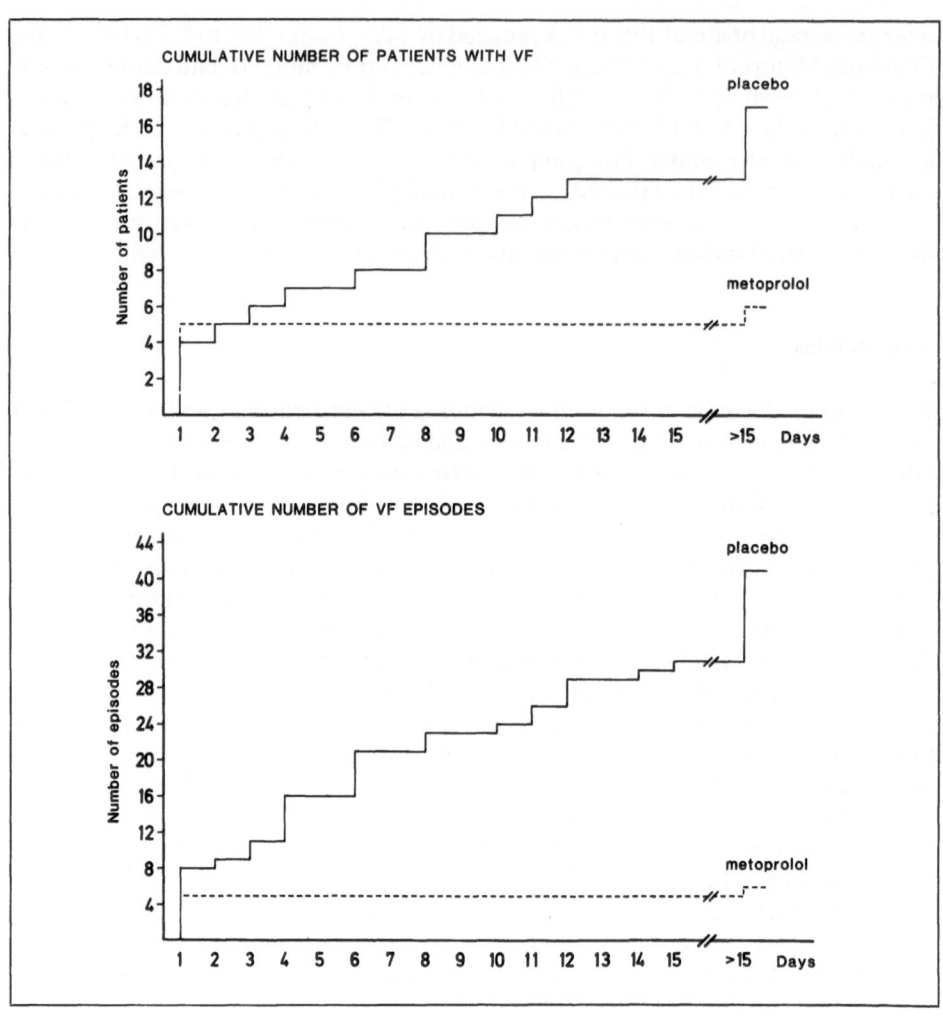

Fig. 2. Cumulative number of patients with ventricular fibrillation in the Göteborg Metoprolol Trial.

Mortality

Beta-blockers were the first to show a significant reduction in mortality among patients with ischemic heart disease. Three large randomized trials were published in 1981, convincingly demonstrating that prophylactic institution of a beta-blocker after acute myocardial infarction reduced long-term mortality. The first postinfarction study to demonstrate this was the Norwegian Timolol Trial [32] showing a 36% reduction in long-term mortality. This was later supported by the Göteborg Metoprolol Trial [10, 27] which also showed a 36% reduction after three months. In this trial, metoprolol was started shortly after a patient's arrival in hospital with suspicion of infarction and was given intravenously. The largest of the three studies published

266

in 1981 was the American BHAT Trial [5]. Propranolol treatment was started about 2 weeks after myocardial infarction with an average follow-up of 35 month during which total mortality was reduced by 26%. There are today 25 randomized trials including more than 23 000 patients which have evaluated the long-term use of various beta-blockers after myocardial infarction [47]. Among patients allocated to placebo treatment, mortality was 9.4% compared to those treated with beta-blocker therapy (7.6%), a 22% reduction in total mortality (p < 0.001). Analysis of patients according to mode of death has demonstrated that sudden death was even more markedly reduced (32%).

The Göteborg Metoprolol Trial in which the beta-blocker was first given intravenously followed by oral treatment, showed reduction in mortality already after three months. This was not found in the Norwegian Timolol and the American BHAT Trials. Most likely, the early start of treatment was important. To further study the potential role of beta-blockers on early infarct mortality, two very large trials were performed, the MIAMI Trial [26, 27] including 5778 patients, and the ISIS Trial with 16 027 patients [22]. These two trials showed that total mortality was reduced by 13% after seven days. The same reduction was found when pooling 27 early intervention trials including about 27 000 patients [48].

Analyses of subgroups in the various studies of beta-blockers have demonstrated that the mortality reduction was more marked among patients with a higher mortality risk, being older and/or sicker at entry. This has been found from the Timolol Trial [32], the BHAT Trial [9, 14], and when pooling all larger postinfarction studies [6]. This observation is not true only for long-term treatment but also during the first two weeks after onset of infarction, as demonstrated in the Göteborg Metoprolol Trial [10, 19] and the MIAMI Trial [26, 27]. The Göteborg Trial demonstrated that mortality reduction was especially marked in patients with elevated heart rate at entry. Among patients with a heart rate above 70 beats/min on arrival to hospital placebo mortality was 12.9% at three months compared to 6.1% in the metoprolol group (p < 0.003). Among patients with a heart rate less than 70 beats/min on arrival to hospital there was no difference between the two treatment groups in three months mortality (placebo 4.9% and metoprolol 5.4%). The very marked difference in mortality in favor of metoprolol was maintained in the patients with elevated entry heart rate over a 2-year follow-up period.

Sudden death was more markedly reduced in the postinfarction studies than nonsudden deaths. Sudden deaths are mainly thought to be due to fatal arrhythmias, ventricular fibrillation being most important. Animal experimental as well as clinical studies have demonstrated that beta-blockers can prevent ventricular fibrillation [19, 37, 38]. In the Göteborg Metoprolol Trial [10, 19], the causes of death were carefully analyzed based upon clinical records and autopsy (performed in all deaths which occurred outside hospitals and in 75% of all deaths). In the metoprolol-treated group there were fewer deaths due to progressive heart failure, arrhythmias, rupture of ventricular wall and fatal late infarction. In the ISIS-1 Trial, atenolol treatment in the early phase of acute myocardial infarction reduced cardiac death classified as cardiac rupture confirmed by necropsy, electromechanical dissociation and primary ventricular fibrillation [23], findings which are in general agreement with those of the Göteborg Metoprolol Trial.

Table 2. Favorable clinical antiischemic effects of beta-blockers on:

- Ischemic chest pain
- ST-segment deviation
- Lactate release
- Infarct development
- Ischemic arrhythmias (including VF)
- Ischemic dysfunction
- Mortality (primary and secondary prevention)

Summary of Clinical Anti-ischemic Effects of Beta-blockers

A number of favorable clinical anti-ischemic effects have been demonstrated in well-controlled clinical studies. As has been reviewed extensively, beta-blockers are well known to prevent and reduce chest pain in patients with stable and unstable angina pectoris, and in threatened and definite acute myocardial infarction. That this is due to antiischemic effects is supported by the fact that beta-blockers have no analgesic property and that other acute antiischemic effects on ST-segments and lactate release have been demonstrated in addition to pain relief in several studies. The acute and long-term effects on reversible myocardial ischemia are not unique for beta-blockers, but have also been demonstrated for, for example, calcium channel blockers and nitrates (see Chapters 14 and 15). What is, however, unique for the beta-blockers among the antiischemic drugs is prevention of severe ischemic events such as infarct development, fatal arrhythmias, progression of heart failure, and death. The favorable clinical effects of beta-blockers are summarized in Table 2.

Mechanisms Behind the Clinical Anti-ischemic Effects of Beta-blockers

The pathophysiology of ischemic heart disease has been discussed in detail in previous chapters. It is today well recognized that ischemic heart disease is mainly due to thromboatherosclerotic narrowing of the coronary arteries, but with dynamic changes over periods of time. There are spontaneous and continuous changes of myocardial metabolic demand (mainly heart rate and pressure development), in coronary vasomotor tone, thrombocyte aggregation, and thrombus formation/lysis. Drugs with antiischemic effects may well influence a number of factors of importance for reversible as well as irreversible myocardial ischemia. The beta-blockers seem to influence several important factors (Table 3).

The most obvious effects of beta-blockers explaining their antiischemic effects are reductions in heart rate, systolic blood pressure, and rate-pressure product. This has been reported in a number of studies of patients with stable and unstable angina pectoris as well as in threatened myocardial infarction [10, 27]. This is in agreement with the observations that beta-blockers are more effective to prevent ischemic events among patients with elevated heart rate and/or systolic blood pressure (see above). In addition, beta-blockers might prevent the influence of locally released noradrenaline in the ischemic myocardium and thereby prevent further ischemic deterioration

Table 3. Possible mechanisms behind clinical antiischemic effects of beta-blockers:

- Reduced global heart work and demand (heart rate; pressure development)
- Reduced regional metabolic demand in ischemic zones (blockade of influence from locally released noradrenaline)
- Improved coronary flow to ischemic zones (flow redistribution from epicardium to endocardium; reduced flow resistance of eccentric stenosis; increased diastolic perfusion time)
- Antithrombotic effects (thrombocyte aggregation reduced by elevated prostacyclin production; reduced plasminogen activator inhibitor)
- Antifibrillatory effects (nonischemic and ischemic)
- Antihypertensive effects (pressure development; regression of hypertrophy)
- Antiatherosclerotic effects (endothelial cell injury prevented via antithrombotic effects)

[1, 41]. A local effect of beta-blockers in the ischemic myocardial region has been investigated by Åblad et al. [1]. They reported that metoprolol being lipophilic was more efficiently distributed to the ischemic myocardium than atenolol in different species of animals. Metoprolol partially restored the electrical and mechanical dysfunction in the ischemic myocardium which was not the case for atenolol under conditions with comparable external work load. The metoprolol effect was thought to be due to improved myocardial cell metabolism in the ischemic region as a result of inhibition of beta-mediated effects of locally released noradrenaline.

There is good evidence that beta-blockade can cause favorable redistribution of blood from nonischemic myocardium to ischemic regions [8, 11, 39]. In stress situations there is an increase in heart work and metabolic demand in all parts of the heart. Beta-blockade will reduce overall heart work and thereby cause a reduction of coronary flow to match demand in the nonischemic myocardium. This reduction in flow will not take place in the ischemic areas and therefore there will be a favorable redistribution from normal to more ischemic zones. In addition, the reduction in heart rate by beta-blockers via the sinus node will cause prolongation of the diastolic perfusion period. In situations where beta-blockers are given intravenously, e.g., in patients with threatened acute myocardial infarction or unstable angina, there is an immediate reduction in heart rate and systolic blood pressure but not in diastolic blood pressure. Therefore, a good diastolic perfusion pressure is maintained at the same time as there is reduced metabolic demand. Guth et al. have demonstrated that acute administration of atenolol has caused an improvement in exercise-induced ischemic dysfunction in conscious dogs, but that this effect was due to the reduction of heart rate [12]. When the bradycardiac effect of atenolol was prevented there was a further depression of regional blood flow and function by atenolol, suggesting an unmasking of a vasoconstrictor effect. For this reason it was postulated that newer bradycardiac agents without effects on contractility and vasoconstrictor tone might be preferable to the beta-blockers for prevention of ischemic attacks [13]. Their conclusions are, however, limited to i.v. administration of a rather high dose of a beta-blocker and exercise stress in conscious dogs with myocardial ischemia. There might be a very marked increase in sympathetic activity to try to overcome the acute beta-blockade and thereby causing marked alpha-constrictor activity.

It has been postulated that chronic beta-blockade can cause inhibition of thrombocyte aggregation in patients with ischemic heart disease. Thus, it was thought that the morning increase in platelet aggregability may be blunted and thereby the morning peak of onset of acute myocardial infarction be reduced. There are, however, also studies with contradictory findings in which long-term beta-blocker treatment has not influenced platelet aggregation [43]. This study demonstrated that metoprolol depressed the numbers of attacks of silent ischemia without alteration of morning increase of platelet aggregability. The decrease in the frequency of silent ischemic episodes was accompanied by a decrease in blood pressure and heart rate, supporting the fact that the major influence of the beta-blocker was through reduction of myocardial oxygen demand.

During the last few years, a number of studies have been done in animals to demonstrate that stress can cause acceleration of atherosclerotic-like changes in the vessels including the coronary arteries, as discussed in a recent publication [3]. In this paper is discussed the fact that in the Stockholm Metoprolol Trial after Myocardial Infarction, long-term treatment over 3 years prevented not only cardiac events but also cerebrovascular complications and complications in the lower extremities. Furthermore, in the recent MAPHY Trial [42], metoprolol treatment of men with hypertension compared to diuretic treatment decreased the incidence of coronary and cerebrovascular complications. In none of the studies could the findings be primarily ascribed to the antihypertensive effect of the beta-blocker. It has been demonstrated recently that beta-blocker treatment of experimental animals prevented endothelial injury, as well as the deposition of platelets at the arterial intima, and increased the prostacyclin production in animals as well as in man [3, 4]. Other possible mechanisms for the antiatherogenic effect of beta-blockers might be prevention of the growth of intimal fibrose tissue [3], and of the binding of low-density lipoprotein to the arterial wall [24, 25]. In view of the very interesting experimental studies there is

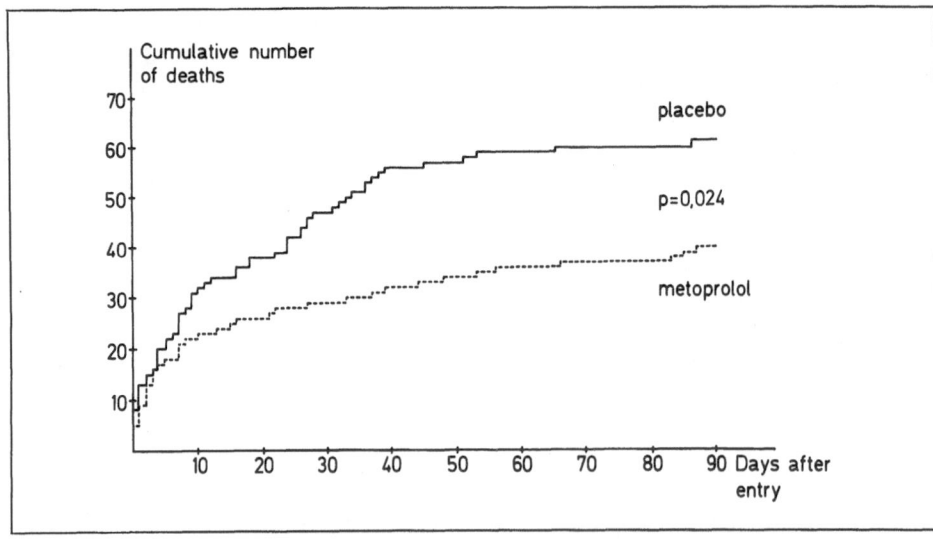

Fig. 3. Cumulative number of deaths among patients randomized to placebo of metoprolol treatment after 90 days in the Göteborg Metoprolol Trial.

an obvious need of direct clinical studies for the antiatherosclerotic possible effects of beta-blockers.

It can be concluded that a number of factors of importance for the development of acute reversible ischemia as well as irreversible ischemic events can be influenced by acute and long-term administration of beta-blockers. The major effect of beta-blockers seems to be the reduction of external heart work, but there are also favorable effects on coronary blood flow distribution, thrombocyte aggregation, thrombus formation and lysis as well as on the progress of thromboatherosclerotic changes.

Type of Beta-blocker and Way of Administration

The present chapter has described the principle effects of beta-blockers on ischemic events in patients with coronary artery disease. There are obvious differences between the beta-blockers regarding degree of beta$_1$-selectivity, lipophilicity and intrinsic stimulatory, membrane-stabilizing and other properties. For most effects described, drugs with intrinsic stimulatory activity seem to be somewhat less effective. This has been clearly described for, e.g. prevention of cardiac deaths after myocardial infarction. The beta$_2$-blocking property may cause side-effects and does not seem to be of importance for favorable effects in patients with ischemic heart disease. The importance of lipophilicity vs hydrophilicity is at present not clear. It seems reasonable to limit results to the actual beta-blocker studied with a proven effect rather than relating this to the whole group of agents.

With the present knowledge of receptor modulation and changes in tissue sensitivity the route of administration is also important. Acute administration to not previously treated persons will cause a marked increase in sympathetic activity in order to overcome the blockade. This might cause specific effects through adrenergic receptors which have not been blocked, e.g., an alpha-receptor-mediated vasoconstrictor response. Long-term chronic beta-blockade will, on the other hand, cause an increase in receptor numbers and tissue sensitivity. This explains the marked withdrawal response which can be seen after long-term treatment.

It has been known for many years that heart rate and systolic blood pressure are reduced immediately after acute administration of a beta-blocker. However, the reduction of diastolic blood pressure and some antiarrhythmic effects seem to develop later. Very recently it has been demonstrated that propranolol and metoprolol can increase biosynthesis of prostacyclin in experimental animals and in man [2, 4]. The increased prostacyclin synthesis induced by beta-blocker treatment has a delayed onset of action and it seems reasonable to assume that there is a relation between prostacyclin synthesis and the antihypertensive effect of beta-blockers. In addition, an increase in prostacyclin synthesis might also contribute to the anti-ischemic and the antiatherosclerotic effects of beta-blockers. Further studies are needed to clarify the cause-effect relationship for different therapeutic effects of the beta-blockers. It is, however, already quite obvious that there are differences between acute and long-term administration of beta-blockers.

Conclusion

Beta-blockers have been in clinical use for more than 25 years. There are a number of favorable therapeutic effects, reported in patients with ischemic heart disease. Beta-blocker administration has been found to reduce ischemic chest pain, ST-segment changes, arrhythmias, to limit infarct development, and to prevent cardiac death. It has recently been suggested that long-term treatment might also cause prevention of thromboatherosclerotic changes in the vascular bed. A major mechanism of action of the beta-blockers is to reduce heart work and metabolic demand. In addition, beta-blockers seem to cause a favorable redistribution of coronary blood flow from normal to more ischemic myocardial zones. Furthermore, there are favorable effects on biosynthesis of prostacyclin, development of endothelial injury, growth of intimal fibrose tissue and binding of low density lipoprotein to the arterial wall. The different actions of beta-blockers might well explain cardioprotective as well as potential vascular protective effects by long-term treatment in man.

References

1. Åblad B, Abrahamsson T, Adler G, Björkman J-A, Bjurö T, Ek L, Ervik M, Sjöquist P-O, Sutherland I, Svensson L (1987) Cardiac antiischemic effect of metoprolol: role of beta-blockade within the ischemic region. J Cardiovasc Pharmacol 10 (Suppl 2):117−125
2. Åblad B, Björkman J-A, Edvardsson N, Hansson G (1987) A beta-adrenergic mechanism in the arterial wall possibly involved in genesis of cardiovascular disease. Fed Proc 46:974 (abstr)
3. Åblad B, Björkman J-A, Gustafsson D, Hansson G, Östlund-Lindqvist A-M, Pettersson K (1988) The role of sympathetic activity in atherogenesis: Effects of beta-blockade. Am Heart J 116:322−327
4. Beckman ML, Gerber JG, Byyny RL, LoVerde M, Nies AS (1988) Propranolol increases prostacyclin synthesis in patients with essential hypertension. Hypertension 12:582−588
5. Beta-blocker Heart Attack Trial Research Group (1982) A randomized trial of propranolol in patients with acute myocardial infarction: I. Mortality results. JAMA 247:1707−1714
6. The Beta-Blocker Pooling Project Research Group (1988) The beta-blocker pooling project (BBPP): subgroup findings from randomized trials ins post infarction patients. Eur Heart J 9:8−16
7. Boyle DMcC, Barber JM, McIlmoyle EL, Salathia KS, Evans AE, Cran G, Elwood JH, Shanks RG (1983) Effect of very early intervention with metoprolol on myocardial infarct size. Br Heart J 49:229−233
8. Buck JD, Hardman HF, Warltier DC, Gross GJ (1981) Changes in ischemic blood flow distribution and dynamic severity of a coronary stenosis induced by beta blockade in the canine heart. Circulation 64:708−715
9. Furberg CD, Byington RP (1983) What do subgroup analyses reveal about differential response to beta-blocker therapy? The beta-blocker heart attack trial experience. Circulation 67 (Suppl I):98−101
10. The Göteborg Metoprolol Trial in Acute Myocardial Infarction (1984) Hjalmarson Å (ed). Am J Cardiol 53:1D−50D
11. Gross GJ, Lamping KG, Warltier DC, Hardman HF (1984) Effects of three bradycardiac drugs on regional myocardial blood flow and function in areas distal to a total or partial coronary occlusion in dogs. Circulation 69:391−399
12. Guth BD, Heusch G, Seitelberger R, Ross J Jr (1987) Mechanism of beneficial effect of beta-adrenergic blockade on exercise-induced myocardial ischemia in conscious dog. Circ Res 60:738−746
13. Guth BD, Heusch G, Seitelberger R, Ross J Jr (1987) Elimination of exercise-induced regional myocardial dysfunction by a bradycardiac agent in dogs with chronic coronary stenosis. Circulation 75:661−669

14. Hawkins CM, Richardsson DW, Vokonas PS (1983) Effect of propranolol in reducing mortality in older myocardial infarction patients. The Beta-blocker heart attack trial experience. Circulation 67 (Suppl I):94–97

15. Herlitz J, Hjalmarson Å, Waldenström J (1984) Relationship between enzymatically estimated infarct size and clinical findings in acute myocardial infarction. Acta Med Scand 215:21–32

16. Herlitz J, Hjalmarson Å, Waldenström J (1984) Relationship between enzymatically estimated infarct size and short- and long-term survival after acute myocardial infarction. Acta Med Scand 216:261–267

17. Herlitz J, Hjalmarson Å, Waldenström J (1984) Relationship between electrocardiographically and enzymatically estimated size in anterior myocardial infarction. J Electrocardiol 17:361–370

18. Herlitz J, Hjalmarson Å, Waagstein F (1989) Treatment of pain in acute myocardial infarction. Br Heart J 61:9–13

19. Hjalmarson Å, Elmfeldt D, Herlitz J, Holmberg S, Málek I, Nyberg G, Rydén L, Swedberg K, Vedin A, Waagstein F, Waldenström A, Waldenström J, Wedel H, Wilhelmsen L, Wilhelmsson C (1981) Effect on mortality of metoprolol in acute myocardial infarction: a double-blind randomized trial. Lancet II:823–827

20. Holland RP, Brooks H (1977) TQ-ST segment mapping: Critical review and analysis of current concepts. Am J Cardiol 40:110–129, 1977

21. The International Collaborative Study Group (1984) Reduction of infarct size with the early use of timolol in acute myocardial infarction. N Engl J Med 310:9–15

22. ISIS-1 (first international Study of Infarct Survival) Collaborative Group (1986) Randomized trial of intravenous atenolol among 16 027 cases of suspected acute myocardial infarction: ISIS-1. Lancet II:57–66

23. ISIS-1 (First International Study of Infarct Survival) Collaborative Group (1988) Mechanisms for the early mortality reduction produced by beta-blockade started early in acute myocardial infarction: ISIS-1. Lancet I:921–923

24. Lindén T, Bondjers G, Camejo G, Bergstrand R, Wiklund O (1989) Affinity of LDL to a human arterial proteoglycan among male survivors of myocardial infarction. Eur J Clin Invest 19:38–44

25. Lindén T, Camejo G, Wiklund O, Warnold I, Bondjers G (1990) Effect of short term beta-blockade on serum lipid levels and on the interaction of LDL with human arterial proteoglycan: a comparison between metoprolol CR and atenolol. J Clin Pharmacol. In press

26. The MIAMI Trial Research Group (1985) Metoprolol in acute myocardial infarction (MIAMI). A randomized placebo-controlled international trial. Eur Heart J 6:199–226

27. MIAMI (1985) Metoprolol in acute myocardial infarction. In: Hjalmarson Å (ed) Am J Cardiol 56:1G–57G

28. Mueller HS, Ayres SM, Religa A, Evans RG (1974) Propranolol in the treatment of acute myocardial infarction. Effect on myocardial oxygenation and hemodynamics. Circulation 49:1078–1087

29. Norris RM, Barratt-Boyes C, Heng MK, Singh BN (1976) Failure of ST segment elevation to predict severity of acute myocardial infarction. Br Heart J 38:85–92

30. Norris RM, Sammel NL, Clarke ED, Smith WM, Williams B (1978) Protective effect of propranolol in threatened myocardial infarction. Lancet II:907–909

31. Norris RM, Brown MA, Clarke ED, Barnaby PF, Geary GG, Logan RL (1984) Prevention of ventricular fibrillation during acute myocardial infarction by intravenous propranolol. Lancet II:883–886

32. Norwegian Multicenter Study Group (1981) Timolol-induced reduction in mortality and reinfarction in patients surviving acute myocardial infarction. N Engl J Med 304:801–807

33. Pelides LJ, Reid DS, Thomas M, Shillingford JP (1972) Inhibition by betablockade of the ST segment elevation after acute myocardial infarction in man. Cardiovasc Res 6:295–301

34. Peter T, Norris RM, Clarke ED, Heng MK, Singh BN, Williams B, Howell DR (1978) Reduction of enzyme levels by propranolol after acute myocardial infarction. Circulation 57:1091–1095

35. Ramsdale DR, Faragher EB, Bennett DH, Bray CL, Ward C, Cruickshank JM, Yusuf S, Sleight P (1982) Ischemic pain relief in patients with acute myocardial infarction by intravenous atenolol. Am Heart J 103:459–467

36. Roberts R, Croft Ch, Gold HK, Hartwell TD, Jaffe AS, Muller JE, Mullin SM, Parker C, Passamani ER, Poole WK, Raabe DS, Rude RE, Stone PH, Turi ZG, Sobel BE, Willerson JT, Braunwald E (1984) Effect of propranolol on myocardial infarct size in a randomized blinded multicenter trial. N Engl J Med 311:218–225
37. Rydén L, Ariniego R, Arnman K, Herlitz J, Hjalmarson Å, Holmberg S, Reyes C, Smedgård P, Swedberg K, Vedin A, Waagstein F, Waldenström A, Wilhelmsson C, Wedel H, Yamamoto M (1983) A double-blind trial of metoprolol in acute myocardial infarction. Effects on ventricular tachyarrhythmias. N Engl J Med 308:614–618
38. Telford AM, Wilson C (1981) Trial of heparin versus atenolol in prevention of myocardial infarction in intermediate coronary syndrome. Lancet II:1225–1228
39. Tomoike H, Ross J Jr, Franklin D, Crozatier B, McKown D, Kemper WS (1978) Improvement by propranolol of regional myocardial dysfunction and abnormal coronary flow pattern in conscious dogs with coronary narrowing. Am J Cardiol 41:689–696
40. Waagstein F, Hjalmarson Å (1976) Double-blind study of cardioselective beta-blockade on chest pain in acute myocardial infarction. Acta Med Scand (suppl) 587:201–208
41. Waldenström AP, Hjalmarson ÅC, Thornell L (1978) A possible role of noradrenaline in the development of myocardial infarction. Am Heart J 95:43–52
42. Wikstrand J, Warnold I, Olsson G, Tuomilehto J, Elmfeldt D, Berglund G (1988) Primary prevention with metoprolol in patients with hypertension. Mortality results from the MAPHY study. J Am Med Assoc 259:1976–1982
43. Willich SN, Pohjola-Sintonen S, Bhatia SJS, Shook TL, Tofler GH, Muller JE, Curtis DG, Williams GH, Stone PH (1989) Suppression of silent ischemia by metoprolol without alteration of morning increase of platelet aggregability in patients with stable coronary artery disease. Circulation 79:557–565
44. Yusuf S, Lopez A, Maw P, Ray N, McMillan S, Sleight P (1979) Value of electrocardiogram in predicting and estimating infarct size in man. Br Heart J 42:286–293
45. Yusuf S, Peto R, Bennet D, Ramsdale D, Furse L, Bray C, Sleight P (1980) Early intravenous atenolol treatment in suspected acute myocardial infarction. Lancet II:273–276
46. Yusuf S, Sleight P, Rossi P, Ramsdale D, Peto R, Furze L, Sterry H, Pearson M, Motwani R, Parish S, Gray R, Bennett D, Bray C (1983) Reduction in infarct size, arrhythmias and chest pain by early intravenous beta blockade in suspected acute myocardial infarction. Circulation 67 suppl I:I32–I41
47. Yusuf S, Peto R, Lewis J, Collins R, Sleight P (1985) Beta blockade during and after myocardial infarction: An overview of the randomized trials. Progr Cardiovasc Dis 27:335–371
48. Yusuf S, Wittes J, Friedman L (1988) Overview of results of randomized clinical trials in heart disease. I. Treatments following myocardial infarction. JAMA 260:2088–2093

Authors' address:
Åke Hjalmarson, M.D.
Division of Cardiology,
Department of Medicine I,
University of Göteborg
Sahlgren's Hospital
Göteborg, Sweden

Treatment of Myocardial Ischemia with Nitrates

Werner Rudolph, Günter Reiniger, Josef Dirschinger, Felicitas Kraus

Klinik für Herz- und Kreislauferkrankungen, Deutsches Herzzentrum München, FRG

Introduction

Nitrates, polyol esters of nitric acid, have served as a cornerstone of anti-ischemic treatment of coronary artery disease for more than 100 years and, today, they continue to be used on a wide-spread basis.

The development of nitrate treatment began with the synthesis of amyl nitrite in 1844 by Balard [4]. Guthrie, in 1859, described the effects of inhaled amyl nitrite on the human circulation: "Flushing of the face, throbbing of the carotids and acceleration of the pulse rate" [34]. In 1867, Brunton first used amyl nitrite for the treatment of angina pectoris [12]. Nitroglycerin, first synthesized by Sobrero in 1846, was subjected to closer investigation in the following years by C. Hering, but about 30 more years elapsed before Murrell first systematically employed this agent for the treatment of angina pectoris [36, 52, 71]. Because of similarities between the chemical structures of nitroglycerin and amyl nitrite, Murrell postulated a similar mechanism of action. In his publication "Nitroglycerin as a remedy for angina pectoris", in 1879, he described that nitroglycerin, as compared with amyl nitrite, led to a similar effect on the pulse, a slower onset of action, a longer duration of action as well as a slower dissipation of the effect. Based on these findings, in addition to their use for treatment of an acute attack, since they enabled successful short-term prophylaxis of angina pectoris, Murrell initiated continuous treatment with nitroglycerin administered several times daily. In 1847, Sobrero also synthesized mannitol hexanitrate which, as was the case with sodium nitrite, was used primarily for the treatment of hypertension for decades [71]. The patent for pentaerythritol tetranitrate, another nitrate, dates back to 1895. It is still used today as an anti-ischemic agent and was introduced for the treatment of coronary artery disease by Bjerlöv in 1943 [9]. Erythritol tetranitrate was tested as an anti-anginal substance as early as 1933: This drug is reported to be the first long-acting nitrate used in humans for therapeutic purposes [21].

The first step toward modern, oral nitrate treatment for long-term prophylaxis was taken in 1940 by Krantz, who tested a number of sugar alcohols for use in diabetics and thereby demonstrated their anhydride forms [45]. In this way, isosorbide dinitrate was synthesized and this agent was found to have a relatively long duration of action. In 1948, Goldberg in a double-blind study confirmed this by documenting a reduction in blood pressure for a period of 4 h [29]. From 1960, isosorbide dinitrate was used in sublingual form for long-term prophylaxis of anginal attacks and, since 1975, after an effect was shown on oral administration, this agent assumed the role of the standard nitrate for treatment of coronary artery disease. The most recent development in nitrate treatment was the synthesis of isosorbide 5-mononitrate, the active primary metabolite of isosorbide dinitrate, which was shown to have more predictable pharmacokinetics than the parent substance. Basically, at the present time, three nitrate derivatives are used in sustained-release and nonsustained-release forms, the effectiveness of which has been sufficiently documented in controlled studies. Nitroglycerin can be administered intravenously, sublingually, buccally, orally or by the transdermal route. For oral long-term treatment, isosorbide dinitrate and isosorbide 5-mononitrate are used primarily, whereas isosorbide dinitrate, as nitroglycerin, can also be given intravenously, sublingually, buccally and transdermally and isosorbide 5-mononitrate can be administered intravenously as well.

Mechanism of Action

The anti-anginal and anti-ischemic effectiveness of nitrates can be attributed to the vasodilating properties. According to the currently-held opinion, the vasodilation is considered to result from NO-mediated stimulation of guanylate cyclase with increased conversion of GTP to c-GMP and, subsequently, relaxation of smooth musculature [23, 40, 46]. There is a differing extent of relaxation in the various vascular beds of the circulation. In the foreground is the vasodilatation in the venous capacitance vessels; only with higher doses, in particular on intravenous administration of the nitrates, can an effect on the peripheral arterial vessels be assumed [13, 55]. In contrast, there is a marked effect on the coronary arteries. Here, the nitrates lead to widening of epicardial vessels and, to the extent that dilatable segments are present, stenotic regions as well, and to dilatation of collateral vessels [11, 49]. The influence on arterioles is regarded as only slight and evanescent [32]. A direct effect on the contractility of the heart by nitrates is not assumed [77]. The increase in heart rate after nitrate administration, in particular in the standing position and at the beginning of treatment, is considered to be reflex-induced in response to the fall in blood pressure.

As a result of the peripheral vasodilatation occuring mainly in the venous capacitance vessels, there is a decrease in the ventricular filling pressures as well as the end-diastolic and end-systolic volumes. Consequently, there is a decrease in the myocardial wall tension and, concomitantly, a reduction in myocardial oxygen requirements [50, 77].

The effects of nitrates on myocardial blood flow result from the changes in left ventricular pressures and volumes, and from the effects on the coronary vessels themselves. In experimental studies it was shown that nitrates lead to a redistribution of blood flow to ischemic regions, in particular subendocardial regions, probably mediated by an increase in collateral flow [3].

Studies in patients in whom total or mean myocardial blood flow was measured have not yielded entirely consistent results but the majority, however, have shown a decrease in blood flow after nitrate administration [5]. Methods enabling assessment of regional perfusion conditions have shown, in addition to a decrease in blood flow in normally-perfused segments indicative of diminished oxygen requirement, an increase in myocardial blood flow in hypoperfused post-stenotic regions [19, 64]. The mechanisms which permit enhancement of the oxygen supply in post-stenotic regions are a decrease in the extravascular component of the coronary artery resistance due to lessening of the wall tension, and a direct dilatation of collateral vessels and, when possible, coronary stenoses as well. Both the elevation of the oxygen supply in the post-stenotic regions due to direct effects on the coronary arteries as well as the reduction in myocardial oxygen requirements due to peripheral vasodilation are capable of exerting a favorable influence on the deranged myocardial oxygen balance.

Nitrate Tolerance

Numerous clinical studies – intercurrently supported by experimental data – have documented that tolerance development during long-term treatment with nitrates is a clinically relevant problem. Even early in the history of their use, there was evidence of tolerance development to nitrates. Stewart, in 1888, described cessation of nitrate-induced headache and attenuation of the blood pressure lowering action in spite of administering increasingly-large doses of nitroglycerin [72]. As early as 1933, Evans and Hoyle questioned the value of nitrates for long-term treatment of angina pectoris [21]. Systematically conducted long-term clinical studies concerned with the influence of nitrates on myocardial ischemia and ventricular function have been available since 1969. Some investigators reported maintained effectiveness while others observed an absence of effect, attenuation or loss of effects not only with respect to the rate of anginal attacks but also with regard to actions on ST-segment depression, exercise capacity and hemodynamic parameters such as pulmonary artery pressure, cardiac output, systemic arterial pressure, and heart rate [66].

Because of varying results, most probably due to discrepant methodological approaches, a long-term effect could not be assumed with certainty and, since in our own studies we had documented a marked anti-ischemic effect after single oral nitrate doses (in contrast to the widely held opinion that extensive first-pass metabolism would preclude a meaningful acute action) in 1977, we began to conduct exactly controlled studies dealing with the question of tolerance; the studies were carried out according to randomized, double-blind, crossover and placebo-controlled protocols [66].

These studies revealed that in patients with angiographically documented coronary artery disease during 8-week treatment phases, each with 20 mg, 40 mg, and 40 mg t.i.d. isosorbide dinitrate in sustained-release form, respectively, as compared to placebo, there was neither a decrease in the rate of anginal attacks, nor in the nitrate consumption, nor in exercise-induced ST-segment depression [63]. The absence of an anti-ischemic effect was also found during long-term treatment with 40 mg isosorbide dinitrate in nonsustained-release form four times daily, while after the initial 40 mg dose there was a marked and significant reduction in exercise-induced ST-segment depression (Fig. 1) [65].

The temporal course of the development of tolerance to 40 mg isosorbide dinitrate in nonsustained-release form four times daily was documented during continuous hemodynamic monitoring of pulmonary artery pressure and systemic arterial pressure. While the first 40 mg-dose led to a marked reduction of the elevated diastolic pulmonary artery pressure reflecting, in turn, the elevated left ventricular end-diastolic pressure, during continuous treatment with the same dose given four times

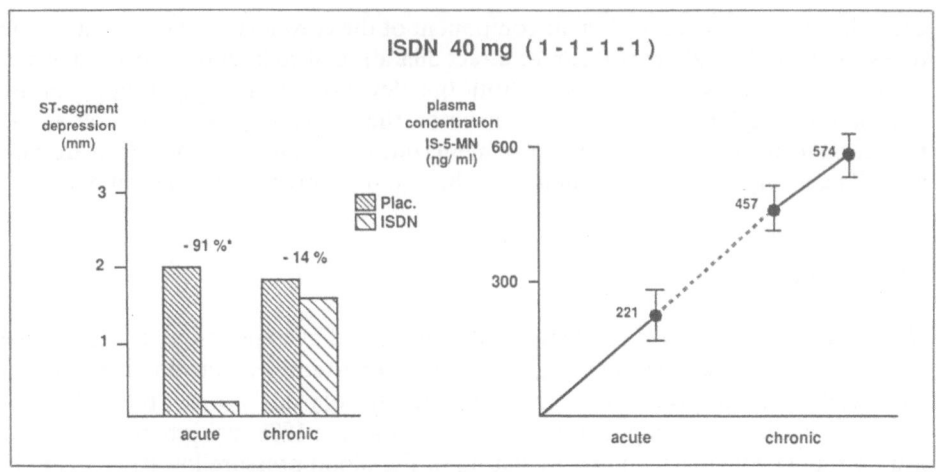

Fig. 1. Exercise-induced ST-segment depression and isosorbide 5-mononitrate plasma concentrations 1 h after administration of 40 mg isosorbide dinitrate (ISDN) at the time of first dose and during long-term treatment with 40 mg ISDN four times daily as compared to placebo (Plac.). After initial administration, as compared to placebo, there is a marked anti-ischemic effect which can no longer be documented after 2 weeks of continuous treatment. During long-term treatment the isosorbide 5-mononitrate plasma concentration at the end of 24 h shows a high residual level. In relation to the prevailing baseline value, the relatively limited increase in nitrate concentration after renewed administration from 457 to 574 ng/ml appears responsible for the loss of action; (* $p < 0.05$).

daily, there was a continuous attenuation of effects and at the end of 7 days, there was a complete loss of action [6].

After the initial application of isosorbide dinitrate ointment, as well, there was a marked anti-ischemic effect for a period of 8 h. However, after 7–10 days of treatment with once-daily application an effect could no longer be detected [58].

In the state of established tolerance to nitrates there are high plasma concentrations such that the increase in the concentration resulting from renewed administration of the dose is relatively limited with respect to the high baseline value. As compared with the initial administration, the absolute increase in concentration is similar but in proportion to the high prevailing baseline value, the relatively limited increase appears responsible for the absent or only diminished response. Most probably because tolerance can be much more rapidly appreciated during use of nitroglycerin patch systems than with isosorbide dinitrate, the phenomenon of tolerance was readily observed by a number of other investigators and is now regarded as a clinically relevant problem. As early as within 8–14 h after application of a patch delivering 10 mg, 15 mg, or 30 mg per 24 h, a markedly attenuated effect and a 24 h loss of action can be detected, even though there are nearly constant plasma concentrations. On application of a second patch at the end of 24 h, with the 15 mg and 30 mg doses no statistically significant effects can be observed and after the second application of a low-dose 10 mg patch there is a markedly attenuated action (Fig. 2) [59, 60, 67].

Based on the studies referred to, it can be concluded that the regularly repeated administration of oral nitrates in high doses in sustained-release or nonsustained-

278

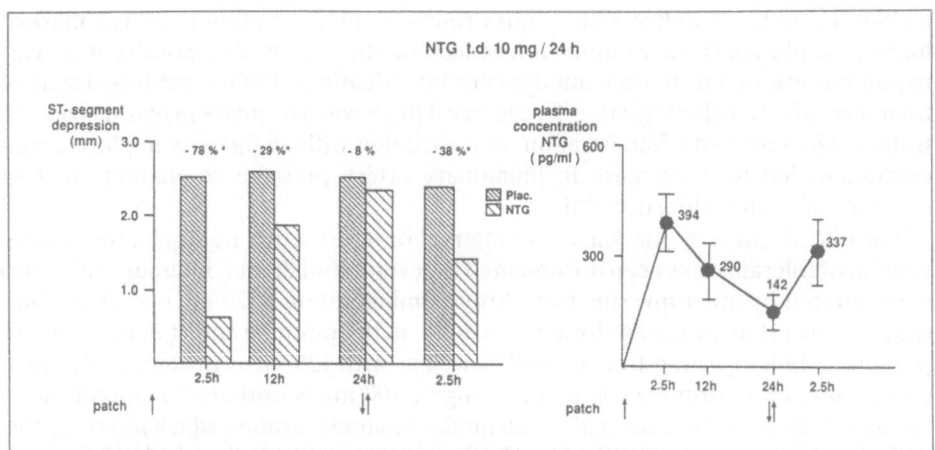

Fig. 2. Exercise-induced ST-segment depression and nitroglycerin (NTG) plasma concentrations during continuous treatment with nitroglycerin patches with a delivery rate of 10 mg/24 h as compared to placebo. At 2.5 h after initial application there is a marked anti-ischemic effect which is clearly attenuated within 12 h and no longer detectable with certainty at 24 h in association with plasma concentrations within the therapeutic range. After renewed patch application, in spite of similar plasma concentrations, as compared with the initial application, only a clearly attenuated anti-ischemic effect can be seen; (* p < 0,05).

release form several times daily results in nitrate accumulation in the plasma and that transdermal application of nitrates similarly leads to nearly constant high plasma concentrations which invariably lead to tolerance development with respect to anti-anginal, anti-ischemic, and hemodynamic actions [67]. The mechanism at the cellular level is currently considered to be the depletion of the intracellular SH-pools. Since the release of NO from the organic nitrates is dependent on the availability of thiol-groups, however, diminished thiol availability and, consequently, insufficient or absent stimulation of guanylate cyclase may result which otherwise is responsible for catalyzation of the formation of cyclic GMP from GPT where the latter, in turn, causes relaxation of vascular smooth musculature [23, 46].

Circumvention of Nitrate Tolerance

In consideration of the marked anti-anginal and anti-ischemic effects after acute administration and the overall good tolerability of nitrates, the search for possibilities to avoid tolerance development seemed particularly meaningful.

Authors, who at the beginning of the 1980s reported a loss of nitrate effect on heart rate and blood pressure while assuming a maintained anti-ischemic effect concluded from their studies that tolerance could be prevented by more frequent administration of higher nitrate dosages [73]. In consideration of our own experience, however, it became apparent that with increasing dosages an effect could only be observed until a new steady-state is achieved. Accordingly, we approached the problem by reducing the number of single doses per day to enable decrementation of the plasma concentration. From low baseline values, after renewed administration there is a marked increase in plasma concentration and, concomitantly, renewed responsiveness. This hypothesis was first tested in hemodynamic investigations. In the established state of tolerance, after a 1-day ingestion-free interval there was a decrease in plasma concentration. The renewed administration, in association with an increase in plasma concentration, led to a decrease in pulmonary artery pressure comparable to that observed after the initial dose [6].

The effectiveness of our concept of interval or intermittent treatment for circumvention of tolerance has been documented in several studies and, intercurrently, also corroborated by other investigators. After administration of 20 mg isosorbide dinitrate in nonsustained-release form twice daily, in the morning and at midday, incorporating a 19 h ingestion-free interval as well as with 120 mg isosorbide dinitrate in sustained-release form once daily or 50 mg or 100 mg isosorbide-5-mononitrate in sustained-release form once daily − all modes of administration which are attractive with respect to patient compliance − both on acute use as well as during long-term treatment, there were comparably marked anti-ischemic and anti-anginal effects with a duration approximately sufficient to provide coverage for the period of maximal physical activities; these regimens are not, however, capable of providing 24 h protection (Fig. 3) [7, 8, 65, 68].

None of these regimens leads to accumulation of the plasma concentration and this is the reason why the incrementation prerequisite to a meaningful response can be achieved after the initial administration as well as during long-term treatment [7, 8, 65, 67, 68]. In general, it can be assumed that attenuation or loss of action can be avoided if the plasma concentrations of isosorbide-5-mononitrate decrement to baseline values between 100 and 200 ng/ml and if after renewed administration they are incremented by more than 2.5-fold (Fig. 4).

In analogy to the experience with orally administered nitrates, the incorporation of a nitrate-free interval with transdermal use was a logical sequel. On application of a nitroglycerin patch system with a delivery rate of 10 mg/24 h and a 12 h patch-free interval, the initial anti-ischemic effect can be reproduced. However, even with this mode of application, over the 12 h active-drug period there is an increasingly attenuated effect associated with nearly constant high plasma concentrations as an expression of the propensity to rapidly developing tolerance (Fig. 5) [60].

This rapid attenuation of action which can be documented to occur within 12 h can be avoided through modification of the delivery rate. On use of four patches delivering 5 mg/24 h each, starting with one, then adding one every 3 h, thereafter removing all patches for a 12 h system-free interval, a marked anti-ischemic effect of nearly constant magnitude can be achieved in association with plasma concentrations continuously increasing to approximately four-fold. After the 12 h patch-free interval, the decrementation of the plasma concentration was sufficient such that on re-application of the same regimen on the next day, the corresponding anti-ischemic effect was reproducible [68].

With the intention of providing a low-nitrate or nitrate-free interval without removing the patch, systems were developed from which two-thirds of the patch content is released during the first 12 h after application. With these patches there are clear fluctuations in the plasma concentration; however, the amplitude does not appear sufficient to prevent completely an attenuation of action within 12 h. When

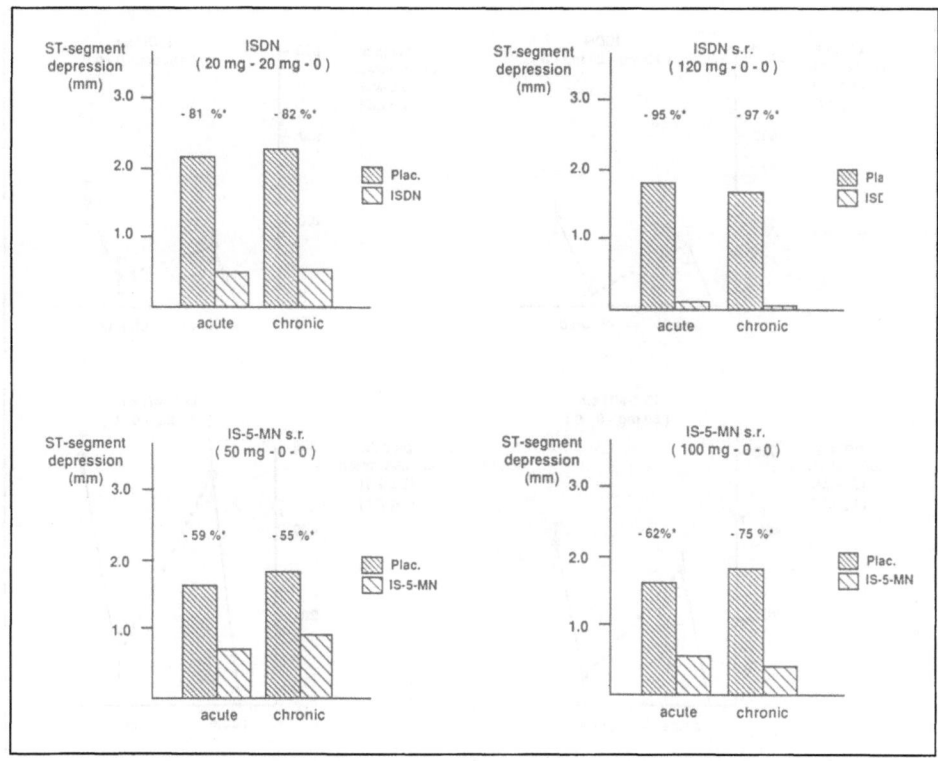

Fig. 3. Exercise-induced ST-segment depression during treatment with 20 mg isosorbide dinitrate (ISDN) twice daily, in the morning and at midday, as well as during once-daily treatment with 120 mg ISDN in sustained-release form, 50 mg isosorbide 5-mononitrate (IS-5-MN) in sustained-release form, and 100 mg IS 5-MN in sustained-release form. Both after initial administration as well as during 3–4 weeks of treatment, there is a comparably marked anti-ischemic effect; (* p < 0.05).

these systems are used with dosages of more than 7.5 mg/24 h, loss of action is incurred [61, 69].

With currently-available commercial patch systems, using a patch with a delivery rate of 10 mg/24 h and incorporating a 12 h patch-free interval, an overall loss of effects can be prevented but not, however, the early attenuation of action. The necessity of removing the patch, nevertheless, defeats to a certain degree the original intention of its proponents, namely that of easy handling.

Studies using SH-donors such as acetyl cystein, methionine, ascorbic acid or ACE-inhibitors with SH groups in attempts to prevent or reverse nitrate tolerance have not led to results which would justify their clinical application [28, 38, 48].

Stable Angina Pectoris

For many years the use of nitrate treatment in stable angina pectoris was limited to amelioration or prophylaxis of acute ischemic pain associated with physical exertion

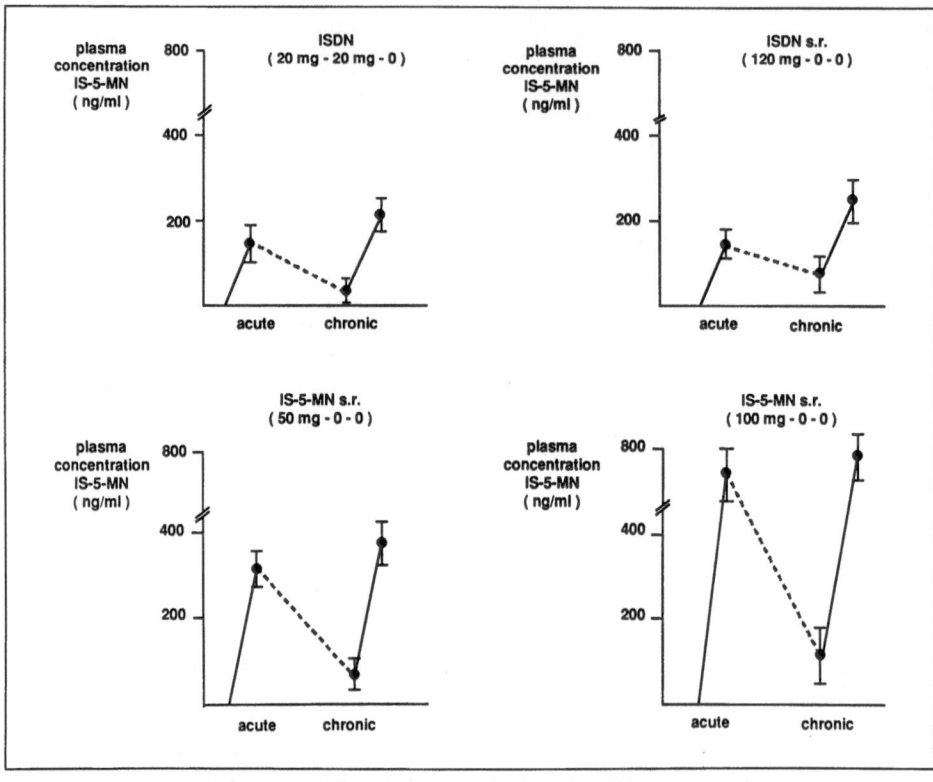

Fig. 4. Isosorbide-5-mononitrate (IS-5-MN) plasma concentrations after initial administration and during long-term treatment with isosorbide dinitrate (ISDN) 20 mg twice daily, in the morning and at midday, and once-daily treatment with 120 mg ISDN in sustained-release form, 50 mg IS 5-MN in sustained-release form, and 100 mg IS 5-MN in sustained-release form. During long-term treatment there is no meaningful accumulation in the IS 5-MN plasma concentration. After renewed nitrate administration there is an increase in the drug level comparable to that after initial administration.

or mental stress. It was only in the 1970s that long-term nitrate treatment was introduced.

For acute treatment or prophylaxis, currently as in the past, nitroglycerin and isosorbide dinitrate are available in rapidly absorbable forms such as capsules, sublingual tablets, and sprays. Isosorbide-5-mononitrate, because of its delayed onset of action, is not useful for acute treatment [64]. Sublingual administration of 0.4 mg nitroglycerin was shown to be effective in 92% of patients treated for anginal attacks. Symptoms were completely relieved in the majority of patients within 3 min [39]. On continuation of the exercise inducing the pain, the onset of action of the 0.4 mg sublingual nitroglycerin dose was observed at 1.6 min and complete relief of symptoms at 3.7 min. On use of 5 mg isosorbide dinitrate sublingually, the onset of action was slightly slower at 2.4 min and the symptoms were completely relieved at 6.2 min [44]. Aerosol forms of nitroglycerin or isosorbide dinitrate for use as a spray have been shown to be equally effective and the onset of action is somewhat more rapid than that of sublingual tablets. After sublingual administration of nitroglycerin or isosor-

282

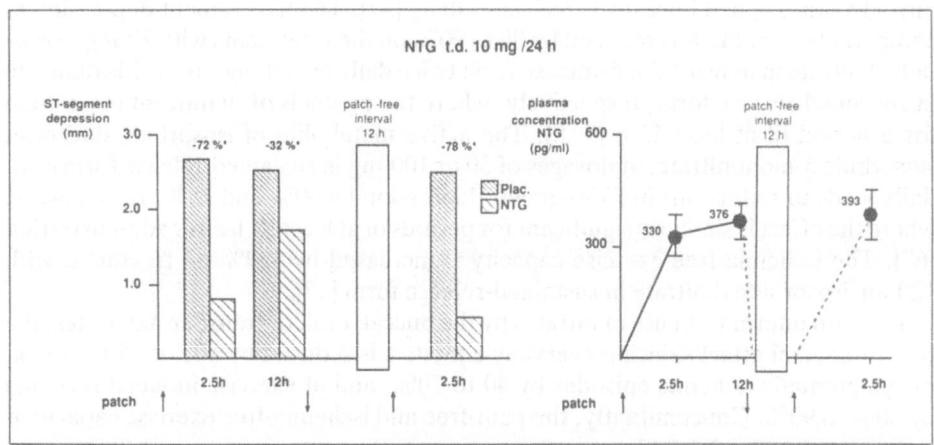

Fig. 5. Exercise-induced ST-segment depression and nitroglycerin (NTG) plasma concentrations during treatment with nitroglycerin patches with a delivery rate of 10 mg/24 h on two consecutive days between which there was a 12 h patch-free interval. At 2.5 h after initial application there was a marked anti-ischemic effect which is clearly attenuated within 12 h in association with nearly constant plasma concentrations. After a 12 h patch-free interval, 2.5 h after renewed application in association with comparable plasma concentrations, there is an undiminished anti-ischemic effect; (* p < 0.05).

bide dinitrate, exercise tolerance is increased by 51% and 55%, respectively, and exercise-induced ST-segment depression reduced by 76% with isosorbide dinitrate. The action of sublingually administered nitroglycerin persists about 30 min and that of sublingually administered isosorbide dinitrate about 60 min [17, 31].

For the acute treatment of an anginal attack, in general, doses of 0.4 to 0.8 mg nitroglycerin or 5 mg isosorbide dinitrate are sufficient. For persistent symptoms the same dose can be repeated after 3 to 5 min. A dosage of more than 2 to 3 capsules or tablets within a period of 15 min should not be exceeded since hypotension and reflex tachycardia can occur with increasing compromise of the myocardial oxygen balance [64].

In addition to acute treatment and prophylaxis it is desirable to provide, as far as possible, freedom from anginal attacks during everyday activities as well as an increase in asymptomatic and ischemia-free exercise capacity through regular drug administration. Although substances with longer durations of action are available, the originally intended, uninterrupted anti-ischemic protection, that is, 24 h a day, cannot be achieved due to the rapid tolerance development.

According to studies in patients with stable angina pectoris which have been carried out with currently prerequisite protocols, the rate of anginal attacks can be reduced by 40%/week with 20 mg isosorbide dinitrate in nonsustained-release form given in the morning and at midday, and by 50%/week with 120 mg isosorbide dinitrate in sustained-release form once daily [7, 8]. Symptomatic ischemic episodes during treatment with 120 mg isosorbide dinitrate in sustained-release form once daily were reduced by 67% as shown by Holter-monitoring and, in our own patients by 44% [35]. On the same treatment, the asymptomatic exercise capacity increased by 153% [47] and the exercise time increase amounted to 34% on treatment with 3 mg nitroglycerin applied buccally three times daily [57]. The ST-segment depression at comparable workloads is reduced by 80 to 90% on the treatments with 20 mg isosorbide dinitrate in nonsustained-release form twice daily or 120 mg isosorbide dinitrate in sustained-release form, respectively, where the action is of significant magnitude for a period of at least 12 h [7, 8]. The active metabolite of isosorbide dinitrate, isosorbide 5-mononitrate, at dosages of 50 or 100 mg in sustained-release form once daily leads to reductions in ST-segment depression by 60% and 75%, respectively, where the effect is similarly significant for periods of at least 12 h after administration [67]. The ischemia-free exercise capacity is increased by 329% on treatment with 120 mg isosorbide dinitrate in sustained-release form [47].

Thus, on intermittent use of nitrates by the buccal, oral, or transdermal routes, the rate of anginal attacks during everyday activities is reduced by 40% to 50%/week, of symptomatic ischemic episodes by 40 to 70%, and of exercise-induced ischemia by 60% to 90%. Concomitantly, the pain-free and ischemia-free exercise capacity is increased by 1.5 to 3.3 fold.

As dosing regimens, we recommend 20 mg isosorbide dinitrate in nonsustained-release form twice daily (in general, administered in the morning and at midday), 120 mg isosorbide dinitrate in sustained-release form once daily, 50 or 100 mg isosorbide-5-mononitrate in sustained-release form once daily or the 12 h application of a nitroglycerin patch releasing 10 mg/24 h with a 12 h patch-free interval [16, 60].

During nightly recumbancy, angina pectoris attacks are relatively rare. Consequently, the nitrate action of the regimens specified, which is still significant 12 h after application at a magnitude of 50% of the maximal effect, in general, is sufficient to provide effective anti-anginal coverage.

To what extent chronic nitrate treatment may prevent structural changes in the myocardium, acute myocardial infarction or sudden cardiac death, or improve the prognosis in patients with stable angina pectoris, is not known. It is conceivable that long-term nitrate treatment can prevent or diminish the extent of structural changes recently reported in association with recurrent myocardial hypoperfusion even though the prognostic and functional relevance of this finding remains unclear [37]. The nitrates could possibly exert a favorable influence on prognosis by virtue of their platelet aggregation inhibiting properties through counteraction of incurrence or growth of wall-adherent thrombi [15]. All of these problems, however, have not been sufficiently investigated.

In practice, such prognostic considerations with regard to nitrate treatment or, with combined treatment as it is generally carried out, are of lesser relevance since patients thought to have a compromised prognosis on the basis of noninvasive examinations, after invasive diagnostic procedures, are subjected to revascularization measures which, for the most part, favorably influence both angina pectoris as well as the prognosis.

Unstable Angina Pectoris

For treatment of unstable angina pectoris, that is, recent-onset, severe exercise-induced angina pectoris, rapidly progressive exercise-induced angina pectoris or angina pectoris at rest, nitrates administered intravenously are considered drugs of first choice. In only a small fraction of patients can unstable angina be attributed to a high-grade fixed stenosis capable of precipitating complaints subsequent to a slight increase in oxygen requirements. Numerous autoptic, angiographic, angioscopic, and bioptic studies have shown that unstable angina pectoris is most frequently due to a thrombotic event, possibly augmented by vasospasm, where the rapidly developing thrombus formation results at the site of an ulcerated or ruptured arteriosclerotic plaque and complete or partial spontaneous lysis is possible [1, 18, 22, 24, 51, 70]. In this regard it is thoroughly justified to make use of the oxygen-consumption reducing, in hypoperfused myocardial regions oxygen-supply enhancing and platelet-aggregation inhibiting properties of nitrates [15]. Other measures directly exerting a stronger influence on the thrombotic event, obviously, should also be considered.

The therapeutic success which can be achieved with nitrates based on reduction of anginal attack frequency or asymptomatic status has been documented in a number of studies. For example, the rate of anginal attacks was reduced by 88% over a 7-day period with a low-dose nitroglycerin infusion of 1.8 mg/h; 63% of the patients were rendered asymptomatic [56]. High-dose treatment with 8.4 mg/h for 6 days led to a 95%-reduction in the rate of anginal attacks and 70% of the patients were asymptomatic [43]. With an isosorbide dinitrate infusion, as well, at a dosage of 3.5 mg/h for 8 days anginal attacks decreased by 88% and 47% of the patients were asymptomatic [20].

Accordingly, it can be assumed that with nitrates the rate of anginal attacks can be reduced by up to 90%, and between 50% to 70% of the patients will be rendered asymptomatic.

The dose should be titrated to reduce mean arterial blood pressure by 10% but not to less than 80 mmHg. If oral administration is employed, the guidelines for use of nitrates in patients with stable angina pectoris should be used.

Based on current knowledge, it must be assumed that the intravenous administration of nitrates in the dosages reported will be associated with rapid tolerance development, that is, the nitrates will lose their effect and continuously increasing doses will be required. The varying dosages specified for the treatment of unstable angina pectoris, as well, permit the assumption that due to prior treatment with orally administered nitrates in different patient collectives preexistent tolerance of differing extent may have been present to account for the nonuniform responsiveness.

With oral nitrate treatment, in our experience, an effect similar to that yielded by intravenous treatment can be achieved. We did not observe any cases in which the unstable angina pectoris was not controlled with oral intermittent nitrate treatment, that is, no patients were seen who responded to intravenous nitrates only, provided that the regimen respected the principles of tolerance avoidance. In spite of the marked effectiveness of nitrate treatment, however, in general, current practice calls for immediate combined treatment with beta-adrenergic receptor blockers, calcium channel blockers, heparin and aspirin as well as a detailed invasive diagnostic investigation for risk assessment.

Asymptomatic (Silent) Ischemia

Nitrates are used for treatment of asymptomatic ischemia with approximately the same success as with their use for the symptomatic form. As can be seen in numerous studies the overwhelming majority of patients with asymptomatic ischemia, with a ratio of about 4 : 1, also have symptomatic ischemia, that is, angina pectoris.

On treatment with 120 mg isosorbide dinitrate in sustained-release form there was a reduction in the incidence of asymptomatic ischemic episodes of 33% and symptomatic ischemic episodes of 67% [35]. In our patients the corresponding values were 49% and 44%, respectively. On treatment with 20 mg isosorbide-5-mononitrate twice daily, the reduction in asymptomatic and symptomatic ischemic episodes was reported at 67% each [2]. The duration of asymptomatic ischemia decreased on treatment with 120 mg isosorbide dinitrate in sustained-release form by 43% and that of symptomatic ischemia by 67% [35]. With isosorbide-5-mononitrate the reduction in the extent of asymptomatic and symptomatic ischemia were both reported at 70% [2].

Assessment of the therapeutic effect with the aid of Holter-monitoring of asymptomatic ischemic episodes is encumbered, as compared with bicycle ergometry and ST-segment registration, because of the high degree of spontaneous variability of transient ischemia during everyday conditions since the incidence and extent of ischemia-precipitating factors cannot be standardized. Accordingly, to confirm a reliable therapeutic effect for a group, as opposed to ergometry, a larger number of patients would be necessary and at least a 48 h monitoring phase without and with treatment [53].

For an individual patient, the suitability of this method for documentation of a therapeutic effect is limited since the spontaneous variability is frequently as great as or greater than the drug effect. Observations of spontaneous 100%-reductions in the ischemic episodes within the usual two 24 h registrations are not uncommon. The duration of registration that would be necessary to enable meaningful assessment in an individual patient would entail time-consuming analyses and incur problems of acceptability for the patient.

The nitrate-dosing regimens for patients with asymptomatic ischemia are the same as those for patients with stable angina pectoris. As Holter-monitoring studies dealing with the circadian distribution of the ischemic episodes occuring during everyday activities have shown, 80% of all ischemic episodes, that is the asymptomatic episodes as well, can be observed during the main period of physical activities between 6 a.m. and 6 p.m. The proportion of nightly episodes is less than 5% (Fig. 6). Thus, for patients with coronary artery disease and asymptomatic ischemia as well, in general an adequate therapy can be provided with a nitrate interval treatment which offers reliable anti-ischemic protection persisting for at least 12 h. The indication for treatment of patients with asymptomatic myocardial ischemia is established on the basis that their prognosis is similarly compromised as in those with symptomatic ischemia. Moreover, the incidence of both sudden cardiac death and myocardial infarction are similar in patients with asymptomatic and in those with symptomatic myocardial ischemia. In this regard the largest body of data can be derived from the CASS study in which ischemia was documented by means of ergometry [75, 76]. The importance of asymptomatic ischemia for the prognosis can

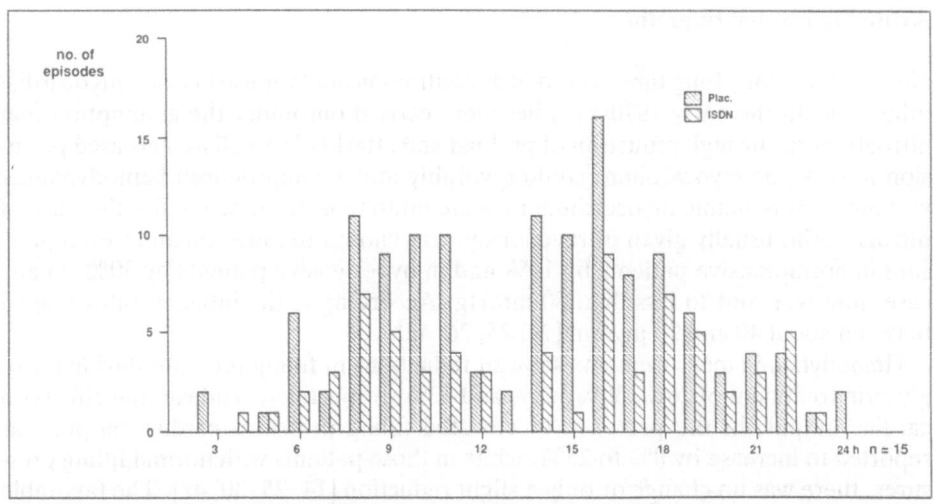

Fig. 6. Circadian distribution of ischemic episodes during everyday activities on placebo or on treatment with 120 mg isosorbide dinitrate (ISDN) in sustained-release form. On placebo there is an increase in the incidence of ischemic episodes in the morning between 6 a.m. and 12 noon, as well as in the afternoon between 2 p.m. and 6 p.m. On treatment with isosorbide dinitrate there is a clear reduction during the approximately 12 h period of action; the double-peaked circadian rhythm, however, is maintained during treatment. Only few episodes were seen during the night between 10 p.m. and 6 a.m.

also be seen in studies showing that in patients with stable angina pectoris and ST-segment depression in the exercise ECG in whom, additionally, asymptomatic and symptomatic ST-segment depression was found on Holter-monitoring during everyday activities, there was a significantly higher rate of death, acute myocardial infarction, and occurrence of unstable angina pectoris, as well as of the necessity for revascularization than in patients without such ST-segment depression [62, 74]. Moreover, in patients with unstable angina pectoris in whom in spite of maximal therapy asymptomatic ischemic episodes were found on Holter-monitoring within a specified observation period, there were more complications than in patients with no ST-segment depression [33]. Patients with prolonged periods of ischemia appear to be at a particularly high risk [54].

However, no studies are available that address directly the question to what extent the prognosis is favorably influenced by anti-ischemic treatment. The indication for treatment is, therefore, established on the hypothesis of a possible improvement in prognosis.

Essentially, however, patients with ischemic periods during everyday activities and, thus, who are at risk should undergo invasive diagnostic procedures to assess the indication for PTCA or surgical intervention.

Acute Myocardial Infarction

Nitrates were for a long time regarded as contraindicated for use in acute myocardial infarction. In the early 1970s, studies were carried out under the assumption that nitroglycerin, through reduction of preload and afterload as well as increased perfusion in ischemic myocardium, could favorably affect compromised hemodynamics and preserve ischemic myocardium in acute infarction. In these studies the dose of nitroglycerin, usually given intravenously, was chosen to lower mean arterial pressure in normotensive patients by 10% and in hypertensive patients by 30%, in any case, however, not to less than 80 mmHg. Accordingly, the infusion rates ranged between about 40 and 90 µg/min [14, 25, 26, 42].

Hemodynamic measurements showed reductions in filling pressure during nitroglycerin treatment of 26%, 30%, 42% and 52%, respectively, and varying effects on cardiac output; in the presence of elevated filling pressures cardiac output was reported to increase by 8% to 25%, while in those patients with normal filling pressures, there was no change or only a slight reduction [14, 25, 30, 42]. The favorable hemodynamic effects achieved with nitroglycerin are more marked with more severe heart failure [27].

The sum of the ST-segment elevation, indicative of the extent of myocardial ischemia, was reduced by 13% to 26% by nitroglycerin; here, as well, the effect was more pronounced in patients with heart failure [10, 25].

Studies dealing with nitrate-induced limitation of infarct size, assessed by creatine kinase determination, showed inconsistent results. In some studies treatment with nitroglycerin was reported to lead to reductions in infarct size from 23% to 27% as compared with controls; in others either no change was found or there was a decrease in infarct size only in certain subgroups [14, 26, 41, 42]. Negative results may be due to late onset of treatment or excessive lowering of blood pressure. The clearest therapeutic effects, as judged by creatine kinase infarct size determination, were achieved when treatment was begun within 4 h of the onset of pain, but a positive effect was seen when treatment was begun within 10 h after the onset of pain [42]. Decisive for the effect of nitrate treatment on infarct size, additionally, was the extent of blood pressure reduction since in patients in whom the mean arterial blood pressure was lowered to less than 80 mmHg by nitroglycerin, infarct size was 122% larger than in those in whom mean arterial pressure was not lowered to less than 80 mmHg [42].

The effect of nitroglycerin on infarct size is also reflected in an improvement of global left ventricular function and a reduction of left ventricular asynergy. The left ventricular ejection fraction in patients with preexistent impairment increased by 11% in patients receiving nitroglycerin as compared to those given only placebo [26] and was 22% better than in the control group, respectively. Patients with previously normal ejection fraction had either no change or a reduction [26, 42]. Parallel to the increase in ejection fraction, the extent of echocardiographically determined left ventricular asynergy decreased on nitroglycerin treatment. There was 40% less asynergy than in control patients who were not seen to have this response [42].

The study of parameters indicative of left ventricular post-infarct remodelling such as the extension and thinning indexes of the infarct area and the left ventricular end-diastolic dimension shows, in control patients, an increase in the infarct extension index of 31% and the end-diastolic dimension of 13% as well as a reduction in the thinning index of 17% while, in patients treated with nitroglycerin, such changes were completely prevented [42].

Infarct-induced complications, as well, were fewer in those treated with nitroglycerin than in control patients. Nitroglycerin led to a reduction in the combined incidence of new onset of heart failure, re-infarction or early death to 15% in those treated early as compared with an incidence of 33% in patients treated late or not treated [26]. There was also a reduction in the syndrome of infarct extension (defined as incurrence of hypotension, heart failure or left ventricular dilatation without new ECG changes or CK elevation) to 2% vs 15% in control patients as well as reductions in the incidence of left ventricular thrombus (to 5% vs 22%), cardiogenic shock (to 5% vs 15%), and re-infarction (to 11% vs 22%), respectively [42]. Of decisive importance, however, is whether these favorable effects are also translated into improvement of prognosis. According to the pooled data of seven studies with a total of 426 patients treated with nitroglycerin and 425 control patients, the mortality in the treated group of 12.5% is clearly lower than the 20.5% in control patients where the greatest reduction was observed in the early phase of myocardial infarction [78]. It should be taken into consideration, however, that the positive effect is predominantly due to the results of one study in 310 patients in which mortality in the nitroglycerin-treated group was lower than in control patients, both in-hospital at 14% vs 26%, 3 months post-infarct at 16% vs 28%, and 12 months post-infarct at 21% vs 31%, respectively [42].

Overall, the conclusion appears justified that nitroglycerin is capable of exerting favorable hemodynamic effects, in particular in patients with congestive heart failure, including limiting infarct size and thereby improving global and regional ventricular function and, especially preventing to a large degree the otherwise to be expected left ventricular remodelling process, as well as lowering the incidence of in-hospital complications and, lastly, reducing mortality. Particularly essential for the effectiveness of treatment appear to be choice of dosage such that mean arterial blood pressure is not lowered to less than 80 mmHg, early initiation of treatment and that the problem of nitrate tolerance is respected with dose incrementation as necessary.

While nitroglycerin has an established role in treatment of acute heart failure arising from myocardial infarction, its role in the scope of the currently standard treatment with thrombolytic agents to preserve myocardium, remains to be defined. Studies to determine if nitrate treatment can elicit an additional effect have not been carried out. It is, however, thoroughly feasible that nitrate treatment concomitant with thrombolytic agents or their combination with PTCA could offer advantages not achievable with thrombolytic treatment alone.

References

1. Ambrose JA, Winters StL, Stern A, Eng A, Teichholz LE, Goldstein J, Fuster V (1985) Angiographic morphology and the pathogenesis of unstable angina. J Am Coll Cardiol 5:609–616
2. von Arnim T, Erath A (1988) Nitrates and calcium antagonists for silent myocardial ischemia. Am J Cardiol 61:15E–18E

3. Bache RJ, Ball RM, Cobb FR, Rembert JC, Greenfield Jr JC (1975) Effects of nitroglycerin on transmural myocardial blood flow in the unanesthetized dog. J Clin Invest 55:1219–1228
4. Balard JM (1844) Premier memoire sur l'alcool amylique. Ann Chim Phys 12:294–330
5. Bernstein L, Friesinger GC, Lichtlen PR, Ross RS (1966) The effect of nitroglycerin on the systemic and coronary circulation in man and dogs. Circulation 33:107–116
6. Blasini R, Froer KL, Blümel G, Rudolph W (1982) Wirkungsverlust von Isosorbiddinitrat bei Langzeitbehandlung der chronischen Herzinsuffizienz. Herz 7:250–258
7. Blasini R, Reiniger G, Brügmann U, Rudolph W (1984) Vermeidung einer Toleranz-entwicklung unter Isosorbiddinitrat durch Intervalltherapie. Herz 9:166–170
8. Blasini R, Brügmann U, Reiniger G, Rudolph W (1985) Langzeittherapie der Belastungs-Angina-pectoris durch einmal tägliche Verabreichung von 120 mg Isosorbiddinitrat in retar-dierter Form. Vergleich von Monotherapie und Kombinationstherapie mit Atenolol und Nifedipin. Herz 10:163–171
9. Bjerlöv H (1943) A new drug against angina pectoris. Svenska Läk-Tidn 40:449
10. Borer JS, Redwood DR, Levitt B, Cagin N, Bianchi C, Vallin H, Epstein St (1975) Reduction in myocardial ischemia with nitroglycerin or nitroglycerin plus phenylephrine administered during acute myocardial infarction. N Engl J Med 293:1008–1012
11. Brown BG, Bolson E, Petersen RB, Pierce CD, Dodge HT (1981) The mechanisms of nitro-glycerin action: stenosis vasodilatation as a major component of the drug response. Circulation 64:1089–1097
12. Brunton TL (1867) Use of nitrite of amyl in angina pectoris. Lancet II:97–98
13. Busse R, Bassenge E (1982) Einfluß von Nitraten, nitratähnlichen Substanzen, Calcium-Antagonisten und β-Rezeptorenblockern auf den peripheren Kreislauf. Herz 7:388–405
14. Bussmann W-D, Passek D, Seidel W, Kaltenbach M (1981) Reduction of CK and CK-MB indexes of infarct size by intravenous nitroglycerin. Circulation 63:615–622
15. DeCaterina R, Gianessi D, Crea F, Chierchia S, Bernini W, Gazzetti P, L'Abbate A (1984) Inhibition of platelet function by injectable isosorbide dinitrate. Am J Cardiol 53:1683–1687
16. Cowan JC, Bourke JP, Reid DS, Julian DG (1987) Prevention of tolerance to nitroglycerin patches by overnight removal. Am J Cardiol 60:271–275
17. Danahy DT, Burwell DT, Aronow WS, Prakash R (1977) Sustained hemodynamic and antian-ginal effect of high dose oral isosorbide dinitrate. Circulation 55:381–387
18. Davies MJ, Thomas AC, Knapman PA, Hangartner JR (1986) Intramyocardial platelet aggre-gation in patients with unstable angina suffering sudden ischemic cardiac death. Circulation 73:418–427
19. Dirschinger J, Fleck E, Redl A, Brandt R, Späth M, Mannes YG, Recke S, Hall D, Rudolph W (1977) Effects of sodium nitroprusside and isosorbide dinitrate on regional myocardial blood flow in patients with coronary artery disease and left ventricular asynergy. Herz 2:71–74
20. Distante A, Maseri A, Severi S, Biagini A, Chierchia S (1979) Management of vasospastic angina at rest with continous infusion of isosorbide dinitrate. Am J Cardiol 44:533–539
21. Evans W, Hoyle C (1933) The comparative value of drugs used in the continuous treatment of angina pectoris. Quart J Med 2:311–338
22. Falk E (1985) Unstable angina with fatal outcome: Dynamic coronary thrombosis leading to infarction and/or sudden death. Autopsy evidence of recurrent mural thrombosis with periph-eral embolization culminating in total vascular occlusion. Circulation 71:699–708
23. Feelisch M, Noack E, Schröder H (1988) Explanation of the discrepancy between the degree of organic nitrate decomposition, nitrite formation and guanylate cyclase stimulation. Eur Heart J 9 [Suppl A]:57–62
24. Fitzgerald DJ, Roy L, Catella F, Fitzgerald GA (1986) Platelet activation in unstable coronary disease. N Engl J Med 315:983–989
25. Flaherty JT, Come PC, Baird MG, Rouleau J, Taylor DR, Weisfeldt ML, Greene HL, Becker LC, Pitt B (1976) Effects of intravenous nitroglycerin on left ventricular function and ST seg-ment changes in acute myocardial infarction. Br Heart J 38:612–621
26. Flaherty JT, Becker LC, Bulkley BH, Weiss JL, Gerstenblith G, Kallman CH, Silverman KJ, Wei JY, Pitt B, Weisfeldt ML (1983) A randomized prospective trial of intravenous nitroglyce-rin in patients with acute myocardial infarction. Circulation 68:576–588
27. Flaherty JT (1987) Role of nitrates in acute myocardial infarction. Am J Cardiol 60:35H–38H
28. van Gilst WH, deGraeff PA, Scholtens E, deLangen CDJ, Wesseling H (1987) Potentiation of

290

isosorbide dinitrate-induced coronary dilatation by captopril. J Cardiovasc Pharmacol 9:254–255

29. Goldberg L (1948) Pharmacological properties of sorbide dinitrate. Acta Physiol Scand 15:173–187
30. Gold HK, Leinbach RC, Sanders CA (1972) Use of sublingual nitroglycerin in congestive failure following acute myocardial infarction. Circulation 46:839–845
31. Goldstein RE, Rosing DR, Redwood DR, Beiser GD, Epstein SE (1971) Clinical and circulatory effects of isosorbide dinitrate. Comparison with nitroglycerin. Circulation 43:629–640
32. Golstein RE (1982) Coronary vascular responses to vasodilator drugs. Progr Cardiovasc Dis 24:419–436
33. Gottlieb SO, Weisfeldt ML, Ouyang P, Mellits ED, Gerstenblith G (1987) Silent ischemia predicts infarction and death during 2 year follow-up of unstable angina. J Am Coll Cardiol 10:756–760
34. Guthrie F (1859) Nitrite of amyl and its derivates. Chem Soc 11:245–252
35. Hausmann D, Nikutta P, Daniel WG, Lichtlen PR (1988) Therapie der stummen Myokardischämie: Wirkung von 1 × 120 mg retardiertem Isosorbiddinitrat bei stabiler Angina pectoris. Z Kardiol 77 [Suppl 1]:83 (abstr.)
36. Hering C (1849) Glonoin. Kirby's N Am J Homeopathy, May
37. Hess OM, Schneider J, Nonogi H, Carroll JD, Schneider K, Turina M, Krayenbuehl HP (1988) Myocardial structure in patients with exercise-induced ischemia. Circulation 77:967–977
38. Horowitz JD, Henry CA, Syrjanen ML, Louis WJ, Fish RD, Smith TW, Antman EM (1988) Combined use of nitroglycerin and N-acetylcysteine in the management of unstable angina pectoris. Circulation 77:787–794
39. Horwitz LD, Herman MV, Gorlin R (1972) Clinical response to nitroglycerin as a diagnostic test for coronary artery disease. Am J Cardiol 29:149–153
40. Ignarro LJ, Lippton H, Edwards JC, Baricous WH, Hyman AL, Kadowitz PJ, Gruetter CA (1981) Mechanism of vascular smooth muscle relaxation by organic nitrates, nitrites, nitroprusside and nitric oxide: Evidence for the involvement of S-nitrosothiols as active intermediates. J Pharmacol Exp Ther 218:739–749
41. Jaffe AS, Geltman EM, Tiefenbrunn AJ, Ambos HD, Strauss HD, Sobel BE, Roberts R (1983) Reduction of infarct size in patients with inferior infarction with intravenous glyceryl trinitrate. A randomized study. Br Heart J 49:452–460
42. Jugdutt BI, Warnica JW (1988) Intravenous nitroglycerin therapy to limit myocardial infarct size, expansion, and complications. Effect of timing, dosage, and infarct location. Circulation 78:906–919
43. Kaplan K, Davison R, Parker M, Przybylek J, Teagarden JR, Lesch M (1983) Intravenous nitroglycerin for the treatment of angina at rest unresponsive to standard nitrate therapy. Am J Cardiol 51:694–698
44. Kattus AA, Alvaro AB, Zohman LR, Coulson AH (1979) Comparison of placebo, nitroglycerin and isosorbide dinitrate for effectiveness of relief of angina and duration of action. Chest 75:17–23
45. Krantz J C Jr, Carr CJ, Forman SE, Cone N (1940) A contribution to the mechanism of the action of organic nitrates. J Pharmacol Exp Ther 70:323–327
46. Kukovetz WR, Holzmann S (1983) Mechanism of nitrate-induced vasodilatation and tolerance. Z Kardiol 72 [Suppl 3]:14–19
47. Lehmann G, Reiniger G, Rudolph W (1988) Wie effektiv ist eine Kombination eines Nitrates und eines Calciumantagonisten bei Patienten mit stabiler Angina pectoris? Z Kardiol 77 [Suppl 1]:119 (abstr.)
48. Levy WS, Katz RJ, Wassermann AG (1989) Methionine reverses tolerance to transdermal nitroglycerin. J Am Coll Cardiol 13:230A (abstr.)
49. Macho P, Vatner SF (1981) Effects of nitroglycerin and nitroprusside on large and small coronary vessels in conscious dogs. Circulation 64:1101–1107
50. McGregor M (1982) The nitrates and myocardial ischemia. Circulation 66:689–692
51. Moise A, Théroux P, Taeymanns Y, Descoings B, Lespérance J, Waters DD, Pelletier GB, Bourassa MG (1983) Unstable angina and progression of coronary atherosclerosis. N Engl J Med 309:685–689

52. Murrell W (1879) Nitroglycerin as a remedy for angina pectoris. Lancet I:80, 113, 225, 642–646
53. Nabel EG, Barry J, Rocco MB, Campbell St, Mead K, Fenton T, Orav EJ, Selwyn AP (1988) Variability of transient myocardial ischemia in ambulatory patients with coronary artery disease. Circulation 78:60–67
54. Nademanee K, Intarachot V, Josephson MA, Rieders D, Mody FV, Singh BN (1987) Prognostic significance of silent myocardial ischemia in patients with unstable angina. J Am Coll Cardiol 10:1–9
55. Noack E (1982) Therapie der Angina pectoris. Pharmakologische Basis für die Therapie mit organischen Nitraten. Herz 7:275–285
56. Page A, Gateau P, Ohayon J, Couppillaud J, LeMinh D, Besse P (1981) Intravenous nitroglycerin in unstable angina. In: Lichtlen PR, Engel HJ, Schrey A, Swan HJC (eds) Nitrates III. Cardiovascular effects. Springer, Berlin Heidelberg New York, pp 371–376
57. Parker JO, van Koughnett KA, Farrell B (1985) Comparison of buccal nitroglycerin and oral isosorbide dinitrate for nitrate tolerance in stable angina pectoris. Am J Cardiol 56:724–728
58. Parker JO, van Koughnett KA, Fung HL (1984) Transdermal isosorbide dinitrate in angina pectoris: effect of acute and sustained therapy. Am J Cardiol 54:8–13
59. Reiniger G, Kraus F, Dirschinger J, Blasini R, Rudolph W (1985) Hochdosierte transdermale Nitroglycerintherapie: Wirkungsverlust innerhalb von 24 Stunden? Herz 10:157–162
60. Reiniger G, Menke G, Boertz A, Kraus F, Rudolph W (1987) Intervalltherapie zur effektiven Behandlung der Angina pectoris mit Nitroglycerin-Pflaster-Systemen. Kontrollierte Studie mit Bestimmung der Nitroglycerin-Plasmaspiegel. Herz 12:68–73
61. Reiniger G, Rudolph W (1987) Diskontinuierliche Wirkstofffreisetzung als Alternative zur Intervalltherapie bei der Behandlung der koronaren Herzerkrankung mit Nitroglycerinpflastern. Herz 12:348–353
62. Rocco MB, Nabel EG, Campbell St, Goldman L, Barry J, Mead K, Selwyn AP (1988) Prognostic importance of myocardial ischemia detected by ambulatory monitoring in patients with stable coronary artery disease. Circulation 78:877–884
63. Rudolph W, Blasini R, Froer KL, Brügmann U, Mannes A, Hall D (1981) Effects of acute and chronic administration of isosorbide dinitrate, sustained-release form, in patients with angina pectoris. In: Lichtlen PR, Engel HJ, Schrey A, Swan HJC (eds) Nitrates III. Cardiovascular effects. Springer, Berlin Heidelberg New York, pp 75–81
64. Rudolph W, Blasini R, Kraus F (1982) Klinische Wirksamkeit der Nitrate bei Belastungs-Angina pectoris. Herz 7:287–295
65. Rudolph W, Blasini R, Reiniger G, Brügmann U (1983) Tolerance development during isosorbide dinitrate treatment: can it be circumvented? Z Kardiol 72 [Suppl 3]:195–198
66. Rudolph W, Blasini R (1984) Nitrate tolerance: A relevant clinical problem? Herz 9:115–122
67. Rudolph W, Dirschinger J, Reiniger G, Beyerle A, Hall D (1988) When does nitrate tolerance develop? What dosages and which intervals are necessary to ensure maintained effectiveness? Eur Heart J 9 [Suppl. A]:63–72
68. Rudolph W, Dirschinger J, Hall D, Reiniger G, Beyerle A (1989) Effectiveness of interval therapy in angina pectoris. Eur Heart J 10 [Suppl A]:50–55
69. Schirnick C, Reifart N (1988) Akute und subchronische Wirkung eines Pflasters mit diskontinuierlicher Nitroglycerinfreisetzung. Z. Kardiol. 77 [Suppl 1]:82 (abstr.)
70. Sherman CT, Litvack F, Grundfest W, Lee M, Hickey A, Chaux A, Kass R, Blanche C, Matloff J, Morgenstern L, Ganz W (1986) Coronary angioscopy in patients with unstable angina pectoris. N Engl J Med 315:913–919
71. Sobrero A (1847) Sur pluriers composes detonants produc avec l'acide nitrique et le sucre, la lactine, la mannite et la glycerine. R Acad Sci (Paris) 24:247–248
72. Stewart DD (1888) Remarkable tolerance to nitroglycerin. Philadelphia Policlinic, p 172
73. Thadani U, Manyari D, Parker JO, Fung HL (1980) Tolerance to the circulatory effects of oral isosorbide dinitrate. Rate of development and cross-tolerance to glyceryl trinitrate. Circulation 61:526–535
74. Tzivoni D, Weisz G, Gavish A, Zin D, Keren A, Stern S (1989) Comparison of mortality and myocardial infarction rates in stable angina pectoris with and without ischemic episodes during daily activities. Am J Cardiol 63:273–276
75. Weiner DA, Ryan TJ, McCabe CH, Luk S, Chaitman BR, Sheffield LT, Tristani F, Fisher LD

(1987) Significance of silent myocardial ischemia during exercise testing in patients with coronary artery disease. Am J Cardiol 59:725–729

76. Weiner DA, Ryan TJ, McCabe CH, Ng G, Chaitman BR, Sheffield LT, Tristani FE, Fisher LD (1988) Risk of developing an acute myocardial infarction or sudden coronary death in patients with exercise-induced silent myocardial ischemia. A report from the coronary artery surgery study (CASS) registry. Am J Cardiol 62:1155–1158

77. Vatner SF, Heyndrickx GR (1975) Mechanism of action of nitroglycerin: coronary, cardiac, and systemic effects. In: Needleman P (ed) Organic Nitrates, Springer, Berlin Heidelberg New York, S 131–161

78. Yusuf S, Collins R, MacMahon St, Peto R (1988) Effect of intravenous nitrates on mortality in acute myocardial infarction. An overview of the randomized trials. Lancet I:1088–1092

Authors' address:
Prof. Dr. W. Rudolph
Deutsches Herzzentrum München
Klinik für Herz- und Kreislauferkrankungen
Lothstraße 11
8000 München 2

Treatment of Myocardial Ischemia with Calcium Antagonists

Wolf Rafflenbeul

Division of Cardiology, Hannover Medical School

Introduction

Calcium is the most common cation in the human body. Its movements through the membrane of excitable cells and its accumulation within the cell are of fundamental importance for the specific activity of tissues. Generally, the concentration of calcium in the cytosol is only a small fraction of the calcium concentration in the plasma, and it is largely independent of the normal variations in plasma calcium [12]. Therefore, a prerequisite for intracellular calcium homeostasis is a very precise modulation mechanism of the entry and efflux rate of calcium together with functioning sarcoplasmic reticulum and mitochondria. Increasing intracellular calcium to abnormally high levels may cause cell injury or cell death. Cell membranes exhibit a low permeability to calcium which enters cells predominantly through specific *calcium channels*.

"Uphill" efflux of calcium from the cytosol requires direct utilization of energy from the specific plasma membrane calcium-activated ATP-ase and uses the sodium electrochemical potential to extrude calcium in exchange for sodium entry (Na^+-Ca^{2+}-exchange). Cellular calcium is either sequestered to mitochondria (which can absorb large amounts of intracellular calcium), the endoplasmatic reticulum, proteins and anions, or to the polar heads of membrane phospholipids. All these mechanisms work in unison to keep the cytosolic calcium concentration at its normally low level and prevent calcium overloading of the intracellular space in the face of an extracellular calcium concentration three to four orders of magnitude higher than the intracellular concentration.

Calcium Channels

Calcium channels are integral surface membrane proteins that regulate the entry of calcium into the cytosol and whose properties are modified by hormones such as catecholamines. They are named "calcium channels" because of their relative selectivity for calcium passage compared to sodium or potassium which might be as high as 10.000-fold. The common mechanism for calcium selectivity are the high affinity binding sites for calcium positioned along a single file pore.

At this time different types of calcium channels have been distinguished in the cardiovascular system:
1. Calcium channels that are opened by depolarization of the cell membrane ("voltage-dependent calcium channels"). Based mainly on differences in the activation threshold, voltage dependence of inactivation, flux rates at the single channel

level (single channel conductance), relative permeability to Ba^{2+}, and variable sensitivity to calcium antagonists, two different voltage-gated calcium channels can be distinguished in heart and smooth muscle cells:

a) The L-type calcium channel is highly voltage-dependent, i.e., it is activated only by relatively strong depolarizations and inactivates relatively slowly (slow calcium channel). The L-type calcium channel, which is the main voltage-dependent calcium entry pathway in heart and vascular smooth muscle, is also the major target for calcium antagonists like nifedipine, diltiazem, and verapamil.

b) The T-type calcium channel which activates with relatively weak depolarizations, but inactivates relatively fast (transient or fast calcium channel). The T-type calcium channels are found in the majority of heart muscle cells (where they are responsible for pacemaker activity), and in most vascular smooth muscle cells. T-type channels seem to be insensitive to calcium antagonists.

2. Particularly in voltage-clamp experiments with vascular smooth muscle cells another type of calcium permeable channel has been described [6, 24] which opens during activation of cell membrane receptors without a change in membrane potential and, therefore, is named receptor-operated calcium channel. It may respond to adrenergic stimulation, releasing ATP from the sympathetic varicosities [6]. Angiotension II may be another source of receptor modulated calcium influx.

As in many other biologic cellular processes the cytosolic calcium concentration plays an important role in numerous cardiovascular functions, including the genesis of action potential, excitation-contraction coupling processes, the inotropic state of the myocardium, and regulation of the tone of vascular smooth muscle in peripheral and coronary arteries. The myocardial and the vascular effects of calcium, particularly initiating contractile activity, are inhibited by calcium antagonists, interfering with the transmembrane calcium flux and the intracellular calcium storage and release mechanisms. Therefore, the specific pathways of calcium ions in myocardium and vascular smooth muscle involved in the functional consequences of ischemia are addressed.

Calcium in Normal Myocardial Function

Within the myocyte calcium is highly compartmentalized and measurements of its concentration in different intracellular compartments are fraught with substantial errors. In ischemia, total cellular calcium is often measured and a substantial rise of intracellular calcium has been directly documented by different methodology [52, 55, 106].

The calcium concentration in the myoplasm is maintained at its low level by a number of pumps that remove calcium, either to the extracellular space or to one of its internal stores. The two main "uphill" calcium extruding pumps are the ATP-dependent surface membrane calcium pump and the sodium-calcium-exchange, which uses the energy of the sodium gradient to pump calcium in exchange to the extracellular space.

Myocardial contraction is initiated by the sudden release of calcium from the sarcoplasmic reticulum – the calcium transient –. The sequestration of calcium into the sarcoplasmic reticulum which is a prerequisite for its triggered release is energy-dependent. Therefore, over several cardiac cycles with ATP declining during ischemia, the calcium transient will decrease (decreasing contractility) and diastolic calcium concentration in the cytosol will increase (increasing diastolic stiffness).

Mitochondria with their large capacity for calcium play an important role in the calcium metabolism of the cell. With prolonged ischemia, mitochondria become loaded with calcium [37] and loss of oxidative phosphorylation might result. In ischemia, troponin – the quantitatively most important myoplasmic calcium buffer – may bind more H^+-ions than calcium, so that even at comparable intracellular calcium concentrations less tension will develop.

Calcium and the Ischemic Myocardium

Ischemia imposes a rapidly progressive damage on the myocytes that is predominantly due to inadequate ATP production because of interrupted or reduced aerobic metabolism. The preexisting stores of high-energy phosphates (phosphocreatine (CP) and ATP) are relatively small and anaerobic glycolysis generates only minor amounts of high-energy phosphates which are not sufficient to maintain contractile force and cellular metabolism.

Calcium homeostasis with cytosolic concentrations below the electrochemical equilibrium requires active energy consuming processes to extrude calcium from the intracellular space. These calcium pumps are ATP-dependent and a failure of these pumps to remove calcium from the myoplasm increases intracellular calcium. The ATP levels which block calcium pumps are lower than those necessary for rigor production. Therefore, contracture precedes the rise in intracellular calcium [1, 16]. Rising intracellular calcium activates other calcium-dependent enzyme systems including ATP-ases which further accelerate the depletion of ATP from the ischemic myocyte. A consequence is the depletion of the already diminished energy resources and an increasing breakdown of cell membranes. These predisposing factors and others, such as cell swelling or structural disorganization of sarcoplasmic reticulum occur with progressive time of ischemia and set the stage for the even more damaging processes that occur during *reperfusion.*

It has been established that calcium influx does increase massively on postischemic reperfusion [10, 47] with calcium overloading of the cell. This "flooding" is not simply due to a nonspecific increase in cell permeability [81] nor does it occur via calcium channels alone. Calcium overloading is most probably due to a progressive opening of different routes of entry into the cytosol. With the sarcolemma still intact calcium will be exchanged with sodium and will reach the intracellular space via the calcium channels. Already a minor increase of intracellular calcium concentration activates a host of calcium-dependent pathways and thereby triggers a cascade of deleterious events, including activation of ATP-ases with further ATP wastage, as well as stimulation of proteases and phospholipases with concomitant membrane disruption (Fig. 1). Although the magnitude and the time-course of calcium rise show considerable variability at this "point of no return" the reperfused myocyte eventually will die.

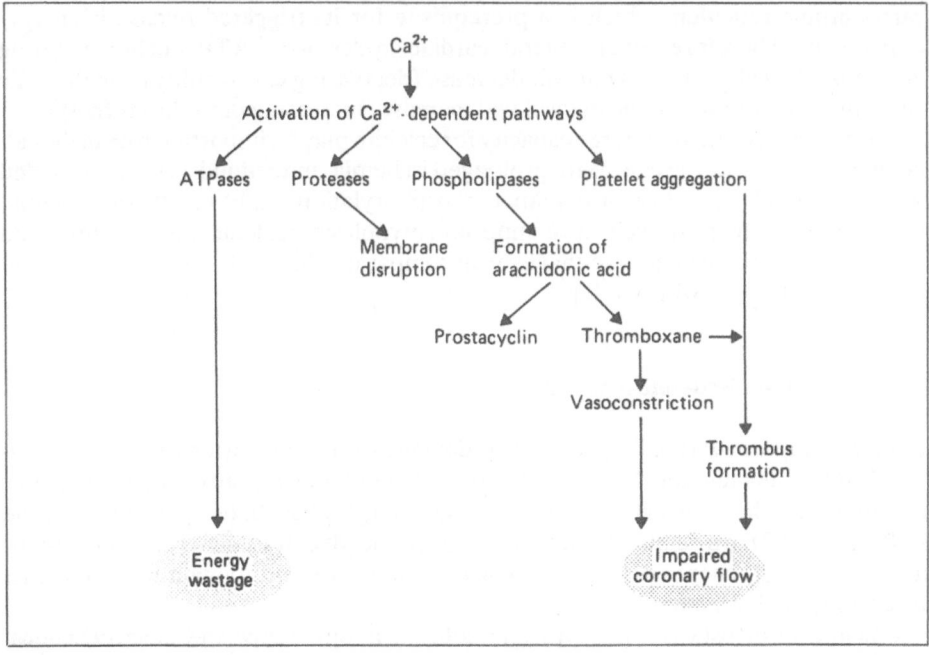

Fig. 1. Cascade of direct consequences of calcium overloading during postischemic reperfusion. from: [71].

Calcium in Vascular Smooth Muscle

Constriction of vascular smooth muscle is also governed by calcium-dependent mechanisms. The sarcolemma of smooth muscle cells contains two distinct calcium entry pathways: as in cardiac muscle, one is activated by membrane depolarization [13] which may be caused by agonists like norepinephrine, serotonin, angiotensin, or histamine. The other type of calcium channel is independent of the electrical state of the cell and is activated by agonist interaction with its receptor ("receptor-operated channel", ROC). The opening of this type of channel is also triggered by chemical neurotransmitters or hormones like norepinephrine, serotonin or histamine [9, 116]. In conjunction with the calcium influx into the cell, calcium is discharged from internal stores, particularly from the sarcoplasmic reticulum (Fig. 2). There is strong evidence of a close interrelation between agonist-induced calcium entry and release of calcium from the sarcoplasmic reticulum, suggestive of a mutual triggering activity [104, 108]. Furthermore, it has been proposed that receptor activation triggers the release of a messenger other than calcium which stimulates the release of calcium from the sarcoplasmic reticulum [58, 116]. As in cardiac muscle, intracellular calcium stores play a major role in supplying sufficient amounts of calcium necessary for active vascular smooth muscle contraction. Their dependence on calcium influx from extracellular fluid explains why calcium antagonists reduce the amount of calcium available for contraction, i.e., relax vascular smooth muscle.

298

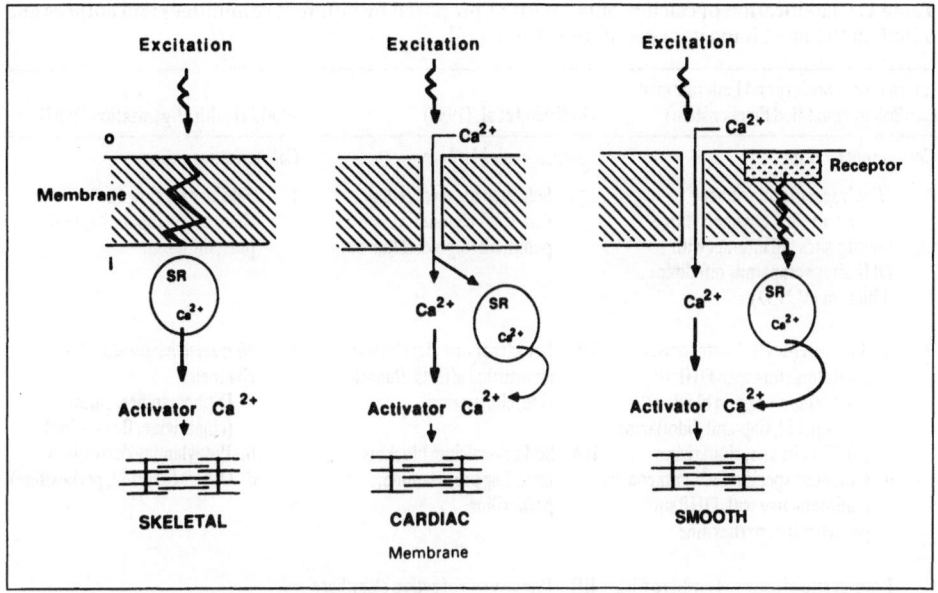

Fig. 2. Schematic illustration of different calcium entry pathways in skeletal, cardiac and smooth muscle cells. See text for details; from [73]).

Calcium Antagonists

Calcium antagonists are compounds which selectively inhibit the inward flow of calcium ions through the calcium-selective channels. Their main pharmacological actions which provide the basis of their use in cardiovascular medicine are depression of myocardial contractility, vasodilation of peripheral and coronary arterial vessels, and in some, negative chronotropic and dromotropic activity in specialized cardiac tissue. At present a uniformly accepted classification of calcium antagonists does not exist, although numerous proposals have been introduced recently emphasizing different properties of the calcium antagonists and reflect the most important actions [25, 33, 73, 103, 105, 118, 119] (Table 1).

With respect to the treatment of myocardial ischemia two major characteristics of calcium antagonists play a predominant role:
1) The *specificity* of inhibiting the slow calcium channel. According to the criteria originally postulated by Fleckenstein [25] the highly selective group of calcium current inhibitors, i.e., calcium antagonists with 90−100% inhibition of the calcium current, include:
   a) the dihydropyridines: nifedipine, nisoldipine, nicardipine, nimodipine, nitrendipine, isradipine, felodipine, and others;
   b) the benzothiazipines, like diltiazem and
   c) the phenylalkylamines: verapamil, desmethoxyverapamil (D888), gallopamil (D600) or tiapamile, and others.

   Besides these groups of compounds with a very high calcium antagonistic activity others demonstrate a weaker inhibitory effect of the slow inward current, act

299

Table 1. Classification of calcium antagonists as proposed by different committees and authors and based on the most important properties. from: [76].

International Society and Federation of Cardiology (modified Fleckenstein)	Godfraind et al. (1986)	World Health Organization (1987)
Calcium antagonists	*Calcium entry blockers*	*Calcium antagonists*
A. *Highly specific* on voltage-dependent Ca^{2+} channels; three binding sites, all interact with DHP site: verapamil, nifedipine, diltiazem (V,N,D)	IA. Selective for myocardial slow Ca^{2+} channel: verapamil, nifedipine, diltiazem	1. *Selective* for slow Ca^{2+} channels: verapamil, nifedipine, diltiazem
B. i. *Less specific* Ca^{2+} antagonists also interacting with DHP site: (a) Sodium-calcium blockers bepridil, tiapamil, lidoflazine (b) Flunarizine, cinnarizine ii. Other less specific Ca^{2+} antagonists noninteractive with DHP site: prenylamine, perhexiline	IB. Selective (vascular) but no myocardial effects: flunarizine, cinnarizine IIA. Sodium-calcium blockers including prenylamine, perhexiline	2. *Non-selective* for slow Ca^{2+}-channels: a. Diphenylpiperazines (cinnarizine, flunarizine) b. Prenylamine derivatives c. Others (bepridil, perhexiline)
C. Primary site of action elsewhere with incidental Ca^{2+} antagonist activity	IIB. Primary site of action elsewhere	

DHP = dihydropyridine binding site.

particularly on vascular smooth muscle with minor or no influence on myocardial slow calcium channels, and may show a simultaneous inhibition of calcium and sodium channels at clinically useful doses. Included in this secondary group are the diphenylalkylamines cinnarizine, flunarizine, fendiline and prenylamine or others like perhexilene. Some of these drugs show a preference for specific organs like the brain which might constitute a valuable property above weak cardiovascular activity.

The potent prototypes of the three chemical groups of highly selective and effective calcium antagonists – nifedipine, verapamil and diltiazem – are investigated most intensively and are widely used in the treatment of myocardial ischemia. Therefore, most of the following data will refer to these compounds.

2) The *tissue selectivity*, i.e., a predominant effect on either the myocardium or on the vasculature or in certain calcium antagonists a quantitative difference with regard to the effect on nodal and conductive tissue in the heart with depression of impulse formation and conduction. In general, based on extensive experimental evidence as well as clinical observations, the dihydropyridines are regarded as more potent smooth muscle relaxants than are other calcium antagonists [30]. Some of them, like nisoldipine, preferentially dilate specific vasculature, namely the coronary bed [31]. In contrast, verapamil and diltiazem have nearly equipotent effects on the conductive tissue and the vascular smooth muscle.

It has to be emphasized that the relative potency of calcium antagonists on myocardium or vasculature is dose-dependent, i.e., the tissue- or organ-selectivity may be attenuated with high concentrations. For example, the relative prefer-

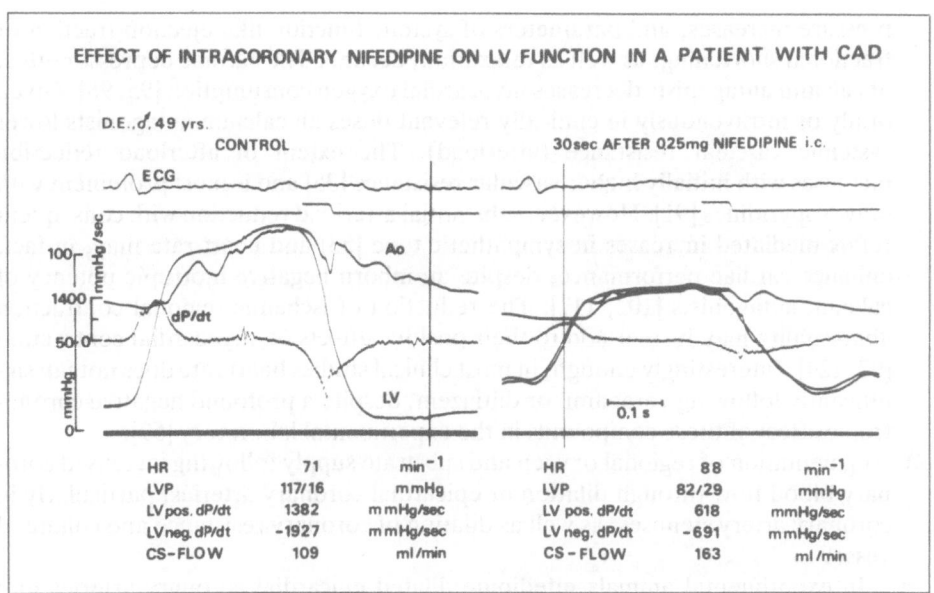

Fig. 3. Effect of 0.25 mg intracoronary nifedipine on left ventricular function in a patient with coronary artery disease. The high concentration of nifedipine causes a substantial negative inotropic effect. from: [101].

ence of nifedipine for vasculature is obscured after intracoronary application of the drug, demonstrating a profound negative inotropic effect (Fig. 3). In addition, in the patient treated with calcium antagonists the net hemodynamic and electrophysiological effect of a specific calcium antagonist is a combination of its direct effects and the reflex-mediated effects superimposed. Furthermore, the anti-ischemic potency of a calcium antagonist may be attenuated by "pro"-ischemic effects like the increase of heart rate following application of dihydropyridines in some patients.

Calcium Antagonists in Ischemia

Based on their predominant pharmacological action – reduction in tension of cardiac muscle and vascular smooth muscle – calcium antagonists may be beneficial in the treatment of regional and global ischemia by one or more of the following mechanisms:
1) Attenuation of oxygen demand through altering cardiovascular hemodynamics with reduction of contractility, systemic vascular resistance (afterload), and/or heart rate.

In cardiac muscle preparations [25] and in the intact heart [65] calcium antagonists exhibit a profound negative inotropic effect. In patients the same negative effect on myocardial contractility can only be demonstrated following intracoronary injection of a calcium antagonist (Fig. 3). It has been demonstrated repeatedly that [2, 51, 94]: systolic pressure falls, left ventricular end-diastolic pressure increases, and parameters of systolic function like ejection fraction or fractional shortening, as well as relaxation, decline. The cardiac depressor effect of calcium antagonists decreases myocardial oxygen consumption [95, 98]. Given orally or intravenously in clinically relevant doses all calcium antagonists lower systemic vascular resistance (afterload). The extent of afterload reduction increases with initially higher vascular resistance [35] and is most prominent with dihydropyridines [91]. However, substantial afterload reduction with consequent reflex-mediated increases in sympathetic tone [53] and heart rate may, in fact, enhance cardiac performance, despite an inborn negative inotropic potency of calcium antagonists [102, 117]. The reduction of ischemic regional contraction abnormality may further add to their positive effects on myocardial contraction [97, 124]. Interestingly enough, in most clinical studies heart rate does not fall significantly following verapamil or diltiazem, despite a profound negative chronotropic effect of these compounds in the experimental laboratory [59].

2) Augmentation of regional oxygen and substrate supply following increased coronary blood flow through dilation of epicardial coronary arteries, particularly in coronary artery stenoses, as well as dilation of coronary resistance and collateral vessels.

In experimental animals nifedipine dilated epicardial coronary arteries in a dose-dependent manner [120]. In isolated human epicardial coronary arteries harvested either from recipient hearts at the time of cardiac transplantation [28] or from victims of sudden non-cardiac death [32], calcium antagonists inhibited the spontaneous and provoked contractile activity of the artery segments. In patients undergoing coronary artery angiography a dilation of epicardial coronary arteries following calcium antagonists has been demonstrated repeatedly [43, 74, 96]. Quantitative measurements [50] revealed an average diameter increase of about 30% following sublingual and intravenous application of different calcium antagonists [49, 74, 85, 87].

More important than the release of vascular tone in normal coronary arteries is the vasodilatory potency of calcium antagonists on coronary artery stenoses. From histological sections of coronary artery stenoses post mortem we have learned that the majority of coronary obstructions is of eccentric nature, i.e., the atherosclerotic plaque involves only a segment of the vessel circumference [26] leaving parts of the vessel wall intact with vascular smooth muscle cells sensitive to vasoconstrictory or dilatory impulses, as demonstrated earlier [83]. Due to the exponential relationship between the degree of obstruction and the pressure drop across its smallest part (Fig. 4) minute changes of the residual lumen result in substantial differences of the transstenotic pressure gradient. Angiographically distinguishable morphological differences of the individual coronary artery lesion are crucial for the dilatory response [86]. Thus, dilation of a high-grade coronary artery obstruction following the application of a calcium antagonist − like that shown in Fig. 5 − may constitute a major anti-ischemic property of the drugs [84, 86].

Concomitant with the dilation of epicardial coronary arteries a decrease of total coronary vascular resistance is observed in concert with the fall in total peripheral vascular resistance [15]. Consequently overall coronary blood flow increases [18]. However, clinically most important is the increase of *regional* myocardial blood

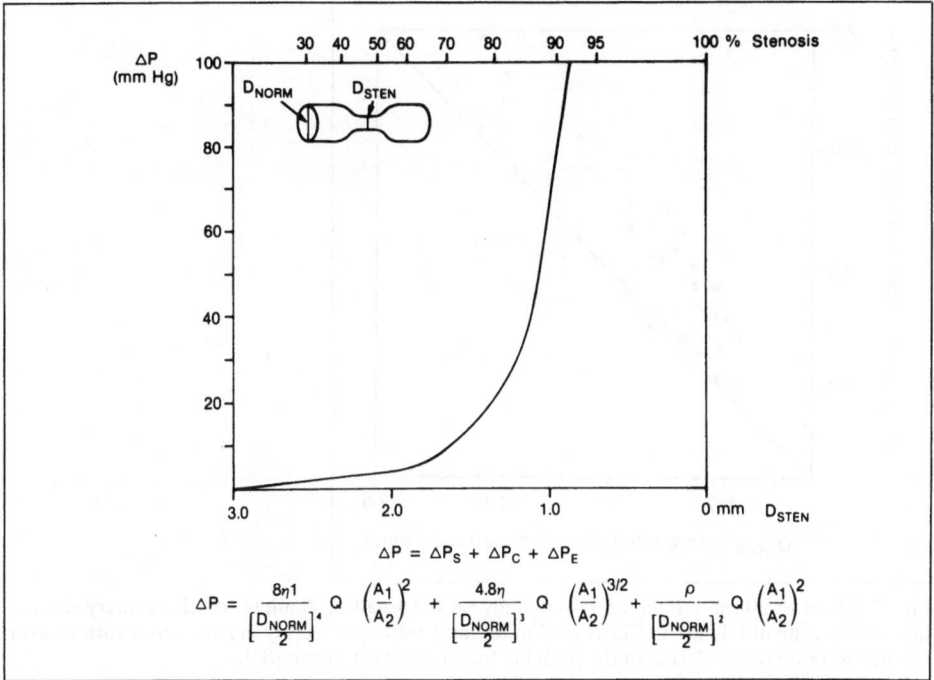

Fig. 4. Relationship between the smallest diameter (D STEN, abscissa) and the pressure gradient (P, ordinate) across the obstructed segment of a simplified coronary artery stenosis. D Norm = 3 mm; length = 5 mm; Q = 80 ml/min; η = 0.035 Poise; ϱ = 1.,056 g/cm³. from: [84].

flow in poststenotic areas following the application of calcium antagonists [23,60]. With pacing-induced angina and ischemic ST-segment shifts, poststenotic myocardial blood flow − measured as total transmural flow − increased to a lower extent than in myocardial areas perfused by an undiseased vessel. Following sublingual application of 20 mg nifedipine regional myocardial blood flow in normally perfused areas tended to fall, probably due to a lower rate-pressure product and reduced myocardial oxygen requirements (Fig. 6). In contrast, in the poststenotic area nifedipine further increases regional flow. Similar results were reported with diltiazem and verapamil [18]. This additional increase of regional coronary blood flow may reflect the sum of multiple beneficial effects of calcium antagonists in myocardial ischemia, including one or more of the following mechanisms:

a) dilation of eccentric epicardial coronary artery stenoses reducing the "conduit" vessel resistance (see above);

b) dilation of coronary collateral vessels with increasing collateral perfusion of ischemic myocardium [38, 48, 123];

c) prevention of inadequate poststenotic alpha-adrenergic coronary constriction on the precapillary level or in coronary collateral vessels [36, 39, 40], and

d) reduction of intraventricular and coronary extravascular pressure enhancing subendocardial blood flow and regional myocardial function [62, 93].

303

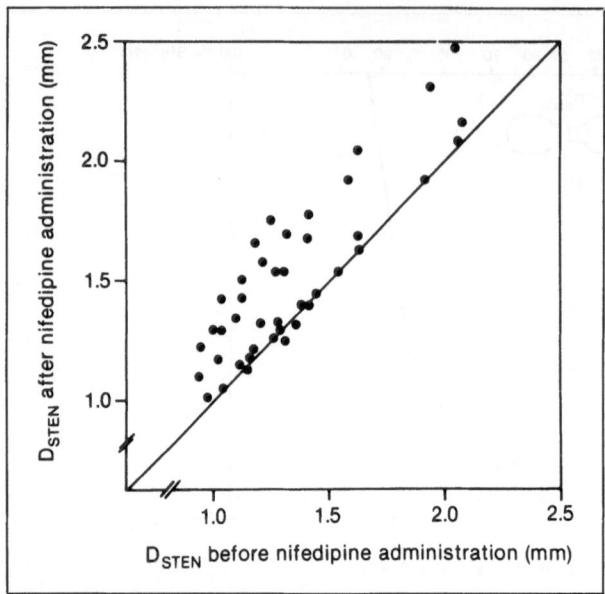

Fig. 5. Effect of 20 mg nifedipine sublingually on the smallest diameter of 42 coronary obstructions. Following nifedipine D STEN (ordinate) increased significantly in 20 stenoses with an average diameter increase of 31% of the initial D Sten (abscissa). from: [86].

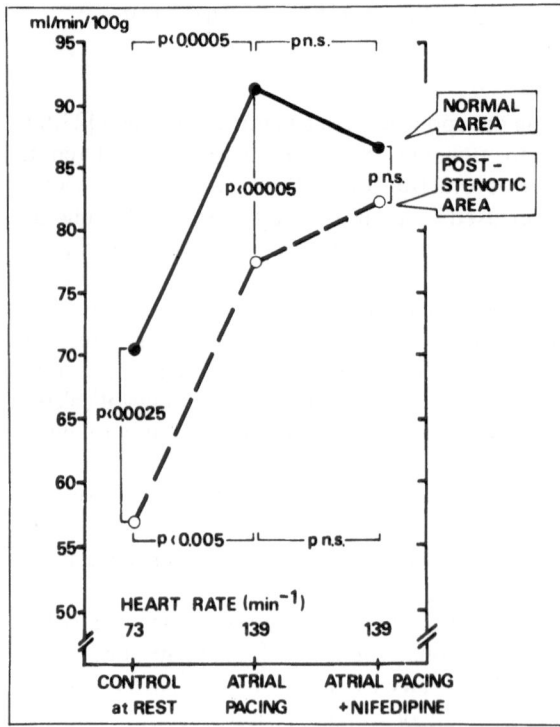

Fig. 6. Increase of regional myocardial blood flow in 11 patients with coronary artery disease during pacing-induced ischemia before and after 20 mg of nifedipine sublingually. from: [23].

3) Myocardial "protection", i.e., reducing myocardial injury during ischemia and reperfusion. The complex mode of protective action may include:

a) Mechanisms which tend to balance the oxygen demand/supply ratio as characterized in 1) and 2).

b) A decrease of ATP- and CP-breakdown and a slowing of catabolite loss [72].

During myocardial ischemia a prolonged depletion of high-energy phosphate stores occurs with an accumulation of catabolites like adenosine, inosine or hypoxanthine. During reperfusion the membrane-permeable catabolites are washed out and ATP has to be restored by the relatively slow resynthesis of purines instead of rephosphorylation of catabolites. Pre-treatment with calcium antagonists before global or regional ischemia and reperfusion has been shown to slow the decline of ATP and phosphocreatine levels and to preserve the levels of major catabolites [19, 54, 68, 75]. The latter mechanism should improve the ATP resynthesis from purine precursors. It is not clear whether the reduced loss of catabolites following calcium antagonists is a direct membrane effect or a secondary consequence of the primary reduction of ischemia-induced cellular damage.

c) Attenuation of mitochondrial calcium overloading [69]. In isolated hearts – eliminating the protective effect of an afterload reduction – pretreatment with calcium antagonists preserved mitochondrial respiration and the capacity to generate ATP as well as ATP and phosphocreatine stores and calcium overload during reperfusion was reduced [7, 37, 69, 88, 90]. Consequently, mechanical properties recovered more completely during reperfusion [99]. Again, these studies demonstrated the critical importance of the time of application of the calcium antagonist, i.e., *before* initiation of ischemia. In another study in which the calcium antagonist was only infused *after* occlusion of the vessel the same protective effects could not be observed [125]. Inhibition of calcium inward transport and preservation of ATP levels will maintain cellular homeostasis of calcium and mitochondrial function.

d) A direct protective effect on the sarcolemma of myocytes and on endothelial cells of coronary vessels. Intravital microscopy permits direct visualization of the coronary microvasculature, and increased vascular permeability to macromolecules during ischemia was demonstrated. This leakage was attenuated following administration of calcium antagonists, indicating maintained microvascular integrity [63, 64, 111]. With addition of verapamil to the perfusate, the enzymatic profile of the myocyte sarcolemma – expressed as the activities of sodium-potassium-ATPase and 5'-nucleotidase and the sodium-calcium exchange rate – could be relatively preserved over a 60 min-period of global myocardial ischemia [17]. Together with the inhibition of calcium accumulation in the sarcolemma [61], this protection from decline of intrinsic biochemistry of the sarcolemma may contribute to the overall myocardial protective effects of calcium antagonists.

e) An attenuated depletion of left ventricular norepinephrine reserve during reperfusion [72] which may limit the damage of the ischemic episode.

f) Less well identified effects of calcium antagonists like the decrease of cytosolic enzyme activity by inhibition of the subcellular leakage of lysosomal enzymes [45] or the inhibition of platelet activity at high concentrations of calcium antagonists may further contribute to the limitation of myocardial injury during ischemia and reperfusion.

305

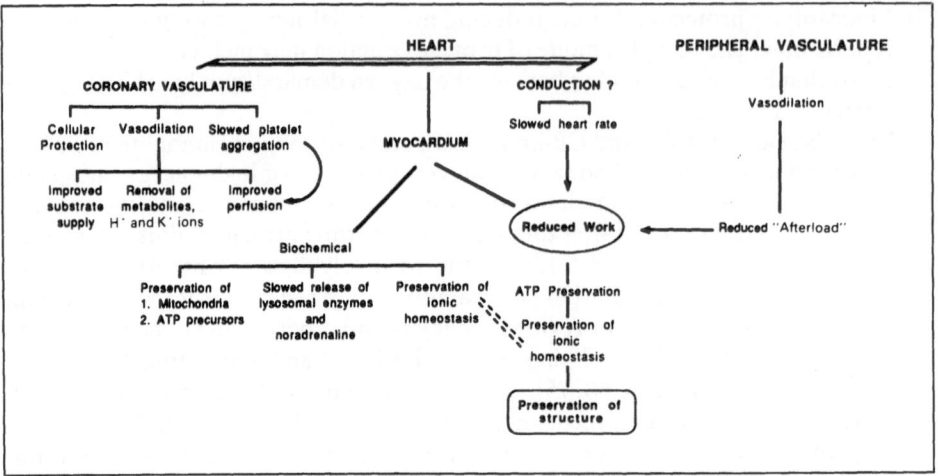

Fig. 7. Summary of the "protective" effects of calcium antagonists in myocardial ischemia. See text for details. from: [73].

In essence, all the "protective" actions of calcium antagonists aim primarily at maintaining a level of energy sufficient to preserve the ionic homeostasis of the ischemic myocyte as summarized in Fig. 7.

The sequence of ischemia- and reperfusion-induced changes in cellular morphology and biochemistry and the interference of calcium antagonists with this cascade of damage emphasize the importance of prophylactic use of calcium antagonists in clinical practice. From numerous experimental studies (for review see: [73]) it seems evident that calcium antagonists must be present at the onset of ischemia to reduce the energy requirements and to establish their protective action. This is of major clinical importance [44].

Clinical Implications

All of the extensively evaluated calcium antagonists demonstrate *in vitro* in appropriately selected preparations hemodynamic effects that are capable of reducing myocardial ischemia and ischemic damage. There is, however, considerable variability between individual calcium antagonists as to their relative potencies for these various actions. In the patient treated with calcium antagonists the net hemodynamic and electrophysiological effect of the individual compound is a combination of its direct effects and reflex-mediated effects.

Myocardial "protection" in the situation of flow-related regional ischemia has been demonstrated in different clinical syndromes.

Chronic Stable Angina Pectoris

In patients afflicted with chronic angina pectoris myocardial ischemia is generally precipitated at a reproducible level of physical exertion. The presence of a high-grade coronary obstruction limits the increase of coronary blood flow necessary to cover myocardial oxygen demand. Numerous placebo-controlled, double-blind studies with a single calcium antagonist demonstrate a 36–60% reduction in frequency of anginal attacks and a 26–39% improvement of exercise time with a daily dose of either 30 to 60 mg nifedipine [8, 20, 21, 46, 66, 100], 120 to 360 mg diltiazem [5, 29, 42, 57, 79, 107, 121] or 120 to 480 mg verapamil [3, 11, 80, 109]. The beneficial effects tended to increase with higher dosages of the individual calcium antagonist. More recently, the effects of calcium antagonists on myocardial ischemia during daily life activities were monitored by ambulatory Holter systems [114]. Compared with long-acting nitrates and different beta-blocking agents the calcium antagonists generally demonstrate an equivalent reduction of ischemia [4, 56].

Variant and Unstable Angina

Chronic exertional angina may deteriorate to angina at minimal exertion or even angina at rest associated with ST-segment changes. In these aggravated (unstable) anginal syndromes and in variant angina (Prinzmetal's angina) inadequate coronary vasoconstriction either superimposed on a preexisting coronary lesion or in angio-graphically "normal" coronary arteries transiently impairs myocardial blood supply. Nifedipine [27, 77] and verapamil [78] have been reported to reduce anginal attacks and ST-segment changes. In the situation of unstable angina with a complex and mul-tifactorial trigger mechanism for myocardial ischemia a combination therapy is advis-able. Different combination strategies have been found to be more effective than a single-agent treatment [67, 110]. In some patients even a triple regimen with nitrates, calcium antagonists, and beta-blockers will not completely prevent attacks of myo-cardial ischemia and an early revascularization procedure is recommended [34].

In variant angina the efficacy of calcium antagonists has been documented in numerous controlled trials [41, 82, 89, 92, 126]. The invariably favorable clinical response with a high percentage of complete protection from ischemic attacks emphasizes the key role of enhanced vasomotor tone as one etiologic factor respon-sible for recurrent ischemia in patients with angina at rest. Spontaneous or drug-induced coronary vasoconstriction can be prevented with calcium antagonists. In addition, the reduction in cardiac work, as well as in platelet aggregation, may further enhance the beneficial effects of calcium antagonists in different clinical syn-dromes of myocardial ischemia.

References

1. Allen DG, Orchard CH (1983) Intracellular calcium concentration during hypoxia and metabolic inhibition in mammalian ventricular muscle. J Physiol (London) 339:107–122
2. Amende I, Simon R, Hood WP, Hetzer R, Lichtlen PR (1983) Intracoronary nifedipine in human beings: magnitude and time course of changes in left ventricular contraction/relaxa-tion and coronary sinus blood flow. J Am Coll Cardiol 2:1141–1145
3. Bala Subramanian V, Bowles M, Penalver H, Davies AB, Raftery EB (1981) Verapamil in ischemic heart disease – an objective study in 100 patients. In: Zanchetti A, Krikler DM (eds). Calcium Antagonism in Cardiovascular Therapy. Excerpta Medica, Amsterdam, pp 118–139

4. Bala Subramanian V (1983) Calcium antagonists in chronic stable angina pectoris. Excerpta Medica, Amsterdam
5. Bala Subramanian V, Khurmi NS, Bowles MJ, O'Hara M, Raftery EB (1983) Objective evaluation of three dose levels of diltiazem in patients with chronic stable angina. J Am Coll Cardiol 1:1144–1153
6. Benham CD, Tsien RW (1987) A novel receptor-operated calcium-permeable channel activated by ATP in smooth muscle. Nature 328:275–278
7. Bier C, Klassen G, Hütter I, Mamer O, Mogensen L, Zborowska-Sluis D (1978) Mitochondrial protection by nifedipine in ischemic myocardium. Circulation 57/58 (Suppl. II) II:99 (abstr.)
8. Boehm C, Christl HL (1983) Comparative studies of efficacy of nifedipine and molsidomin for the treatment of stable angina pectoris. Therapiewoche 33:4604–4613
9. Bolton TB (1979) Mechanism of action of transmitters and other substances on smooth muscle. Physiol Rev 59:606–621
10. Bourdillon PD, Poole-Wilson PA (1982) The effects of verapamil, quiescence and cardioplegia on calcium exchange and mechanical function in ischemic rabbit myocardium. Circ Res 50:360–368
11. Brodsky SJ, Cutler SS, Weiner DA, McCabe CH, Ryan TJ, Klein MD (1982) Treatment of stable angina of effort with verapamil: a double-blind, placebo-controlled, randomized crossover study. Circulation 66:569–574
12. Carafoli E (1984) How calcium crosses plasma membranes including the sarcolemma. In: Opie LH (ed). Calcium Antagonists and Cardiovascular Disease. Raven Press, New York, pp 29–41
13. Casteels R, Droogmans G (1982) Membrane potential and excitation – contraction coupling in smooth muscle. Fed Proc 41: 2879–2884
14. Cheung JY, Bonventre JV, Malis CD, Leaf A (1986) Calcium and ischemic injury. New Engl J Med 314:1670–1676
15. Chew CC, Hecht HS, Collett JT, McAllister RG, Singh BN (1981) Influence of severity of ventricular dysfunction on hemodynamic responses to intravenously administered verapamil in ischemic heart disease. Am J Cardiol 47:917–922
16. Cobbold PH, Bonrue PK (1984) Aequorin measurements of free calcium in single heart cells. Nature 312:444–446
17. Daly MJ, Elz JS, Nayler WG (1985) The effect of verapamil on ischemia-induced changes to the sarcolemma. J Mol Cell Cardiol 17:667–674
18. Daniel WG, Engel HJ, Lichtlen PR (1984) Effects of Ca-antagonists on regional myocardial blood flow. In: Althaus U, Burckhardt D, Vogt E (eds) Ca-Antagonism. International Symposium on Calcium-Antagonism. Universimed, Frankfurt, pp 152–162
19. De Jong JW, Harmsen E, De Tombe PP, Keijzer E (1982) Nifedipine reduces adenine nucleotide breakdown in ischemic rat heart. Eur J Pharmacol 81:89–94
20. Deanfield J, Wright C, Fox K (1983) Treatment of angina pectoris with nifedipine: importance of dose titration. Br Med J 286:1467–1470
21. Ebner F (1975) Survey of the results of world wide clinical trials with nifedipine. In: Hashimoto K, Kimura E, Kobayashi T (eds) First International Nifedipine Symposium. University Tokyo press, Tokyo, pp 282–290
22. Elz JS, Panagiotopoulos S, Nayler WG (1989) Reperfusion-induced calcium gain after ischemia. Am J Cardiol 63:7E–13E
23. Engel HJ, Lichtlen PR (1981) Beneficial enhancement of coronary blood flow by nifedipine. Comparison with nitroglycerin and beta-blocking agents. Am J Med, 71:658–666
24. Finkel AS, Hirst GDS, van Helden DF (1984) Some properties of excitatory junction currents recorded from submucosal arterioles of guinea-pig ileum. J Physiol 351:87–98
25. Fleckenstein A (1983) History of calcium antagonists. Circ Res 52:I3–I16
26. Freudenberg H, Lichtlen PR (1981) Das normale Wandsegment bei Koronarstenose – eine postmortale Studie. Z Kardiol 70: 863–869
27. Gerstenblith G, Ouyang P, Achuff SC, Bulkley BH, Becker LC, Mellits ED, Baughman KL, Weiss JL, Flaherty JT, Kallman CH, Llewellyn M, Weisfeldt ML (1982) Nifedipine in unstable angina: a double-blind randomized trial. N Engl J Med 306:885–889
28. Ginsburg R, Bristow MR, Harrison DC, Stinson EB (1980) Studies with isolated human

coronary arteries: some general observations, potential mediators of spasm, role of calcium antagonists. Chest 78 [Suppl]:180−186

29. Go M, Hollenberg M (1984) Improved efficacy of high dose versus medium and low dose diltiazem therapy for chronic stable angina pectoris. Am J Cardiol 53:669−673
30. Godfraind T, Finet M, Socrates Lima J, Miller RC (1984) Contractile activity of human coronary arteries and human myocardium in vitro and their sensitivity to calcium entry blockade by nifedipine. J Pharmacol Exp Ther 230:514−518
31. Godfraind T (1986) Calcium entry blockade and excitation-contraction coupling in the cardiovascular system (with an attempt of pharmacological classification). Acta Pharmacol Toxicol 58 [Suppl II]:5−30
32. Godfraind T, Egleme C, Finet M, Debande B, Jaumin P (1987) Comparison of nifedipine and nisoldipine on human arteries and human cardiac tissues in vitro. In: Hugenholtz PG, Meyer J (eds) Nisoldipine 1987, Springer, Berlin − Heidelberg − New York, pp 36−44
33. Godfraind T (1987) Classification of the calcium antagonists. Am J Cardiol 59:11B−23B
34. Gottlieb SO, Weisfeldt ML, Ouyang P, Mellits ED, Gerstenblith G (1987) Silent ischemia predicts infarction and death during two year follow-up of unstable angina. J Am Coll Cardiol 10:756−760
35. Guazzi MD, Fiorentini C, Olivari MT, Bartorelli A, Necchi G, Polese A (1980) Short- and long-term efficacy of a calcium antagonistic agent (nifedipine) combined with methyldopa in the treatment of severe hypertension. Circulation 61:913−919
36. Gunther S, Muller JE, Mudge GH, Grossman W (1981) Therapy of coronary vasoconstriction in patients with coronary artery disease. Am J Cardiol 47:157−162
37. Henry PD, Shuchleib R, Davis J, Weiss ES, Sobel BE (1977) Myocardial contracture and accumulation of mitochondrial calcium in ischemic rabbit heart. Am J Physiol 233:H677−H680
38. Henry PD, Shuchleib R, Clark RE, Perez JE (1979) Effects of nifedipine on myocardial ischemia: analysis of collateral flow, pulsatile heat and regional muscle shortening. Am J Cardiol 44:817−824
39. Heusch G, Deussen A (1984) Nifedipine prevents sympathetic vasoconstriction distal to severe coronary stenoses. J Cardiovasc Pharmacol 6:378−383
40. Heusch G, Guth BD, Seitelberger R, Ross J (1987) Attenuation of exercise-induced myocardial ischemia in dogs with recruitment of coronary vasodilator reserve by nifedipine. Circulation 75:482−490
41. Hill JA, Feldman RL, Pepine CJ, Conti CR (1982) Randomized double-blind comparison of nifedipine and isosorbide dinitrate in patients with coronary arterial spasm. Am J Cardiol 49:431−438
42. Hossack KF, Pool PE, Steele P, Crawford MH, De Maria AN, Cohen LS, Ports TA (1982) Efficacy of diltiazem in angina on effort: a multicenter trial. Am J Cardiol 49:567−572
43. Hugenholtz PG, Michels HR, Serruys PW, Brower RW (1981) Nifedipine in the treatment of unstable angina, coronary spasm and myocardial ischemia. Am J Cardiol 47:163−173
44. Hugenholtz PG, Serruys PW, Fleckenstein A, Nayler W (1986) Why calcium antagonists will be most useful before or during early myocardial ischemia and not after infarction has been established. Eur Heart J 7:270−278
45. Ichihara K, Haneda T, Onodera S, Abiko Y (1987) Inhibition of ischemia-induced subcellular redistribution of lysosomal enzymes in the perfused rat heart by the calcium entry blocker diltiazem. J Pharmacol Exp Ther 242:1109−1113
46. Jenkins RM, Nagle RE (1982) The symptomatic and objective effects of nifedipine in combination with betablocker therapy in severe angina pectoris. Postgrad Med J 58:697−700
47. Jennings RB, Schaper J, Hill ML, Steenbergen G, Reimer KA (1985) Effects of reperfusion late in the phase of reversible ischemic injury. Circ Res 56:262−278
48. Jolly SR, Gross GJ (1980) Improvement in ischemic myocardial blood flow following a new calcium antagonist. Am J Physiol 239:H163−H171
49. Jost S, Rafflenbeul W, Mogwitz B, Nellessen U, Hecker H, Ahr G, Lichtlen PR (1987) Coronary vasomotility with different intravenous doses of nisoldipine. In: Hugenholtz PG, Meyer J (eds) Nisoldipine 1987. Springer, Berlin, Heidelberg, New York, pp 165−170
50. Jost S, Rafflenbeul W, Lichtlen PR (1989) Assessment of the vasomotility of epicardial coronary arteries with quantitative coronary angiography. Z Kardiol 78 [Suppl 6]:143−148

51. Kaltenbach M, Schulz W, Kober G (1979) Effects of nifedipine after intravenous and intracoronary administration. Am J Cardiol 44:832–836

52. Kihara Y, Grossmann W, Morgan JP (1989) Direct measurement of changes in intracellular calcium transient during hypoxia, ischemia and reperfusion of the intact mammalian heart. Circ Res 65:1029–1044

53. Koch G (1980) Beta-receptor and calcium blockade in ischemic heart disease: effects on systemic and pulmonary hemodynamics and on plasma catecholamines at rest and during exercise. In: Puech P, Krebs R (eds) 4th International Adalat Symposium. Excerpta Medica, Amsterdam, pp 131–142

54. Lange R, Ingwall J, Hale SL, Alker KJ, Braunwald E, Kloner RA (1984) Preservation of high-energy phosphates by verapamil in reperfused myocardium. Circulation 70:734–741

55. Lee HC, Mohabir R, Smith N, Franz MR, Clusin WT (1988) Effects of ischemia on calcium-dependent fluorescence transients in rabbit hearts containing indo 1: correlation with monophasic action potentials and contraction. Circulation 78:1047–1059

56. Liang CS, Coplin B, Wellington K (1985) Comparison of antianginal efficacy of nifedipine and isosorbide dinitrate in chronic stable angina: a long term, randomized, double blind, crossover study. Am J Cardiol 55:9E–14E

57. Lindenberg BS, Weiner DA, McCabe CH, Cutler SS, Ryan TJ, Klein MD (1983) Efficacy and safety of incremental doses of diltiazem for the treatment of stable angina pectoris. J Am Coll Cardiol 2:1129–1133

58. Lodge NJ, Van Breemen C (1988) Calcium pathways mediating agonist-activated contraction of vascular smooth muscle and EDRF release from endothelium. In: Morad M, Nayler W, Kazda S, Schramm M (eds) The Calcium Channel: Structure, Function and Implications. Springer, Berlin, Heidelberg, New York. pp 283–292

59. Low RI, Takeda P, Mason DT, DeMaria AN (1982) The effects of calcium channel blocking agents on cardiovascular function. Am J Cardiol 49:547–553

60. Malacoff RF, Lorell BH, Mudge GH, Holman BL, Idoine J, Bifolck L, Cohn PF (1982) Beneficial effect of nifedipine on regional myocardial blood flow in patients with coronary artery disease. Circulation 65 [Suppl. I]:I32–I37

61. Mas-Oliva J, Nayler WG (1980) The effect of verapamil on the calcium-transporting and calcium ATPase activity of isolated cardiac sarcolemmal preparations. Br J Pharmacol 70:617–624

62. Matsuzaki M, Guth BD, Tajimi T, Kemper WS, Ross J (1985) Effect of the combination of diltiazem and atenolol on exercise-induced regional myocardial ischemia in conscious dogs. Circulation 72:233–243

63. Mayhan WG, Joyner WL (1984) The effect of altering the external calcium concentration and a calcium channel blocker, verapamil, on microvascular leaky sites and dextran clearance in the hamster cheek pouch. Microvasc Res 28:159–179

64. Mc Donagh PF, Roberts DJ (1986) Prevention of transcoronary macromolecular leakage after ischemia-reperfusion by the calcium entry blocker nisoldipine. Circ Res 58:127–136

65. Millard R, Grupp G, Grupp IL, Disalvo J, De Pover A, Schwartz A (1983) Chronotropic, inotropic and vasodilator actions of diltiazem, nifedipine and verapamil. A comparative study of physiological responses and membrane receptor activity. Circ Res 52 [Suppl I]:29–39

66. Mueller HS, Chahine RA (1981) Interim report of multicenter double-blind, placebo-controlled studies of nifedipine in chronic stable angina. Am J Med 71:645–657

67. Muller JE, Morrison J, Stone PH, Rude RE, Rosner B, Roberts P, Pearle D, Toorey ZG, Schneider JF, Serfas DH (1984) Nifedipine therapy for threatened and acute myocardial infarction: a randomized double-blind placebo-controlled comparison. Circulation 69:740–747

68. Nayler WG, Gran A, Slade A (1976) A protective effect of verapamil on hypoxic heart muscle. Cardiovasc Res 10:650–657

69. Nayler WG, Ferrari R, Williams A (1980) Protective effect of pretreatment with verapamil, nifedipine and propranolol on mitochondrial function in the ischemic and reperfused myocardium. Am J Cardiol 46:242–248

70. Nayler WG, Sturrock WJ (1985) Inhibitory effect of calcium antagonists on the depletion of cardiac norepinephrine during postischemic reperfusion. J Cardiovasc Pharmacol 7:581–587

310

71. Nayler WG, Elz JS (1986) Reperfusion injury: laboratory artifact or clinical dilemma? Circulation 74:215–221
72. Nayler WG, Panagiotopoulos S, Elz JS, Sturrock WJ (1987) Fundamental mechanisms of action of calcium-antagonists in myocardial ischemia. Am J Cardiol 79:75B–83B
73. Nayler W (1988) Calcium Antagonists. Academic Press, London, San Diego
74. Nellessen U, Rafflenbeul W, Jost S, Daniel W, Hecker H, Lichtlen PR (1987) Koronarweitenänderung nach sublingualer und intravenöser Nifedipinapplikation in Korrelation zu Plasmaspiegeln. Z Kardiol 76:329–339
75. Nigdikar SV, Bowditch J, Dow JW (1986) Calcium antagonists and adenine nucleotide metabolism in rat heart. Cardiovasc Res 20:604–608
76. Opie LH (1987) Calcium channel antagonists, part I: fundamental properties: mechanisms, classification, sites of action. Cardiovasc Drugs Ther 1:411–430
77. Otsu F, Kishida H (1987) Antianginal efficacy of nisoldipine in patients with unstable angina pectoris: evaluation on Holter ECG. In: Hugenholtz PG, Meyer J (eds) Nisoldipine 1987. Springer, Berlin Heidelberg New York Tokyo, pp 115–122
78. Parodi O, Maseri A, Simonetti I (1979) Management of unstable angina at rest by verapamil. A double-blind cross-over study in coronary care unit. Br Heart J 41:167–174
79. Petru MA, Crawford MH, Sorenson SG, Chaudhuri TK, Levine S, O'Rourke RA (1983) Short- and long-term efficacy of high-dose oral diltiazem for angina due to coronary artery disease: a placebo-controlled randomized, double-blind crossover study. Circulation 68:139–147
80. Pine MD, Citron D, Bailley DJ, Butman S, Plasencia GO, Landa DW, Wong RK (1982) Verapamil versus placebo in relieving stable angina pectoris. Circulation 65:17–22
81. Poole-Wilson PA, Harding DP, Bourdillon PDV, Tones MA (1984) Calcium out of control. J Mol Cell Cardiol 16:175–187
82. Previtali M, Salerno JA, Pancipoli C, Guasti L, Chimienti M, Montemantini C (1986) Short term effectiveness of nifedipine, diltiazem and verapamil in Prinzmetal's variant angina. Evaluation by Holter monitoring and ergometrine testing. In: Lichtlen PR (ed) 6th International Adalat Symposium, pp 280–286
83. Rafflenbeul W, Urthaler F, Russell R, Lichtlen P, James TN (1980) Dilatation of coronary artery stenoses after isosorbide dinitrate in man. Br Heart J 43:546–549
84. Rafflenbeul W, Lichtlen PR (1982) Zum Konzept der "dynamischen" Koronarstenose. Z Kardiol 71:439–444
85. Rafflenbeul W (1983) Dilatation of coronary artery stenoses with diltiazem i.v. In: Fleckenstein A, Hashimoto K, Herrmann M, Schwartz A, Seipel L (eds) New Calcium Antagonists – Recent Developments and Concepts. Fischer, Stuttgart, pp 181–190
86. Rafflenbeul W, Lichtlen P (1983) Quantitative coronary angiography – evidence of a sustained increase in vascular smooth muscle tone in coronary artery stenosis. Z Kardiol 72 [Suppl. III]:87–91
87. Rafflenbeul W, Jost S, Berger C, Lichtlen P (1989) Wirkung von Kalziumantagonisten und Betarezeptorenblockern auf die Koronarweite. Z Kardiol 78 [Suppl 5]:16–19
88. Reimer KA, Löwe JE, Jennings RB (1977) Effects of the calcium antagonist verapamil on necrosis following temporary coronary artery occlusion in dogs. Circulation 55:581–587
89. Rizzon P, Scrutinio D, Mangini SG, Lagioia R, de Toma L (1986) Randomized placebo-controlled comparative study of nifedipine, verapamil and isosorbide dinitrate in the treatment of angina at rest. Eur Heart J 7:67–76
90. Robb-Nicholson C, Currie WD, Wechsler A (1978) Effects of verapamil on myocardial tolerance to ischemic arrest. Circulation 58 [Suppl I]:I119–I123
91. Rosendorff C (1984) Calcium channel blockers and hypertension. In: Opie LH (ed) Calcium antagonists and cardiovascular disease. Raven Press, New York, pp 323–331
92. Rosenthal SJ, Ginsberg R, Lamb IH, Baim DS, Schroeder JS (1980) Efficacy of diltiazem for control of symptoms of coronary arterial spasm. Am J Cardiol 46:1027–1032
93. Ross Jr. J (1989) Mechanisms of regional ischemia and antianginal drug action during exercise. Progr Cardiovasc Dis 31: 455–466
94. Rouleau JL, Chatterjee K, Ports TA, Doyle MB, Hiramatsu B, Parmley W (1983) Mechanism of relief of pacing-induced angina with oral verapamil: reduced oxygen demand. Circulation 67:94–100

95. Schanzenbächer P, Liebau G, Deeg P, Kochsiek K (1982) Koronarsinusfluß und myokardialer Sauerstoffverbrauch nach intrakoronarer Nifedipininjektion bei Patienten mit koronarer Herzkrankheit. Z Kardiol 71:393–396

96. Schulz W, Krauss G, Kaltenbach M, Kober G (1981) Einfluß von intrakoronarem und intravenösem Nifedipin auf die allgemeine und lokale Gefäßweite von epikardialen Koronararterien bei stabiler Angina pectoris – ein antianginöser Wirkaspekt? Z Kardiol 70:809–815

97. Serruys PW, Brower RW, ten Katen HJ, Bom AH, Hugenholtz PG (1981) Regional wall motion from radiopaque markers after intravenous and intracoronary injections of nifedipine. Circulation 63:584–591

98. Serruys PW, Hooghondt TEH, Reiber JHC, Slager C, Brower RW, Hugenholtz PG (1983) Influence of intracoronary nifedipine on left ventricular function, coronary vasomobility, and myocardial oxygen consumption. Br Heart J 49:427–433

99. Sherman LG, Liang CS, Boden WE, Hood WB (1981) The effect of verapamil on mechanical performance of acutely ischemic and reperfused myocardium in the conscious dog. Circ Res 48:224–230

100. Sherman LS, Liang CS (1983) Nifedipine in chronic stable angina: a double-blind placebo controlled crossover trial. Am J Cardiol 51:706–711

101. Simon R (1984) Kalziumantagonisten: Wirkung auf periphere und koronare Hämodynamik. Z Kardiol 73 [Suppl 2]:79–88

102. Singh BN, Roche AHG (1977) Effects of intravenous verapamil on hemodynamics in patients with heart disease. Am Heart J 94: 593–599

103. Singh BN (1986) The mechanism of action of calcium antagonists relative to their clinical applications. Br J Clin Pharmacol 21:109 S–119 S

104. Somlyo AP (1985) Excitation-contraction coupling in smooth muscle. Circ Res 57:497–507

105. Spedding M (1985) Calcium antagonists subgroups. Trends Pharmacol Sci 6:109–114

106. Steenbergen C, Murphy E, Levy L, London RE (1987) Elevation in cytosolic free calcium concentration early in myocardial ischemia in perfused rat hearts. Circ Res 60:700–707

107. Strauss WE, McIntyre S, Parisi AF, Shapiro W (1982) Safety and efficacy of diltiazem hydrochloride for the treatment of stable angina pectoris: report of a cooperative clinical trial. Am J Cardiol 49:560–566

108. Sturek M, Thayer SA, Miller RJ (1988) Intracellular calcium release activates calcium-permeable ion-channels in coronary artery smooth muscle cells. Biophys J 53:561 (abstr.)

109. Tan ATH, Sadick N, Kelley DT, Harris PJ, Freedman SB, Bautovich G (1982) Verapamil in stable effort angina: effects on left ventricular function evaluated with exercise radionuclide ventriculography. Am J Cardiol 49:425–430

110. Theroux P, Taeymans Y, Morissette D, Bosch X, Pelletier GB, Waters DD (1985) A randomized study comparing propranolol and diltiazem in the treatment of unstable angina. J Am Coll Cardiol 5:717–722

111. Tilton RG, Larson KB, Cole PA, Williamson JR (1984) Diltiazem prevents coronary vascular resistance and permeability changes following no-flow ischemia. Fed Proc 43:335 (abstr.)

112. Tsien KRW, Bean BP, Hess P, Lansman JB, Nilius B, Nowycky MC (1986) Mechanism of calcium channel modulation by beta-adrenergic agents and dihydropyridine calcium agonists. J Mol Cell Cardiol 18:691–710

113. Tsien RW (1989) Calcium channels as molecular transducers in heart, smooth muscle, and sympathetic neurons. Circulation 80 [Suppl II]:IIA–IIB

114. Tzivoni D, Keren A, Gavish A, Benhorin J, Stern S (1987) Guiding anti-ischemic therapy by Holter monitoring. In: v. Arnim Th, Maseri A (eds) Silent ischemia, Steinkopff, Darmstadt, pp 177–183

115. Urquhart J, Patterson RE, Bacharach SL, Green MV, Speir EH, Aamodt R, Epstein SE (1984) Comparative effects of verapamil, diltiazem, and nifedipine on hemodynamics and left ventricular function during acute myocardial ischemia in dogs. Circulation 69:382–390

116. van Breemen C, Siegel B (1980) The mechanism of alpha-adrenergic activation of the dog coronary artery. Circ Res 46: 426–429

117. van den Brand M, Remme WJ, Meesters GT, Tiggelaar-de Widt J, de Ruiter R, Hugenholtz PG (1975) Hemodynamic effects of nifedipine in patients catheterized for coronary artery dis-

ease. In: Braasch W, Kroneberg G (eds) Second International Adalat Symposium, Springer, Berlin Heidelberg New York Tokyo, pp 145–152

118. van Zwieten PA (1985) Calcium antagonists – terminology, classification and comparison. Drug Res 35:298–301
119. Vanhoutte PM (1987) The expert committee of the World Health Organisation on classification of calcium antagonists: the viewpoint of the raporteur. Am J Cardiol 59:3A
120. Vatner SF, Hintze TH (1982) Effects of a calcium-channel antagonist on large and small coronary arteries in conscious dogs. Circulation 66:579–588
121. Wagniart P, Ferguson RJ, Chaitman BR, Achard F, Benacerraf A, Delanguenhagen B, Morin B, Pasternac A, Bourassa MG (1982) Increased exercise tolerance and reduced electrocardiographic ischemia with diltiazem in patients with stable angina pectoris. Circulation 66:23–28
122. Ware AJ, Johnson PC, Smith M, Salzman EW (1986) Inhibition of human platelet aggregation and cytoplasmic calcium response by calcium antagonists: studies with aequorin and Quin 2. Circ Res 59:39–42
123. Warltier DC, Meils CM, Gross GJ, Brooks HL (1981) Blood flow in normal and acutely ischemic myocardium after verapamil, diltiazem and nisoldipine (BAY K5552), a new dihydropyridine calcium antagonist. J Pharmacol Exp Ther 228:296–302
124. Weintraub WS, Akizuki S, Agarwal JB, Bodenheimer MM, Banka VS, Helfant RH (1982) Comparative effects of nitroglycerin and nifedipine on myocardial blood flow and contraction during flow-limiting coronary stenosis in the dog. Am J Cardiol 50:281–288
125. Weishaar R, Ashikawa K, Bing RJ (1979) Effect of diltiazem, a calcium antagonist, on myocardial ischemia. Am J Cardiol 43: 1137–1143
126. Winniford MD, Johnson SM, Mauritson DR, Rellas JS, Redish GA, Willerson JT, Hillis LD (1982) Verapamil therapy for Prinzmetal's variant angina: comparison with placebo and nifedipine. Am J Cardiol 50:913–918

Authors' address:
Wolf Rafflenbeul, M.D.
Division of Cardiology
Hannover Medical School
Konstanty-Gutschow-Str. 8
3000 Hannover 61, FRG

Combination Anti-Anginal Therapy: Rationale and Results

Hugh A. McCann and John Ross, Jr

Division of Cardiology, Department of Medicine, University of California, San Diego, La Jolla, California, USA

Introduction

It was not until the early 1900s that Osler in his Lumelian lectures on angina pectoris [65] observed the frequently poor correlation between the clinical symptoms and the underlying pathology of coronary artery disease. This presaged our current conception that myocardial ischemia is a complex, dynamic entity resulting from a number of anatomic and physiologic factors which combine to overwhelm the autoregulatory mechanisms of the coronary circulation. With the advancement of research techniques and the availability of methods for investigating coronary anatomy, regional perfusion, and myocardial function during life in both animals and in man, we have come to appreciate the dynamic balance which exists between myocardial O_2 supply and demand. However, understanding of the pathophysiologic mechanisms involved, and therefore the actions of antianginal drugs, has been limited by the lack of quantitative methods for accurately quantifying the degree of ischemia in man (the transmural coronary blood flow distribution and/or coronary perfusion pressure) and its effects on the myocardium.

The term "myocardial ischemia" refers, of course, to a state of diminished myocardial blood flow, resulting in O_2 deprivation and inadequate removal of metabolites. It is usually a consequence of coronary atherosclerosis developing over decades which eventually causes transient myocardial ischemia, with or without associated angina pectoris, and often results ultimately in irreversible damage with myocardial cell death. The term ischemia is often defined in relative, rather than absolute terms, since the requirements of the myocardium vary under different conditions. This supply-demand balance is readily appreciated when one considers the variety of provocative tests used in clinical cardiology and research designed to upset this balance and thereby precipitate myocardial ischemia. The increased use of ambulatory ECG monitoring to objectively assess the frequency and duration of ischemic episodes in a wide variety of patients with clinically evident or occult coronary artery disease (documented by coronary angiography) has led clinicians to appreciate that: 1) there is not a direct correlation between symptoms and severity of disease [17, 75]; 2) coronary vasomotor phenomena can contribute to ischemia [52]; 3) the presence or absence of angina pectoris may not give an adequate picture of prognosis [44]; and 4) the "total ischemic burden" may provide a useful indicator of the risk for coronary events and arrhythmias [79].

This work was supported by Ischemic Heart Disease SCOR Grant HL 17682 awarded by the National Heart, Lung and Blood Institute.

Although the issue of how and when to manage ischemia in the absence of all symptoms remains unsettled, the implication of the above observations is that treating symptomatic ischemic episodes alone may not be sufficient, since angina pectoris can be an unreliable underestimator of disease activity and risk. This possibility is supported by the fact that myocardial infarction is a common initial event, that up to one-third of myocardial infarctions occur undetected [43, 45], and that advanced coronary artery disease can be observed postmortem in relatively or entirely asymptomatic patients, and in patients experiencing sudden cardiac death [70, 81]. Thus, management is usually aimed at the detection of true ischemia and assessment of its severity in patients known or suspected to have coronary artery disease [74], and at treatment of the ischemia with drugs or revascularization in order to relieve symptoms more effectively, prevent myocardial injury, and prolong life.

Pathophysiology

The heart is an aerobic organ, relying almost entirely on oxidative metabolism under normal conditions. Under such circumstances, autoregulatory mechanisms maintain coronary blood flow in proportion to need [58]. The O_2 delivered to the myocardium, (the product of the coronary blood flow and the coronary A-VO$_2$ difference), is not usually measured clinically, and changes in coronary blood flow which normally are directly related to energy use are taken as an indicator of the O_2 supply [8]. Major factors affecting the coronary blood flow, in addition to the myocardial O_2 consumption (MVO_2), include the presence of fixed coronary stenosis, the coronary vasomotor tone, the aortic diastolic pressure (if it is reduced), the left ventricular end-diastolic pressure, the level of vasodilator reserve, and the presence or absence of collateral channels.

When progressive coronary stenosis is produced experimentally, a transstenotic pressure gradient develops associated with failure of coronary flow to increase appropriately on demand [30, 31]. As the coronary resistance-vessels dilate further, the poststenotic perfusion pressure falls as well, and as the stenosis becomes critical (i.e., greater than 70% reduction of diameter) the relationship between stenosis area and the resistance to flow becomes exponential [30].

The importance of coronary vasoconstriction has been emphasized by Maseri et al. [51, 52, 53], and more recently Selwyn and Deanfield [18, 77] have demonstrated that episodes of myocardial ischemia (both symptomatic and asymptomatic) in patients with coronary artery disease occur very frequently at rest or at low exercise levels, implying that such imbalances are due to a decrease in O_2 supply rather than increased demand. Moreover, atherosclerotic coronary arteries have been shown during coronary arteriography to react by constriction to cold stimulus [59], isometric hand grip exercise [12], and supine bicycle exercise [23]. Such vasoconstriction is most likely related to smooth muscle constriction due in part to increased alpha-adrenergic tone to the coronary arteries and to impaired endothelial function [29]; the latter is suggested by dose-dependent paradoxical vasoconstriction in response to local infusions of acetylcholine in diseased coronary arteries, consistent with impaired release of endothelial relaxing factor [50]. Finally, and beyond the scope of this discussion, the expanding use of coronary angiography in the setting of unstable angina and acute myocardial infarction has facilitated assessment of the role of plaque rupture and thrombosis in progressive unstable syndromes [2], and prompted experimental studies suggesting a role for platelet aggregation with the release of vasoconstrictor substances such as thromboxane A2 [90].

An important factor which has received increasing attention is the vasodilator reserve [46]. Autoregulation controls vascular tone in response to normal alterations in MVO_2 and coronary perfusion pressure to maintain coronary flow within the normal range, and therefore increases in the O_2 demand and/or reductions in perfusion pressure will compromise the coronary reserve [46, 58]. In addition to the effects of a coronary stenosis on distal perfusion pressure, some patients without coronary stenosis have impaired vasodilator reserve and objective evidence of myocardial ischemia, suggesting an abnormal vasodilator reserve in small coronary arteries or arterioles [15].

Regional Function during Ischemia and its Correlates

Experimentally, within 10−15 s after coronary occlusion regional dysfunction becomes marked [87], and with various degrees of coronary stenosis regional dysfunction provides a very sensitive marker of the degree of ischemia [25], the contractile defect appearing sooner and being more sensitive than body surface ECG abnormalities [5]. Regional dysfunction also persists longer than ECG changes after exercise-induced ischemia [33], reflecting post-exercise dysfunction or "stunning" [86].

In human subjects, within seconds of the onset of myocardial ischemia, abnormalities of diastolic function can be demonstrated, and systolic function promptly declines with depression of the contractile state of the ventricle, followed by disturbances in cation fluxes between ischemic and normal tissue resulting in characteristic ECG changes [62]. It is noteworthy, however, that these pathophysiologic functional changes occur prior to the onset of ECG changes and characteristic chest pain, and indeed they may be painless [38]. Also, several investigators have shown prolonged disturbances in myocardial perfusion and function in patients following exercise-induced ischemia, lasting two- to five-times longer than chest pain or ECG changes [71, 78]. All of these findings confirm the clinical relevance of the animal models.

The regional nature of myocardial ischemia has certain unique features which contrast with the effects of global ischemia [72]. These can be magnified when ischemia is induced by exercise, in that there is augmented cardiac output and enhanced inotropic state of normally perfused regions of the left ventricle, which in turn tend to increase the aortic pressure and thereby enhance the coronary perfusion pressure. In the ischemic zone, however, the initially increased O_2 demands due to increased heart rate, blood pressure, and contractility shift the relationship of coronary flow to perfusion pressure upward [58] so that the lower limit of autoregulation occurs at a higher pressure [39, 58]. Contractile performance in an area supplied by a stenosed coronary artery deteriorates promptly after exercise onset, as tachycardia leads to reduced subendocardial blood flow [27, 72, 89], and it is affected by the altered loading conditions and by complex interactions with the remainder of the left ventricle [72]. The tachycardia limits the diastolic perfusion time as well as increases the MVO_2, thereby quickly exhausting the vasodilator reserve of the subendocardium in

317

the ischemic zone [4, 27]. Regional ischemia can be induced experimentally by tread-mill exercise [54], by rapid atrial pacing [47], or by isoproterenol infusion [24].

Regional Flow-Function Relations

Using a conscious canine model of chronic single-vessel coronary stenosis produced by an ameroid constrictor we have been able to produce regional myocardial ischemia during exercise (resembling that seen in patients, with single-vessel coro-nary stenosis) accompanied by a severe regional contractile dysfunction, a subnormal blood pressure response to exercise, and characteristic ECG changes. The model of chronic coronary stenosis with reproducible regional ischemia induced by exercise allows study of both regional myocardial function and perfusion before and after antianginal drugs, as discussed subsequently. During exercise with critical coronary stenosis, subendocardial myocardial blood flow falls while subepicardial flow rises [27] and the relation between the steady-state level of subendocardial flow and regional wall contraction is nearly linear. At the onset of exercise, there is a transient period of mismatch between the rapidly increasing hemodynamic responses and the limited coronary vascular reserve. After several minutes, regional contraction reaches a depressed steady-state level, accompanied by reduced subendocardial and mid-wall flows implying that the reduced contraction is *matched* to the reduced blood flow [73]. Moreover, with exercise during incremental gradations of coronary stenosis to produce varying steady-state levels of ischemia, the nearly linear relation between subendocardial blood flow *per beat* and regional myocardial function falls very close to the relation observed under resting conditions, showing *perfusion-con-traction matching* under the altered conditions of exercise [73].

The relative importance of any single component in the pathophysiology of angina pectoris can vary from patient to patient and from minute to minute within the same patient. Such a diverse pathophysiologic spectrum often cannot be treated with an empiric approach, but rather consideration should be given to each of the pathophysiologic factors and its relative importance in a particular clinical setting. Since any single anti-anginal agent modifies only a few of the determinants of the O_2 supply-demand mismatch, the addition of other agents which have distinct but often additive anti-ischemic properties may act in a complementary fashion. These factors are summarized in Tables 1 and 2. Agents which decrease O_2 demand per minute (beta blockers) by affecting heart rate also result in improved subendocardial blood flow (O_2 supply) per beat (Table 1). Vasodilator agents decrease O_2 demand by reducing systolic wall stress, sometimes with a heart rate reduction, but they also can increase O_2 supply by reducing resistance at a stenosis site and/or reducing alpha adrenergically mediated coronary vasoconstriction (Table 2) [73]. Of course, alpha adrenergic tone is unopposed and enhanced after beta blockade. It is such effects which form the rationale for combined anti-anginal therapy. Moreover, the reflex circulatory reactions which often limit the therapeutic effects of any single vaso-dilator agent may be attenuated by a beta blocking agent, thus potentiating its anti-ischemic actions. Drug-induced adverse hemodynamic effects, (e.g., reflex tachycar-dia or negative inotropy) can sometimes be attenuated by the addition of a second, carefully chosen agent.

318

Table 1. Anti-anginal Mechanisms of Beta-Blockade on Ischemic Region during Steady Exercise

	Actions	Mechanism(s)	Effect
1.	↓Heart rate ↓Arterial pressure	↓O_2 Cons.$_{endo}$	↑O_2 Avail/beat ↑Contraction
2.	↓Heart rate	↑Diastolic perf. Time/min.	↑MBF_{endo} ↑↑MBF_{endo}/beat
3.	↓Heart rate ↓Arterial pressure ↓Contractility	↓O_2Cons.$_{epi}$ ↓Transmural steal	↓MBF_{epi} ↑MBF_{endo} ↑endo/epi

Cons = Consumption. Perf = perfusion. Avail = available. MBF = myocardial blood flow.
epi = subepicardium. endo = subendocardium.
(Reproduced by permission of author and publisher from [73].)

Table 2. Anti-anginal Mechanisms: Vasodilators and α-Blockade on Ischemic Region during Exercise

	Actions	Mechanism(s)	Effect
1.	Relax coronary smooth muscle	↓Vascular resistance of arterioles and/or collaterals ↑Caliber of large coronary arteries	↑$MBF_{transmural}$ Diminished degree of coronary stenosis
2.	↓LVED pressure ↓Heart size	↓Subendocardial tissue pressure	↑MBF_{endo}
3.	↓Arterial pressure ↓Heart size (↓Heart rate, variable)	↓O_2 Cons/minute	↑O_2 Avail/beat

Cons = Consumption. Perf = perfusion. Avail = available, MBF = myocardial blood flow,
endo = subendocardium.
(Reproduced by permission of author and publisher from [73].)

Actions of Anti-anginal Drugs

The medical therapy of ischemic heart disease has made great strides in the past 10 years. The agents currently available are traditionally categorized by whether their predominant effect is to reduce myocardial O_2 demand, or to increase myocardial blood flow and thereby increase O_2 supply. From our understanding of the pathophysiology, slowing of the heart rate should often be a major goal of therapy, particularly with exercise or stress-induced angina. While lowering blood pressure will decrease afterload and thereby diminish myocardial O_2 demands, it may also reduce coronary perfusion pressure and impair flow in the bed supplied by the stenosed artery.

319

Nitroglycerin

Until the early 1970s, nitroglycerin was the mainstay of therapy for patients with angina pectoris. At that time concerns about first pass metabolism [60] almost caused it to be abandoned, but many clinical trials since then have shown nitroglycerin therapy to be effective in improving exercise performance and reducing ST depression in patients with angina [67, 68, 73, 85], and long-acting nitrates often remain the underlying therapy to which other agents are added. It is likely that nitroglycerin acts by uptake of the compound into the vascular wall where it is reduced to nitric acid, leading to the production of S-nitrosothiols by interaction with sulphydryl donors [41]. This stimulates guanylate-cyclase, generating cyclic guanosine monophosphate (c-GMP), which is associated with inhibition of calcium influx into the cells and/or increased calcium sequestration by the sarcoplasmic reticulum, thereby producing relaxation of vascular smooth muscle and vasodilation [1]. Nitroglycerin acts strongly on the veins of the peripheral circulation to produce dose-related venodilation [3, 22], and with higher doses there is lowering of systemic vascular resistance. Nitroglycerin also dilates large coronary arteries and arterioles [10]. By reducing pressure, left ventricular size and afterload (and hence wall stress), nitroglycerin reduces O_2 demands in the normal and ischemic zones [48], but it also increases blood flow in collateral coronary vessels [32] as well as dilates stenotic lesions [10], thereby leading to improved regional O_2 supply.

Beta-Blocking Agents

Beta adrenergic blocking drugs have antianginal, antihypertensive, and antiarrhythmic properties, and they may also limit infarct size and prevent sudden death [76]. Their primary modes of action in relieving ischemia are: slowed heart rate, which decreases O_2 demand, and increased subendocardial myocardial blood flow per beat (Table 1). Thus, the lessened ischemia is due *both* to more time available for recovery of metabolism per beat and to increased subendocardial blood flow. This leads to augmented contractile function in the ischemic zone during exercise after propranolol both in the chronic coronary stenosis canine model [49], and in patients with exercise-induced ischemia [6]. Oral atenolol was also shown experimentally to improve blood flow to the subendocardium in the ischemic zone [56], with partial correction of the transmural maldistribution of blood flow in that region; the improvement of ischemic regional wall dysfunction was also associated with a reduction of the usual enhancement of contraction in normal regions during exercise. Thus, beta-blockade reduces O_2 consumption in the outer wall layers in the ischemic region (where fibers have been shown to contract even with severe dyskinesia [26]), thereby reducing blood flow requirements per minute which, in turn, leads to increased outer wall coronary vascular resistance with improved distal coronary perfusion pressure and a reduction in exercise-induced transmural "steal" [27]. Aortic pressure was lower after atenolol than during control exercise; thus, the diastolic perfusion pressure difference from aorta to the distal coronary bed, a major determinant of collateral blood flow [11], is reduced by beta-blockade [13]. It is therefore unlikely that changes in collateral flow play a significant role in the anti-ischemic effect. Finally, in a chronic coronary stenosis canine model, in which the heart rate was maintained by atrial pacing at the rate which occurred during exercise prior to beta-blockade, regional myocardial function failed to improve with beta-blockade and the blood flow to the ischemic region actually diminished below that observed during control exercise [34]. This emphasizes the singular importance of the anti-tachycardiac effects of beta-blockers on regional ischemia during exercise, as summarized in Table 1.

Calcium Channel Blockers

Calcium channel blockers constitute a group of organic compounds of diverse chemical structure, which produce coronary vasodilation and have negative inotropic, chronotropic, and dromotropic effects. The dihydropyridines (e.g., nifedipine) are far more potent coronary vasodilators relative to their negative inotropic, chronotropic and dromotropic effects. The phenylalkylamines (e.g., verapamil) and benzothiazepines (e.g., diltiazem) are roughly equipotent in producing coronary vasodilation and a negative chronotropic effect, but the latter is less potent in producing a negative inotropic effect [16]. In patients with coronary artery disease it is possible that the vasodilation of large arteries induced by most of these agents can reduce the degree of stenosis in non-circumferential, eccentric lesions [9]; however, this is unlikely to be a factor in the chronic animal model with fixed constriction produced by an ameroid constrictor. In that model, regional myocardial function in the ischemic zone was shown to increase during severe exercise-induced ischemia with diltiazem [54] nifedipine [37] and verapamil [64]. However, this response occurred at the same or reduced arterial pressure, indicating that vasodilation with lowered vascular resistance occurred in the native coronary bed and/or coronary collateral channels. With diltiazem, the increased subendocardial and transmural blood flows in the ischemic and normal zones were mediated, in part, by a modest reduction in heart rate [54]. But the improvement of contraction and blood flow produced by diltiazem was equal to that produced by atenolol, which caused a much larger reduction in heart rate, implicating an additional vasodilator action of diltiazem (Fig. 1). Moreover, in contrast to β-blockade, diltiazem and nifedipine caused no depression of contraction in normal regions during exercise [37, 54]. Experimentally, these effects on regional blood flow could also be demonstrated with a nifedipine dose below that necessary to produce appreciable systemic hemodynamic effects, suggesting that coronary vasodilator reserve was recruited [37]. These effects of calcium channel blockers are summarized in Table 2.

The comparative effects of nitroglycerin, nifedipine, and metoprolol on regional myocardial function were studied in humans with isolated left anterior descending disease by Pfisterer et al. [69]. They divided patients into two groups, those with exercise-induced ischemia (LAD stenosis) and those with previous transmural myocardial infarction and no ischemic changes during thallium imaging (LAD occlusion). A regional anti-ischemic effect evidenced by improved regional function in the patients with ischemia (LAD stenosis) was observed with all three agents. Although the anti-ischemic effects were similar with all agents, the authors pointed out the different underlying hemodynamic mechanisms, reflected by the effects of each agent on normal regions of myocardium and global left ventricular function. These findings serve

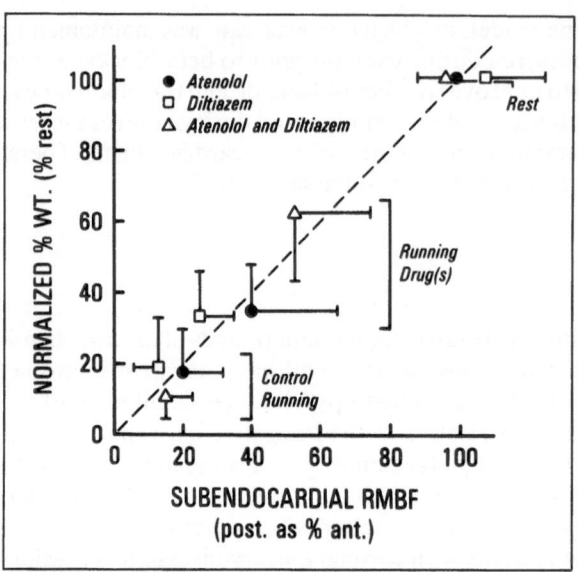

Fig. 1. Regional flow-function relations at rest and during exercise in a canine model of chronic coronary stenosis. Subendocardial regional myocardial blood flow values (RMBF) in the posterior (post.) ischemic wall (expressed as a percentage of that in the control anterior (ant.) wall) are plotted against normalized regional systolic wall thickening (%WT) in the ischemic zone (expressed as a percentage of the resting value).

The reductions in flow and function during control running prior to drug administration are similar for atenolol (closed circles), diltiazem (open squares), and the combination of these two drugs (open triangles). Significant improvement in both regional flow and regional function during exercise is evident with either drug alone, compared to control exercise, but when both drugs were given there was substantial further improvement of flow and function during running. (Reproduced by the permission of the American Heart Association and the authors [55].)

to confirm in man the effects observed with these agents in the canine chronic coronary stenosis model.

Combination Anti-anginal Therapy

In view of the varied mechanisms of action of antianginal agents, it might be expected that combinations of such drugs would have additive effects. Indeed, in the chronic coronary stenosis animal model, the combination of beta-blockade (atenolol) plus calcium channel blockade (diltiazem) was found to be complimentary, with both subendocardial myocardial blood flow and regional wall function in the ischemic zone during exercise improving twofold over the benefit observed with either agent alone (Fig. 1), [55]. When isosorbide dinitrate was added to this combination, regional dysfunction during milder ischemia in this animal model could be entirely prevented [35].

Clinical Studies on Combined Therapy

The many reported clinical studies of combination antianginal therapy have used different end points to assess efficacy, including the double product at peak exercise [82], exercise duration prior to angina [84], global and regional ejection fraction at rest and exercise as assessed by radionuclide ventriculography [66], and the frequency and duration of ischemic changes on ambulatory ECG monitoring. This endpoint variability, coupled with the varying doses of drugs administered in different temporal sequences and by different routes, render it difficult to directly compare any given combination with another.

Effects on Ischemia

Clinical studies on combination therapy are summarized in recent reviews [66, 84]. We will review briefly some of these studies, usually carried out in patients with persistent angina pectoris despite beta-blocker and nitrate therapy. When nifedipine was added to such therapy in a double-blind, placebo-controlled trial, patients exercised for significantly longer periods, and fewer were limited by angina [63]. Findlay et al. [20], in a placebo-controlled blind study of 15 patients, found that atenolol plus verapamil decreased the resting ejection fraction from 60% to 53%, but ST-segment depression during exercise, episodes of angina, and nitroglycerin use were significantly improved over use of either drug alone. Placebo-controlled studies on the addition of diltiazem (90 mg, q.i.d.) to propranolol and nitrates in patients with refractory angina have shown increased elimination of angina associated with improved exercise tolerance and reduced ST-segment changes [7, 83]. Miller et al. studied the combination of the beta-blocker nadolol (once-daily dose; low lipid solubility) and diltiazem in 18 patients with persistent angina on medical treatment; there was no difference in exercise time, but the number of patients with angina and ST-segment depression during exercise was significantly reduced [57]. It has also been reported that the combination of the calcium antagonist nicardipine with propranolol is superior to propranolol alone in reducing pacing-induced ischemia [36].

Egstrup studied the total ischemic burden (ambulatory monitoring and exercise testing) in 42 patients with coronary artery stenosis >75%; this burden was reduced by metoprolol but not by nifedipine alone, and there was a trend for further improvement (NS) with the combination [19]. The Mixed Angina Study Group investigated in a double-blind crossover study the efficacy of adding nifedipine to a regimen of a beta-blocker and/or long acting nitrates [82], subdividing 72 patients into those with "classic exertional angina" (17 patients) and "mixed" angina (55 patients, defined as exertional angina provoked by a variable threshold, or angina occurring occasionally at rest). Both exercise testing and 24 h ambulatory ECG monitoring were employed. It was found that patients with mixed angina were more likely to benefit symptomatically from the addition of nifedipine, suggesting that such patients may have more episodes of ischemia due to transient increases in coronary vasomotor tone superimposed on a fixed atherosclerotic obstruction. Although there was overall improvement in the mixed group, the individual responses were somewhat variable. Interestingly, in this study the authors also noted that the reduction in ischemia seen in the mixed group occurred only with painful episodes; there was no effect of combined therapy on the frequency of silent episodes [82].

An additional benefit of adding a calcium channel blocker to beta-blockade is that diltiazem attenuates the coronary vasoconstriction which may accompany beta-blocker therapy [40, 88].

Effects on Left Ventricular Function

Combined beta-blocker and calcium channel blocker therapy can have potentially deleterious effects on cardiac function in patients with impaired left ventricular function due to coronary heart disease. Several clinical studies have addressed this issue, assessing the safety of combinations of nifedipine and metoprolol [61], verapamil and metoprolol [80], and atenolol and nifedipine [14]. The latter study concluded that chronic oral administration of nifedipine in combination with atenolol precipitates left ventricular failure only in patients with a resting ejection fraction lower than 30% and a left ventricular end-diastolic pressure greater than 32 mmHg [14]. However, extensive infarction was usually present, and left ventricular failure was commonly apparent with atenolol therapy alone. Silke et al. [80] randomized patients with angiographically proven coronary artery disease and mildly depressed ejection fractions to intravenous verapamil or metoprolol, and combination therapy. Mild further depression of ejection fraction was found to be due mainly to the effects of metoprolol alone, with little or no added effect of verapamil in such patients. However, Johnston et al. [42] reported that propranolol plus verapamil improved the ejection fraction during exercise compared to either drug alone.

Adverse Effects

Strauss and Parisi [84] have recently reviewed in detail the complications of combined therapy. Increased fatigue has been noted, with lack of change in exercise time despite reduced angina pectoris. The combination of nifedipine with a beta-blocker does not appear to cause sinus bradycardia or significant conduction defects. Diltiazem plus a beta-blocker has shown only occasional minor conduction defects in controlled trials, with occasional sinus bradycardia, and rarely CHF. With verapamil plus a beta blocker, the incidence of significant sinus bradycardia or conduction defects (such as 2° heart block) is about 10%; also CHF or hypotension are most common with this combination, and occasionally these effects can be severe [84]. A few studies have documented exacerbation of angina pectoris by calcium antagonist plus beta-blocker combinations, especially when their interaction served to increase ventricular wall stress or cause profound bradycardia. Strauss and Parisi noted that most of the above effects occurred when a calcium channel blocker was added to high-dose beta-blockade, suggesting that the dose of beta-blocker should have been lowered, or therapy started with the calcium channel antagonist [63].

It may be concluded from the studies on combined calcium channel and beta-blocker therapy that this approach should be used selectively. Thus, as emphasized in a recent review by Packer [66], patients treated with multiple drugs may experience more side effects than when optimal treatment with a single agent is employed. Therefore, it was recommended that combination therapy be employed only in those patients with persistent symptoms despite maximal therapy with one or the other drug [66].

Recently, Frishman et al. reported on their experience with the combination of diltiazem and nifedipine in a selected group of patients with stable angina pectoris [21]. The combination of two calcium antagonists has some pharmacokinetic basis. Thus, the binding sites for each calcium antagonist are different, but the sites interact with one another allosterically. Diltiazem has been shown to stimulate the binding of both verapamil and nifedipine [28]. As monotherapy, diltiazem proved to be more effective than nifedipine in reducing the frequency of angina and the amount of nitroglycerin consumed, as well as in increasing exercise duration on a treadmill; however, combination therapy was effective in patients who did not satisfactorily respond to diltiazem or nifedipine alone [21].

In conclusion, myocardial ischemia is a multifactorial syndrome with different pathophysiological mechanisms operating to various degrees in any individual setting. Laboratory and clinical evidence for differing anti-ischemic mechanisms of antianginal drugs is reviewed. Since the mechanisms for ischemia may vary from patient to patient and over time in individual patients, it could be expected that combinations of anti-anginal drugs with differing modes of action would have additive effects. Persistent angina pectoris or signs of ischemia on maximal monotherapy should serve as the major impetus for considering the combination of a calcium antagonist and a beta blocker in patients without heart failure. It is suggested that the combination of verapamil and beta blocker, in particular, is clinically contraindicated in those patients with severe resting left ventricular dysfunction. Benefits of combination therapy in patients with mixed angina appear promising, but their role in silent ischemia appears to require further study. Whether or not combination therapy can reduce ischemic events and ultimately improve patient prognosis also remains to be determined.

References

1. Abrams J (1985) Hemodynamic effects of nitroglycerin and long-acting nitrates. Am Heart J 110:216–224
2. Ambrose JA, Winters SL, Arora RR, Eng A, Riccio A, Gorlin R, Fuster V (1986) Angiographic evaluation of coronary artery morphology in unstable angina. J Am Coll Cardiol 7:472–478
3. Armstrong PW, Moffat JA, Marks GS (1982) Arterial-venous nitroglycerin gradient during intravenous infusion in man. Circulation 66:1273–1276
4. Bache RJ, Cobb FR (1977) Effect of maximal coronary vasodilation on transmural perfusion during tachycardia in the awake dog. Circ Res 41:648–653
5. Battler A, Froelicher VF, Gallagher KP, Kemper WS, Ross J Jr (1980) Dissociation between regional myocardial dysfunction and ECG changes during ischemia in the conscious dog. Circulation 62:735–744
6. Battler A, Ross J Jr, Slutsky R, Pfisterer M, Ashburn W, Froelicher V (1979) Improvement of exercise-induced left ventricular dysfunction with oral propranolol in patients with coronary heart disease. Am J Cardiol 44:318–324
7. Boden W, Bough E, Reichman M, Rich V, Young P, Korr K, Shulman R (1985) Beneficial effects of high-dose diltiazem in patients with persistent effort angina on β-blockers and

nitrates: a randomized, double-blind, placebo-controlled cross-over study. Circulation 71:1197−1204

8. Braunwald E, Sobel BE (1988) Coronary blood flow and myocardial ischaemia. In: Braunwald E (ed) Heart Disease, 3rd ed. WB Saunders Co, Philadelphia, pp. 1191−1221

9. Brown BG, Bolson EL, Dodge HT (1984) Dynamic mechanisms in human coronary stenosis. Circulation 70:917−922

10. Brown BG, Bolson EL, Petersen RB, Pierce DC, Dodge HT (1981) The mechanisms of nitroglycerin action: stenosis vasodilation as a major component of the drug response. Circulation 64:1089−1097

11. Brown BG, Gundel WB, Gott VL, Covell JW (1974) Coronary collateral blood flow following acute coronary occlusion: a diastolic phenomenon. Cardiovasc Res 8:621−631

12. Brown BG, Lee AB, Bolson EL, Dodge HT (1984) Reflex constriction of significant coronary stenosis as a mechanism contributing to ischemic left ventricular dysfunction during isometric exercise. Circulation 70:18−24

13. Buck JD, Hardman HF, Warltier DC, Gross GJ (1981) Changes in ischemic blood flow distribution and dynamic severity of a coronary stenosis induced by beta blockade in the canine heart. Circulation 64:708−715

14. deBuitleir M, Rowland E, Krikler DM (1985) Hemodynamic effects of nifedipine given alone and in combination with atenolol in patients with impaired left ventricular function. Am J Cardiol 55:15E−20E

15. Cannon RO III, Watson RM, Rosing DR, Epstein SE (1983) Angina caused by reduced vasodilator reserve of the small coronary arteries. J Am Coll Cardiol 1:1359−1373

16. Chan P, Heo J, Gariban G, Askenase A, Segal BL, Iskandrian AS (1988) The role of nitrates, beta blockers and calcium antagonists in stable angina pectoris. Am Heart J 116:838−848

17. Cohn PF (1980) Silent myocardial ischemia in patients with a defective anginal warning system. (Editorial) Am J Cardiol 45:697−702

18. Deanfield JE, Selwyn AP, Chierchia S, Maseri A, Ribeiro P, Krikler S (1983) Myocardial ischemia during daily life in patients with stable angina. Its relation to symptoms and heart rate changes. Lancet II:753−758

19. Egstrup K (1988) Randomized double-blind comparison of metoprolol, nifedipine, and their combination in chronic stable angina: effects on total ischemic activity and heart rate at onset of ischemia. Am Heart J 116:971−978

20. Findlay IN, MacLeod K, Gillen G, Elliott AT, Aitchinson T, Dargie HJ (1987) A double-blind placebo controlled comparison of verapamil, atenolol, and their combination in patients with chronic stable angina pectoris. Br Heart J 57:336−343

21. Frishman W, Charlap S, Kimmel B, Teicher M, Cinnamon J, Allan L, Strom J (1988) Diltiazem, nifedipine and their combination in patients with stable angina pectoris: effects on angina, exercise tolerance and the ambulatory electrocardiographic ST segment. Circulation 77:774−786

22. Fung H-L, Sutton SG, Kamiya A (1984) Blood vessel uptake and metabolism of organic nitrates in the rat. J Pharmacol Exp Ther 228:334−341

23. Gage JE, Hess OM, Murakami T, Ritter M, Grimm J, Krayenbuehl HP (1986) Vasoconstriction of stenotic coronary arteries during dynamic exercise in patients with classic angina pectoris: reversibility by nitroglycerin. Circulation 73:865−876

24. Gallagher KP, Kumada T, Battler A, Kemper WS, Ross J Jr (1982) Isoproterenol-induced myocardial dysfunction in dogs with coronary stenosis. Am J Physiol 242:H260−H267

25. Gallagher KP, Matsuzaki M, Koziol JA, Kemper WS, Ross J Jr (1984) Regional myocardial perfusion and wall thickening during ischemia in conscious dogs. Am J Physiol 247:H727−H738

26. Gallagher KP, Osakada G, Hess OM, Koziol JA, Kemper WS, Ross J Jr (1982) Subepicardial segmental function during coronary stenosis and the role of myocardial fiber orientation. Circ Res 50:352−359

27. Gallagher KP, Osakada G, Matsuzaki M, Kemper WS, Ross J Jr (1982) Myocardial blood flow and function with critical coronary stenosis in exercising dogs. Am J Physiol 243:H698−H707

28. Glossman H, Ferry DR, Goll A, Strussing J, Schober M (1985) Calcium channels basic properties as revealed by radioligand binding studies. J Cardiovasc Pharmacol 7:S20−S30

29. Gordon JB, Zebede J, Wayne RR, Mudge GH, Ganz P, Selwyn AP (1986) Coronary constriction with exercise: possible role for endothelial dysfunction and alpha tone. Circulation 74:II–481 (abstr.)
30. Gould KL, Lipscomb K (1974) Effects of coronary stenoses on coronary flow reserve and resistance. Am J Cardiol 34:48–55
31. Gould KL, Westcott RJ, Albro PL, Hamilton GW (1978) Noninvasive assessment of coronary stenoses by myocardial imaging during pharmacologic coronary vasodilation. II. Clinical methodology and feasibility. Am J Cardiol 41:279–287
32. McGregor M (1982) The nitrates and myocardial ischemia. Circulation 66:689–692
33. Grover-McKay M, Matsuzaki M, Ross J Jr (1987) Dissociation between regional myocardial dysfunction and subendocardial ST segment elevation during and after exercise-induced ischemia in dogs. J Am Coll Cardiol 10:1105–1112
34. Guth BD, Heusch G, Seitelberger R, Ross J Jr (1987) Mechanism of beneficial effect of beta-adrenergic blockade on exercise-induced ischemia in conscious dogs. Circ Res 60:738–746
35. Guth BD, Tajimi T, Seitelberger R, Lee JD, Matsuzaki M, Ross J Jr (1986) Experimental exercise-induced ischemia: drug therapy can eliminate regional dysfunction and oxygen supply-demand imbalance. J Am Coll Cardiol 7:1036–1046
36. Hanet C, Pouleur H, Harlow BJ, Rousseau MF (1988) Effects of longterm combined dosing with nicardipine and propranolol on coronary hemodynamics, myocardial metabolism and exercise tolerance in patients with angina pectoris-comparison with monotherapy. Am Heart J 116:431–439
37. Heusch G, Guth BD, Seitelberger R, Ross J Jr (1987) Attenuation of exercise-induced myocardial ischemia in dogs with recruitment of coronary vasodilator reserve by nifedipine. Circulation 75:482–490
38. Hirzel HO, Leutwyler R, Krayenbuehl HP (1985) Silent myocardial ischemia: Hemodynamic changes during dynamic exercise in patients with proven coronary artery disease despite absence of angina pectoris. J Am Coll Cardiol 2:275–284
39. Hoffman JIE (1984) Maximal coronary flow and the concept of coronary vascular reserve. Circulation 70:153–159
40. Hossack KF, Brown BG, Stewart DK, Dodge HT (1984) Diltiazem-induced blockade of sympathetically mediated constriction of normal and diseased coronary arteries: lack of epicardial coronary dilatory effect in humans. Circulation 70:465–471
41. Ignarro LJ, Lippton H, Edwards JC, Baricos WH, Hyman AL, Kadowitz PJ, Gruetter CA (1981) Mechanisms of vascular smooth muscle relaxation by organic nitrates, nitrites, nitroprusside, and nitric oxide. Evidence for the involvement of S-nitrosothiols as active intermediates. J Pharmacol Exp Ther 218:739–749
42. Johnston DL, Gebhardt VA, Donald A, Kostuk WJ (1983) Comparative effects of propranolol and verapamil-alone and in combination on left ventricular function and volumes in patients with chronic exertional angina: a double-blind, placebo-controlled, randomized, crossover study with radionuclide ventriculography. Circulation 68:1280–1289
43. Kannel WB (1976) Some lessons in cardiovascular epidemiology from Framingham. Cardiovasc Epidemiol 37:269–282
44. Kannel WB (1989) Detection and management of patients with silent myocardial ischemia. Am Heart J 117:221–226
45. Kannel WB, McNamara PM, Feinleib M, Dawber TR (1970) The unrecognized myocardial infarction. Fourteen-year follow-up experience in the Framingham study. Geriatrics 75–87
46. Klocke FJ (1987) Measurements of coronary flow reserve: defining pathophysiology versus making decisions about patient care. Circulation 76:1183–1189
47. Kumada T, Gallagher KP, Battler A, White F, Kemper WS, Ross J Jr (1982) Comparison of post-pacing and exercise-induced myocardial dysfunction during collateral development in conscious dogs. Circulation 65:1178–1185
48. Kumada T, Gallagher KP, Miller M, McKown M, White F, McKown D, Kemper WS, Ross J Jr (1980) Improvement by isosorbide dinitrate of exercise-induced regional myocardial dysfunction in the dog. Am J Physiol 239:H399–H405
49. Kumada T, Gallagher K, Shirato K, McKown D, Miller M, Kemper WS, White F, Ross J Jr (1980) Reduction of exercise-induced regional myocardial dysfunction by propranolol: studies in a canine model of chronic coronary artery stenosis. Circ Res 46:190–200

50. Ludmer PL, Selwyn AP, Shook TL, Wayne RR, Mudge GH, Alexander W, Ganz P (1986) Paradoxical vasoconstriction induced by acetylcholine in atherosclerotic coronary arteries. N Engl J Med 315:1046–1051
51. Maseri A (1980) Pathogenic mechanisms of angina pectoris: expanding views. Br Heart J 43:648–660
52. Maseri A, Chierchia S (1982) Coronary artery spasm: demonstration, definition, diagnosis and consequences. Prog Cardiovasc Dis 25:169–192
53. Maseri A, Chierchia S, L'Abbate A (1980) Pathogenic mechanisms underlying the clinical events associated with atherosclerotic heart disease. Circulation 62(V):V3–V13
54. Matsuzaki M, Gallagher KP, Patritti J, Tajimi T, Kemper WS, White FC, Ross J Jr (1984) Effects of a calcium entry blocker (diltiazem) on regional myocardial flow and function during exercise in conscious dogs. Circulation 69:801–814
55. Matsuzaki M, Guth B, Tajimi T, Kemper WS, Ross J Jr (1985) Effect of the combination of diltiazem and atenolol on exercise-induced regional myocardial ischemia in conscious dogs. Circulation 72:233–243
56. Matsuzaki M, Patritti J, Tajimi T, Miller M, Kemper WS, Ross J Jr (1984) Effect of beta blockade on regional myocardial flow and function during exercise. Am J Physiol 247:H52–H60
57. Miller WE, Vittitoe J, O'Rourke RA, Crawford MH (1988) Nadolol versus diltiazem and combination for preventing exercise-induced ischemia in severe angina pectoris. Am J Cardiol 62:373–376
58. Mosher P, Ross J Jr, McFate PA, Shaw RF (1964) Control of coronary blood flow by an autoregulatory mechanism. Circ Res 14:250–259
59. Mudge GH Jr, Grossman W, Mills RM Jr, Lesch M, Braunwald E (1976) Reflex increase in coronary vascular resistance in patients with ischemic heart disease. N Engl J Med 295:1333–1337
60. Needleman P, Blehm DJ, Karkey AB, Johnson EM, Lang S (1971) The metabolic pathway in the degradation of glyceryl trinitrate. J Pharmacol Exp Ther 179:347–353
61. Nelson GIC, Silke B, Ahuja RC, Hussain M, Forsyth D, Taylor SH (1984) The effect on left ventricular performance of nifedipine and metoprolol singly and together in exercise-induced angina pectoris. Eur Heart J 5:67–79
62. Nesto RW, Kowalchuk GJ (1987) The ischemic cascade: Temporal sequence of hemodynamic, electrocardiographic, and symptomatic expressions of ischemia. Am J Cardiol 59:23C–30C
63. Nesto RW, White HD, Ganz P, Koslowski J, Wynne J, Holman BL, Antman E (1985) Addition of nifedipine to maximal beta-blocker-nitrate therapy: effects on exercise capacity and global left ventricular performance at rest and during exercise. Am J Cardiol 55:3E–8E
64. Osakada G, Kumada T, Gallagher KP, Kemper WS, Ross J Jr (1981) Reduction of exercise-induced regional myocardial dysfunction by verapamil in conscious dogs. Am Heart J 101:707–712
65. Osler W (1910) The Lumelian lectures on angina pectoris. Lancet (Lectures 1–3) I:698–702; II:829–844; III:973–977
66. Packer M (1989) Combined beta-adrenergic and calcium-entry blockade in angina pectoris. N Engl J Med 320:709–718
67. Parker JO (1987) Nitrate therapy in stable angina pectoris. N Engl J Med 316:1635–1642
68. Pepine CJ (1986) Clinical aspects of silent myocardial ischemia in patients with angina and other forms of coronary heart disease. Am J Med 80:25–34
69. Pfisterer M, Glaus L, Burkart F (1983) Comparative effects of nitroglycerin, nifedipine and metoprolol on regional left ventricular function in patients with one-vessel coronary disease. Circulation 67:291–301
70. Roberts WC (1975) The coronary arteries in coronary heart disease: morphologic observations. Pathol Annu 5:149–282
71. Robertson WS, Feigenbaum H, Armstrong WF, Dillon JC, O'Donnell J, McHenry RW (1983) Exercise echocardiography: a clinically practical addition in the evaluation of coronary artery disease. J Am Coll Cardiol 2:1085–1091
72. Ross J Jr (1986) Perspective: assessment of ischemic regional myocardial dysfunction and its reversibility. Circulation 74:1186–1190

73. Ross J Jr (1989) Mechanisms of regional ischemia and antianginal drug action during exercise. Prog Cardiovasc Dis 31:455–466

74. Ross J Jr, Brandenburg R, Dinsmore R, Friesinger GC, Hultgren HH, Pepine CJ, Rapaport E, Ryan TJ, Weinberg SL, Williams JF (1987) Guidelines for coronary angiography. A report of the American College of Cardiology/American Heart Association Task Force on Assessment of diagnostic and therapeutic cardiovascular procedures (Subcommittee on Coronary Angiography). J Am Coll Cardiol 10:935–950

75. Ross EM, Roberts WC (1986) Severe atherosclerotic coronary artery disease, healed myocardial infarction, and chronic congestive heart failure: analysis of 81 patients studied at necropsy. Am J Cardiol 57:44–50

76. Scheidt S (1987) Ischemic heart disease: a patient specific therapeutic approach with emphasis on quality of life considerations. Am Heart J 114:251–257

77. Selwyn AP, Fox K, Eves M, Oakley D, Dargie H, Shillingford J (1978) Myocardial ischemia in patients with frequent angina pectoris. Br Med J 2:1594–1596

78. Shea MJ, Deanfield JE, Wilson RA, deLandsheere CM, Jonathan A, Selwyn AP (1985) The distribution and frequency of prolonged ischemia in patients with angina pectoris. (Abstract) J Am Coll Cardiol 5:412

79. Sheps DS, Heiss G (1989) Sudden death and silent myocardial ischemia. Am Heart J 117:177–186

80. Silke B, Verma SP, Nelson GIC, Hussain M, Forsyth D, Frais MA, Taylor SH (1985) The effects on left ventricular performance of verapamil and metoprolol singly and together in exercise-induced angina pectoris. Am Heart J 109:1286–1292

81. Spain DB, Brades VA (1970) Sudden death from coronary heart disease. Chest 58:107–110

82. Stone PH, Ware JH, DeWood MA, Gore JM, Eich RH, Pietro DA, Parisi AF, Nesto RW, Boden WE, Sharma SC, Vlay SC, Ennis LE, Gianelly RE, Turi ZG, McCall NT, Curtis DG, Chierchia S, Maseri A, Braunwald E (1988) The efficacy of the addition of nifedipine in patients with mixed angina compared to patients with classic exertional angina: a multicenter, randomized, double-blind, placebo-controlled clinical trial. Am J Heart 116:961–970

83. Strauss W, Parisi A (1985) Superiority of combined diltiazem and propranolol therapy for angina pectoris. Circulation 71:951–957

84. Strauss WE, Parisi AF (1988) Combined use of calcium-channel and beta-adrenergic blockers for the treatment of chronic stable angina. Ann Intern Med 109:570–581

85. Taylor SH (1986) The role of transdermal nitroglycerin in the treatment of coronary heart disease. Am Heart J 112:197–207

86. Thaulow E, Guth BD, Heusch G, Gilpin E, Schulz R, Kroeger K, Ross J Jr (1989): Characteristics of regional myocardial stunning after exercise in dogs with chronic coronary stenosis. Am J Physiol. 257, H113–H119

87. Theroux P, Franklin D, Ross J Jr, Kemper WS (1974) Regional myocardial function during acute coronary artery occlusion and its modification of pharmacologic agents in the dog. Circ Res 35:896–908

88. Tilmant PY, LaBlanche JM, Thieuleux FA, Dupuis PA, Bertrand ME (1983) Detrimental effect of propranolol in patients with coronary arterial spasm countered by combination with diltiazem. Am J Cardiol 52:230–233

89. Tomoike H, Franklin D, McKown D, Kemper WS, Guberek M, Ross J Jr (1978) Regional myocardial dysfunction and hemodynamic abnormalities during strenuous exercise in dogs with limited coronary flow. Circ Res 42:487–496

90. Willerson JT, Campbell WB, Winniford MD, Schmidt J, Apprill P, Firth BG, Ashton J, Smitherman T, Bush L, Buja LM (1984) Conversion from chronic to acute coronary artery disease: speculation regarding mechanisms. (Editorial) Am J Cardiol 54:1349–1354

Author's address:
Dr. John Ross, Jr., M.D.
University of California, San Diego
Mail Code M013-B,
School of Medicine
La Jolla, CA 92093

Combined Treatment of Stable Effort-Induced Angina and Mixed Angina

Kenneth Egstrup

Department of Cardiology, Odense University Hospital, Denmark

Introduction

The principles for medical management of transient myocardial ischemia in patients with angina pectoris are based on drugs that affect either myocardial oxygen demand, myocardial oxygen supply or both, as the development of symptomatic or asymptomatic myocardial ischemia is caused by an imbalance between myocardial oxygen demand and myocardial oxygen supply. Traditionally, a fixed coronary artery stenosis has been assessed in most patients with stable angina pectoris as the cause of transient myocardial ischemia through an increase in myocardial oxygen demand and with a fixed coronary flow reserve. Hitherto, this concept has been supported by exercise testing, which often is very reproducible with the same level of work load and with the same level of myocardial ischemia in individual patients. However, with the use of ambulatory electrocardiographic monitoring it has been noted that transient myocardial ischemia during daily life in patients with stable effort-induced angina pectoris occurs at a heart rate level which is substantially lower than the heart rate at exercise-induced myocardial ischemia [6, 8, 13, 21, 62, 75, 85]. This observation makes the assessment of transient myocardial ischemia during daily life more complex and a reduction in myocardial oxygen supply, which is normally tolerated during exercise testing, seems to play an essential role even in classic exertional angina. It has in this context been observed, that patients with stable angina may exhibit episodic vasoconstriction superimposed on fixed atherosclerotic lesions induced by increased alpha-adrenergic stimulation by a cold pressor test [50]. The concept of the "dynamic" coronary artery stenosis has been furthermore supported by quantitative angiographic studies, where a significant reduction was noted in the luminal area of a localized coronary obstruction during dynamic and isometric exercise [26], and this reduction in luminal area was influenced by intracoronary nitroglycerin or intracoronary propranolol [27]. As medical therapy involves the use of nitrates, beta-adrenoceptor blocking drugs, calcium antagonists and their combination, the choice of pharmacologic agents is dependent on a number of factors such as suspected pathophysiology, clinical manifestation of the disease, other concomitant illness and possible side effects.

The main goals in the management of angina pectoris are to reduce symptoms and to improve prognosis. The primary objective for the success of the treatment has for many years been the reduction in frequency and duration of angina and nitroglycerin consumption. However, with the use of ambulatory monitoring, angina pectoris has been identified as an unreliable indicator of total ischemic activity in chronic stable angina as the majority of ischemic episodes are not accompanied by symptoms, and it has even been noted that ischemic episodes, even those which are silent, are often very prolonged [1, 10, 12, 14, 60]. Furthermore, it has been demonstrated that angina is subjective and may not necessarily identify episodes with transient myocardial ischemia [20]. Based on these observations and the increasing evidence that silent myocardial ischemia may be associated with an adverse prognosis in patients with coronary artery disease [16, 22, 23, 29], objective measurement of total ischemic activity seems necessary.

The aims of this presentation are mainly to focus on the effects of medical therapy on total ischemic activity and to discuss the principles for monotherapy and combined therapy; furthermore, to evaluate if the medical history presented by the patients can be used to identify different pathophysiologic mechanisms and to predict the therapeutic effect of antianginal therapy.

Concept of Effort-Induced Angina and Mixed Angina

Based on the medical history, patients with stable angina pectoris are often divided into those with effort-induced angina and those with mixed angina, as it is suspected that alterations in coronary vasomotor tone may contribute to transient ischemic episodes in mixed angina and that myocardial ischemia in effort-induced angina is dependent only on fixed atherosclerotic lesions. This assumption is based on the definition of patients with mixed angina as patients with stable angina pectoris, who have a recognizable threshold of exertion, often demonstrated during exercise testing, and who also develop occasionally angina at rest or at levels of exertion that are normally well tolerated [45–47]. This is in contrast to patients with effort angina, with a reproducible threshold of angina caused by an increase in myocardial oxygen demand in the presence of a suggested fixed reduction of the coronary artery flow reserve.

Traditionally, beta-adrenoceptor blocking drugs are recommended for patients with effort angina and vasodilators such as calcium antagonists or nitrates are recommended for those with mixed angina [25, 28, 79].

However, the question is if there is a rational basis for this division into effort-induced angina and mixed angina, and for the traditional guidelines to medical therapy. To assess this question, it is necessary to evaluate one of the main determinants of myocardial oxygen demand, the heart rate.

In a study [17] including 16 patients with effort-induced angina and 25 patients with mixed angina, the heart rate was assessed at the onset of transient ischemia during daily life by means of ambulatory electrocardiographic monitoring and compared with the heart rate at the onset of exercise-induced angina/ischemia. The heart rate at the onset of exercise-induced ischemia was the same in both effort-induced angina and mixed angina and also maximal working capacity was the same, indicating that the limitation of maximal coronary flow reserve was the same in both groups of patients. Careful analysis of the heart rate at the onset of both symptomatic and asymptomatic ischemia obtained by ambulatory monitoring during unrestricted daily life showed that the heart rate at the onset of ischemia was significantly lower in both effort-induced angina and mixed angina as compared with exercise-induced myocardial ischemia. This observation suggests that reduced myocardial oxygen supply may play a major role in both subtypes of patients and the distinction between effort angina and mixed angina therefore seems arbitrary and not based on general differences in pathophysiology. This assessment has been confirmed by others [7] where objective signs and determinants of myocardial oxygen demand were the same in a group of 17 patients with classic exertional angina and 55 patients with mixed angina [76]. The question of whether the classification of patients with stable angina into subsets of those with mixed angina and those with effort angina has clinical applications for the selection of antianginal therapy, will be discussed later in this chapter.

Assessment of Total Ischemic Activity

As has been stated before, angina pectoris seems to be an unreliable parameter of total ischemic activity; as it has been documented by ambulatory monitoring, that silent ischemia accounts for the majority of transient ischemic episodes in most patients with stable angina pectoris. It has been noted that silent ischemia accounts for about 75% of all ischemic episodes in patients with stable angina pectoris and that these episodes may be very prolonged and even associated with high-grade ST-segment depression [6, 13, 21]. The assessment of total ischemic activity may be of importance for the management of patients with angina pectoris as silent ischemia seems to be associated with an adverse prognosis in patients with coronary artery disease. This adverse prognosis of asymptomatic ischemia has been identified in asymptomatic persons with documented coronary artery disease [9], in patients with unstable angina [29], in patients after coronary artery bypass surgery [16], and in patients after myocardial infarction [48, 80]. However, it must be emphasized that no prospective study has been performed where a reduction or elimination of total ischemic activity has been related to reduced cardiac morbidity and mortality.

Of great interest for the management of angina patients is the observation that both symptomatic and asymptomatic ischemic episodes have a circadian variation with the highest frequency of transient myocardial ischemia in the morning hours [13, 21, 62, 68]. The same daily rhythm has been observed for the onset of myocardial infarction [52] and the same pattern is also found for sudden cardiac death [42]. The mechanisms underlying this circadian variation are not understood in detail but may have a relationship to a similar daily rhythm in plasma catecholamine level, heart rate, blood pressure, platelet aggregability and even changes in coronary artery tone [15, 49, 81, 86]. In several large post-myocardial infarction trials beta-receptor blockade has been found to reduce cardiac mortality and reinfarction, and it is therefore interesting to note that treatment with beta-receptor blocking drugs such as metoprolol and atenolol reduces the frequency of both symptomatic and asymptomatic ischemic episodes, mostly in the morning hours when the frequency is highest [35, 51].

Ambulatory monitoring seems therefore useful to evaluate the effect of medical therapy on transient ischemic episodes out of hospital, especially those not accompanied by symptoms. However, it must be emphasized that this evaluation must be based on the knowledge of the natural variability of transient myocardial ischemia, which makes this method less suitable for the assessment of individual patients as changes in ischemic activity may be caused by day-to-day variability in transient ischemic episodes rather than be the result of pharmacologic treatment [53]. In this context, treatment must be further based on an individual evaluation by medical history, results of exercise testing, scintigraphic studies, and coronary anatomy.

Pharmacotherapy in Stable Angina Pectoris

Monotherapy with Nitrates

Nitroglycerin is still the backbone of the medical therapy of angina pectoris, and the beneficial effects of nitrates have been documented in numerous studies. Their effectiveness has been demonstrated by exercise testing [66] and also by ambulatory monitoring where hourly administration of sublingual nitroglycerin reduced the frequency of silent ischemia. This beneficial effect has been confirmed in other studies where both transdermal nitroglycerin and slow-release nitrate formulations have been used either alone or in combination with other antianginal drugs [67, 70, 72, 84].

The side effects of nitrates are related to vasodilatation, especially headache, which is thought to be due to dilatation of the cerebral vessels. A limitation for the use of nitrates is the development of nitrate tolerance, where increasing doses are required to maintain the antianginal effect. The mechanism of nitrate tolerance seems to be related to decreased production of cyclic guanosine monophosphate [57], and nitrate tolerance only develops during sustained therapy with high constant plasma levels of nitrates. It has therefore been recommended to use intermittent therapy to avoid the occurrence of nitrate tolerance [11, 56].

Nitrates can be combined with beta-adrenoceptor blocking drugs, with calcium antagonists or both, and the effect seems additive [33, 41].

Monotherapy with Beta-Adrenoceptor Blocking Drugs

The main property of beta-receptor blockade is to decrease myocardial oxygen demand and it is well established that beta-blockers reduce anginal attack rate, nitroglycerin consumption, and increase exercise capacity [40, 44, 82]. As angina pectoris is only an indicator of a minor part of ischemic episodes, detailed studies have been performed to evaluate the effect of beta-receptor blockade on total ischemic activity.

In a small number of patients it has been shown that propranolol, atenolol, and labetalol reduced both frequency and duration of all ischemic episodes [43, 54, 63, 64]. In a detailed study [18] of 42 patients with severe angina despite medical therapy and with angiographically documented coronary artery stenosis, it was found that metoprolol significantly reduced the frequency of total ischemic activity (Fig. 1), silent ischemia and duration of ischemia. The ischemic burden, which combines frequency, duration and magnitude of all ischemic episodes, was also reduced significantly (Fig. 2). The heart rate at the onset of ischemic episodes, both painful and without pain, was significantly reduced during therapy with metoprolol compared with placebo, indicating that beta-blockade may act by reducing myocardial oxygen demand even in asymptomatic ischemic episodes. It has therefore been suggested that this reduction in myocardial oxygen demand, mediated through a reduction in

334

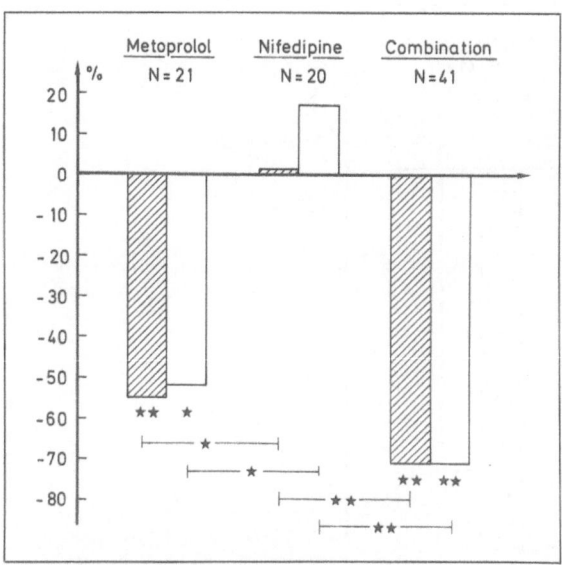

Fig. 1. Changes (%) in frequency of total ischemia and asymptomatic ischemic episodes during treatment with metoprolol, nifedipine, and their combination compared with placebo. Numbers of episodes at baseline (placebo) = 0%. Hatched bars denote all ischemic episodes and open bars denote silent ischemia.
* p < 0.05 and ** p < 0.01 compared with baseline (placebo). (Reproduced from [18] with the permission of American Heart Journal.)

heart rate, could make a transient reduction in myocardial oxygen supply tolerable and that metoprolol could blunt the beta-adrenoceptor mediated increase in heart rate and thus reduce ischemic activity [18, 35]. A direct dilatatory effect on the coronary stenosis has been demonstrated for other beta-blockers [27] so that a beneficial influence of beta-blockade on myocardial oxygen supply cannot be excluded, although this vasodilatatory effect has not been observed in experimental animal studies where the improvement in exercise variables was related to the reduction in heart rate [30]. On the other hand, even an adverse effect of beta-blockade has been suggested through increased coronary vasoconstriction mediated by the unmasking of alpha-receptor tone [38]. However, this possible increase in coronary artery tone seems not to be of clinical relevance as no adverse effects has been demonstrated in any of the studies assessing the effect of beta-blockade on total ischemic activity out of hospital.

Monotherapy with Calcium Channel Blockers

Nifedipine: The effect of nifedipine monotherapy in angina pectoris has given variable results. In a small study [43] it was found that nifedipine reduced objective signs of myocardial ischemia in patients with severe angina pectoris. This is in contrast to the effect noted in a double-blind comparison with metoprolol in patients with stable

335

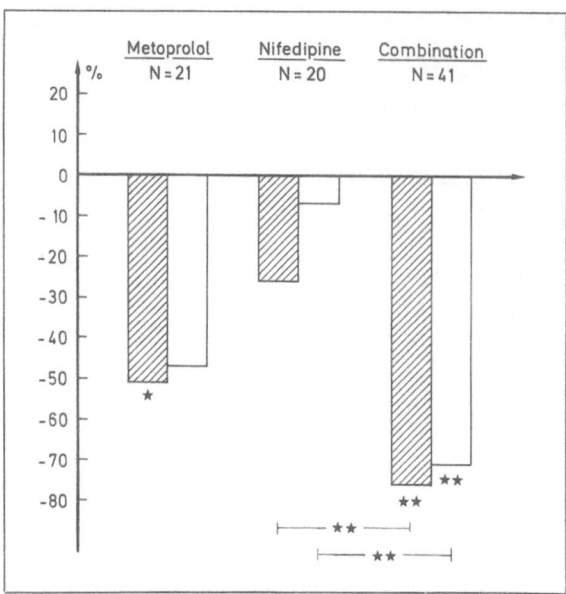

Fig. 2. Changes (%) in ischemic burden for total ischemia and asymptomatic ischemia during treatment with metoprolol, nifedipine, and their combination compared with placebo. Ischemic burden at baseline (placebo) = 0%. Hatched bars denote all ischemic episodes and open bars denote silent ischemia.
* p < 0.05 and ** p < 0.01 compared with baseline (placebo). (Reproduced from [18] with the permission of American Heart Journal.)

angina where exercise parameters and ischemic activity detected by ambulatory monitoring were used as the study objectives. Nifedipine was without a significant effect on ischemic variables and it was even noted that the number of silent ischemic episodes increased (Fig. 1) [18]. The failure of nifedipine to relieve myocardial ischemia could not be ascribed to reflex tachycardia as a detailed evaluation of the heart rate at the onset of ischemia during daily life and during exercise testing did not reveal a significant difference as compared with placebo treatment. This observation is in accordance with others [24]. A possible explanation could be a coronary steal phenomenon during therapy with nifedipine as a dilatation of resistance vessels induces increased flow to non-ischemic myocardium. This effect has been described to occur during exercise testing [71]. If the patients were angiographically divided into two groups with either good or no/bad coronary collateral flow, it was noted that nifedipine reduced significantly the frequency of both symptomatic and asymptomatic ischemia in patients with no/bad coronary collateral flow, whereas an increase in ischemic activity was noted in the group of patients with good coronary collateral flow [19]. This may explain the conflicting results that exist concerning the antianginal potency of nifedipine.

Due to its pharmacologic action it should be supposed that nifedipine is more effective in patients where alterations in coronary vasomotor tone contribute significantly to the occurrence of myocardial ischemia. In accordance with this, it has been noted that nifedipine is effective in patients with variant angina [37] and in patients with angina at rest [67]. Therefore, the effect of nifedipine has been assessed in a group of patients with mixed angina and the effect was compared with that in a group of patients with effort-induced angina [17]. In neither of these two groups of patients could a beneficial effect of nifedipine be noted on ischemic variables detected either by exercise testing or during ambulatory monitoring. In fact, an insignificant increase in total ischemia (24%) and silent ischemia (35%) was noted in the patients with mixed angina (Figs. 3 and 4). This observation indicates that the antiischemic action of nifedipine is the same in both effort angina and mixed angina, which further is in accordance with observations of the heart rate at the onset of ischemia. This suggests that the pathophysiological mechanisms of effort-induced angina and mixed angina are mainly the same and the clinical identification of patients with mixed angina is no specific indication for the use of nifedipine in monotherapy.

Verapamil and Diltiazem: These drugs have been shown to be effective in the treatment of angina pectoris with a prolongation of exercise time, time to angina, and time to the appearance of ischemic ST-segment changes [39, 59, 61]. Furthermore, it has been noted that verapamil reduced symptomatic as well as asymptomatic ischemic episodes in effort-induced angina [3, 4, 36] and verapamil has even been shown to be effective in variant angina [37] and unstable angina [58]. In most studies the effect of verapamil and diltiazem seems equal in reducing anginal attack rate and nitroglycerin consumption, as well as in improving exercise variables, but so far there is no report concerning the effect of diltiazem on silent myocardial ischemia.

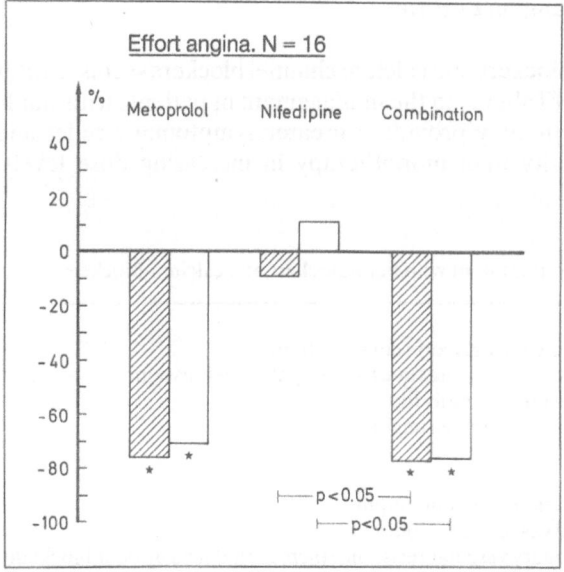

Fig. 3. Changes (%) in number of total and silent ischemic episodes/24 h during treatment with metoprolol, nifedipine, and their combination compared with baseline (placebo). Number of episodes at baseline = 0%. Hatched bars denote total ischemic events and open bars denote silent ischemia.
* $p < 0.05$ in comparison to baseline (placebo). (From [17].)

337

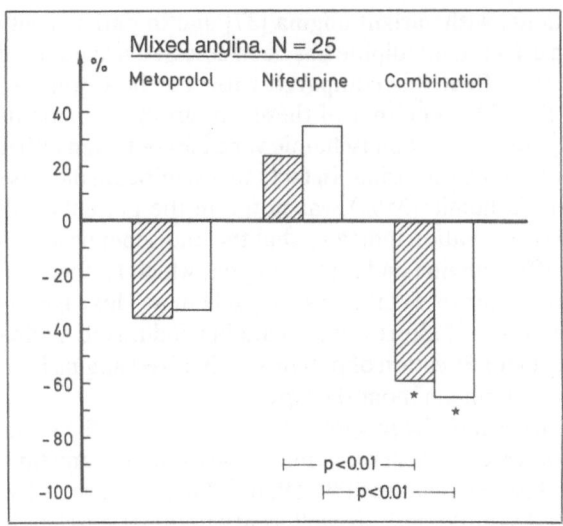

Fig. 4. Changes (%) in number of total and silent ischemic episodes/24 h during treatment with metoprolol, nifedipine, and their combination compared with baseline (placebo). Number of episodes at baseline = 0%. Hatched bars denote total ischemic events and open bars denote silent ischemia.
* p < 0.05 in comparison to baseline (placebo). (From [17].)

Combined Treatment in Stable Angina Pectoris

Combination therapy with beta-blockers and calcium channel blockers seems to offer an advantage over monotherapy (Table 1) in the management of patients with stable angina. Firstly, combined therapy may provide a greater symptomatic relief and reduction in total ischemic activity than monotherapy in increasing dose levels.

Table 1. Favorable effects of combined treatment with beta-blockers and calcium-blockers

---

I. Improved antiischemic effects.
   1) Decrease in angina frequency and nitroglycerin consumption.
   2) Increased excercise time, time to angina, and time to ST-segment depression.
   3) Decreased total ischemic activity during daily life.
   4) Effective in effort-induced angina and mixed angina.

II. Positive hemodynamic profile.
   1) Reduction in reflex tachycardia induced by nifedipine.
   2) Prevention of exercise-induced vasoconstriction.
   3) Reduction of systemic and coronary vascular resistance increase induced by beta-blockade.
   4) Increased antegrade collateral flow by decreased preload and decreased end-diastolic pressure.

III. Reduced side effects due to lower doses of each drug and possible positive hemodynamic interplay.

338

Secondly, the side effects may be lower than during monotherapy as the dose level of the individual drug can be lower and side effects can even be reduced through a beneficial hemodynamic interplay [2, 34, 69, 77, 83]. Thirdly, combined therapy gives a more physiological effect on the hemodynamic parameters as the combination of calcium antagonists and beta-blockers may prevent the increase in vascular resistance seen during beta-blockade and may prevent the reflex tachycardia induced by nifedipine and may increase antegrade coronary collateral flow which could prevent a coronary steal phenomenon. Combined treatment with verapamil and beta-blockers should be used with caution, especially in patients with impaired left ventricular function or with constant or intermittent conduction defects [65, 74]. When a combination therapy is considered, the combination of a beta-blocker with either nifedipine or diltiazem should be considered as the first choice. Hitherto, there is only limited information concerning the use of combination therapy with diltiazem and a beta-blocker, whereas the combined therapy with nifedipine and beta-blockers is well described.

Combined therapy with metoprolol and nifedipine has been carefully evaluated in 42 patients with severe angina pectoris despite optimal medical treatment [18]. The patients were required to have typical stable angina pectoris of at least 6 months' duration. All patients were aged under 70 years and none had experienced a myocardial infarction in the preceding three months. Patients with unstable angina, severe hypertension, and overt heart failure were excluded. Also excluded were patients with bundle branch block, those taking any medication that could influence the evaluation of the ST-segment, and those with gross left hypertrophy, severe valvular heart disease, or those not in sinus rhythm. In all patients selective coronary arteriography was performed and luminal narrowing $\geq 75\%$ in one or more of the major coronary arteries was considered significant. According to this definition 18 patients had single-vessel disease, 14 patients had two-vessel disease, and 10 patients had three-vessel disease.

If the patient was included in the study, all prophylactic antianginal medication was withdrawn allowing only sublingual nitroglycerin for the relief of pain but not for prophylactic use. All patients single-blindly received placebo therapy for 1 week: 5 days in hospital and the last 2 days out of the hospital. The data obtained during the end of the placebo period provided baseline data. The patients were then allocated to double-blind treatment with metoprolol 50 mg twice daily for 1 week and 100 mg twice daily for the next week, or nifedipine 10 mg three-times daily for 1 week and 20 mg three-times daily in the following week. In the third week, either nifedipine 10 mg three-times daily was added to high-dose metoprolol, or metoprolol 50 mg twice daily was added to high-dose nifedipine. The combination therapy thus consisted of two different dose levels. A double-dummy technique was used to ensure double-blind conditions.

Ambulatory electrocardiographic monitoring was performed for 36 h during the end of placebo, monotherapy, and combination therapy periods, and the number, duration, and magnitude of transient ischemia was noted for each patient. To define symptomatic ischemic episodes each patient kept a detailed diary and was instructed to activate the event button on the recorder, when experiencing anginal chest pain. A maximal bicycle exercise test was performed before starting ambulatory monitoring.

339

Frequency of Ischemia

During the placebo period 196 ischemic episodes were noted, of these 74% were not accompanied by symptoms. Metoprolol reduced significantly the number of both symptomatic and asymptomatic ischemic episodes, whereas nifedipine therapy was without any statistically significant effects (Fig. 1). The combination therapies significantly reduced both total ischemic events (71% reduction) and silent ischemic episodes (71% reduction). Combination therapy was significantly superior to nifedipine monotherapy and with a tendency to improve compared with metoprolol monotherapy.

Duration of Ischemia

Duration of total ischemia was reduced by metoprolol compared with placebo, whereas no significant reduction in duration of silent ischemia was noted during therapy with metoprolol. Monotherapy with nifedipine was without a statistically significant effect. Combination therapy produced a highly significant reduction in both total ischemic duration and duration of asymptomatic ischemia, which was superior to nifedipine and metoprolol.

Ischemic Burden

Frequency, duration, and magnitude of transient ischemia may have individual or combined prognostic implications, therefore the term ischemic burden combined frequency, duration, and magnitude of transient ischemia into a simple index. The combined therapy significantly reduced ischemic burden both in those associated with symptoms and in those without symptoms, and the effect was more pronounced than with metoprolol monotherapy (Fig. 2).

Dose Levels

In this study two dose levels were used with metoprolol 50 mg twice daily and nifedipine 20 mg three-times daily vs metoprolol 100 mg twice daily and nifedipine 10 mg three-times daily. If the data were analyzed according to the dose levels no significant difference could be obtained regarding frequency of ischemia, duration of ischemia, and ischemic burden, which indicates that none of those two dose levels was more preferable than the other.

Heart Rate at the Onset of Ischemia

The heart rate reduction at the onset of ischemia (Fig. 5) was the same during combination therapy as during metoprolol monotherapy, so that the main anti-ischemic effects seems to be a reduction in myocardial oxygen demand. Furthermore, there

340

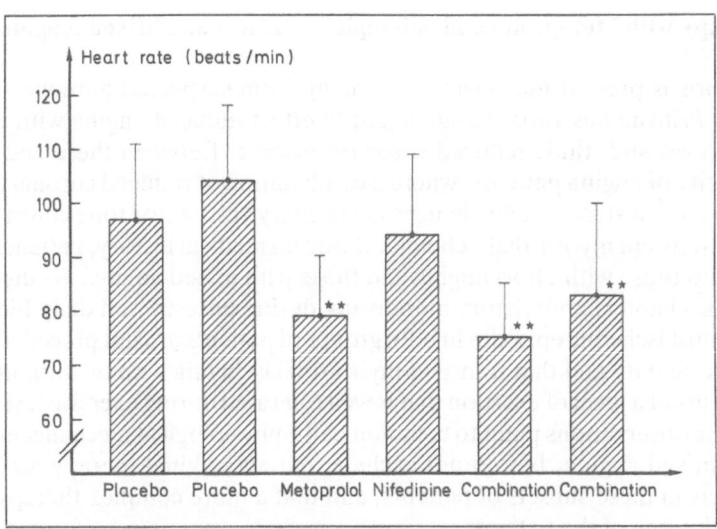

Fig. 5. Heart rate at the onset of all ischemic episodes during placebo, metoprolol, nifedipine, and combination therapy period. The two groups noted for placebo and combination represent patients randomized to metoprolol and nifedipine, respectively.
** p < 0.01 compared with placebo. (Reproduced from [18] with the permission of American Heart Journal.)

seems to be an additional effect of calcium channel blockade with nifedipine, which is noted by a further decrease in ischemic variables compared with metoprolol mono-therapy. This effect may be mediated through a reduction in peripheral vascular resistance that results in increased left ventricular contractility [73, 78]. Nifedipine may also improve left ventricular compliance, facilitate antegrade collateral flow and even attenuate myocardial ischemia due to recruitment of coronary vasodilator reserve [31]. In the presence of atrioventricular abnormalities or significantly reduced left ventricular function, combined therapy with a beta-blocker and nifedipine may be a better choice than combined therapy with diltiazem and a beta-blocker [5, 32].

Exercise Testing

The combined therapy with metoprolol and nifedipine produced a significant increase in exercise time, time to the onset of pain and the onset of ST-segment depression. This study supports the conclusion that combined therapy with meto-prolol and nifedipine has an additive antianginal effect and this beneficial effect is achieved without an increase in side effects. It should be mentioned that this study was performed in a highly selected group of patients with severe angina and known severe coronary artery disease. Therefore, medical therapy may obtain an even greater reduction in ischemic activity in patients with less severe disease. On the other hand, combined therapy may not be a first consideration in patients with less active disease.

341

Combination Therapy with Metoprolol and Nifedipine in Effort and Mixed Angina

Stable angina pectoris is present in a spectrum ranging from suspected pure vaso-spastic angina as in Prinzmetals variant angina [55] to effort-induced angina with a suspected fixed stenosis and, thus, reduced coronary reserve. Between these end-points are the majority of angina patients, where a combination of reduced coronary flow reserve due to a fixed stenosis and changes in coronary vasomotor tone contri-bute to episodes of transient myocardial ischemia. Based on medical history, patients are often divided into those with effort angina and those with mixed angina. As dis-cussed earlier in this chapter, ambulatory monitoring during unrestricted daily life demonstrates that most ischemic episodes in both groups of patients are not preceded by increased cardiac activity and that transient myocardial ischemia with or without symptoms often occurs at a level of exertion that is well tolerated during exercise test-ing in hospital. These observations point to common pathophysiological mechanisms in both effort and mixed angina, but do not exclude that antianginal therapy may have different effects in these subsets of patients, and that a more complex therapy may be preferable in some of the patients.

Only a few studies have addressed this question. In 16 patients with stable effort-induced angina and 25 patients with mixed angina the effect on total ischemic activity was assessed in a double-blind randomized study using metoprolol, nifedipine, and their combination [17]. All patients fulfilled the following criteria: severe angina pec-toris despite medical treatment, documented coronary artery disease, stable angina of at least 6 months duration, no electrocardiographic changes or medication, which could influence the evaluation of the ST-segment, and all patients were in sinus rhythm.

Effort Angina

It was found that metoprolol and the combination therapy of metoprolol and nifedipine produced a significant reduction in the frequency of the total number of ischemic episodes and silent ischemic episodes (Fig. 3). The same beneficial effect was noted on the duration of ischemia, where a significant reduction was noted for total ischemic duration and duration of asymptomatic ischemia. There was no signif-icant difference between metoprolol monotherapy and the combined therapy with metoprolol and nifedipine. Nifedipine was without any significant effect on ischemic variables. As expected, heart rate at the onset of ischemia and at 1 mm ST-segment depression was significantly lower during metoprolol monotherapy and the com-bined therapy compared with placebo and nifedipine monotherapy. The heart rate at the onset of ischemia during daily life was also lower compared with the correspond-ing values during exercise testing.

Mixed Angina

In this group of patients the same pattern was observed: metoprolol and the combina-tion therapy were effective in reducing frequency and duration of both total and silent ischemic episodes. The treatment effect was less pronounced as compared with the effect in effort angina. The combination therapy seems superior to monotherapy with metoprolol, but the difference did not reach the level of significance, which may depend on the number of patients included in the study. Notable was the lack of antianginal effect of nifedipine in monotherapy, where even an increase in both total and silent ischemia (24% vs 35%) were noted. This increase in ischemic activity was not related to reflex tachycardia as the heart rate at the onset of ischemia during ambulatory monitoring and during exercise testing was the same as for placebo treatment. Heart rate variables during ambulatory monitoring and exercise testing showed the same reduction as in the group of patients with effort angina during treatment with metoprolol and combined treatment with metoprolol and nifedipine. These data indicate that the pathophysiology of transient ischemia during daily life is the same in both effort angina and mixed angina, and the subdivision of these types of angina is not a guideline for the specific use of a beta-blocker or a calcium antagonist. On the other hand, mixed angina seems to identify a group of patients with a higher level of ischemic activity despite the fact, that coronary anatomy was the same in both patients with effort angina and those with mixed angina. This may have clinical applications as a combined therapy seems more rational and effective in this group of patients. This conclusion is supported by a recently performed study [76] where the addition of nifedipine to treatment with beta-blockers and/or nitrates reduced the frequency of symptomatic ischemic episodes detected by ambulatory monitoring in a group of patients with mixed angina. The combination therapy was also found to be more effective based on the evaluation of angina frequency, nitroglycerin consumption, and exercise variables in patients with effort-induced angina.

Conclusion

Rational therapy in stable angina pectoris implies objective assessment of total ischemic activity as the symptom angina is an unreliable parameter of transient myocardial ischemia during unrestricted daily life. Studies with ambulatory electrocardiographic monitoring indicate that myocardial ischemia out of hospital is often not preceded by an increase in the major determinants of myocardial oxygen demand, suggesting that reduced myocardial oxygen supply plays a major role in angina patients during daily life. Based on clinical history patients with stable angina are often divided into those with effort-induced angina and those with mixed angina as it is suggested that patients with mixed angina have a higher component of changes in vasomotor tone than those with effort angina. However, it has not been possible to identify differences in pathophysiological mechanisms between patients with effort-induced angina and mixed angina, although patients with symptoms suggestive of mixed angina may have a higher level of ischemic activity despite no observed differences in coronary anatomy. This group of patients may be especially suitable for a more complex therapy with a combination of beta-blockers and calcium antagonists. The combined therapy may have further advantages as the hemodynamic interplay appears beneficial. By this effect the side effects may be reduced and even lower doses of each of the two drugs may be used.

References

1. Allen RD, Gettes LS, Phalan C, Avington MD (1976) Painless ST-segment depression in patients with angina pectoris. Chest 69:467−473
2. Anderson JL, Wagner JM, Datz FL, Christian PE, Bray BE, Taylor AT (1984) Comparative effects of diltiazem, propranolol and placebo on exercise performance using radionuclide ventriculography in patients with symptomatic coronary artery disease: results of a double-blind, randomized, cross-over study. Am Heart J 107:698−706
3. Bala Subramanian V, Bowles MJ, Davies AB, Raftery EB (1982) Combined therapy with verapamil and propranolol in chronic stable angina. Am J Cardiol 49:125−132
4. Bala Subramanian V, Bowles MJ, Khurmi NS, Davies AB, Raftery EB (1982) Randomized double-blind comparison of verapamil and nifedipine in chronic stable angina. Am J Cardiol 50:696−703
5. Braun S, Terdiman R, Bennfeld D, Laniado S (1985) Clinical and hemodynamic effects of combined propranolol and nifedipine therapy versus propranolol alone in patients with angina pectoris. Am Heart J 109:478−485
6. Cecchi AC, Dovellini EV, Marchi F, Pucci P, Santoro GM, Fazzini PF (1983) Silent myocardial ischemia during ambulatory electrocardiographic monitoring in patients with effort angina. J Am Coll Cardiol 1:934−939
7. Chierchia S, Glazier JJ, Gerosa S (1987) A single-blind, placebo-controlled study of effects of atenolol on transient ischemia in "mixed" angina. Am J Cardiol 60:36A−40A
8. Cocco G, Braun S, Strozzi C, Leishman B, Chu D, Rochat N (1982) Asymptomatic myocardial ischemia in patients with stable and typical angina pectoris. Clin Cardiol 5:403−408
9. Cohn PF (1983) Prognosis and treatment of asymptomatic coronary artery disease. J Am Coll Cardiol 1:959−964
10. Cohn PF (1984) Time for a new approach to management of patients with both symptomatic and asymptomatic episodes of myocardial ischemia. Am J Cardiol 54:1357−1359
11. Cowan JC (1987) Prevention of tolerance to nitroglycerin patch by overnight removal. Am J Cardiol 60:271−276
12. Deanfield JE, Kensett M, Wilson RA, Shea M, Horlock P, de Landsheere CM, Selwyn AP (1984) Silent myocardial ischaemia due to mental stress. Lancet II:1001−1005
13. Deanfield JE, Selwyn AP, Chierchia S, Maseri A, Ribeiro P, Krikler S, Morgan M (1983) Myocardial ischaemia during daily life in patients with stable angina: its relation to symptoms and heart rate changes. Lancet II:753−758
14. Deanfield JE, Shea MJ, Selwyn AP (1985) Clinical evaluation of transient myocardial ischemia during daily life. Am J Med 79 (Suppl 3A):18−24
15. Decousus HA, Croze M, Levi FA, Jaubert JC, Perpoint BM, De-Bonadona JF, Reinberg A, Queneau PM (1985) Circadian changes in anticoagulant effect of heparin infused at a constant rate. Br Med J 290:341−344
16. Egstrup K (1988) Asymptomatic myocardial ischemia as a predictor of cardiac events after coronary artery bypass grafting for stable angina pectoris. Am J Cardiol 61:248−252
17. Egstrup K (1989) Effects of metoprolol, nifedipine and their combination on total ischemic activity in effort and mixed angina. A randomized double-blind study. Am J Noninvas Cardiol 3:290−296
18. Egstrup K (1988) Randomized double-blind comparison of metoprolol, nifedipine, and their combination in chronic stable angina: Effects on total ischemic activity and heart rate at onset of ischemia. Am Heart J 116:971−978
19. Egstrup K (1990) Relationship between the antiischemic effect of nifedipine during daily life and coronary collateral flow. Submitted to Am Heart J
20. Egstrup K (1987) The sensitivity of the symptom angina pectoris as a marker of transient myocardial ischaemia in chronic stable angina. Acta Med Scand 222:301−306
21. Egstrup K (1988) Transient myocardial ischaemia during ambulatory monitoring out of hospital in patients with chronic stable angina pectoris. Acta Med Scand 224:311−318
22. Eriksen J, Enge J, Forfang K, Storstein O (1976) False-positive diagnostic tests and coronary angiographic findings in 105 presumably healthy males. Circulation 54:371−376

23. Eriksen J, Thaulow E (1984) Follow-up of patients with asymptomatic myocardial ischemia. In: Rutishauser W, Roskamm H (eds) Silent Myocardial Ischemia. Springer, Berlin Heidelberg New York, pp 156–164
24. Fox KM, Deanfield JE, Selwyn AP, Krikler S, Wright CA (1982) Treatment of chronic stable angina pectoris with nifedipine. In: Kaltenbach M, Neufeld HN (eds) Fifth International Nifedipine Symposium. Amsterdam: Exerpta Medica:197–204
25. Frishman WH (1983) Multifactorial actions of beta-adrenergic blocking drugs in ischemic heart disease. Current concepts. Circulation 67 (Suppl I):11–18
26. Gage JE, Hess OM, Murakami T, Ritter M, Grimm J, Krayenbuehl HP (1986) Vasoconstriction of stenotic coronary arteries during dynamic exercise in patients with classic angina pectoris: reversibility by nitroglycerin. Circulation 73:865–876
27. Gaglione A, Hess OM, Corin WJ, Ritter M, Grimm J, Krayenbuehl HP (1987) Is there coronary vasoconstriction after intracoronary beta-adrenergic blockade in patients with coronary artery disease? J Am Coll Cardiol 10:299–310
28. Gorlin R (1982) Role of coronary vasospasm in the pathogenesis of myocardial ischemia and angina pectoris. Am Heart J 103:598–603
29. Gottlieb SO, Weisfeldt ML, Ouyang P, Mellits ED, Gerstenblith G (1986) Silent ischemia as a marker for early unfavourable outcomes in patients with unstable angina. N Engl J Med 314:1214–1219
30. Guth BD, Heusch G, Seitelberger R, Ross J (1987) Mechanism of beneficial effect of beta-adrenergic blockade on exercise-induced myocardial ischemia in conscious dogs. Circ Res 60:738–746
31. Heusch G, Guth BD, Seitelberger R, Ross J (1987) Attenuation of exercise-induced myocardial ischemia in dogs with recruitment of coronary vasodilator reserve by nifedipine. Circulation 75:482–490
32. Higginbotham MB, Morris KG, Coleman RE, Cobb FR (1986) Comparison of nifedipine alone with propranolol alone for stable angina pectoris including hemodynamics at rest and during exercise. Am J Cardiol 57:1022–1028
33. Hossack KF, Elridge JE, Buckner K (1986) Comparison of acute hemodynamic effects of nitroglycerin versus diltiazem and combined acute effects of both drugs in angina pectoris. Am J Cardiol 58:722–726
34. Humen DP, O'Brian P, Purves P, Johnson D, Kostuk WJ (1986) Effort angina with adequate beta-receptor blockade: comparison with diltiazem alone and in combination. J Am Coll Cardiol 7:329–335
35. Imperi GA, Lambert CR, Coy K, Lopez L, Pepine CJ (1987) Effects of titrated beta blockade (metoprolol) on silent myocardial ischemia in ambulatory patients with coronary artery disease. Am J Cardiol 60:519–524
36. Johnson SM, Mauritson DR, Corbett JR, Woodward W, Willerson JT, Hillis LD (1981) Double-blind, randomized, placebo-controlled comparison of propranolol and verapamil in the treatment of patients with stable angina pectoris. Am J Med 71:443–451
37. Johnson SM, Mauritson DR, Willerson JT, Hillis LD (1981) Comparison of verapamil and nifedipine in the treatment of variant angina pectoris: preliminary observations in 10 patients. Am J Cardiol 47:1295–1300
38. Kern MJ, Ganz P, Horowitz JD, Gaspar J, Barry WH, Lorell BH, Grossman W, Mudge GH (1983) Potentiation of coronary vasoconstriction by beta-adrenergic blockade in patients with coronary artery disease. Circulation 67:1178–1185
39. Khurmi NS, Bowles MJ, O'Hara MJ, Bala Subramanian V, Raftery EB (1984) Long term efficacy of diltiazem assessed with multistage graded exercise tests in patients with chronic stable angina pectoris. Am J Cardiol 54:738–743
40. Kritzer G, Warr TA, Strong ML, Froelicher VF (1983) Effect of atenolol on treadmill performance in patients with angina pectoris. Clin Pharm 2:236–242
41. Liang CS, Coplin B, Wellington K (1985) Comparison of antianginal efficacy of nifedipine and isosorbide dinitrate in chronic stable angina: a long-term, randomized, double-blind, crossover study. Am J Cardiol 55:9E–14E
42. Ludmer PL, Alymer GZ, Muller JE (1986) Circadian variation of sudden cardiac death. J Am Coll Cardiol 7:209A (abstr.)

43. Lynch P, Dargie H, Krikler S, Krikler D (1980) Objective assessment of antianginal treatment: a double-blind comparison of propranolol, nifedipine and their combination. Br Med J 2:184−187
44. Manyari DE, Kostuk WJ, Carruthers SG, Johnston DJ, Purves P (1983) Pindolol and propranolol in patients with angina pectoris and normal or near-normal ventricular function. Am J Cardiol 51:427−433
45. Maseri A (1980) Pathogenetic mechanisms of angina pectoris: expanding views. Br Heart J 43:648−660
46. Maseri A, Chierchia S (1981) A new rationale for the clinical approach to the patient with angina pectoris. Am J Med 71:639−644
47. Maseri A, Chierchia S, Kaski JC (1985) Mixed angina pectoris. Am J Cardiol 56:30E−33E
48. Mickley H, Pless P, Egstrup K, Rokkedal J, Møller M (1988) Silent ischemia is a predictor of severe angina pectoris after first myocardial infarction. Eur Heart J 9:314 (abstr.)
49. Millar-Craig MW, Bishop CN, Raftery EB (1978) Circadian variation of blood pressure. Lancet I:795−797
50. Mudge GH Jr, Grossman W, Mills RM Jr, Lesch M, Braunwald E (1976) Reflex increase in coronary vascular resistance in patients with ischemic heart disease. N Engl J Med 295:1333−1337
51. Mulcahy D, Cunningham D, Crean P, Wright C, Keegan J, Quyyumi A, Park A, Fox K (1988) Circadian variations of total ischaemic burden and its alteration with anti-anginal agents. Lancet II:755−759
52. Muller JE, Stone PH, Turi ZG, Rutherford JD, Czeisler CA, Parker C, Poole KW, Passamani E, Roberts R, Robertson T, Sobel BE, Willerson JT, Braunwald E, MILIS Study Group (1985) Circadian variation in the frequency of onset of acute myocardial infarction. N Engl J Med 21:1315−1322
53. Nabel EG, Barry J, Rocco MB, Campbell S, Mead K, Fenton T, Orav EJ, Selwyn AP (1988) Variability of transient myocardial ischemia in ambulatory patients with coronary artery disease. Circulation 78:60−67
54. Oakley GDG, Fox KM, Dargie HJ, Selwyn AP (1979) Objective assessment of treatment in severe angina. Br Med J 1:1540
55. Oliva PB, Potts DE, Pluss RG (1973) Coronary arterial spasm in Prinzmetal angina. Documentation by coronary arteriography. N Engl J Med 288:745−751
56. Parker JO (1987) Effect of intervals between doses on the development of tolerance to isosorbide dinitrate. N Engl J Med 316:1440−1444
57. Parker JO (1985) Nitrate tolerance. Am J Cardiol 56:281−311
58. Parodi O, Simonetti I, L'Abbate A, Maseri A (1982) Verapamil versus propranolol for angina at rest. Am J Cardiol 50:923−928
59. Patton JW, Vlietstra RE, Frye RL (1984) Randomized, placebo-controlled study of the effect of verapamil on exercise hemodynamics in coronary artery disease. Am J Cardiol 53:674−678
60. Pepine CJ (1986) Clinical aspects of silent myocardial ischemia in patients with angina and other forms of coronary heart disease. Am J Med 80 (Suppl 4C):25−34
61. Petru MA, Crawford MH, Sorensen SG, Chandhuri TK, Levine S, O'Rourke RA (1983) Short and long term efficacy of high-dose oral diltiazem for angina due to coronary artery disease: a placebo-controlled, randomized, double-blind, cross-over study. Circulation 68:139−147
62. Quyyumi AA, Mockus LJ, Wright CA, Fox KM (1984) Mechanisms of nocturnal angina pectoris: importance of increased myocardial oxygen demand in patients with severe coronary artery disease. Lancet I:1207−1209
63. Quyyumi AA, Wright C, Mockus L, Fox KM (1984) Effect of partial agonist activity in beta-blockers in severe angina pectoris. Br Med J 289:951−953
64. Quyyumi AA, Wright C, Mockus L, Shackell M, Sutton GC, Fox KM (1985) Effects of combined alpha and beta adrenoceptor blockade in patients with angina pectoris. A double-blind study comparing labetalol with placebo. Br Heart J, 53:47−52
65. Reddy PS, Uretsky BF, Steinfeld M (1984) The hemodynamic effects of intravenous verapamil in patients on chronic propranolol therapy. Am Heart J 107:97−101
66. Reichek N (1983) Role of nitroglycerin in effort angina. Am J Med 74:33−39
67. Rizzon P, Scrutinio D, Mangini SG, Lagioia R, de Toma L (1986) Randomized placebo-con-

346

trolled comparative study of nifedipine, verapamil and isosorbide dinitrate in the treatment of angina at rest. Eur Heart J 7:67−76

68. Rocco MB, Barry J, Campbell S, Nabel EF, Cook E, Goldman L, Selwyn AP (1987) Circadian variation of transient myocardial ischemia in patients with coronary artery disease. Circulation 75:395−400

69. Rocha P, Baron B, Delestrain A, Pathe M, Cazon JL, Kahn JC (1986) Hemodynamic effects of intravenous diltiazem in patients treated chronically with propranolol. Am Heart J 111:62−68

70. Schang SJ, Pepine CJ (1977) Transient asymptomatic S-T segment depression during daily activity. Am J Cardiol 39:396−402

71. Schulz W, Jost S, Kober G, Kaltenbach M (1985) Relation of antianginal efficacy of nifedipine to degree of coronary arterial narrowing and to presence of coronary collateral vessels. Am J Cardiol 55:26−32

72. Shell WE, Kivowitz CF, Rubins SB, See J (1986) Mechanisms and therapy of silent myocardial ischemia: the effect of transdermal nitroglycerin. Am Heart J 112:222−229

73. Sherman LG, Liang CS (1983) Nifedipine in chronic stable angina: a double-blind placebo-controlled cross-over trial. Am J Cardiol 51:706−711

74. Silke B, Verma SP, Nelson GI, Hussian M, Forsyth D, Frais MA, Taylor SH (1985) The effect on left ventricular performance of verapamil and metoprolol singly and together in exercise-induced angina pectoris. Am Heart J 109:1286−1293

75. Stern S, Tzivoni D (1973) Dynamic changes in the ST-T segment during sleep in ischemic heart disease. Am J Cardiol 32:17−20

76. Stone PH, Ware JH, de Wood MA, Gore JM, Eich RH, Pietro DA, Parisi AF, Nesto RW, Boden WE, Sharma SC, Vlay SC, Ennis LE, Gianelly RE, Turzi ZG, McCall NT, Curtis DG, Chierchia S, Maseri A, Braunwald E (1988) The efficacy of the addition of nifedipine in patients with mixed angina compared to patients with classic exertional angina: A multicenter, randomized, double-blind, placebo-controlled clinical trial. Am Heart J 116:961−970

77. Strauss WE, Parisi AF (1985) Superiority of combined diltiazem and propranolol therapy for angina pectoris. Circulation 71:951−957

78. Terris S, Bourdillon PD, Cheng DT, Pitt B (1986) Direct cardiac and peripheral vascular effects of intracoronary and intravenous nifedipine. Am J Cardiol 58:25−30

79. Thadani U, Davidson C, Singleton W, Taylor SH (1980) Comparison of five beta-adrenoreceptor antagonists with different ancillary properties during sustained twice daily therapy in angina pectoris. Am J Med 68:243−250

80. Theroux P, Waters DD, Halphen C, Debaisieux JC, Mizgala HF (1979) Prognostic value of exercise testing soon after myocardial infarction. N Engl J Med 301:341−345

81. Thorton MB, Deegan T (1974) Circadian variations of plasma catecholamine, cortisol and immunoreactive insulin concentrations in supine subjects. Acta Clin Chem 55:389−394

82. Uusitala A, Arstila M, Bae EA, Härkönen R, Keyriläinen O, Rytkönen U, Schjelderup-Mathiesen PM, Wendelin H (1986) Metoprolol, nifedipine and the combination in stable effort angina pectoris. Am J Cardiol 57:733−737

83. Vetrovec GW, Parker VE (1986) Alternative medical treatment for patients with angina pectoris and adverse reactions to beta-blockers. Usefulness of nifedipine. Am J Med 81:20−27

84. Winniford MD, Gabliani G, Johnson SM, Mauritson DR, Fulton KL, Hillis LD (1984) Concomitant calcium antagonist plus isosorbide dinitrate therapy for markedly active variant angina. Am Heart J 108:1269−1273

85. Wolf E, Tzivoni D, Stern S (1974) Comparison of exercise tests and 24-hour ambulatory electrocardiographic monitoring in detection of ST-T changes. Br Heart J 36:90−95

86. Yasue H (1980) Pathophysiology and treatment of coronary arterial spasm. Chest 78:216−221

Author's address:
Kenneth Egstrup, M.D.
Department of Cardiology
Odense University Hospital
DK-5000 Odense C, Denmark

Subject Index

A

α-adrenergic coronary constriction 205, 208−210, 254, 256, 269, 271, 303, 316, 318, 335

α-adrenoceptor 156, 158, 201, 211

α_1-adrenoceptor 157, 158, 165, 202, 203, 208

α_2-adrenoceptor 206, 208, 210, 211, 254

α_2-makroglobulin 87

absolute ischemia 7

ACE-inhibitors 281

acetate 63

acetylcholine 203, 316

acetyl CoA 65

acidosis 47, 50, 51, 140, 141, 149, 154, 158, 159

action potential 138, 139, 149, 159, 296

action potential amplitude 138, 140, 141, 161

action potential duration 47, 48, 138, 141, 152, 154, 159, 161

acyl carnitine 43, 52, 153, 154, 158, 160, 161, 163−165

acyl CoA 43, 52, 161

adenosine 44, 176, 188, 200, 204, 205, 241

adenylate cyclase 94, 155, 159

aica-ribose 176

afterload 4, 128, 288, 301, 302, 306, 319, 320

afferent nerve fibers 231−235

ameroid constrictor 250−252, 318, 321

amylnitrate 4, 275

anaerobic glycolysis 297

anesthesia 115, 120, 234, 235

angina pectoris 1, 2, 4, 5, 8, 27, 50, 188, 211, 224, 247, 261, 275, 277, 284, 286, 315, 316, 318, 320, 324, 331, 333, 335

angioplasty (PTCA) 11, 26, 74, 98, 176, 241, 287, 289

angiotensin (AGT) 205, 210, 296, 298

anisoylated streptokinase 102

anisoylated streptokinase plasminogen complex 99

anoxia 38, 41

anticoagulants 83

antithrombin III 84, 87, 90, 97

arrhythmias 48, 52, 54, 88, 99, 137, 138, 155, 186, 190, 206, 221, 241, 261, 265, 267, 268, 272, 315

aspirin 83, 89, 91−93, 101, 224, 285

asynergy 288

atenolol 251, 252, 257, 261, 263−265, 267, 269, 320−324, 333, 334

atherosclerosis 1−3, 5, 83, 84, 89, 201, 203, 205, 218, 219, 221−223, 249, 257, 268, 270, 302, 315, 316, 331

ATP 6, 38−40, 44−46, 48, 49, 51, 52, 54, 66, 68, 72, 138, 153, 161, 162, 176, 177, 183−185, 188, 192, 296, 297, 305, 306

atrial fibrillation 265

automaticity 49, 149, 154, 156

autoregulation 199, 200, 249, 315, 317

B

β-adrenoceptor 155, 156, 158, 285

β-blocker 83, 96, 115, 156, 247, 250−254, 257, 258, 261−263, 265−272, 285, 307, 318−322, 324, 331, 332, 334, 335, 338, 339, 343

β-oxidation 154, 161, 162

basement membrane 13, 23

biopsy 22, 29, 31, 192

blood pressure 271, 275−277, 279, 285, 288, 289, 317−319, 321, 333

border zone 125−131, 141, 142, 148, 151, 152

bradycardia 252, 254, 256−258, 324

bradykinin 87, 234−239

Brunton 3, 275

C

C1-inhibitor 87

C-11 acetate 63−66, 68, 74

C-11 butanol 61

C-11 dioxide 64, 67

C-11 glucose 63

C-11 palmitate 63−70, 72, 74

caffeine 149

calcium 12, 43, 46−54, 86, 140, 141, 149, 153−159, 161, 177, 181−183, 190, 295−298, 305

calcium antagonists 115, 153, 220, 223, 224, 247, 250, 256, 257, 285, 295, 296, 298, 299, 301−303, 305, 306, 321, 331, 332, 335, 338, 339, 341, 343

calcium channel 295−300

calcium overload 51, 52, 182, 190, 297, 305